Introduction
to
Liquid State Chemistry

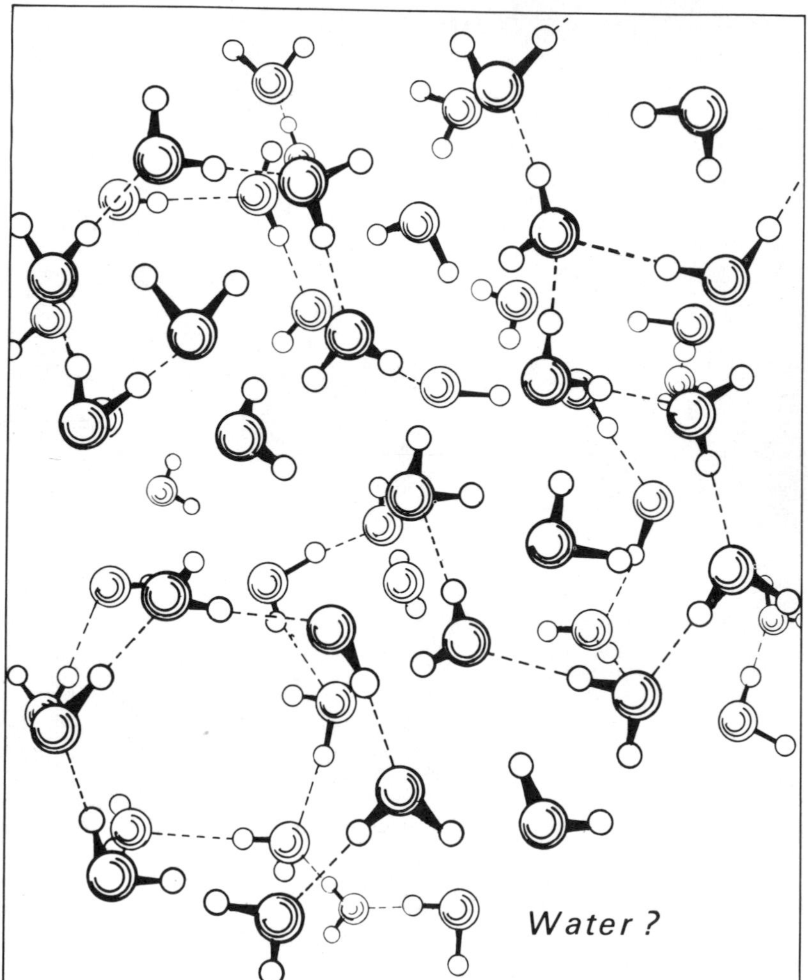
Water ?

Introduction to Liquid State Chemistry

Y. Marcus

*The Hebrew University of Jerusalem
Jerusalem, Israel*

A Wiley–Interscience Publication

JOHN WILEY & SONS
London · New York · Sydney · Toronto

Library of Congress Cataloging in Publication Data:

Marcus, Yizhak.
Introduction to liquid state chemistry.

'A Wiley–Interscience publication.'
Includes indexes.
1, Liquids. I. Title.
QD541.M338 541'.042'2 76-40230

IBSN 0 471 99448 0

Typeset in IBM Journal by
Preface Ltd., Salisbury, Wilts.
Printed in Great Britain by
The Pitman Press, Bath, Avon.

Contents

Preface

Chemists have been working with liquids since the dawn of chemistry, so that much of chemistry is indeed 'liquid state chemistry' although this phrase is not in common use. Conversely, both 'solid state physics' and 'solid state chemistry' are recognized fields of science in which many advances have been achieved over the past decades. But, although in recent years the liquid state has been the subject of intense study, the several books that have been published treat mainly the physics of this state. Liquid argon is an admirable substance upon which test liquid state theories, almost as good as the old stand-by of hard spheres, but it is relatively little interest to chemists. Crude petroleum, sea water, aqueous acetone or molten cryolite, which are of more practical interest, cannot be handled by the theories of physics, unless simple models and far-reaching approximations and assumptions are introduced. It is the proper domain of the liquid state chemist to provide these, and with this the present book is concerned.

This introduction to liquid state chemistry applies the tools of statistical mechanics and thermodynamics to the equilibrium properties of liquids and liquid mixtures, so that reactions and interactions phenomenologically described can also be understood and made the basis for predictions regarding systems not yet studied. There are several broad areas of liquid state chemistry which are currently subject to extensive research but do not sufficiently share their concepts and approaches. These areas comprise *non-polar liquids*, such as form regular solutions, but may also have non-ideal entropies of mixing (e.g. petroleum, certain polymer solutions); *electrolyte solutions*, ranging from dilute aqueous solutions to molten hydrate salts, with organic solvents and inorganic non-aqueous solvents included; *molten salts*, which range from ideally ionic fluids to molecular melts and liquid or glassy network systems; *non-aqueous solvents* used by organic chemists as reaction media, ranging from the rather inert to those with highly pronounced donor or acceptor properties. and *liquid metals and alloys*, and their solutions in other liquids (e.g. in liquid ammonia or in molten salts). Discussions of these have rarely been brought together in one volume from a unified point of view based on common models, and extended as the need arises to cover specific requirements.

The great advances that have been made in the last two decades in deriving approximations of integral equations relating to correlation functions in fluids, and the simulation by computers of the equilibrium and dynamic behaviour of assemblies of particles, have revived the interest of liquid state scientists in theoretical aspects. One of the major problems is the statement of the correct potential functions describing the interactions in the various types of fluids, and of approximating these according to well understood simplified models. However, this approach often disguises the lack of knowledge of the actual interactions by providing a too simplified model which is only capable of describing the properties of highly artificial model liquids rather than those of real liquids. Although the chemist is involved with real liquids, and usually with mixtures of real liquids, the theories cannot as yet provide detailed information about those properties which are required in order to make the best use of these liquids in practical applications. One must, therefore, revert to the phenomenology of these liquids and liquid mixtures, but this field has been somewhat neglected in recent years. In this book there is strong emphasis on the phenomenological aspects of liquids and liquid mixtures, with the understanding that not everything can be generalized and forced into the framework of simple phenomenological expressions, successful though these may be in describing some of the properties of some of the liquids and mixtures of interest. This is the reason that in many cases several phenomenological approaches are described in this book side-by-side, without the presumption of a decision on which is best among them.

This book provides also a considerable amount of factual knowledge so that the properties discussed and described by theoretical or phenomenological expressions are well illustrated. In many ways, the book provides a convenient source of information, in so far as its scope and size permit. The more extensive tables of properties are relegated to appendixes, so as not to break the continuity of the presentation. Among the properties displayed, much space has been devoted to solubilities and phase relationships, since these are often among the prime considerations of chemists in their decision on the suitability of a liquid or a liquid mixture for their purposes.

The book opens with a consideration of the liquid state as a disordered condensed phase, with thermal motion producing randomness, internal equilibrium and isotropic properties, and intermolecular forces producing cohesion and in some cases local order. The thermodynamic and statistical mechanical concepts and relationships used are briefly outlined, as are the phenomenological studies. Detailed considerations of transport properties and irreversible processes, as well as spectroscopic studies of liquids and mixtures, are excluded, although their results are frequently drawn upon to decide a point of discussion.

Pure liquids are discussed from two approaches: in the first, their physics is pursued as far as is profitable, treating the liquid either as a very dense gas (distribution function theories) or as a highly disordered solid (lattice theories); in the second, the structure and properties of actual liquids are described and compared with the results of the theoretical studies. The liquids are classified into atomic liquids (inert gases and metals); molecular liquids, with increasing degrees of interaction, culminating with water; and ionic liquids, that is molten salts, with varying degrees of ionicity. The relationships among these various kinds of liquids may be demonstrated by a schematic diagram.

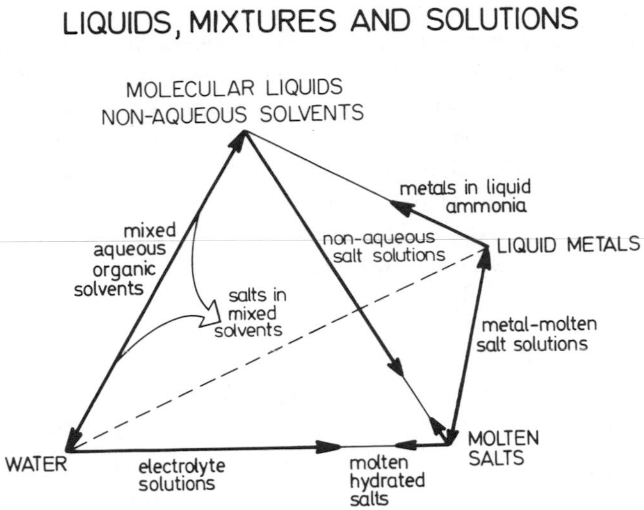

Liquid mixtures and solutions are subsequently discussed. Ideally, theories of the liquid state should provide potential functions which can be applied to dissimilar particles, leading to correlation functions involving all pairs of particles, the same and different, and finally to partition functions and free energies which are functions of the state of the system and the composition. Valiant attempts have been made in this direction, but a chemist is assisted more if it is permitted to take the macroscopic properties of the components of the mixture as granted, and information is then provided on how the properties of the mixture depend on those of the components at the given conditions and as a function of the composition. Heavy reliance is therefore put on the concept of excess functions. The systems are classified mainly according to the intermolecular forces: dispersion forces only (e.g. regular solutions); superposed dipole interactions (solutions of polar and hydrogen-bonded liquids); ion–dipole interactions and ion–ion interactions in a dielectric medium (electrolyte solutions); ionic interactions with no medium (molten salts); and electron–ion or –molecule interactions (solutions of metals). These interactions are usually regarded as 'physical interactions' since they are non-specific and non-directed, although in certain cases it is best to regard them as 'chemical', leading to definite association products, solvates or complexes.

It is hoped that practising chemists employing liquids as their reaction media, as well as advanced students of chemistry, will find this book a helpful introduction to the chemistry of the liquid state. They will gain an appreciation of the importance of this field of science and of its relationships to other fields of knowledge. They will then be able to apply a deeper comprehension to the problems of the liquid state in their work, and to extend their knowledge by studying more advanced texts and the current literature.

The author is indebted to many sources for the material in this book. These have been cited in the lists of references at the ends of tables, in figure-captions, and at the ends of chapters. The more subtle influences on his mental picture of the chemistry of the liquid state, which has found expression in this book, is more difficult to acknowledge. The author has read, and it is hoped at least partly digested, the contents of many books. Following the preface is a list of such general references, which may also be of help to the reader. In his capacity as a university teacher the author had the opportunity of presenting some of the material in this book to his students, and he has profited from this experience too. Fruitful discussions with his graduate students and his colleagues, and in particular with Professor Arieh Ben-Naim, are acknowledged. The financial support of the Alexander von Humboldt Stiftung in the German Federal Republic, and the hospitality of Professor Franz Baumgärtner in his institute in the Kernforschungszentrum Karlsruhe, where a major part of this book was written, are also deeply appreciated. Last but not least, the loving patience of his wife and daughters, taxed by long absences caused by the need to concentrate on the writing of this book, is very gratefully recognized.

A Selected List of General References

Liquids and Liquid Mixtures, by J. S. Rowlinson, Butterworth, London, 2nd ed., 1969.

Liquid State, edited by D. Henderson, constituting *Physical Chemistry, An Advanced Treatise, Vol. 8A*, Academic Press, New York, 1971.

The Liquid State, by F. Kohler, Verlag Chemie, Weinheim, 1972.

Introduction to Statistical Thermodynamics, by T. L. Hill, Addison–Wesley, Reading, Mass., 1960.

Simple Dense Fluids, edited by H. L. Frisch and Z. W. Salsburg, Academic Press, New York, 1968.

Introduction to Liquid State Physics, by C. Croxton, Wiley, London, 1975.

Lattice Theories of The Liquid State, by J. A. Barker, Pergamon Press, Oxford, 1963.

Melting and Crystal Structure, by A. R. Ubbelohde, Clarendon Press, Oxford, 1965.

Water and Aqueous Solutions, by A. Ben-Naim, Plenum Press, New York, 1974.

Water, Vol. 1, edited by F. Franks, Plenum Press, New York, 1972.

Molten Salts, edited by M. Blander, Wiley–Interscience, New York, 1964.

Fused Salts, edited by B. R. Sundheim, McGraw-Hill, New York, 1964.

Liquid Metals, by N. H. March, Pergamon Press, Oxford, 1968.

Mixtures, by E. A. Guggenheim, Clarendon Press, Oxford, 1952.

The Molecular Theory of Solutions, by I. Prigogine, North-Holland, Amsterdam, 1957.

The Solubility of Nonelectrolytes, by J. H. Hildebrand and R. L. Scott, Reinhold, New York, 3rd ed., 1950.

Regular and Related Solutions, by J. H. Hildebrand, J. M. Prausnitz and R. L. Scott, Van Nostrand Reinhold, New York, 1970.

The Physical Chemistry of Electrolyte Solutions, by H. Harned and B. B. Owen, Reinhold, New York, 3rd ed., 1958.

Electrolyte Solutions, by R. A. Robinson and R. H. Stokes, Butterworth, London, 2nd ed., 1959.

Glossary of Symbols

Roman type letters are used for chemical species, *italicized* and Greek letters for physical quantities expressible in numerical terms, and **heavy type** letters for universal physical constants. Mathematical operators have their usual meaning (ln denotes natural, log denotes decadic logarithm), with the addition of ⟨ ⟩ for average quantity over a system and $\rightarrow$ over a symbol designating a vector. Extensive thermodynamic quantities are normally in capital letters, molar quantities in lower case, while partial molar quantities have generally a subscript indicating the relevant component (occasionally a bar $\bar{}$ over the symbol if confusion would otherwise ensue). Superscripts for processes are normally in capital letters, superscripts and subscripts for states in lower case. Generally accepted and widely used symbols are preferred, however, to the systematic application of these generalizations, if in conflict.

The following list includes the main symbols used, their SI units and the equation or page (prefix p.) where the symbol is first defined or applied.

List of Principal Symbols – Latin Letters

A	Helmholz (free) energy (J)
AN	acceptor number (p. 109)
a	molar Helmholz energy (J mol^{-1})
a	distance between centres of cells (m) (p. 78)
a	mean ionic diameter (m) (p. 230)
a_i	(relative) activity of component i (4.44)
$a(k)$	atomic scattering intensity (p. 50)
B_i	ith virial coefficient (B_2, B_3, etc.) (m^3 mol^{-1})$^{i-1}$ (1.44)
B	coefficient of linear term in Jones-Dole viscosity equation (6.19)
b	field-dependence coefficient of dielectric constant (m^2 V^{-2}) (1.32)
b	bound atom scattering length (m) (p. 50)
b	parameter in ionic association theory (6.109)
b	coefficient of the term in log $\gamma_\pm$ linear in the ionic strength (dm^3 mol^{-1}) (6.88)
$b(i,j)$	set of graphs that contain at least one bridge point (p. 63)
b_j	coefficient of Redlich-Kister equation (4.65)
b'_j, b''_j	coefficient of Margules equation (4.66), (4.67)

$b_{i(jk)}$	coefficient in Gibbs energy equation of molten salt mixtures ($J\ mol^{-1}$) (p. 274)
CAV_{ij}	contribution of dielectric cavity to potential energy between ions i and j (p. 241)
COR_{ij}	contribution of core repulsion to potential energy (p. 241)
C_p	heat capacity at constant pressure (JK^{-1})
C_2	total number of contact points of solute (p. 175)
C_{ij}	interaction parameter ($J\ m^3$) (p. 154)
$\mathbf{c}$	speed of light in vacuum, $2.9979 \times 10^8\ m\ s^{-1}$
c_i	concentration of component i (M, $dm^3\ mol^{-1}$)
c_p	molar heat capacity at constant pressure ($J\ K^{-1}\ mol^{-1}$) (1.20)
c_V	molar heat capacity at constant volume ($J\ K^{-1}\ mol^{-1}$) (p. 12)
c_σ	orthobaric molar heat capacity ($J\ K^{-1}\ mol^{-1}$) (p. 13)
c_{ij}	scaling factor of potential in molten salts (7.24)
D	bond dissociation energy (J) (p. 32)
D	electric displacement ($C\ m^{-2}$) (p. 219)
D	distribution ratio (p. 229, 316)
d	density, mass per unit volume ($kg\ m^{-3}$) (p. 10)
d	molecular size parameter (m) (p. 69)
d_{ij}	cation-anion distance in molten salts (m) (p. 260)
DN	donor number ($kcal\ mol^{-1}$, 1 cal = 4.184 J) (p. 109)
E	energy (J), of electronic state, etc., internal energy
E	electric field intensity ($V\ m^{-1}$)
E_A	electron affinity ($J\ molecule^{-1}$)
E_F	Fermi energy (J) (p. 35)
E_κ	activation energy for conductance ($J\ mol^{-1}$) (3.14)
E_η	activation energy for viscosity, fluidity ($J\ mol^{-1}$) (3.14)
$\mathbf{e}$	charge of proton, 1.6021×10^{-19} C
e	molar expansibility ($m^3\ K^{-1}\ mol^{-1}$) (p. 236)
$\mathbf{F}$	Faraday constant, $9.6487 \times 10^4\ C\ mol^{-1}$
f	intermolecular force (N) (p. 21)
f	scaling factor for intermolecular potential (p. 25)
$f(r)$	direct correlation function (p. 52)
f_i	rational activity coefficient of component i (4.43)
GUR_{ij}	contribution of liquid structure to potential energy (p. 240)
G	Gibbs (free) energy (J)
G	affinity between particles (m^3) (2.6)
G	cavity surface-concentration of particles (p. 68)
G_j	group contribution to solubility parameter ($J^{1/2}\ cm^{3/2}$)
δG^{HI}	Gibbs energy change for hydrophobic interactions ($J\ mol^{-1}$) (5.102)
g	scaling factor for molecular diameter (p. 25)
g	dipole correlation parameter (1.35, 3.4)
$g(r)$	pair correlation function (2.5)

$g^{(3)}$	triplet correlation function (p. 72)
g_i	molar Gibbs energy of component i (J mol^{-1})
g_p	coefficient in power series of Gibbs energy of electrolyte mixtures (6.118)
H	enthalpy (J)
H	Hamiltonian function (1.71)
h	Planck's constant 6.6256 x 10^{-34} J s
h	molar enthalpy (J mol^{-1})
$h(r)$	total correlation function (p. 45)
I	ionization potential (J $molecule^{-1}$)
I	ionic strength (mol kg^{-1} or M, mol dm^{-3}) (6.78)
$I(\theta)$	intensity of beam of radiation diffracted through angle θ
K	equilibrium constant (units according to reaction and concentration scales)
K	factor representing the reciprocal of the dielectric constant of a molten salt (C^2 J^{-1} m^{-1}) (3.7, 7.7)
K	scattering parameter (m^{-1}) (p. 317)
K_a	association constant for ion-pair (M, mol dm^{-3}) (6.105)
K_ϕ	equilibrium constant in terms of volume fraction
$K_{B(A)}$	Henry's law constant (B = solute, A = solvent) (Pa) (4.35)
k	Boltzmann's constant, 1.3805 x 10^{-23} J K^{-1}
k	force constant (N m^{-1})
k	wave number (m^{-1}) (2.18)
k_F	Fermi wave number of electrons in metals (m^{-1})
k_p	packing factor (6.44)
k_s	salting (Setchenov) coefficient (M^{-1}, dm^3 mol^{-1}) (6.55)
k'_{ij}	deviation parameter (5.49)
L	length of edge of box containing particles (m) (p. 54)
L_i	partial molal relative heat content of component i (J mol^{-1}) (p. 212)
l_c	average intermolecular distance at critical state (m) (1.1)
(l)	liquid state
M	molar mass (kg mol^{-1})
m	mass of electron, 9.1066 x 10^{-31} kg
m_i	molality of component i (mol (kg $solvent)^{-1}$)
N	Avogadro's number, 6.0225 x 10^{23} molecule mol^{-1}
N	number of molecules
n	number of moles (mol)
n	refractive index (n_D when measured with sodium D line light) (1.30)
n	number of metallic bonds formed between two metal atoms (p. 309)
$n^{(1)}$	singlet distribution function (2.2)
$n^{(2)}$	pair distribution function (2.3)
$n^{(i)}$	*i*-particle distribution function (p. 61)
$n(r)$	radial distribution function

P	pressure (Pa)
P	molar polarization (m^3 mol^{-1}) (1.33)
P	probability that neighbouring sites are occupied (5.13)
P_σ	(saturated) vapour pressure (Pa)
p	momentum (kg m s^{-1})
p	probability that a neighbour of a molecule of type 1 is a molecule of type 2 (p. 139)
p_i	vapour pressure of component i (Pa) (4.32)
$\vec{p}$	momentum vector (Cartesian coordinates) (p. 37)
Q	(total) partition function (1.72)
$Q(b)$	integral function concerning ion-pairs (6.109)
q	internal partition function (p. 38)
q	quadrupole moment (C m^2) (p. 27)
q	geometric parameter for polymers (p. 167)
q	critical distance for association (m) (p. 243)
q_n	collective coordinate (2.47)
q_{ij}	electrical potential energy divided by kT (6.91)
R	gas constant, 8.3144 J K^{-1} mol^{-1}
R	specified distance between particles (m)
R	molar refractivity (m^3 mol^{-1}) (1.30)
R	width, relative to σ, of square-well pontential (1.40)
R_{ij}	mole ratio in dilute reciprocal molten salt mixture (7.63)
r	distance between particles (m)
r	critical index of surface tension (p. 16)
r	ratio of (molar or free) volumes of components of mixture
$r_c(r_{c+},r_{c-})$	(crystal) radius of ion (cation, anion) (m)
r_i	reaction field correction to dipole interaction (5.59)
$r_p(r_s)$	limit of primary (secondary) hydration of ions (m) (p. 215)
r_e	effective radius (m) (p. 215)
r^0	equilibrium distance between two neighbouring molecules (m)
$\vec{r}$	distance vector (Cartesian coordinates) (Fig. 2.1, p. 44)
$\vec{r}^N$	set of N molecules at specified distances r from origin (p. 37)
S	entropy (J K^{-1})
S	overlap integral (p. 32)
S	structuredness (3.5)
S_{comm}	communal entropy (J K^{-1}) (1.3)
$S(k),S(K)$	structure factor (pp. 50, 317)
S_t	generalized limiting slope of properties of electrolyte solutions (units depend on property) (6.87)
S_γ	limiting slope of activity coefficient ($kg^{1/2}$ $mol^{-1/2}$) (6.86)
T	temperature (K)
T_m	melting point (K)
T_b	normal boiling point (K)
T_t	triple point (K)
T_c	critical point (K)

T_0	ideal glass transition point (vanishing configurational entropy) (K) (p. 5)
T_g	glass transition temperature (K) (p. 6)
t	centigrade temperature (°C)
t	ordinal number of a step in a sequence (p. 55)
t_s	generalized partial molar quantity of an electrolyte (7.87) (units depend on property)
U	internal energy (J)
U	potential energy of a system of particles (J)
u	molar internal energy (J mol^{-1})
$u(r)$	mutual pair-interaction energy of molecules (J $molecule^{-1}$)
$\bar{u}(r)$	angle averaged mutual interaction energy (J $molecule^{-1}$)
V	volume (m^3)
v	molar volume (m^3 mol^{-1}) (p. 11)
v_f	free volume (m^3 mol^{-1}) (p. 75)
W	work (J)
W_i	mass of a component i (kg)
w	interaction (exchange) energy (J $molecule^{-1}$) (p. 81) (4.121)
w_i	mass fraction of component i (4.1)
x	mole fraction of a specified component in a binary mixture
x'	equivalent fraction in molten salt mixture (7.5)
x_i	mole fraction of component i
x_{ij}	fraction of particles i and j which are neighbours
Y_m	generalized thermodynamic property of a mixture (units depend on property)
y	generalized molar thermodynamic function (units depend on properties) (p. 7)
y	mole fraction of an electrolyte in a binary electrolyte mixture (6.117)
y_i	partial molar generalized thermodynamic property (4.12)
y_i	molar (concentration) activity coefficient (4.57)
$y(i,j)$	set of graphs (i,j)-irreducible and free from bridge points (p. 63)
Z	atomic number
Z	configurational partition function (1.75)
Z	coordination number for nearest-neighbours (2.28)
Z'	coordination number for next-nearest-neighbours (p. 60)
Z	compressibility factor (p. 90)
Z^z	generalized ion of charge z
z	charge of an ion (algebraic, in units of e)
z	composition descriptor (4.74)

List of Symbols – Greek Letters

α	polarizability (m^3 $molecule^{-1}$)
α	critical index of heat capacity (p. 16)

α	ratio of site occupancy probabilities (5.15)
α	degree of dissociation of an electrolyte (6.105)
α_j	(regular mixture) interaction coefficient (J mol^{-1}) (4.69)
α_j	solvent-sorting coefficient (6.65)
α_P	isobaric thermal expansibility (K^{-1}) (1.10)
α_σ	orthobaric thermal expansibility (K^{-1}) (1.12)
α_{ij}	non-randomness parameter in Renon's equation (p. 173)
β	resonance integral (p. 33)
β	critical index of molar volume (p. 16)
β_j	temperature coefficient of interaction parameter (J K^{-1} mol^{-1}) (4.69)
Γ	surface excess of matter (mol m^{-2}) (1.23)
γ	critical index for pressure-derivative with volume (p.16)
γ	coupling parameter (2.60)
γ_i	molal activity coefficient of component i (4.56)
γ_j	interaction coefficient of excess heat capacity (J K^{-1} mol^{-1}) (4.69)
Δ	change in a thermodynamic property for a process
δ	square root of cohesive energy density (solubility parameter) ($J^{1/2}$ $cm^{-3/2}$) (p. 10) (5.31)
δ	critical index of pressure (p. 16)
δ	Kihara hard-core diameter (m) (p. 26)
δ	perturbation parameter relating to energies (4.103)
δ	size parameter in molten salts (7.14)
δ_i	molar dielectric decrement caused by component i (M^{-1}, dm^3 mol^{-1}) (p. 227)
ϵ_0	permittivity of vacuum 8.8542×10^{-12} C^2 J^{-1} m^{-1}
ϵ	(relative) dielectric constant
ϵ	depth of potential-well (J $molecule^{-1}$) (1.40, 1.42)
ϵ	self interaction parameter (p. 307)
ϵ_h	minimum energy required for a molecule to enter a hole (J $molecule^{-1}$) (p.84)
ζ	fraction of adduct (p. 187)
η	viscosity (Pa s^{-1})
η	reduced density $\pi/6\rho\sigma^3$ (p. 66)
η	electronegativity factor (J mol^{-1}) (p. 304)
θ	angle in polar coordinates
θ	critical index of $(\partial^2 p/\partial T^2)_\sigma$ (p.16)
θ	perturbation parameter pertaining to energies (4.105)
θ_E	characteristic Einstein temperature (K) (p. 85)
θ_{ai}	fraction of contact points of type a in component i (p. 175)
κ	specific conductance (Ω^{-1} m^{-1})
κ	reciprocal thickness of ionic atmosphere (m^{-1}) (6.70)
κ_S	adiabatic compressibility (Pa^{-1}) (1.17)
κ_T	isothermal compressibility (Pa^{-1}) (1.13)

Λ	ternary (reciprocal) ion mixing enthalpy in molten salts (J mol^{-1}) (7.50)
Λ	momentum partition function (reciprocal, per degree of freedom) (1.73)
Λ	interaction parameter in Wilson's equation (5.42)
λ	wavelength of radiation (m)
λ	scaling parameter (pp.68, 232)
λ	equivalent conductivity (Ω^{-1} m^2 mol^{-1})
λ	dispersion contribution to solubility parameter ($J^{1/2}$ $cm^{-3/2}$) (p.179)
λ	coulomb interaction energy (J mol^{-1}) (7.20).
λ_i	absolute activity of component i (4.17)
λ_{ij}	short-range interaction parameter (6.95)
μ	chemical potential (J $molecule^{-1}$) (1.83)
μ	dipole moment (Debye, 3.33564 x 10^{-30} C m)
μ_i	chemical potential of component i (J mol^{-1}) (4.14)
ν	vibration frequency (s^{-1})
$\nu(\nu_+, \nu_-)$	number of ions (cations, anions) into which electrolyte is dissociated
ν_{CT}	vibration frequency for charge transfer absorption band
ρ	number density (molecules m^{-3})
ρ	perturbation parameter pertaining to sizes (4.104)
ρ	size factor (p.304)
ρ	electrical resistivity (Ω m) (p.317)
ρ_0	number density of close-packed system $2^{1/2}\sigma^{-3}$ (molecule m^{-3})
ρ_i	perturbation parameter (7.9)
σ	molecular diameter (m)
σ	surface tension (N m^{-1}) (p.13)
σ	scattering cross-section (p.50)
σ	perturbation parameter pertaining to sizes (4.106)
σ	degree of order in a mixture (4.122)
$\sigma(\kappa a)$	parameter of Gibbs energy of solvent (6.81)
$\sigma_M(\sigma_X)$	softness parameter of cation (anion) (7.27, 7.28)
τ	induction and orientation contribution to solubility parameter ($J^{1/2}$ $cm^{-3/2}$) (p.179)
$\tau(\kappa a)$	parameter of Gibbs energy of electrolyte (6.75)
Φ	intermolecular virial function (J) (p.40)
Φ	local volume fraction (5.39)
ϕ	volume fraction of a component (4.6)
ϕ	osmotic coefficient (6.11)
$\phi(i,j)$	interaction path (p. 63, Fig. 2.8)
$\phi(r)$	potential function for polarizable ions (J $molecule^{-1}$) (7.24)
ϕ	angle in polar coordinates
ϕ_i	volume fraction of component i (4.8)
ψ	wave function
ψ	RT times a generalized dimensionless function of electrolyte mixtures (J mol^{-1}) (6.124)

ψ_a	electrostatic potential due to ionic atmosphere (V) (6.72)
ψ_j	electrostatic potential caused by ion j (V)
$\psi(r/d)$	potential function for molten salts (J m molecule^{-1}) (7.8)
ψ_{12}	binary interaction parameter (J mol^{-1}) (5.56)
ξ	random number between 0 and 1 (p.55)
ξ	coupling parameter (p. 67)
ξ	fraction of self-associated species (dimer) (p.184)
χ	interaction parameter for z-regular solutions (J mol^{-1}) (5.45)
χ_i	electronegativity of i (kJ mol^{-1})$^{1/2}$ (p.304)
ω	acentric factor (3.1, 5.46)

Superscripts

$\circ$	for standard state
∞	for (infinitely) dilute solutions
$*$	of pure component
$*$	of reference, for conformal substances (p.25)
$'(\prime\prime)$	designation of phases in equilibrium
$'$	equivalent fraction composition scale in molten salts (7.5)
$\hat{}$	equimolar binary mixture
$\tilde{}$	reduced (divided by the relevant standard quantity)
E	excess function of mixing (p.144)
F	of fusion (negative sign : of freezing)
g	gaseous state
l	liquid state
M	change on mixing
s	solid state
tr	of transfer
V	of vaporization

Subscripts

c	at the critical state
h	of hydration
i	of component i
i(j,k)	of molecule i (j or k)
m	of a mixture
r	reduced, i.e. divided by the corresponding quantity for the critical, or another specified, state (p. 13)
s	of surface
u	of unionized part of electrolyte (p. 242)
σ	orthobaric state (liquid–vapour equilibrium line) (p. 9)

Chapter 1

Introduction to Liquids

A. The Liquid State

The world picture of the ancients, as well as classical science, recognized the existence of matter in three well defined states of aggregation: solid, liquid and gas. A closer examination, and attempts at a definition of these states of aggregation, brings out the complications, i.e. the need for quantitative criteria for placing matter under given conditions in one of these states. This also leads to the recognition of the existence of intermediate states, such as glasses, liquid crystals, amorphous solids, supercritical fluids, etc.

Operationally, the criteria usually employed in a qualitative sense for classification into one of the states of aggregation, utilizing common experience rather than exact measurement, are the properties of compressibility and fluidity used in combination. Matter that under the given conditions has low compressibility is considered to exist as a condensed phase, i.e. to be either liquid or solid, while if it cannot support sheer stresses and has a high fluidity, it is considered to be *fluid*, i.e. either liquid or gas. A combination of the criteria then classifies uniquely the state. Corollaries of the criteria are that solids retain their shape and have a free surface; liquids, while having a free surface, take the shape of the containing vessel; and gases fill the container completely, lacking a free surface.

The molecular criteria for classification utilize two different properties, again on a rather qualitative basis. Condensed phases are those where the intermolecular distances are similar to the molecular sizes, and fluids arc disordered aggregates of molecules ('molecules' here includes atoms and ions). A liquid, then, is a disordered aggregate of molecules in great proximity.

In order to introduce quantitative criteria it is necessary to consider the range of existence of matter in the liquid state. Since matter is examined ordinarily at atmospheric pressure of air in the laboratory, roughly 10^5 Pa, the liquid range is commonly considered to extend from the freezing point, or the melting point T_m, to the normal boiling point T_b. One may also say, however, that at normal laboratory room temperature the liquid range extends from as high pressures as may be attained down to the saturated vapour pressure P_σ. It is therefore obvious that the boundaries of the liquid range are given by both the temperature and pressure-variables, as shown in Fig. 1.1. It is now apparent that the $P_\sigma(T)$ curve determines the boundary between the liquid (L) and vapour (V) states, and represents the lowest pressure at a given temperature and the highest temperature at a given pressure where the liquid can exist under equilibrium conditions. The normal boiling point T_b ($P = 1.01325 \times 10^5$ Pa) is therefore not a unique limit. The melting point refers to the equilibrium liquid under the combined pressure of its vapour and of air to a total of one atmosphere (1.01325×10^5 Pa), from which it is seen that the liquid is saturated by dissolved air. If the air is removed, and the liquid is left under its own vapour pressure, the equilibrium systems of liquid and vapour solidifies at the triple point T_t, where gas (vapour), liquid and solid are in equilibrium. The almost vertical $P(T)$ curve terminating at the triple point, obtained from the Clausius–Clapeyron equation, is the demarcation line between solid (S) and liquid (L). Since both are condensed phases of low compressibility, the volume change on melting is small, and therefore the slope of the $P(T)$ equilibrium curve $\mathrm{d}P/\mathrm{d}T = \Delta S^F/\Delta V^F$ is nearly infinite, and the difference between T_m and T_t is quite small. For example, water has $T_m = 273.15$ K and $T_t = 273.16$ K, $\Delta T = T_t - T_m = 0.01$ K (exceptionally > 0, since ordinarily $\Delta T < 0$ because $\Delta V^F > 0$).

The upper limit of existence of the liquid state, distinct from the gaseous (vapour) and solid states, depends on the extension of the $P(T)$ curves to high values of these variables. Experience so far has not shown a limit to the $P(T)$ boundary between the solid and liquid states, nor has theory indicated any requirement of its existence, that is, compression to the highest pressures obtainable at present, somewhat above 10^9 Pa, does not remove the distinction between liquid and solid phases that exist in equilibrium. This is, however, not so for the boundary between the liquid and vapour, where the well known *critical point* (c in Fig. 1.1) sets the limit to the distinction between these two states of aggregation. At $P < P_c$ and $T < T_c$, where P_c and T_c are the critical pressure and temperature, there is a distinction between liquid and vapour; the two may coexist in equilibrium (along the saturation line $P_\sigma(T)$) with a boundary surface separating them. At $P > P_c$ or $T > T_c$ the matter is in the fluid state, but no distinction between liquid and vapour can be made. Above T_c, for instance, a gas cannot be

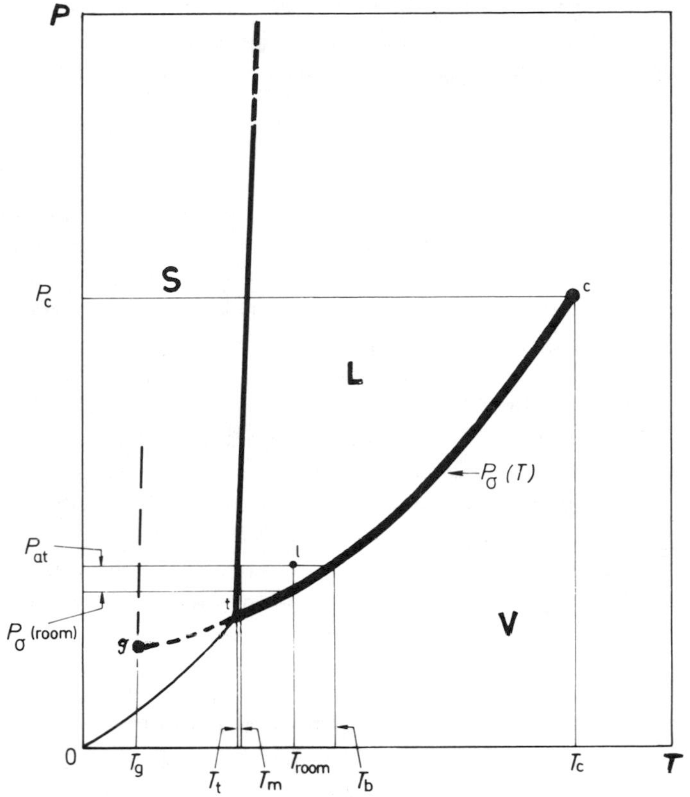

Figure 1.1. A schematic pressure-temperature diagram for a substance that is liquid under laboratory conditions (point l). The heavy line g–t–c is the lower demarcation line of the liquid state L against the vapour state V – the saturated vapour pressure curve $P_\sigma(T)$. The almost vertical line rising from point t (the dashed line rising from point g) is the demarcation line between the liquid (supercooled liquid) state L and the solid state S. The temperatures of glass transition T_g, triple point T_t, melting T_m, the room T_{room}, normal boiling T_b, and critical point T_c, are indicated. The pressures of the saturated vapour P_σ, the atmosphere P_{at}, and the critical point P_c are also indicated.

liquified (i.e. a fluid with a bounding surface cannot be produced) by compression to any pressure, hence the appellation critical temperature.

A fluid possesses a set of *critical constants*, T_c, P_c, the critical molar volume v_c, density d_c, concentration ρ_c (number density, particles per unit volume, $\rho_c = \mathbf{N}v_c^{-1}$) etc. Critical temperatures of fluids vary very considerably, from $T_c = 5.3$ K for helium to an estimated[1] 23 000 K for tungsten. Critical pressures vary somewhat less, but critical volumes vary relatively little; v_c ranges from 4×10^{-5} to 4×10^{-4} m^3 mol^{-1}; as do critical densities, d_c ranges from 31 kg m^{-3} for hydrogen to 4700 kg m^{-3} for mercury, but for most liquids it is in the range 200 to

500. It is significant to note that the average distances $l_c = (v_c/N)^{1/3}$ between the molecules or atoms in the critical state range only from 0.4 to 0.9 nm. Also, there exists an almost strict proportionality between this distance and the molecular diameter σ (see section 1C(a))

$$l_c/\sigma = 1.50 \pm 0.16 \tag{1.1}$$

for such diverse fluids as the noble gases, inorganic gases of small molecules such as hydrogen, nitrogen, carbon dioxide, ammonia, hydrocarbons, substances such as methanol, and metals such as mercury. Since the potential energy for attraction between the molecules decreases with the sixth power of the distance (section 1C(b)), this energy therefore decreases by $(l_c/\sigma)^6$, i.e. just over an order-of-magnitude when the inter-molecular distance increases from contact in the liquid, at σ, to the critical state of the liquid, at l_c. A fluid can therefore be said to cease to exist as a liquid when the intermolecular separation exceeds $(3/2)\sigma$ and the attractive potential energy falls below about (1/10) that at contact. This, then, is the molecular criterion for the end of the (upper) liquid range of existence, while the critical point serves as the thermodynamic criterion.

The lower limit of liquid existence is generally taken, as mentioned above, as T_t. However, the phenomenon of supercooling is known and a liquid may exist in the *supercooled state* at $T < T_t$, with a smooth transition to the *glassy state* in some cases. The latter may be defined as a supercooled liquid with a viscosity exceeding 10^{12} kg m^{-1} s^{-1}, which is *not* in internal configurational equilibrium. In the supercooled state, the liquid is in fact in internal equilibrium, although it is metastable with respect to some crystalline phase. For example, calcium nitrate tetrahydrate is almost indefinitely stable as a liquid at room temperature, $T_{room} < T_m = 315.9$ K.

It has been argued[2] that supercooled liquids should be considered as legitimate representatives of the liquid state. The fact that a phase exists with a lower Gibbs energy than the liquid, so that a spontaneous transformation (crystallization) is possible, although it may be slow, is not an inherent property of the liquid state and according to some views may be disregarded. Indeed, all liquids (and solids for that matter) are in principle metastable with regard to the gas phase, and only the experimental conditions, i.e. the presence of a gravitational field and their being kept in closed containers, prevent their complete evaporation, in time, at any temperature. Thus the restrictive experimental conditions: freedom from dust, smooth walls of vessel, freedom from vibration, rapid cooling, etc., required to bring a liquid to the supercooled state do not detract from its 'legitimacy'.

If this view is accepted, the liquid range is extended considerably below T_t, and the question arises where the limit is, if there is indeed a limit before absolute zero. When a liquid is cooled, say from T_1 to T_0, it loses configurational and vibrational entropy to the extent

$$\Delta S(\mathrm{l}) = \int_{T_1}^{T_0} C_p(\mathrm{l})\,\mathrm{d}(\ln T),$$

and the amount of disorder, which is a characteristic property of the liquid, decreases. If a crystalline, that is ordered, solid exists in the corresponding temperature range, from T_1 down to T_0, it will also lose entropy ΔS(s) according to a similar expression, noting however that since $C_p(s) < C_p(l)$ less entropy is lost. At the melting point, the entropy of the liquid exceeds that of the solid by ΔS^F, the entropy of fusion. The temperature T_0 is therefore the lowest temperature for which

$$\int_{T_m}^{T_0} [C_p(s) - C_p(l)]\, d(\ln T) \leqslant \Delta S^F \tag{1.2}$$

that is, the entropy of the (disordered) liquid is larger than that of the (ordered) solid. Since it is inconceivable that a liquid can be less disordered than the corresponding crystal and have a lower entropy, T_0 must be the lower limit of existence of the liquid state. The operation represented in eq. (1.2) is illustrated in Fig. 1.2. It is seen that c_p of the (supercooled) liquid drops suddenly to a much lower value, so that the (single hatched) area between the liquid and crystal

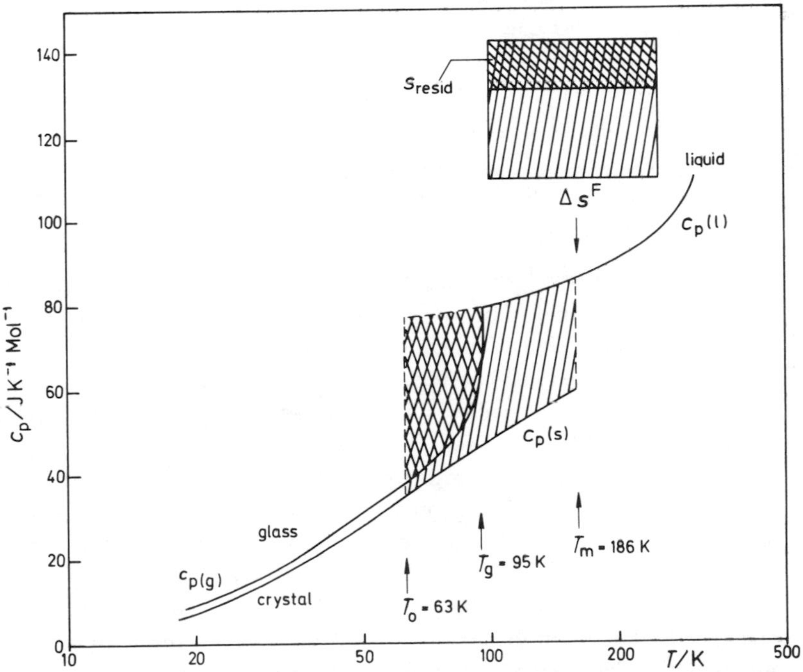

Figure 1.2. The molar heat capacity at constant pressure of ethanol plotted against the temperature (logarithmic scale) for the liquid $c_p(l)$, the glass $c_p(g)$, and the solid $c_p(s)$. Since entropy is represented as an area in this diagram, a rectangle with area Δs^F is drawn above T_m to represent the entropy of fusion. It equals in area that between the curves $c_p(l)$ and $c_p(s)$ and the verticals T_0 and T_m (hatched plus cross-hatched). The cross-hatched area corresponds to the experimentally unrealizable extension of the supercooling of the liquid below T_g, and to the residual entropy of the ensuing glass.

$c_p(\ln T)$ curves is less than the area represented by Δs^F. The lowest temperature by which this drop should have occurred is T_0; actually it occurs at a few to a few tens of kelvins higher, at T_g. This is the experimental *glass transition temperature*, whereas T_0 may be called the ideal glass transition temperature. The difference $T_g - T_0$ corresponds to a residual entropy (the cross-hatched area) left in the disordered phase, the glass, at absolute zero of temperature.

The ideal glass transition temperature T_0 is the lowest limit below which the liquid no longer possesses configurational entropy. Below it, therefore, the liquid is indistinguishable from the crystalline solid in this respect, and T_0 may therefore be construed as a critical temperature. Practically, the liquid state extends down to T_g, where it ceases to be at internal thermodynamic equilibrium since it cannot take up all the configurations available. This extension is shown in Fig. 1.1 by the heavy dashed line terminating at point g.

The configurational entropy is here seen to be a distinguishing criterion of the liquid. For monatomic fluids, or those made up of non-polar spherical molecules, a large part of the configurational entropy is the *communal entropy,*[3] which is essentially a translational entropy. In the liquid, each of the N molecules can be found anywhere in the total volume V, and they are indistinguishable from one another in this respect. This contributes a factor $V^N/N!$ to the partition function. In the solid, the molecules are confined to *cells* of volume V/N, and can be distinguished by their position in the lattice, so that the factor in the partition function is $(V/N)^N$. The communal entropy is obtained from the ratio of these factors

$$S_{\text{comm}} = \mathbf{k} \ln[V^N/N!]/[(V/N)^N] = N\mathbf{k} \tag{1.3}$$

or $s_{\text{comm}} = N\mathbf{k} = \mathbf{R}$. Liquids with non-spherical shapes and non-radial forces have orientational, conformational and other contributions to the configurational entropy, beyond the positional contribution which constitutes the communal entropy. Loss of configurational entropy below T_g (practically) or T_0 (ideally) is therefore loss of positional degeneracy, and constraint of the molecules in specified positions and orientations.

Summarizing, the liquid state can be characterized by a set of criteria, thermodynamic and molecular. The former set the limits for existence of matter in the liquid state as $T_g < T < T_c$ (or if the requirement of internal equilibrium is relaxed, $T_0 < T < T_c$) and $P_g < P < P_c$ (where P_g is rarely known). The latter relate to the average intermolecular distances $l \leqslant (3/2)\sigma$ and to the configurational entropy, $s_{\text{conf}} \geqslant \mathbf{R}$. These are the quantitative expressions of the definition (p. 2) that a liquid is 'a disordered aggregate of molecules in great proximity'.

B. Bulk Properties of Liquids

A detailed description of the properties of specific groups of liquids and of specific liquids is deferred until Chapter 3. Here general properties will be dealt with, and it should be remembered that in this book only the equilibrium properties of liquids are discussed. This will be done under the headings of bulk properties, molecular-statistical properties, and structure. The first heading deals with the phenomenology of material in bulk in the liquid state. The second deals with the relationship of these properties to the molecular properties of the material (sections 1C, 1D(a)). The third deals with the molecular distribution function, and the remnant structure in the liquid (section 2A).

The bulk properties of liquids are mainly thermodynamic in nature, although some static electrical and optical properties (the dielectric constant and the refractivity) should also be mentioned. Transport and relaxation properties, which are also of great importance, are time-dependent, and hence outside the scope of this book.

a. Thermodynamic properties of phase transitions

Material in the liquid state shows two *first-order transitions:* first-order transitions: freezing* and evaporation, symbolized by superscripts F and V. In these transitions (Fig. 1.3), the Gibbs energy G of the material and its first derivatives with respect to the variables of state, the temperature T and the pressure P, show discontinuities. The second derivatives become infinite

$$-(\partial^2 G/\partial T^2)_P^{\mathrm{F\ (or\ V)}} = (\partial S/\partial T)_P^{\mathrm{F\ (or\ V)}} = T^{-1} C_{\mathrm{p}}^{\mathrm{F\ (or\ V)}} = \infty \tag{1.4a}$$

$$-(\partial^2 G/\partial P^2)_T^{\mathrm{F\ (or\ V)}} = -(\partial V/\partial P)_T^{\mathrm{F\ (or\ V)}} = V\kappa_T^{\mathrm{F\ (or\ V)}} = \infty \tag{1.4b}$$

Equation (1.4a) means that at the transition temperature at a given pressure the system can absorb (release) heat by liquid evaporation (freezing) without change of temperature, i.e. with an infinite heat capacity C_{p}. Similarly, eq. (1.4b) means that the system can absorb (release) mechanical compression (expansion) work without changing the pressure, i.e. with an infinite isothermal compressibility κ_T.

The first-order transitions are characterized by definite differences between the molar properties of the liquid and solid or gas: $\Delta y^{\mathrm{F}} = y^{\mathrm{l}} - y^{\mathrm{s}}$ is the difference on freezing between the generalized property y of liquid and solid, and similarly $\Delta y^{\mathrm{V}} = y^{\mathrm{g}} - y^{\mathrm{l}}$ for evaporation. A property such as Δh^{F} is properly called the molar heat change on fusion, commonly abbreviated to the (latent) heat of fusion. The quantities Δy^{F} are normally positive, notable exceptions being Δv^{F} for water, rubidium nitrate, antimony, and a few other substances. The corresponding quantities Δy^{V} are also positive, except $\Delta c_{\mathrm{p}}^{\mathrm{V}}$ which is always negative. These property changes of fusion and evaporation are functions of the temperature and of the pressure, and change along the coexistence curves, solid + liquid and liquid + vapour (gas) respectively. The changes are moderate, and over a short range of (T,P)

*Commonly, the superscript F denotes fusion rather than freezing. In this section attention is focused on the liquid, hence the proper transition to consider is freezing, with, for example, $\Delta h^{\mathrm{Freezing}} = -\Delta h^{\mathrm{Fusion}}$.

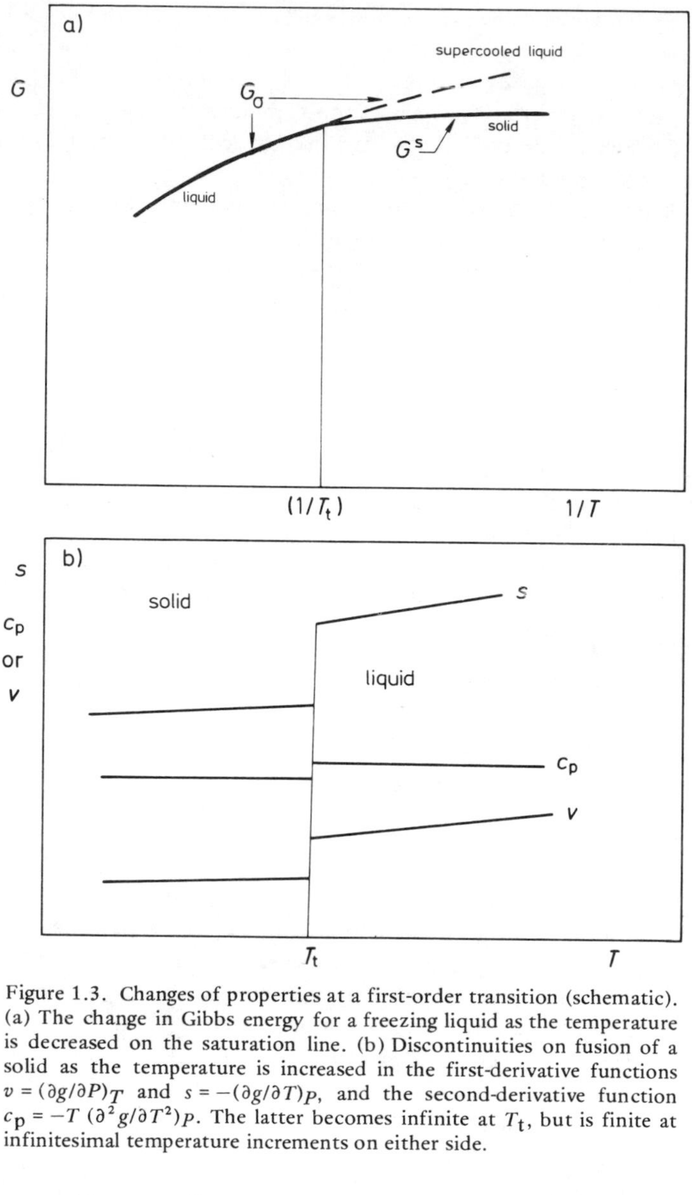

Figure 1.3. Changes of properties at a first-order transition (schematic). (a) The change in Gibbs energy for a freezing liquid as the temperature is decreased on the saturation line. (b) Discontinuities on fusion of a solid as the temperature is increased in the first-derivative functions $v = (\partial g/\partial P)_T$ and $s = -(\partial g/\partial T)_P$, and the second-derivative function $c_p = -T\,(\partial^2 g/\partial T^2)_P$. The latter becomes infinite at T_t, but is finite at infinitesimal temperature increments on either side.

the functions Δy^F and Δy^V can be considered as constants. Nearing the critical point, Δy^V approaches zero, and

$$\lim_{T \to T_c} \Delta y^V \; (= \Delta h^V, \Delta v^V, \Delta s^V, \Delta c_p^V, \Delta \kappa_T^V, \text{etc.}) = 0 \tag{1.5}$$

At a given external pressure, these first-order phase transitions occur at a definite temperature, as required by the phase rule (one component, one restraint – fixed pressure – two phases in equilibrium; variance = 1 + (2 – 1) – 2 = 0). At atmospheric pressure, 1.01325×10^5 Pa, $T^F \equiv T_m$ for freezing (fusion, melting) and $T^V \equiv T_b$ for evaporation (boiling).

The temperature and pressure for a two-phase equilibrium system are related by the Clapeyron equation

$$dP/dT = \Delta s^{F\ (\text{or } V)}/\Delta v^{F\ (\text{or } V)} \tag{1.6}$$

Consider now the liquid-vapour equilibrium system, which exists along the *saturation line*, also called the orthobaric liquid curve (line t–c in Fig. 1.1), designated by subscript σ. If the virial equation of state to the second term is used for the vapour, $v^g = \mathbf{R}TP^{-1} + B_2$, and if the volume of the liquid is neglected, so that $\Delta v^V \simeq v^g$, and if Δs^V is replaced by $\Delta h^V/T$, see below, then

$$(\partial P/\partial T)_\sigma \simeq P\Delta h^V/\mathbf{R}T^2(1 + B_2/v^g) \sim P\Delta h^V/\mathbf{R}T^2 \tag{1.7}$$

The second approximate equality is the familiar Clausius-Clapeyron equation. The magnitude of the approximation can be seen by examining the case of benzene. At 353 K ($\sim T_b$), $v^g = 29.0 \times 10^{-3}$ m^3, $B_2 = -0.96 \times 10^{-3}$ m^3 and v^l = 0.096×10^{-3} m^3; neglecting B_2/v^g introduces and error of 3%.

Most of the properties of a liquid are measured when it is in equilibrium with a small quantity of vapour, somewhere along the σ line. This equilibrium state exists, as long as $P < P_c$, $T < T_c$, between the *bubble point*, where the first minute quantity of vapour is produced in a liquid if, for example at a given temperature T, the pressure is lowered below the vapour pressure $P_\sigma(T)$, and the *dew point*, where the first drop of liquid is produced in a vapour if, for example at a given pressure P, the temperature is lowered below $T_\sigma(P)$.

Along the saturation line $\Delta g^V = 0$ because of the equilibrium between the two phases, hence

$$\Delta s^V = \Delta h^V/T(p_\sigma) \tag{1.8}$$

The heat of vaporization may be obtained from the vapour pressure curve, using (1.7), or by direct experiment, calorimetrically. It may also be estimated from (1.8), using the empirical *Trouton's rule*, applicable to simple, non-associating liquids

$$\Delta s^V \text{ (at } T_b) = 85 \pm 10 \text{ J K}^{-1} \text{ mol}^{-1} \tag{1.9}$$

This rule compares the entropies of vaporization of liquids at a constant (atmospheric) pressure, but since the critical pressures of liquids vary widely (p. 20), this pressure does not constitute a 'corresponding state' (see section 1C). Since the critical molar volumes of liquids vary so much less, comparison at a constant volume of vapour, a suitable multiple of the critical volume, seems more appropriate. *Hildebrand's rule* states that liquids have equal entropies of vaporization at equal vapour volumes; $\Delta s^V = 84.2$ J K^{-1} mol^{-1} at $v^g = 0.05$ m^3 mol^{-1}.

Trouton's rule applies to liquid metals, as far as boiling points are known, with few exceptions (gallium and bismuth have *ca.* 30% high values of Δs^V), as it does to inorganic or organic liquids. Since the observed boiling points vary over an enormous range, so do perforce the derived heats of vaporization, according to (1.8)*.

The corresponding energies of vaporization, $\Delta u^{V} = \Delta h^{V} - P\Delta v^{V}$, are a measure of the internal cohesive forces, in particular when divided by the molar liquid volume to give the *cohesive energy density*, $\Delta u^{V}/v^{l}$, which is set equal to δ^2 (cf. section 5A(b)). The quantity δ, called the solubility parameter, is important for describing the properties of mixtures, and is further discussed in sections 5A(c) and 8A(b). A similar measure is the *internal pressure*, $P_i = (\partial u/\partial V)_T = T(\partial P_\sigma/\partial T)_V - P$, where the last term, $-P$, is negligible for liquids. The temperature-derivative of the vapour pressure, occurring in the second equation, is discussed below, section 1B(b).

The process of melting is not characterized by a uniformity in entropy change as is the case for the process of evaporation, although some regularities can be discerned for certain restricted groups of substances. Entropies of melting $\Delta s^{F} = \Delta h^{F}/T$ in analogy with (1.8) are, as a rule, not measured under 'corresponding' conditions, since T_m or T_t is not a constant fraction of T_c†. Solids melting to monatomic liquids, or to liquids consisting of monatonic ions, show $\Delta s^{F}/\mathbf{R}$ values ranging from near 0.8 (alkali metals) to 1.7 (noble gases) with typical values 1.3 ± 0.2 covering liquid metals, ionic melts, and small spherical molecules, where no new rotational levels are excited on melting. Solids melting to liquids with more complicated molecules show higher entropies of melting (cf. p. 6). The corresponding volume change of melting is usually positive, and $\Delta v^{F}/v^{s}$, the volume increase on fusion relative to the volume of the solid, is in the range 0.05 to 0.25, with a typical value of 0.15 for monatomic (atomic or ionic) substances. Negative values of Δv^{F} are not ruled out and occur when peculiarities of packing are shown by the solids. No valid generalizations concerning the energies, enthalpies or heat capacity changes on melting can be made.

Since the equilibrium between solid and liquid phases is attained over a severely restricted temperature range at accessible pressures (and since no critical point for solid-liquid transitions has been found), the variation of the above Δy^{F} functions with temperature and with pressure are of relatively little importance, and not well known.

b. Mechanical and thermal properties

The density d of liquid is its mass per unit volume, and the reciprocal of its specific volume. At standard room temperature, 298 K, it ranges from 0.6×10^3 kg m^{-3} for pentane to 13.6×10^3 kg m^{-3} for mercury. At other

*The reverse statement should perhaps be preferred: since the intermolecular forces lead to a wide range of heats of vaporization, a wide range of boiling temperatures results from Trouton's rule.

†The relationship $T_t/T_c = 0.56$ applies within 1% to the heavier noble gases Ne, Ar, Kr and Xe, and to a few other fluids, notably BF_3. Higher values are rare ($C(CH_3)_4$ has a ratio 0.59, SiF_4 and SF_6 have ratios 0.70); most inorganic and organic fluids, with both simple and complicated molecules, show ratios in the range 0.25 to 0.45, while metals have ratios even lower than 0.25.

temperatures the range is wider: liquid hydrogen has a density of 0.070×10^3 kg m^{-3} at T_b = 20 K, while liquid gold has d = 17.1 x 10^3 kg m^{-3} at T_m = 1336 K. The *molar volumes* $v = M/d$ are important properties of liquids which range even wider than the densities, from 5.4×10^{-6} m^3 mol^{-1} for liquid beryllium at T_m = 1556 K to values for high-molecular-weight organic liquids, such as perhydrosqualene 5.2×10^{-4} m^3 mol^{-1} or organometallic liquids such as tetra-dodecyltin 8.9×10^{-4} m^3 mol^{-1} at room temperature. The molar volumes, and densities, are temperature- and pressure-dependent, and the coefficients of these functions are important quantities.

The *isobaric thermal expansibility*

$$\alpha_P = V^{-1}(\partial V/\partial T)_P = v^{-1}(\partial v/\partial T)_P = -d^{-1}(\partial d/\partial T)_P \tag{1.10}$$

is readily measurable for liquids with low vapour pressures. Over a limited temperature range the density often shows linear behaviour (t is the centigrade temperature)

$$d = a - bT = a' - bt \tag{1.11}$$

and in this instance α_p itself is mildly temperature-dependent ($\alpha_p \simeq ba^{-1}$ $(1 + ba^{-1}T)$). Actually it is α_σ which is measured, since the pressure of a saturated liquid which is heated is not constant. At low vapour pressures the difference between α_p and α_σ is negligible; at higher vapour pressures

$$\alpha_P = \alpha_\sigma + \kappa_T(\partial P/\partial T)_\sigma \tag{1.12}$$

must be applied (see below). Thermal expansibilities are ordinarily in the range (0.7 to 1.5) x 10^{-3} K^{-1} for organic liquids, (0.15 to 1.0) x 10^{-3} K^{-1} for molten inorganic salts, and (0.07 to 0.3) x 10^{-3} K^{-1} for liquid metals.

The *isothermal compressibility*

$$\kappa_T = - V^{-1}(\partial V/\partial P)_T = - v^{-1}(\partial v/\partial P)_T = d^{-1}(\partial d/\partial P)_T \tag{1.13}$$

is usually obtained from volume measurement at high pressures, up to 3 x 10^8 Pa. The *Tait equation*, originally[4] written

$$(1 - V/V^o)/P = A/(B + P) \tag{1.14a}$$

but usually employed in the modified form

$$(1 - V/V^o) = C \log(B + P)/(B + P^o) \tag{1.14b}$$

where $V^o = V(P^o = 1.01325 \times 10^5$ Pa), is often used. The constant C is temperature-independent, so that B is left to carry the burden of the temperature-dependence. Of interest is $\kappa_T^o = \lim(P \to 0)\kappa_T$, which does not differ much from $\kappa_T(P^o = 1.01325 \times 10^5$ Pa), and is obtained from (1-14b) as

$$\kappa_T^o = (\ln 10)C/(B + P^o) \tag{1.15}$$

The value of C for a variety of organic liquids is *ca.* 0.2, for water it is 0.315, while B is in the range (0.4 to 3.0) x 10^8 Pa. Isothermal compressibilities are thus in the range (0.15 to 2.0) x 10^{-9} Pa^{-1}, the lower limit referring to some molten salts and liquid metals with strong cohesive forces, the higher one to bulky organic liquids. The pressure dependence at high pressures of κ_T can be quite marked, and accurate measurements are not easy.

There are two other ways to obtain values of κ_T: from the expansibility and the thermal pressure coefficient

$$\kappa_T = \alpha_P (\partial P/\partial T)_V^{-1} \tag{1.16}$$

(for the latter, see below), an expression obtained from the relation between the pressure coefficient of α_p and the temperature coefficient of κ_T: $(\partial \alpha_p/\partial P)_T = V^{-1}(\partial^2 V/\partial T \partial P) = -(\partial \kappa_T/\partial T)_P$, and from the *adiabatic compressibility*, κ_s. This quantity is readily obtained from the speed of sound at moderately high frequencies ($10^6 - 10^8$ s^{-1}), u

$$\kappa_S = -V^{-1}(\partial V/\partial P)_S = d^{-1}u^{-2} \tag{1.17}$$

and is related to the isothermal quantity by the ratio of specific heats: $\kappa_T/\kappa_S = c_p/c_v$. From this follows

$$\kappa_T = \kappa_S + Tv\alpha_P^2/c_p \tag{1.18}$$

where the second term constitutes a correction term which can amount to as much as 20%. Lack of knowledge of c_p is the main obstacle to the use of (1.18).

The *isochoric thermal pressure coefficient* $(\partial P/\partial T)_V$ can be measured directly, but this has not (except mainly for Hildebrand's works[5]) been done except for a few liquids. It can, of course, be derived from eq. (1.16), if the data are available. A value such as 9×10^5 Pa K^{-1}, that for diethyl ether at 293 K, is typical for $(\partial P/\partial T)_V$, which is however quite strongly temperature-dependent, $(\partial^2 P/\partial T^2)_V < 0$ and becomes $-\infty$ at T_c. Indeed, at T_c, $(\partial P/\partial T)_V = (\partial P/\partial T)_\sigma$, and for $T > T_c$ the thermal pressure coefficient for the fluid becomes almost temperature-independent. The thermal pressure coefficient along the saturation line, $(\partial P/\partial T)_\sigma$, is the normal vapour pressure curve, which has already been expressed by means of the Clapeyron and Clausius–Clapeyron equations (1.6) and (1.7).

The vapour pressure is often empirically expressed with high accuracy by the *Antoine equation*[6]

$$\log P_\sigma = a - b/(T + c) \tag{1.19}$$

The constant c equals -43 K for many liquids, and the thermal coefficient $(\partial P/\partial T)_\sigma = Pb(\ln 10)(T + c)^{-2}$, from which, in combination with eq. (1.7), the heat of vaporization is obtainable: $h^V \simeq \mathbf{R}b(\ln 10)(1 + c/T)^{-2}$. The *molar heat capacity* at constant volume, c_v, is difficult to measure except near T_c, where $(\partial P/\partial T)_V$ is small, so that the pressure developed by heating at constant volume is not too large. The molar heat capacity at constant pressure c_p is measurable in principle only via c^{1+g}, which is measured directly, and suitable correction terms. The quantity c^{1+g} refers to a liquid confined in the calorimeter vessel so that a fraction x^g of the total amount is in the gaseous phase above the liquid.

$$c_p = c^{1+g} + T(\partial P/\partial T)_\sigma v(2\alpha_p - \kappa_T(\partial P/\partial T)_\sigma) - x^g T(v^g - v)\,(\partial^2 P/\partial T^2)_\sigma \tag{1.20}$$

where quantities without superscript refer to the liquid. The last term is negligible if x^g is made sufficiently small, and $c_\sigma = c^{1+g} + T(\partial P/\partial T)_\sigma$ differs from the measured quantity by a small amount only. The difference $c_p - c_\sigma$ amounts to *ca.* 0.1% at low vapour pressures and is commonly neglected. Heat capacities are usually represented as power series of temperature, in second or third order. The *specific heat* (capacity) of water, c_p/M is exceptionally high, 4179.7 J K^{-1} kg^{-1} at 298.15 K, but the molar heat capacity is not exceptional. This quantity depends on molecular size and complexity, as well as on intermolecular interactions, and, for ordinary inorganic and organic liquids, is in the range 40 to 200 J K^{-1} mol^{-1}, with liquid metals having definitely lower heat capacities (mercury at 298 K has $c_p = 27.98$ J K^{-1} mol^{-1}).

The *surface tension* σ of a liquid is the force acting normal to the surface per unit length of the surface, tending to decrease the surface area. The saturation surface tension, σ_σ, between the liquid and the vapour at equilibrium is usually meant, and the value at unit atmospheric pressure between the liquid and air is usually measured. The difference is negligible, and the symbol σ without suffix will be used. Over a limited temperature range, the surface tension is linear with the temperature, and in analogy with the density, eq. (1.11)

$$\sigma = a - bT = \sigma_0(1 - b't) \tag{1.21}$$

where t is the centigrade temperature. Over a wider temperature range for many liquids

$$\sigma = \sigma_0 v_c^{-2/3} T_c [1 - (T/T_c)]^{(1.27\pm0.02)} \tag{1.22}$$

holds where $\sigma_0 = 44$ J $mol^{2/3}$ m^{-4} K^{-1}. Therefore the reduced surface tension $\sigma_r = \sigma v_c^{2/3} T_c^{-1}$ is a universal function of the reduced temperature, $\sigma_0(1 - T_r)^{1.27}$, and is the same at corresponding states. Surface tensions of organic liquids at room temperature are in the range (1 to 5) x 10^2 N m^{-1}, whilst water has σ(298 K) = 7.181 x 10^2 N m^{-1}, i.e. a higher surface tension. Liquid ionic salts and liquid metals have still higher surface tensions, commonly in the range (0.5 to 2.0) x 10^3 N m^{-1} and (4 to 8) x 10^3 N m^{-1} respectively at, say, 700 – 1 300 K.

The Gibbs adsorption equation applies to a pure liquid in equilibrium with its vapour at constant volume

$$S_s dT + \Gamma d\mu + d\sigma = 0 \tag{1.23}$$

where S_s is the excess entropy at the surface per unit of area, Γ the excess of matter at the surface per unit of area and μ the chemical potential, provided that the mathematical surface between liquid and vapour is planar. If this surface is chosen at the location where $\Gamma = 0$, then

$$S_s(\Gamma = 0) = -\,d\sigma/dT \tag{1.24a}$$

$$U_s(\Gamma = 0) = -\ Td\sigma/dT \tag{1.24b}$$

$$A_s(\Gamma = 0) = \sigma \tag{1.24c}$$

where U_s and A_s are the surface energy and Helmholtz energy per unit area respectively. The Helmholtz energy per unit surface area is seen to equal the surface tension.

The surface tension is related to the internal forces in the liquid, and is expected to bear a relationship to the internal energy or cohesive energy density (p. 10). In fact a rough proportionality exists between $\sigma v^{-1/3} = A_s v^{-1/3}$ and $\Delta u^V/v$ for non-polar liquids. Formerly, a quantity called the *parachor*, defined as $\sigma^{1/4}M/(d^l - d^g) \approx \sigma^{1/4}v$, was used extensively to compare the structure of liquids. The comparison of the parachors of different liquids is in fact a comparison of their molar volumes at a corresponding 'state' of unit surface tension. It has been argued that in this manner a better account is taken of the intermolecular forces than by a comparison made at a constant reduced temperature (roughly the normal boiling point). A particular advantage of the parachor is its additivity of contributions from atoms and groups, and such structural features in organic molecules as double bonds, rings, etc. However, the definition of the parachor is empirical, and in spite of many attempts in the past to ascribe to it a theoretical significance, none has been found.

c. *Properties at the critical point*

The critical data of fluids are quantities which have to be accounted for by any theory of the liquid state. If isotherms in the $P(V)$ diagram of a fluid are considered (Fig. 1.4), experiment shows that at high temperatures they are nearly hyperbolic (ideal gas behaviour is approached, $Pv = \mathbf{R}T = const.$ at a constant temperature). At low temperatures these isotherms have a horizontal section, from the bubble point at a certain small volume to the dew point at a certain high volume (p. 9). The length of the horizontal section decreases as the temperature is increased, until it vanishes at the inflexion point of the *critical isotherm*. The P, V, T data therefore furnish experimental values of the *critical constants* P_c, V_c and T_c for each fluid. The *equation of state* $f(P, V, T) = 0$ of a fluid permits the critical constants to be obtained, and for many fluids (*conformal fluids*) an empirical relationship exists between them

$$P_c v_c/T_c \simeq (0.290 \pm 0.005)\mathbf{R} \tag{1.25}$$

so that only two independent parameters are required to yield the equation of state in terms of the reduced variables (cf. eq. 1.43)

$$f'(P_r, v_r, T_r) \equiv f'(P/P_c, v/v_c, T/T_c) = 0 \tag{1.26}$$

The classical van der Waals' equation can be written in terms of the reduced variables as $(P_r + 3v_r^{-2})(3v_r^{-2} - 1) = 8T_r$, yielding the relationship $P_c v_c/T_c = (3/8)\mathbf{R} = 0.375\mathbf{R}$, rather than the experimental value, eq. (1.25). The classical van der Waals'

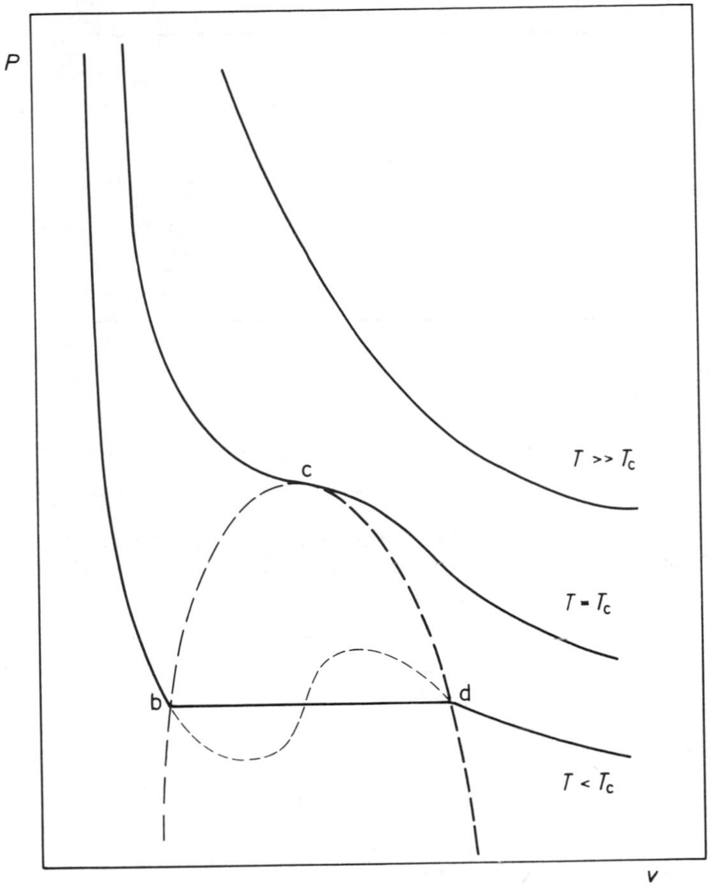

Figure 1.4. Isotherms on a $P(v)_T$ diagram of a fluid showing the one-phase region, and below the dashed line the two-phase region. $T = T_c$ denotes the critical isotherm. A $P(v)$ curve calculated from the van der Waals' equation of state, $(P + av^{-2})(v - b) = RT$, is shown as the continuous line for $T < T_c$, the portion between b (the bubble point) and d (the dew point) on the dashed coexistence curve b–c–d being dotted, since the fluid splits to two phases.

equation thus fails to quantitively reproduce the observed relationship between the critical constants, although it yields isotherms of the correct general shape.

The thermodynamic requirements at the critical point can be related to the volume- and temperature-derivatives of the Helmholtz energy a of the fluid. With the supposition (found out to be incorrect) that a is an analytic function of T and v, this parameter may be expanded to a Taylor series about its value at the critical point

$$a = a_c + \sum_{n=1}^{\infty} \sum_{m=0}^{n} (\partial^n a/\partial v^m \partial T^n)_c \Delta v_c^m \Delta T_c^n / m!(n-m)! \tag{1.27}$$

Table 1.1. Critical indices.

Function X	Index	v.d.W. fluid	Experimental[a]
C_V	α^+	0	-0.2 ± 0.1
	α^-	0	-0.1 ± 0.1
Δv_c	β	1/2	0.35 ± 0.01
$-(\partial P/\partial v)_T$	γ^+	1	1.3 ± 0.2
	γ^-	1	1.0 ± 0.3
ΔP_c	δ[b]	3	4.8 ± 0.4
$(\partial^2 P/\partial T^2)_\sigma$	θ	0	$\sim 0\ (<0)$
σ	r	1½	1.29 ± 0.02

[a]The range of values found for typical liquids is given.
[b]$\delta = \lim(\Delta v_c \to 0)[\ln \Delta P/\ln \Delta v_c]$.

where $\Delta v_c = v - v_c$ and $\Delta T_c = T - T_c$. A sufficient condition for a critical point is

$$(\partial a/\partial v)_T < 0; \quad (\partial^2 a/\partial v^2)_T = (\partial^3 a/\partial v^3)_T = 0; \quad (\partial^4 a/\partial v^4)_T > 0 \qquad (1.28)$$

A fluid obeying the van der Waals' equation obeys (1.27) and (1.28) also. A corollary of (1.27) and of $(\partial a/\partial v^l)_T = (\partial a/\partial v^g)_T$, which is a condition for the coexistence of the two phases at $T < T_c$, is that $\Delta v_c^l + \Delta v_c^g = 0$, or that $v_c = \frac{1}{2}(v^l + v^g)$. Near to the critical point, where $\Delta v_c/v_c \ll 1$, this is equivalent to the *law of rectilinear diameters*, which expresses the densities of the coexisting phases

$$d_c = \tfrac{1}{2}(d^l + d^g)_T + k\Delta T_c \qquad (1.29)$$

and is an empirical rule valid over a long temperature range T up to T_c, k being a constant.

The rate of approach of the thermodynamic quantities to their value at the critical point is described by means of the *critical indices*[7]. These are the limits $\lim(\Delta T_c \to 0)[\ln X(v, T)/\ln \Delta T_c]_v$, where X is the function considered. The indices may be different for positive and negative values of ΔT_c, marked by the superscript sign, and for measurements at a given v along orthobaric states of a single fluid or along the critical isochore. Table 1.1 shows the values of these indices for a van der Waals' (v.d.W.) fluid and obtained experimentally.

Certain inequalities have been established theoretically between these indices, namely: $\alpha^- + \beta(\delta + 1) \geqslant 2$; $\alpha^- + 2\beta + \gamma^- \geqslant 2$; $\gamma^- - \beta(\delta - 1) \geqslant 0$ and $\alpha^- + \beta - \theta \geqslant 0$. The experimental values are 1.9 to 2.3; 1.7 to 2.2; −0.4 to 0.1; and >0, respectively; i.e. near enough to the theoretical values to satisfy them. However, the discrepancy between the van der Waals' value of $\beta = 0.50$ and the experimental value $\beta = 0.35$ is real and points to the non-analyticity of $a(V, T)$ at the critical point, and to the non-validity of (1.27).

d. Static optical and electrical properties

Attention will be limited here to two bulk properties which are intimately connected with molecular properties and liquid structure, namely the refractive index and the static (relative) dielectric constant. The former may be considered as the limiting case of the latter when measurements are carried out at such high frequencies (visible light, usually the sodium D line of 589.26 nm wavelength, $5.0876 \times 10^{14}\ s^{-1}$ frequency) that the molecules cannot reorient themselves sufficiently fast, and only electronic reorientation occurs.

The *refractive index* is the ratio of the speed of light in vacuum to that in the liquid, and can be measured readily to 10 p.p.m. (the quantity is dimensionless). It is frequency-dependent to some extent and, as mentioned, the value measured with the sodium D line light, n_D, is usually used. The values range from *ca.* 1.3 to 1.7, and are temperature-dependent, $dn_D/dT \sim 4.5 \times 10^{-4}\ K^{-1}$ applying to many organic liquids. The refractive index is not very pressure-sensitive, the dependency dn_D/dP being hardly measurable.

A quantity that relates the refractive index to the density, and moreover is additive for the contributions of atoms, groups and structural features of liquids, is the *molar refractivity*, given by the Lorentz–Lorentz expression

$$R = (n_D^2 - 1)(n_D^2 + 2)^{-1} v \qquad (1.30)$$

Since the quantity $(n_D^2 - 1)/(n_D^2 + 2)^{5/3}$ has been shown[8] to be proportional to the cohesive energy density of liquids, the molar refractivities are a means of comparing molar volumes, adjusted for internal forces (cf. discussion of parachors, p. 14). The molar refractivity is nearly independent of the temperature, the pressure, and the state of aggregation, making it a molecular property (section 1*c*) rather than a bulk liquid property.

The (relative) *dielectric constant*, ϵ, is the ratio of the force acting between two charges at a given distance apart in a vacuum, to the force acting between them in the dielectric medium. The lower limit of ϵ for non-polar liquids is n_D^2, the square of the refractive index, roughly 1.8 (the quantity is dimensionless). The dielectric constant increases with the polarity of the liquid, and a liquid, the molecules of which have a high dipole moment, has also a high dielectric constant. For example ethylene carbonate, $\mu = 4.87$ Debye (1 Debye = 3.33564×10^{-30} C m), has ϵ 89.6 at 313 K. However, both cooperative and antagonistic effects between the dipoles of neighbouring molecules are common, and the correlation between the bulk property, ϵ, and the molecular property, μ, is generally very poor. The liquid with practically the highest dielectric constant at room temperature, *N*-methylformamide with $\epsilon = 182.4$ at 298 K, has a dipole moment $\mu = 3.86$ Debye, exceeded by several others with a lower dielectric constant. Indeed, water with $\epsilon = 78.4$ at 298 K, which is very high, has only a modest dipole moment, 1.84 Debye. On the other hand, triethylene glycol $HO(C_2H_4O)_3H$ with the very high dipole moment, $\mu = 5.58$ Debye, has only a moderate dielectric constant, $\epsilon = 23.7$ at 293 K.

The temperature-dependence of the dielectric constant is marked and usually negative. For polar liquids, the negative trend is due partly to the destruction of the cooperative effect. For liquid water, for instance, ϵ decreases from 87.7 at the melting point, 273.2 K, to 55.7 at the normal boiling point, 373.2 K, and to 9.7 near the critical temperature, 643 K. The temperature-derivative of $\ln \epsilon$ is the quantity needed for theoretical applications and is typically $-(3 \text{ to } 7) \times 10^{-3}\ \mathrm{K}^{-1}$. There are, however, liquids, notably carboxylic acids, which have a small positive temperature-derivative (e.g. $d(\ln \epsilon)$ (acetic acid)$/dT = +1.1 \times 10^{-3}\ \mathrm{K}^{-1}$). This arises from the destruction of the antagonistic effects of the dipoles of the pairs of associated molecules in these liquids by increasing temperatures, as a result of decreasing association.

The pressure-dependence of the dielectric constant of liquids is not so well known. The derivative $(\partial(\ln \epsilon)/\partial(P\ (\mathrm{bar})))_T$ is of the order of 10^{-4} (1 bar = 10^5 Pa), and is not too different from the corresponding compressibility κ_T. In fact, a close relationship exists between the pressure-dependence of the density and of the dielectric constant. An expression similar to the Tait equation (1.14b) holds, which may be written as

$$(\partial(\ln \epsilon(T, P))/\partial P)_{T,1} = A(T)\epsilon(T,1)\kappa_T/C(T)d(T, 1) \tag{1.31a}$$

where $C(T)$ is the pressure-independent constant obtained from (1.15), $\epsilon(T, 1)$ and $d(T, 1)$ are the dielectric constant and density at the given temperature and unit pressure, and $A(T)$ is a pressure-independent constant specific to each liquid. Many liquids obey the alternative[9] relationship

$$(\partial(\ln \epsilon(T, P))/\partial P)_{T,1} = [(\epsilon(T, 1) - 1)/\epsilon(T, 1)]\ [D + 2 + (1 - D)\epsilon(T, 1)]\kappa_T/3 \tag{1.31b}$$

where $D = 0.9743$ is a temperature-, pressure- and specific liquid-independent constant.

The electric field-dependence of the dielectric constant can be expressed by Grahame's equation[10]

$$\epsilon(E) = n_D^2 + (\epsilon(E = 0) - n_D^2)/(1 + bE^2) \tag{1.32}$$

where E is the electric field, and b a field-independent constant, which is usually positive. Typical values of b are $1.1 \times 10^{-16}\ \mathrm{V}^{-2}\ \mathrm{m}^2$ for water, $8 \times 10^{-17}\ \mathrm{V}^{-2}\ \mathrm{m}^2$ for diethyl ether, and $-8.4 \times 10^{-14}\ \mathrm{V}^{-2}\ \mathrm{m}^2$ for nitrobenzene. The large negative value for the latter is due to disruption of the antiparallel structure of the dipoles of adjacent molecules by the high field. High electric fields decrease ordinarily the dielectric constant, and lead to *dielectric saturation*, ϵ approaching n_D^2 as E tends to infinity. Hence ionic liquids have low dielectric constants, not much higher than n_D^2, i.e. near 3. (The high temperatures characteristic for these liquids, i.e. molten salts, also contribute to lowering ϵ.)

Corresponding to the molar refractivity, a quantity called the *molar polarization* is calculated from the Clausius–Mosotti equation

$$P = (\epsilon - 1)(\epsilon + 2)^{-1} v \tag{1.33}$$

which, again, is practically temperature- and pressure-insensitive, but is field-dependent, $\lim (E \to \infty)\, P = R$ from (1.30), (1.32) and (1.34). It too is more of a molecular rather than a bulk property, and is directly related to the dipole moment and to the molecular structure. The dielectric constant is related to the molecular properties, the *polarizability* $\alpha = (3/4\pi)\mathrm{N}^{-1} R$ and the *dipole moment* μ, by the Debye equation

$$P = (4\pi/3)\mathrm{N}(\alpha + \mu^2/3\mathrm{k}T) \qquad (1.34)$$

This equation is valid for gases and dilute solutions of polar substances in non-polar solvents, where the mutual interactions of the dipoles are negligible. In pure polar liquids these interactions are dominant: head-to-tail interactions lead to an antagonistic effect, head-to-head interactions to a synergistic effect, on the dielectric constant. The liquid becomes associated, and the measure of the association is the parameter g, which has to multiply $\mu^2/3\mathrm{k}T$ in eq. (1.34) in order to account for the hindered rotation of the dipoles. According to Onsager and to Kirkwood the ratio $(\epsilon - 1)/(\epsilon + 2)$ is replaced by $(\epsilon - 1)(2\epsilon + 1)/9\epsilon$, and for high values of ϵ the latter quantity is approximated by $(2/9)\epsilon$. Then the relationship

$$g = \mathrm{k}Tv[\epsilon - 9(n_\mathrm{D}^2 - 1)/2(n_\mathrm{D}^2 + 2)]/2\pi\mathrm{N}\mu^2 \qquad (1.35)$$

results, where μ is obtained by gas-phase measurements from eq. (1.34). The parameter g is near unity for 'normal' liquids, which obey Trouton's rule, eq. (1.9), but is as high as 2.7 for water, 3.0 for ethanol and 4.1 for hydrogen cyanide. The parameter g is further discussed in section 3B(d).

e. *Summary*

The bulk properties of liquids may, in principle, be used to classify them, but a much better classification is made according to the intermolecular forces, discussed in the next section. It will there become evident that it is useful to distinguish between the following classes (examples in parentheses): simple fluids (argon), non-polar molecular liquids (n-hexane), polar molecular liquids (acetone), hydrogen-bonded molecular liquids (water), ionic liquids or molten salts (potassium chloride), and liquid metals (lead). The bulk properties of the above representatives of these classes are summarized in Table 1.2.

Apart from the quantities that refer to the first-order phase transitions and the critical data, properties at one chosen temperature are compared. It is impossible to find representatives of all classes which can be compared at one given temperature: simple fluids and ionic liquids just do not coexist as liquids at any temperature because of the large discrepancy between the relevant intermolecular forces. Nor is it convenient to choose a given reduced temperature $T_\mathrm{r} = T/T_\mathrm{c}$, since data are usually not available at the narrow interval $T_\mathrm{r}^\mathrm{t} = 0.556$ (triple point) and $T_\mathrm{r}^\mathrm{b} = 0.579$ (normal boiling point) of argon for the other liquids because they have either a lower reduced boiling point or else the temperatures are, in practice too high. The temperature chosen corresponds, therefore, to T_r values decreasing from left to right in Table 1.2. Some trends may be pointed out in Table 1.2, while admitting that no sharp dividing line with respect to the classes of liquids can be drawn on the basis of the bulk properties. As will be shown in section C, the groups listed above, the representatives of which appear in the Table 1.2 from left to right, correspond to increasing strength and/or range of the *intermolecular forces*. Along with this

Table 1.2. Bulk properties of some liquids representing different classes.

Property	Argon	n-Hexane	Acetone	Water	Potassium chloride	Lead
T_m (K)	83.8[a]	177.8	178.5	273.2	1043	600.5
T_b (K)	87.3	341.9	329.4	373.2	1680	2024
T_c (K)	150.7	507.3	508.2	647.4	3090	5400
$10^{-6}P_c$ (Pa)	4.90	3.01	4.70	22.12	13.6	85
$10^4 v_c$ (m^3)	0.746	3.70	2.08	5.54	4.3	1
$10^{-3}\Delta h^F$ (J mol^{-1})	1.176	13.08	5.69	6.00	26.10	4.97
Δs^F (J K^{-1} mol^{-1})	14.0	73.6	31.9	22.0	25.9	8.3
$10^6 \Delta v^F$ (m^3)	3.54	11.92		-1.50	7.23	0.7
$10^{-3}\Delta h^V$ (J mol^{-1})	6.52	28.85	29.09	40.65	162.3	191.4
Δs^V (J K^{-1} mol^{-1})	74.7	84.4	93.5	109.0	96.7	94.5
T (K) for the following	83.8	298.2	298.2	298.2	1073	1073
$10^{-3}d$ (kg m^{-3})	1.416	0.6548	0.7844	0.9970	1.5102	10.056
$10^5 v$ (m^3 mol^{-1})	2.82	13.16	7.41	1.807	4.93	2.06
$10^3 \alpha_p$ (K^{-1})	4.38	1.36	1.43	0.257	0.270	0.131
$10^{10}\kappa_T$ (Pa^{-1})	20.3	16.06	12.39	4.53	3.84	0.35[b]
$10^{-3}P_\sigma$ (Pa)	68.9	20.17	30.7	3.167	0.092	0.0046
$10^{-2}(\partial P/\partial T)_\sigma$ (Pa K^{-1})	80	8.75	5.69	1.888	0.017	9×10^{-4}
$10^{-5}(\partial P/\partial T)_V$ (Pa K^{-1})	21.5	9.56	11.07			
$10^{-8}\Delta u^V v^{-1}$ (J m^{-3})	2.05	2.20	3.80	22.97	31.1	94.2
c_V (J K^{-1} mol^{-1})	19.5			74.3	51.9[c]	23.7[b]
c_p (J K^{-1} mol^{-1})	41.9	195.4	125	75.30	73.6[c]	29.2
$10^{-2}\sigma$ (N m^{-1})	1.34	1.79	2.27	7.18	9.88	46.4
n_D		1.3749	1.3560	1.3325	1.387	
ϵ		1.88	20.70	78.54	1.9	

[a] T_t rather than T_m; [b] at T_m rather than at 1073 K; [c] at 1133 K rather than at 1073 K.

increase, in general, T_m, T_b, T_c, Δh^V, $\Delta u^V v^{-1}$ and σ increase too, while α_p, κ_T, $(\partial P/\partial T)_\sigma$ decrease. The quantities Δs^V and v_c are remarkably insensitive to the classification, and d, v, c_p, n_D and ϵ depend on the molecular complexity and size rather than on classification. The intermolecular forces operating between molecules of representatives of these classes are discussed in detail in the next section: those for argon on p. 24, for n-hexane on p. 26, for acetone on pp. 28 and 29, for water on pp. 28 and 30 to 31, for potassium chloride on p. 27, and for lead on pp. 35 to 37. Detailed discussions of the properties and structure of these liquids are deferred until Chapter 3. The above examples, though typical, do not represent the range of properties encountered in members of these classes, and as stated above, the bulk properties cannot serve well for the purpose of classification.

Certain other bulk properties have not been included although very relevant: these include transport and spectroscopic properties, which have been excluded from the discussion. Obviously, electrical conductivity is a key property of ionic liquids, and vibrations characteristic of intermolecular hydrogen bonds are a key property of hydrogen-bonded liquids. Properties such as viscosity and dielectric relaxation times are very important for the dynamical description and classification of liquids, but, however, are excluded from the present context, which deals only with equilibrium properties.

C. Intermolecular Forces

The finite compressibility and the relatively high density, which characterize liquids in general, point to the existence of repulsive and attractive intermolecular forces. The same forces that are known in the gaseous form of a fluid may be imagined to play a role also in the liquid form, except for the necessarily more important role of the repulsive forces, because of the short average distances, and excepting the far greater importance that many-body interactions must have. Indeed, it is surprising that two-body interactions can form any basis for a discussion of intermolecular forces in a liquid. However, it has been shown[11] that an effective two-body interaction potential can be defined for liquids which has the same general properties as the corresponding two-body potential in gases. It is therefore profitable to discuss the latter as a basis for the interactions in liquids.

In order to discuss the interactions in liquids, it is necessary first to assume that the mutual interaction energy U of a liquid collection of N molecules is independent of the intramolecular energy levels. It is thus assumed that the interactions between neighbouring molecules are not sufficiently strong to excite new vibrational or electronic levels or to influence the occupation of rotational levels. In terms of the partition functions (section 1D), the internal degrees of freedom of the molecules can be factored out, to leave the interaction or configurational part. Further, although it is the intermolecular *forces* of attraction and repulsion which lend the liquid its properties, it is preferable to discuss the mutual *potential energy, u,* of molecules. As Fig. 1.5 shows for the two molecules a and b, the potential energy $u(\mathrm{a}, \mathrm{b}, R, \theta_\mathrm{a}, \theta_\mathrm{b}, \phi_\mathrm{a}, \phi_\mathrm{b})$ is both distance (r)- and orientation (θ, ϕ)-dependent. This leads to ambiguities when the force between the molecules, $f = -(\partial u/\partial r)_{\theta,\phi}$ is discussed, since it ignores the orientation. The

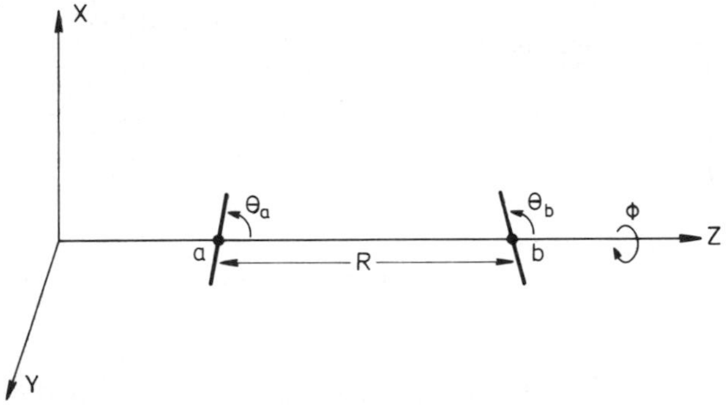

Figure 1.5. The relative orientation of two molecules a and b which are a distance R apart along the Z axis in a Cartesian coordinate system X, Y, Z. The angles θ_a and θ_b refer to the Y, Z plane, the angles ϕ_a and ϕ_b (only the general sense of ϕ is shown) to the X, Y plane. The five quantities R, θ_a, θ_b, ϕ_a and ϕ_b completely determine the mutual space relationship between the two molecules.

instantaneous potential energy depends on this orientation but the thermal motion averages this dependency out in such a way that only a dependence on the distance r between the molecules remains; however, a new dependence on the temperature T is thereby introduced

$$\bar{u}(\mathrm{a}, \mathrm{b}, r, T) = \frac{\iint u \exp(-u/\mathrm{k}T)\mathrm{d}\omega_\mathrm{a}\mathrm{d}\omega_\mathrm{b}}{\iint \exp(-u/\mathrm{k}T)\mathrm{d}\omega_\mathrm{a}\mathrm{d}\omega_\mathrm{b}} \tag{1.36}$$

where $\omega = \sin\theta \, \mathrm{d}\theta \, \mathrm{d}\phi$, and $\bar{u}$ is the orientation-averaged potential energy.

When more than the two molecules a and b interact, the potential energy can be summed over 2-, 3-, ... up to n-body interactions[11] and all N interacting molecules

$$\begin{aligned} U &= (1/2!) \sum_i \sum_j u_{ij} + (1/3!) \sum_i \sum_j \sum_k u_{ijk} + \ldots \\ &\quad + (1/n!) \sum_i \ldots \sum_n u_{i\ldots n} \qquad (1.37) \\ &= (1/2) \sum_i \sum_j \left[u_{ij} + (2/3!) \sum_k u_{ijk} + \ldots + (2/n!) \sum_{i+2} \ldots \sum_n u_{i\ldots n} \right] \end{aligned}$$

The sums are to be understood to run from index 1 to index N excluding identical indices (i.e. $i \neq j$). In the second equation, the summation over the indices i and j is performed last. The quantity u_{ij}^{eff}, an effective two-body potential energy, is now defined formally as $U = (1/2) \Sigma_i \Sigma_j u_{ij}^{\mathrm{eff}}$, and may be written as

$$u_{ij}^{\mathrm{eff}} = u_{ij} \left\{ 1 + (1/3) \sum_k (u_{ijk}/u_{ij}) + \ldots \right\} \tag{1.38}$$

The effective two-body potential energy is thus seen to be directly related to the direct two-body potential u_{ij} known from interactions in gases, and may be obtained from it if the correction terms can be estimated (section 2B(b)). The first step in the determination of the interaction energy of a liquid collection of N molecules, U, is therefore the determination of the relevant two-body potential energies u_{ij}.

a. Simple two-body interactions

The simplest two-body interaction exists between imaginary hard spheres (Fig. 1.6(a)). *Hard spheres* are defined by the following potential energy functions

$$\begin{aligned} u^{\mathrm{hs}}(r) &= \infty \quad \text{for} \quad r < \sigma \\ &= 0 \quad \text{for} \quad r \geqslant \sigma \end{aligned} \tag{1.39}$$

These are characterized by an infinitely strong repulsion when an attempt is made to bring the spheres closer together than a distance σ (which may be taken as their diameter), They show, however, no attractive forces whatsoever and are therefore a very poor model for the liquid. They serve only as a first approximation, when the repulsive forces between molecules, or their sizes, need to be emphasized.

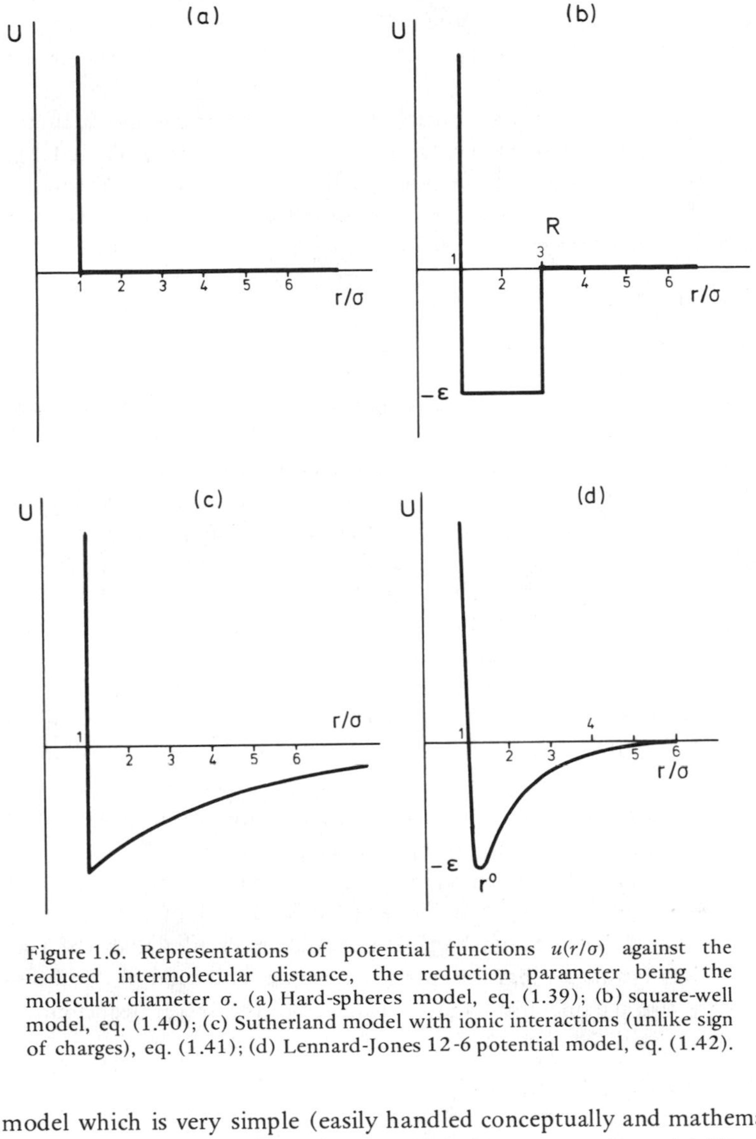

Figure 1.6. Representations of potential functions $u(r/\sigma)$ against the reduced intermolecular distance, the reduction parameter being the molecular diameter σ. (a) Hard-spheres model, eq. (1.39); (b) square-well model, eq. (1.40); (c) Sutherland model with ionic interactions (unlike sign of charges), eq. (1.41); (d) Lennard-Jones 12-6 potential model, eq. (1.42).

A model which is very simple (easily handled conceptually and mathematically) but a fair approximation to the real potential is the *square-well* model (Fig. 1.6(b)). The potential energy functions have the form

$$\begin{aligned} u^{\mathrm{SW}}(r) &= \infty \quad \text{for} \quad r < \sigma \\ &= -\epsilon \quad \text{for} \quad \sigma \leqslant r \leqslant R\sigma \\ &= 0 \quad \text{for} \quad r > R\sigma \end{aligned} \tag{1.40}$$

This is a three-parameter model (ϵ, σ and R), which retains hard-sphere repulsive properties but allows the molecules to attract each other with a constant potential energy, $-\epsilon$, when near each other, up to a distance of R times σ (in Fig. 1.6(b), $R = 3$). Although the square-well potential can reproduce roughly the macroscopic properties of a liquid, no physical picture of a model with this behaviour can be set up.

A model which has been used successfully to represent the interactions in one class of liquids (ionic liquids, i.e. molten salts, cf. KCl in Table 1.2) is the *Sutherland potential*, where again the repulsive forces are represented by the properties of hard spheres but the attractive forces are related to coulombic interactions between charges, Fig. 1.6(c)

$$\begin{aligned} u^{su}(r) &= \infty \quad \text{for} \quad r < \sigma \\ &= Cr^{-1} \quad \text{for} \quad r \geqslant \sigma \end{aligned} \tag{1.41}$$

(This is a restricted form of the Sutherland potential, which in general permits any negative exponent of r.)

The model has two parameters (C and σ), of which the parameter C has the function of the product of the charges on molecules (ions) a and b, and the reciprocal of a dielectric constant: $C = (1/K)z_a z_b e^2$. This potential is of interest mainly if z_a and z_b differ algebraically in sign, since only then is the interaction an attraction. The energy of a configuration where z_a and z_b are of the same sign is too high, and the Boltzmann coefficient makes this a rare configuration which may be ignored as far as the present level of approximation is concerned. Nothing can be said about the 'dielectric constant' K, so that C is in fact a free parameter. The main feature of this potential function, however, is its long effective range, since u decreases only as the first power of the distance r. This means that interactions are important over many molecular diameters, and perforce many molecules are always involved. These, in a way, contribute to the quantity K (just as in an ionic solid, distant ions contribute to the Madelung constant).

The *Lennard-Jones 12-6* potential function[12], Fig. 1.6(d), is also a simple function, easily handled mathematically, which is a much better approximation to the actual potential between uncharged non-polar molecules than the foregoing ones. It has the advantages that the attractive part has a sound physical basis and that with only two parameters, ϵ and σ, it describes the actual behaviour of gases and liquids fairly well. The potential has the form

$$u^{LJ}(r) = 4\epsilon[(\sigma/r)^{12} - (\sigma/r)^6] \tag{1.42a}$$

where the reciprocal exponent 12 of the distance, with a positive coefficient, represents the repulsive forces, while the reciprocal exponent 6, with the negative coefficient, represents the attractive forces. These exponents, 12 and 6, are but a particular case of a family of m, n potentials, but since r^{-n} with $n = 6$ has a physical justification (see below), and since $m = 2n$ provides a particularly convenient form of the potential, the $m = 12$ and $n = 6$ case is the most commonly used of the Lennard-Jones potentials. From (1.42a) can be seen that the mutual interaction energy vanishes when the molecules approach each other to the distance σ, their diameter; $u(\sigma) = 0$. The minimal potential energy results when the distance is the equilibrium distance $r^0 = 2^{1/2}\sigma = 1.12246\sigma$, and $u(r^0) = -\epsilon$. Equation (1.42a) may therefore be written in terms of r^0 rather than of σ

$$u^{LJ}(r) = \epsilon[(r^0/r)^{12} - 2(r^0/r)^6] \tag{1.42b}$$

A two-parameter potential, such as the Lennard-Jones 12-6 potential, eq. (1.42), has the advantage that it can serve as a universal function for *conformal fluids* that differ only in the scaling factors $f = \epsilon/\epsilon^*$ and $g = \sigma/\sigma^*$ (or r^0/r^{0*}). The potential function of any of these fluids is then $u(r) = fu^*(r/g)$, where u^* is the 'LJ 12-6' function (1.42) in the reference parameters ϵ^* and σ^*. Pressures, volumes and temperatures have to be multiplied by the scaling factors fg^{-3}, g^3 and f, respectively. Since this applies also to the critical pressures, volumes and temperatures, eq. 1.26, it follows that for the whole family of conformal liquids

$$(P_c v_c/T_c) = (fg^{-3}P_c^* g^3 v_c^*/fT_c^*) = P_c^* v_c^*/T_c^* \tag{1.43}$$

For the product $P_c^* v_c^*/T_c^*$ the empirical value $(0.290 \pm 0.005)\mathbf{R}$, given in eq. (1.25), may be used, and argon, with the product equalling 0.292, may be designated as the reference fluid. It should be emphasized that any other potential function with two parameters, one referring to the energy (such as ϵ, the equilibrium attractive energy) and the other to the distance (such as r^0, the equilibrium intermolecular distance), may serve for the purpose of defining the conformal fluids in place of the 'LJ 12-6' potential.

The justification of any model for describing the intermolecular forces lies in its ability to reproduce the macroscopic properties of the fluid. For the gaseous form, this means mainly the second virial coefficient, $B_2(T)$ in (1.44) and the higher virial coefficients, $B_3(T)$ . . . as far as they are known, and such transport properties as gas viscosities. The virial coefficients can be obtained from highly accurate (P, v, T) data

$$Pv/\mathbf{R}T = 1 + B_2(T)v^{-1} + B_3(T)v^{-2} + \ldots \tag{1.44}$$

For dilute gases, the series may be truncated after the second term, but for dense gases as many as seven terms may be required in order to represent the behaviour adequately.

The second virial coefficient $B_2(T)$ is related to the two-body interaction potential $u(r)$ by the simple integral equation:

$$B_2(T) = -2\pi\mathrm{N}\int_0^\infty [\exp(-u(r)/\mathrm{k}T) - 1]\,r^2\,dr \tag{1.45}$$

However, the quantity $B_2(T)$ is not sufficiently sensitive to small variations in $u(r)$ to permit unique values of the potential function to be obtained from experimentally available $B_2(T)$ values. The same applies also to the gas viscosity

$$\eta(T) = M^{1/2}T^{-1/2}\epsilon\sigma^{-2}\,\Omega(u(r)) \tag{1.46}$$

where M is the molecular mass, and $\Omega(u(r))$ is a complicated function of the potential. The measured viscosity is insufficiently sensitive to the precise form of the function $u(r)$. Other experimental approaches are not much better: equilibrium properties such as the third virial coefficient $B_3(T)$, or the Joule–Thompson coefficient $(\partial h/\partial P)_T$, or transport properties such as the thermal diffusion coefficient, although more sensitive to the form of $u(r)$, are also sensitive to three-body and higher interactions, and again cannot lead to unique values of $u(r)$ for the two-body interaction. Although the 'LJ 12-6' potential, eq. (1.42), is fairly successful for describing the properties of fluids with spherical non-polar molecules (e.g. those of argon in Table 1.2, including $B(T)$), improved models have their merits if they do not introduce too many additional free parameters. (The Lennard-Jones function can, indeed, be treated as a four-parameter function if the exponents m and n are allowed to vary.) Of the many modifications, the *Kihara core-model*[13] seems to be the most successful, since it can be applied to non-spherical molecules. This assumes a hard core inside each molecule, of diameter δ (so that $u(r<\delta)=\infty$), which is surrounded by a soft envelope, so that at a certain intermolecular distance $\sigma>\delta$ the potential energy is zero, while at any distance $r\geqslant\delta$, the potential function is

$$u(r) = 4\epsilon\{[(\sigma-\delta)/(r-\delta)]^{12} - [(\sigma-\delta)/(r-\delta)]^{6}\} \tag{1.47}$$

For the particular case where the third parameter, the core diameter δ, vanishes, eq. (1.47) reduces to the 'LJ 12-6' potential function (1.42). Realistic estimates[11] of δ/σ vary from 0.084 for argon or krypton, to 0.194 for nitrogen, 0.655 for benzene and 0.900 for n-hexane.

For given values of ϵ and σ, the Kihara model potential has the same general form as shown in Fig. 1.6(d) for the 'LJ 12-6' model, but the potential well becomes narrower (the repulsion 'softer') as the parameter δ increases. The three-parameter $(\epsilon, \sigma, \delta)$ equation (1.47) was found to be numerically very successful for describing several macroscopic properties: the second virial coefficient, gas viscosity, thermal diffusivity, etc. For calculating the thermodynamic properties of fluids such as n-hexane, Table 1.2, structural data (cf. Chapter 2) are required.

b. Dispersion forces and multipole interactions

The models presented above have little physical significance in as far as the origin of the repulsive and the attractive forces is concerned. Very little is known concerning the repulsive forces. They may be ultimately traced to the Pauli exclusion principle for the electrons and the coulombic repulsion of the positive nuclei of the atoms. No physical theory is available however, and the form $(\sigma/r)^{12}$ is chosen merely for convenience. Indeed, an exponential form $A\exp(-Br)$ may be a better representation of the r-dependence of the repulsive potential energy than an inverse power representation but has no better physical theoretical explanation. The attractive forces, however, are much better understood theoretically.

Uncharged non-polar molecules attract one another only on account of quantum-mechanical reasons, and not on account of interactions of the classical type. The dispersion forces, as pointed out first by London[14], may be traced back to the quantum-mechanical zero-point vibrations of the molecules, that is to instantaneous non-isotropic charge distributions inside the molecules. Although the time-average of these instantaneous dipole moments is zero, at any given moment such a dipole induces a corresponding dipole in a neighbouring molecule, leading to a net attractive force between these instantaneous dipoles and induced dipoles. The interaction leading to the perturbance of the electronic motion in one molecule by another is related to the perturbance of this motion by light, as measured by the dependence of the refractive index on the frequency of the light, the light dispersion; hence the name *dispersion forces* for this interaction. The potential energy for the interaction of two spherical molecules is

$$u(\mathrm{a, b}, r) = -(3/2)\alpha_a\alpha_b \mathrm{h}\nu_a\nu_b(\nu_a + \nu_b)^{-1} r^{-6} \tag{1.48a}$$

and when the two molecules are identical

$$u(r) = -(3/4)\alpha^2 \mathrm{h}\nu r^{-6} \tag{1.48b}$$

where α is the polarizability, and ν is the characteristic frequency ($(1/2)\mathrm{h}\nu$ is the zero-point energy). This equation is valid for spherical molecules, and is in fact the leading term of an expanded series of even reciprocal powers of r, although the coefficients of r^{-8}, r^{-10} . . . are not well established theoretically. The energy $\mathrm{h}\nu$ can be approximated by the ionization potential of the molecule, I, so that the dispersion forces between molecules of a fluid can be well represented in terms of the independently measureable quantities α and I

$$u(r) = -Cr^{-6} \quad \text{with} \quad C = (3/4)\alpha^2 I \tag{1.49}$$

Dispersion forces are a major contribution to the attractive potential even for polar molecules. As will be seen presently, a permanent dipole leads to an induction energy (the permanent-dipole interacting with an induced-dipole), which is relatively unimportant, and to a dipole–dipole interaction energy, which is temperature-dependent. Unless the permanent-dipole moment is very large, the dipole–dipole interaction is less than the dispersion interaction. All three interactions depend on the reciprocal sixth power of the intermolecular distance. Other interactions, arising from net charges on the molecules (ions) and from higher multipoles (quadrupoles etc.) play a role under particular circumstances.

Contributions of multipoles, permanent or induced, to the interaction energy can be understood in terms of the spatial relationship shown in Fig. 1.5. A charge on the molecule a will be designated by z_a, a permanent dipole by μ_a, and a quadrupole moment by q_a. The dipole moment is a vector and only ideal or point dipoles are considered: $\mu_a = \lim (\vec{l} \to 0) \vec{l} \cdot z_a \mathrm{e}$ where the charge $z_a\mathrm{e}$ is permitted to grow indefinitely at constant μ_a, as $\vec{l}$ becomes very small compared to the intermolecular distance.

The quadrupole moment may be represented by $q_a = (1/2)\, z_a \mathrm{e}(\vec{l} \cdot \vec{l} - \vec{l}^2 \mathrm{U})$ where U is a unit tensor. The quadrupole moment vanishes for an isotropic spherical charge distribution but must be considered for a linear charge distribution. The following equations give the interaction energies arising from the more important of these multipole interactions. Both the angle-dependent instantaneous values, and the orientation-independent average values, according to eq. (1.35), are shown

(i) charge–charge interaction

$$u(z_a, z_b) = \bar{u}(z_a, z_b) = z_a z_b e^2 r^{-1} \tag{1.50}$$

(ii) charge–dipole interaction

$$u(z_a, \mu_b) = -z_a e \mu_b \cos\theta_b r^{-2} \tag{1.51a}$$

$$\bar{u}(z_a, \mu_b) = -(1/3kT) z_a^2 e^2 \mu_b^2 r^{-4} \tag{1.51b}$$

(iii) charge–quadropole interaction

$$u(z_a, q_b) = (1/4) z_a e q_b (3\cos^2\theta_b - 1) r^{-3} \tag{1.52a}$$

$$\bar{u}(z_a, q_b) = -(1/20kT) z_a^2 e^2 q_b^2 r^{-6} \tag{1.52b}$$

(iv) dipole–dipole interaction

$$u(\mu_a, \mu_b) = -\mu_a \mu_b (2\cos\theta_a \cos\theta_b - \sin\theta_a \sin\theta_b \cos(\phi_a - \phi_b)) r^{-3} \tag{1.53a}$$

$$\bar{u}(\mu_a, \mu_b) = -(2/3kT) \mu_a^2 \mu_b^2 r^{-6} \tag{1.53b}$$

(v) charge–induced-dipole interaction

$$u(z_a, \text{ind. } \mu_b) = \bar{u}(z_a, \text{ind. } \mu_b) = -(1/2) z_a^2 e^2 \alpha_b r^{-4} \tag{1.54}$$

(vi) dipole–induced-dipole interaction

$$u(\mu_a, \text{ind. } \mu_b) = -(1/2) \mu_a^2 \alpha_b (3\cos^2\theta_a + 1) r^{-6} \tag{1.55a}$$

$$\bar{u}(\mu_a, \text{ind. } \mu_b) = -\mu_a^2 \alpha_b r^{-6} \tag{1.55b}$$

It is seen that the presence of net charges leads to potential energies which decrease with r more slowly than the dispersion energy, eq. (1.49), i.e. r^{-1} for charge–charge interactions (cf. eq. (1.41)) and r^{-4} for the thermally averaged orientation-independent charge–dipole and charge–induced-dipole interactions. These are thus of longer range than the dispersion energy, and exceed it in importance beyond, say, $r = 3\sigma$. If net charges are absent, the dipole–dipole and dipole–induced-dipole are the leading interactions, varying with r^{-6} just as the dispersion energy. In practice, higher terms in the London formula for the dispersion energy, as well as multipole interactions leading to terms in r^{-8}, r^{-10}, etc. are ignored. The coefficient of the r^{-6} term in the case where permanent dipoles (but no net charges) are present then becomes

$$C = (3/4)\alpha^2 I + (2/3kT)\mu^4 + \mu^2 \alpha \tag{1.56}$$

instead of the first term only as in (1.49). The temperature-dependence should be noted. Quadrupole interactions contribute only to terms in r^{-8} etc. if net charges are absent. The Stockmayer–Keesom model for interactions of polar molecules thus utilizes in fact the 'LJ 12-6' potential (1.42), and equates the parameter ϵ with $(1/4)C\sigma^{-6}$, where C is given by eq. (1.56).

Table 1.3. Contributions to the total attractive interaction energy $-u_t{}^a$ from dispersion, induction and dipole interactions.

Substance	$-u_t r^6$	$u(\text{disp.})/u_t$	$u(\mu_a,\mu_b)/u_t$ ^{b}	$u(\mu_a, \text{ind. } \mu_b)/u_t$
Ar	53	1.000	–	–
CH_4	106	1.000	–	–
CO	68	0.999	0.00005	0.00084
HCl	129	0.814	0.144	0.042
NH_3	187	0.497	0.449	0.054
H_2O	247	0.190	0.769	0.041

[a]The units are 10^{-79} J m^6. [b]The temperature T = 293 K when applying eq. (1.56).

It was however pointed out some time ago[14] that the dispersion interaction as given by (1.49) provides nearly the correct attractive energy for non-polar fluids, and a major portion of this energy for polar ones, as shown in Table 1.3. For very polar liquids, such as acetone or water (Table 1.2), these two-body potentials are not adequate and additional interactions must be considered.

c. *Other kinds of interactions*

The interactions discussed above are quite general and dispersion interactions occur even for the most 'inert' of molecules, the rare gases, leading to their liquification below the critical temperature as they are compressed. Interactions of this type may be designated as *physical interactions*, to differentiate them from *chemical interactions*, which are more specific, but of very short range and strongly orientation-dependent. In such interactions the internal energy levels of the interacting molecules are no longer unaffected and a permanent dislocation of electrons, relative to their state in the separate molecules, occurs. Because of this, no simple distance- and orientation-dependence of the intermolecular potential can be written and in general the two-body interaction potential is not known. Several kinds of such interactions will be described here, more or less in qualitative terms: strong dipole interactions, hydrogen bonds, charge-transfer complexes, and metallic bonds. A feature that these interactions have in common, except the metallic bonding, is that they depend on a definite stoichiometry of the reactants, usually 1:1, so that for pure substances dimers result. This specific, chemical, interaction, is then superimposed on the non-specific physical interactions previously considered.

Strong dipole interactions occur when very polar molecules become adjacent to each other, the interaction potential energy calculated from eq. (1.53a) then becoming large compared to their thermal, i.e. average kinetic, energy, provided that the orientation is favourable. The molecules may then find a mutual configuration at a distance, r^o, considerably shorter than their usual repulsion envelope (the sum of their van der Waals' radii, i.e. σ for an unfavourable configuration), further enhancing the interaction. The potential energy for two identical molecules aligned head-to-tail becomes

$$u\ (\text{aligned dipoles}) = -\mu^2\,(r^o)^{-3} \tag{1.57}$$

and when $-u$(aligned dipoles)$\gg kT$, a dimer results. This potential energy is not comparable with (1.53b) for thermally averaged dipole interactions, but can be taken to be $\mathbf{N}^{-1}$ times the molar energy of dimerization, except for some energy that may be consumed by restricted rotation and other changes in the internal energy levels brought about by the confining orientation. The molar energy of dimerization can then be related via the van't Hoff equation to the equilibrium constant for the dimerization ($-\mathbf{R}(\mathrm{d}(\ln K_{\mathrm{dim}})/\mathrm{d}(1/T)) + \mathbf{R}T$, the last term correcting for the pressure–volume work, as two gas molecules are consumed to form the dimeric gas molecules). The equilibrium constant for dimerization and its temperature-dependence may be derived from P, V, T data, that is from the measured second virial coefficient, B_{meas}, provided a suitable second virial coefficient for the monomeric polar molecules can be estimated[15]. For this, the Berthelot equation has been proposed

$$B_{\mathrm{Bert}} = (9/128)\mathbf{R}T_{\mathrm{c}}P_{\mathrm{c}}^{-1}(1 - 6T_{\mathrm{c}}^{2}T^{-2}) \tag{1.58}$$

requiring the knowledge of the critical constants. The dimerization constant is then obtained as $K_{\mathrm{dim}} = (B_{\mathrm{Bert}} - B_{\mathrm{meas}})P/\mathbf{R}T$. Agreement of the dimerization energy as obtained from the experimental P, V, T data and from the molecular properties, the dipole moment μ and a crudely estimated r^0, has been found[15] for several polar substances, such as acetone (Table 1.2). There is no definite indication, however, that this dipole interaction is the major interaction responsible for the observed behaviour, since the parameter $|r^0|$ is adjustable, and agreement was found also for cases where other interactions, such as hydrogen-bond formation, are known to occur. Such dipole interactions may, however, play an important role in certain gas-phase self associations, as well as for the self association of polar solutes in inert solvents (cf. section 5B(d)).

The subject of *hydrogen-bond formation* should be treated on a quantum-mechanical basis. It is generally accepted that when two molecules approach each other, one containing a suitable functional group A–H and the other a suitable atom B, a hydrogen bond is formed, A–H . . . B. Suitable functional groups A–H are O–H and N–H (also X–H where X is a halogen or a sulphur atom, and rarely C–H, where the carbon atom must be combined with strong electron-withdrawing groups, as in chloroform, Cl_3CH) and suitable atoms B are F, O and N and sometimes also S and π-electron systems such as in aromatic solvents. Furthermore, the full potential energy decrease is gained only when the A–H . . . B configuration is colinear. On the other hand, A and B may be the same atoms, and A–H and B may belong to identical molecules. The wave function for the hydrogen bond may be written as the combination

$$\psi_{\mathrm{H\text{-}bond}} = a\psi(\mathrm{A{-}H\ B}) + b\psi(\mathrm{A^{-}H^{+}\ B}) + c\psi(\mathrm{A^{-}\ H{-}B^{+}}) \tag{1.59}$$

and the relative contributions of the various structures A–H B (no bond), A^- H^+ B (ionic interaction), and A^- $H{-}B^+$ (charge-transfer covalent bond), with the normalizing constraint that $a^2 + b^2 + c^2 = 1$, depend on the distances $r_{\mathrm{A,H}}$ and $r_{\mathrm{B,H}}$ (and on the angle θ_{AHB} if the three atoms are not colinear).

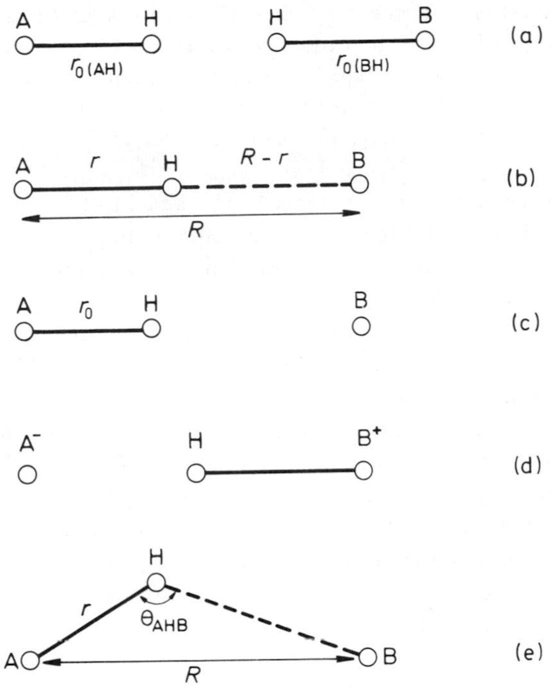

Figure 1.7. Schematics of the hydrogen bond. (a) Bond distances of hydrogen atoms bonded to A atoms, r_{0}(AH), and to B atoms, r_{0}(BH), in isolated groups. (b) The equilibrium position of the hydrogen bond, the A–H distance stretched to r, the B–H distance much longer than either r or r_{0}(BH). (c) The unbonded contribution to the hydrogen bond structure. (d) The charge transfer contribution to the hydrogen bond structure. (e) The bent hydrogen bond, a state of higher energy than (b).

A potential function that relates the distance $\Delta r = r - r_0$ to the distance R (Fig. 1.7), and permits the calculation of the 'energy of the hydrogen bond', which is directly related to the enthalpy of association through hydrogen-bonding, and the frequency shift of the A–H vibration due to the hydrogen bond, has been developed[16] and is being applied to many cases.[17] The function has essentially three terms: one describing the stretching of the A–H bond from its state in isolated A–H (r_0) to its equilibrium state (r); the second the correspondingly much longer stretching for B–H, from $r_{0\,\mathrm{BH}}$ to $R - r$; and the third the general interaction between A and B, repulsion at short R, and attraction through dispersion forces at long R, and designated u_{AB}. For the special case where A and B are identical

$$u_{\text{H-bond}} = D\{1 - \exp[-k_0 r_0 (\Delta r)^2/2Dr]\} - D\exp[-k_0 r_0 (R - r - r_0)^2/2D(R - r)] + u_{\mathrm{AB}} \qquad (1.60)$$

Here D is the bond-dissociation energy of A—H, and k_0 is the force constant of this bond, which is proportional to the vibration frequency ν_0 of the unperturbed bond. Knowing the parameters D, r_0 and k_0 and the dependence of u_{AB} on R, the dependence of r, or of Δr, on R can be obtained from the condition $(\partial u_{\text{H-bond}}/\partial r)_{\text{equil}} = 0$. The frequency shift $\Delta\nu = \nu_{\text{H-bond}} - \nu_0$ can be obtained from $\nu_{\text{H-bond}} = \nu_0(k_{\text{H-bond}}/k_0)$ and $k_{\text{H-bond}} = (\partial^2 u_{\text{H-bond}}/\partial r^2)_{\text{equil}}$. A comparison with experimental data then fixes R, r and $u_{\text{H-bond}}$ or the association enthalpy. If A and B are not identical, and if the hydrogen bond is not colinear, modifications are introduced, also the dependence of u_{AB} on R must be specified, and requires some more parameters.

Typical quantities referring to the hydrogen bond are given below for the case A = B = oxygen (c.f. water in Table 1.2). The distance R varies from 0.25 nm for short (strong) hydrogen bonds to 0.28 nm for weak ones (for water the distance is 0.276 nm in the liquid and in ice), while r_0 = 0.097 nm. The equilibrium distances r for O—H and $R - r$ for H . . . O are however always different, $r < (R - r)$, that is the hydrogen bond in dense fluids is always asymmetrical (but this is not so in the isolated ion HF_2^-, i.e. $[F - - H - - F]^-$, where it is symmetrical, as it also is in certain solids). Typical enthalpies of association to dimers range from 12 to 25 kJ mol^{-1}, except for carboxylic acids in the gaseous phase where the enthalpy of dimerization is *ca.* 30 kJ mol^{-1} (but where evidently two bent hydrogen bonds are formed).

The wave function of a hydrogen bond includes a contribution from a state in which (negative) charge is transferred from B to A, the third term in eq. (1.59). Such *charge transfer* processes can lead to bonding even when no protons act as intermediates, as in molecular complexes. These may be of several types, e.g. a donor with a non-bonding lone-pair of electrons (n-type) interacts with an acceptor with a vacant orbital (v-type), or a donor with a bonding π-orbital (bπ-type) interacts with an acceptor with antibonding σ-orbitals (aσ-type) or π-orbitals (aπ-type). An example of the first kind of interaction (n-v) is $R_3N \cdot BCl_3$, for the second (bπ—aσ) $C_6H_6 \cdot I_2$ and for the third (bπ—aπ) $C_6H_6 \cdot (NC)_2C{=}C(CN)_2$. The wave function for the ground state of the 1 : 1 complex between a donor D and an acceptor A may be written[18] as

$$\psi_N = a\psi(D, A) + b\psi(D^+{-}A^-) \tag{1.61}$$

where normalization is made according to $a^2 + b^2 + 2abS = 1$, $S = \int\psi(D,A)\,\psi(D^+ - A^-)\mathrm{d}(\text{all space})$ is the overlap integral, of the order of 0.1. For weak complexes $a \gg b$ (for $C_6H_6 \cdot I_2$, $b^2 \sim 0.06$). An excited state exists with higher energy, such that

$$\psi_V = -b^*\psi(D, A) + a^*\psi(D^+ - A^-) \tag{1.62}$$

with $a^{*2} + b^{*2} - 2a^*b^*S = 1$ and $a^* \approx a$ and $b^* \sim b$.

The ground state ψ_N is therefore mostly an unbonded state (D, A), and the excited state ψ_V possesses mostly dative bonding $(D^+ - A^-)$. The solution of the Schrödinger equation leads to the energy difference between the excited and the ground state, expressible as the frequency ν_{CT} of the charge-transfer absorption band

$$h\nu_{CT} = E_V - E_N = [(E_1 - E_0)^2 + 4\beta_0\beta_1]^{1/2} (1 - S^2)^{-1} \tag{1.63}$$

here $\beta_0 = E_{01} - E_0 S$ and $\beta_1 = E_{01} - E_1 S$ are resonance integrals, and if $E_1 - E_0 = \Delta$ is used, then approximately (for $S^2 \ll 1$ and $\Delta^2 \gg 4\beta_0\beta_1$)

$$E_N = E_0 - \beta_0^2/\Delta + \ldots$$

$$E_V = E_1 + \beta_1^2/\Delta + \ldots$$

$$h\nu_{CT} = \Delta + (\beta_0^2 + \beta_1^2)/\Delta + \ldots$$

$$b/a = -\beta_0/\Delta \quad \text{and} \quad b^*/a^* = -\beta_1/\Delta \tag{1.64}$$

That is, all the quantities of interest may be expressed in terms of the eigenvalues of the energies of the unbonded state E_0, of the dative bond state E_1 (or their difference Δ), the overlap integral S and the resonance integral β_0 (since $\beta_1 = \beta_0 - \Delta S$). the difference Δ can be estimated from the ionization energy I_D of the donor and the electron affinity E_A of the acceptor, and β_0 may then be obtained from the measured ν_{CT}.

Figure 1.8 shows the relationship between the energetic terms. The energy of the system in the states 0 (unbonded), 1 (complete charge transfer and dative bond), N (actual ground state) and V (excited state) is plotted against the distance parameter r, and it is assumed that the configuration of nuclei and electrons continuously adjusts itself to give the minimal energy for a given state and distance as r varies. At infinite distance, the separated molecules D and A have zero mutual potential energy, but energy must be expended to produce the ions D^+ and A^-, namely $I_D^V - E_A^V$, the difference between the ionization potential of the donor and the electron affinity of the acceptor, both however being constrained in the ionized state to have the same configuration as in the free neutral state. The difference $I_D - I_D^V$ should be small, and is estimated to range from 0 to -0.8×10^{-19} J molecule^{-1}, whereas values of I_D range from 12 to 22 $\times 10^{-19}$ J molecule^{-1} for many common donors. No information concerning E_A^V is available, but it is not expected to differ much from E_A, for which, however, not very accurate data are known either. The common acceptor iodine has $E_A \sim 2.7 \times 10^{-19}$ J molecule^{-1}, tetracyanoethylene has $E_A \sim 4.2 \times 10^{-19}$ J molecule^{-1}.

As D and A are brought closer together they interact normally in the 0 state by van der Waals' forces, and have a mutual potential energy of u_0. However, the mixing-in of the wave function ψ_1 into ψ_0 produces a stabilization of $-\beta_0^2/\Delta$, eq. (1.64), so that the energy of formation of the ground state becomes $-u_N = -u_0 - |\beta_0^2/\Delta|$ at an equilibrium distance r_0 which is somewhat shorter, because of the stronger attraction, than that of the uncomplexed system D + A. The quantity Δ may be estimated from

$$\Delta = E_1 - E_0 = I_D^V - E_A^V + u_0(r_0) - u_1(r_0) \tag{1.65}$$

where u_1 is the mutual potential energy of the ions D^+ and A^-, given mainly by the

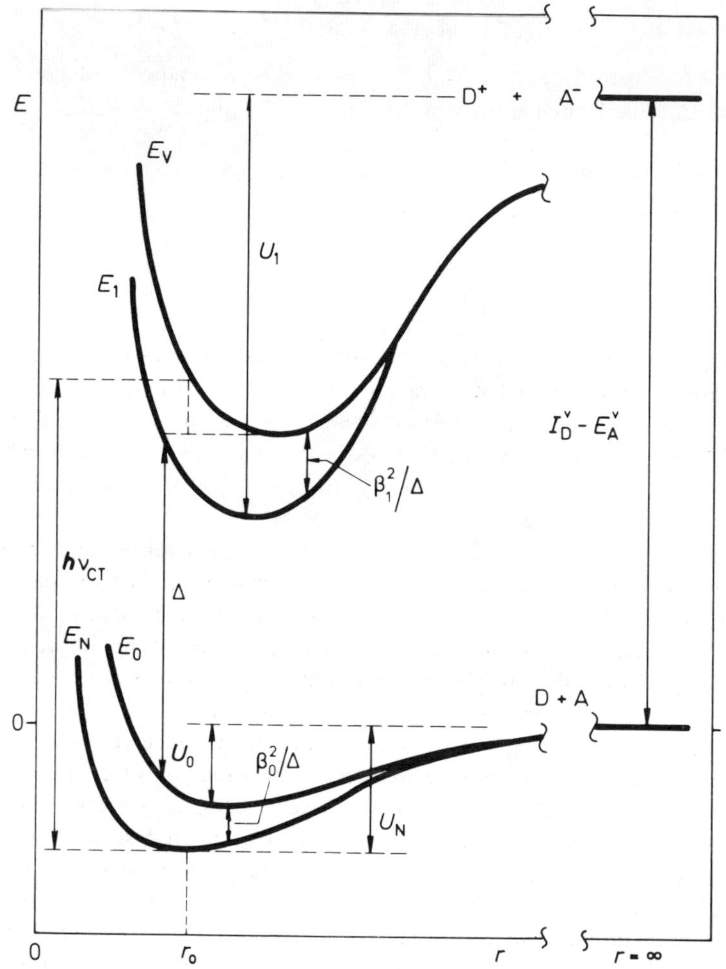

Figure 1.8. Energy diagram of a charge transfer complex between donor D and acceptor A. E_0 is the potential curve of the pure state D + A, E_1 that of the pure state $D^+ + A^-$, E_N of the ground state and E_V of the excited state of the bonded complex. r_0 is the equilibrium distance and $h\nu_{CT}$ is the charge transfer absorption energy. u_0 is the dispersion–attraction and repulsion energy between non-bonded D and A, β_0^2/Δ is the stabilization afforded by the mixing-in of the E_1 state. u_1 is the mainly coulomb energy between the ions D^+ and A^-, β_1^2/Δ is the destabilization produced by mixing in of the E_0 state. I_D^V is the ionization energy of the donor, E_A^V the electron affinity of the acceptor, both at the configuration of the complex. u_N is the formation energy of the complex at its equilibrium state.

coulomb interaction $-e^2/r_0$, and modified by dispersion and repulsion terms, as well as the dative bond energy in the D^+—A^- state, which can all be neglected. The values of u_1 range from 3.7 to 6.7 x 10^{-19} J molecule^{-1}. The stabilization energy β_0^2/Δ may now be estimated from (1.64) using the observed frequency ν_{CT} of the charge transfer absorption band

$$h\nu_{CT} \simeq \Delta + 2\beta_0^2/\Delta - 2\beta_0 S + \Delta S^2 \qquad (1.66)$$

recalling that the overlap integral S has values of the order of 0.1 in weak complexes (and may be as high as 0.4 in strong complexes, such as $R_3N \cdot I_2$). Thus, for iodine complexes with aromatic hydrocarbons, Δ ranges from 5.9 to 7.2, β_0 from −0.8 to −1.6; with alcohols and ethers Δ ranges from 4.2 to 4.6, β_0 from −2.4 to −4.8; and with aliphatic amines Δ ranges from 0.5 to 1.6, β_0 from −2.7 to −4.8 (all x 10^{-19} J per complex). The extent of charge transfer, or the mixing-in of ψ_1 into ψ_0, is measured by $b^2 + abS = a^2(-\beta_0/\Delta)\cdot[(-\beta_0/\Delta) - S]$ and is *ca.* 0.08 for the aromatic iodine complexes, *ca.* 0.2 for the ether—iodine complexes and *ca.* 0.6 for the amine—iodine complexes.

The energy of formation of the complex, $-u_N$, may be used to estimate the heat of formation of the complex in the gas phase $-\Delta h^0$ (gas phase CT complex) $= -\mathbf{N}u_N - \mathbf{R}T$, which is measurable for appropriate systems. The estimation of the extent of complexation in solution, however, requires the knowledge of solvation enthalpies and of the entropy change of complexation, which are not available from the theory delineated above.

Whereas the interactions discussed above — strong dipole interactions, hydrogen-bonding and donor—acceptor charge transfer complexing — are related to the association of molecules in definite 1 : 1 stoichiometries, the interaction between metal atoms is much more complicated because of the effect of the free-electron gas on the *metallic bond* (cf. lead in Table 1.2). This means, that with fluid metals, special interatomic forces come into play in the many-body problem, which are absent in the two body situation. Therefore, the many-body problem can be solved also in terms of sums of two-body, etc. interactions, eqs. (1.36) to (1.38), but only if cognizance is taken of the prevailing electron gas, and the problem must be attacked in a different way.

In a fluid metal, the density of free electrons is very high — of the order of 10^{30} m^{-3}. The positive charges of the ionized atoms are buried in this electron cloud and interact with it. The electrons occupy energy states up to the maximal energy, the *Fermi energy* E_F. Very few electrons have higher energies, attained through scattering interactions, but the maximum is quite sharp. To this maximum corresponds a maximal momentum $p_F = (2\mathrm{m}E_F)^{1/2}$, where m is the mass of the electron. Since only two electrons can be placed in a cell of size $\mathbf{h}^3$ in phase space, because of the uncertainty principle and Pauli's exclusion principle, the volume in phase space occupied by N electrons is $(\mathbf{h}^3 N/2) = (4/3)\pi p_F^3 V$, where V is the volume, so that the electron density, $\rho_0 = N/V = (8\pi/3\mathbf{h}^3)\cdot(2\mathrm{m}E_F)^{3/2}$, from which E_F may be computed. In the presence of a positive charge the electrons are perturbed and their momentum depends on the position relative to the charge. The maximal perturbed momentum is $p_F(\vec{r}) = [2\mathrm{m}(E_F - U(\vec{r})]^{1/2}$, where $U(\vec{r})$ is the potential energy. The electronic charge displaced around the positive ion is

$$\rho(\vec{r}) - \rho_0 = -(4\pi/\mathrm{h}^3)(2\mathrm{m})^{3/2}\, E_{\mathrm{F}}^{1/2}\, U(\vec{r}) \tag{1.67}$$

(when the difference $[E_{\mathrm{F}} - U(\vec{r})]^{3/2} - E_{\mathrm{F}}^{3/2}$ is linearized). Poisson's equation is now used to give $\nabla^2 U = -4\pi \mathrm{e}^2 (\rho(\vec{r}) - \rho_0) = q^2 U$ with $q^2 = 16\pi^2 \mathrm{e}^2 \mathrm{h}^{-3} (2\mathrm{m})^{3/2}\, E_{\mathrm{F}}^{1/2}$, which finally leads to the semiclassical solution[19] for the pair interaction potential energy function for an ion with charge $z\mathrm{e}$ at a distance r from an ion with the electrostatic potential $z\mathrm{e}r^{-1}$

$$u(r) = -z\mathrm{e}^2 r^{-1} \exp(-qr) \tag{1.68}$$

However, in wave-mechanical terms[20], the electronic charge displaced around the positive ion obeys the proportionality

$$[\rho(\vec{r}) - \rho_0] \propto r^{-2} \mathrm{j}[4\pi \mathrm{h}^{-1} (2\mathrm{m}E_{\mathrm{F}})^{1/2} r]$$

where

$$\mathrm{j}(x) = x^{-2} [\sin x - x \cos x] \tag{1.69}$$

At large r this becomes

$$u(r) \propto [\rho(\vec{r}) - \rho_0] \propto r^{-3} \cos[4\pi \mathrm{h}^{-1} (2\mathrm{m}E_{\mathrm{F}})^{1/2} r] \tag{1.70}$$

since $u(r)$ is proportional to the charge displacement. This means that instead of the

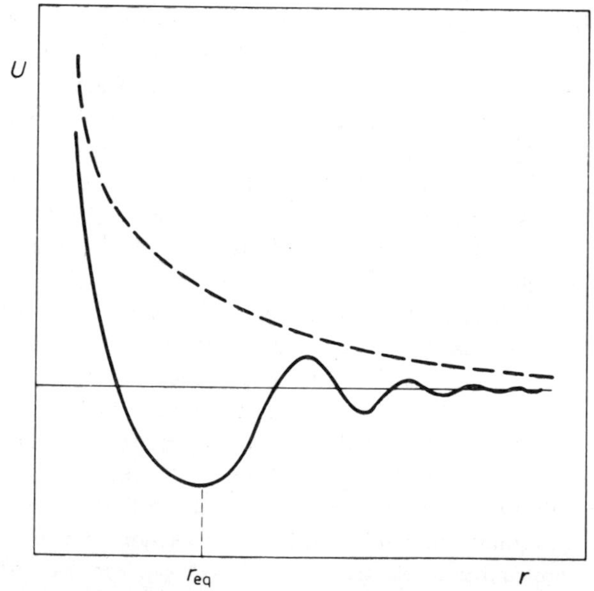

Figure 1.9. Schematics of the potential energy between two metal ions in a Fermi electron gas. Dashed curve, according to semiclassical theory, eq. (1.68); continuous curve, according to wave-mechanical theory, eq. (1.70), leading to an equilibrium configuration at r_{eq}.

semiclassical dependence of $u(r)$ on r^{-1} exp $(-qr)$, eq. (1.68), a dependence on r^{-3} $\cos(kr)$ is obtained, i.e. an oscillating function instead of a smoothly varying function, Fig. 1.9. Thus a function (1.70) with a minimum in u at some r is obtained, leading to a stable state for the fluid metal, which would be impossible with the continuously decreasing potential function (1.68). It is also seen that (1.70) represents long-range forces: compare with (1.41) or (1.50) on the one hand and with (1.42) or (1.49) on the other.

D. Appendix

a. Some statistical thermodynamics

It will be useful at this stage to state some of the concepts and equations of statistical thermodynamics, without going into details or into proofs of the relationships presented. These may be found in appropriate textbooks on this subject[21]. Consider a system of N particles occupying a total volume V at a temperature T. The values of V and of N are considered so large that no boundary effects need be taken into account. Such a system is called a *canonical ensemble.* The particles are indistinguishable from one another, but can be described by their coordinates, $\vec{r}_1, \vec{r}_2, \ldots \vec{r}_N$ (Fig. 2.1) and by their momenta $\vec{p}_1, \vec{p}_2, \ldots \vec{p}_N$. Indistinguishability means that the system is not changed when two particles described by the coordinates (r_i, p_i) and (r_j, p_j) in *phase space* are interchanged. The system is characterized by its *Hamiltonian*

$$H = \tfrac{1}{2} \sum_{i}^{N} \vec{p}_i^{\,2} M^{-1} + U(\vec{r}^{\,N}) \tag{1.71}$$

where M is the molecular mass (so $\tfrac{1}{2}\vec{p}_i^{\,2}M^{-1}$ is the kinetic energy of molecule i) and $U(\vec{r}^{\,N})$ is the potential energy, which is a function of the N coordinates $\vec{r}_i$, symbolized as $(\vec{r}^{\,N})$. The system is then described by the phase integral $\iint \exp(-\beta H)\, d(\vec{r}^{\,N}) d(\vec{p}^{\,N})$, where β is identified as $(\mathrm{k}T)^{-1}$.

In quantum-mechanical statistics (so-called semiclassical statistics, which will be employed here) the phase integral must be modified for the uncertainty principle, which states that $d\vec{r}d\vec{p}$ cannot be known more accurately than to $\mathbf{h}^3$, producing a factor $\mathbf{h}^{-3N}$, and for the indistinguishability, producing a factor $(N!)^{-1}$. This leads to the *total partition function*

$$Q = \mathbf{h}^{-3N}(N!)^{-1} \iint \exp(-H/\mathrm{k}T) d(\vec{r}^{\,N}) d(\vec{p}^{\,N}) \tag{1.72}$$

Integration can be carried out over the momenta separately. A quantity

$$\Lambda = \mathrm{h}/(2\pi M \mathrm{k}T)^{1/2} \tag{1.73}$$

is introduced, so that Λ^{-3N} is the momentum partition function, and

$$Q = \Lambda^{-3N}(N!)^{-1} \int \exp(-U(\vec{r}^{\,N})/\mathrm{k}T) d(\vec{r}^{\,N}) = \Lambda^{-3N}(N!)^{-1} Z \tag{1.74}$$

where Z is the *configurational partition function.* Whereas Λ^{-3N} depends on the temperature but not on volume, Z is a function of both (and of N)

$$Z(N, V, T) = \int \exp(-U(\vec{r}^N)/kT)d(\vec{r}^N) \quad (1.75)$$

A link to thermodynamics is obtained when it is noted that the Helmholz (free) energy of the system is

$$A = -kT \ln Q \quad (1.76)$$

The other thermodynamic functions are then obtained by straightforward differentiations. The internal energy is

$$E = kT^2 (\partial(\ln Q)/\partial T)_V \quad (1.77)$$

the entropy is

$$S = k(\ln Q + T(\partial(\ln Q)/\partial T)_V) \quad (1.78)$$

and the pressure is

$$P = kT(\partial(\ln Q)/\partial V)_T = kT(\partial(\ln Z)/\partial V)_T \quad (1.79)$$

since, as seen, Λ^{-3N} is volume-independent and only Z depends on it.

For the special case of an ideal gas, $U = 0$ by definition, since no interactions exist between the particles that may lead to a potential energy in the system, and hence $\exp(-U/kT) = 1$. The integration $\int d(\vec{r}^N)$ therefore yields N repeated integrations over the whole volume, i.e. V^N, so that $Z = V^N/(N!)$. From (1.79) then results

$$P = kT(\partial(\ln V^N(N!)^{-1})/\partial V)_T = NkTV^{-1} \quad (1.80)$$

which is the ideal gas equation of state $Pv = \mathbf{R}T$ where $\mathbf{R} = \mathbf{N}k$ and $v = (\mathbf{N}/N)V$ is the molar volume. Furthermore, when $U = 0$ at constant volume, only Λ^{-3N} remains a temperature-dependent factor of Q, so that (1.77) yields

$$E = kT^2 (\partial(\ln \Lambda^{-3N})/\partial T)_V = (3/2)NkT \quad (1.81)$$

which is an expression of the equipartition of the (kinetic) energy in an ideal gas.

For molecules with internal degrees of freedom of vibration, rotation and possibly electronic excitation, it is necessary to add an *internal partition function* q. However it will be assumed that the internal properties of the molecules and the configuration are independent of each other, so that q may be factored out in (1.72), and henceforth $U = U(\vec{r}^N)$ will mean the configurational potential energy, and

$$Q = (q/\Lambda^3)^N Z \quad (1.82)$$

Consider now an open system (a *grand canonical ensemble*), i.e. a system at constant V and T consisting of N particles, to which another particle of the same kind is added[22] at a constant chemical potential. The change in the Helmholtz energy of the system (per particle, not the molar quantity as is usual in thermodynamics) is

$$\mu(T, V) = A(T, V, N+1) - A(T, V, N) = (\partial A/\partial N)_{T,V} \tag{1.83}$$

From (1.75), (1.76), (1.82) and (1.83)

$$\exp(-\mu/kT) = \frac{q^{N+1} N!\Lambda^{3N}}{q^N (N+1)!\Lambda^{3(N+1)}} \frac{\int \exp(-U(\vec{r}^{N+1})/kT)d(\vec{r}^{N+1}))}{\int \exp(-U(\vec{r}^N)/kT)d(\vec{r}^N)} \tag{1.84}$$

If the coordinates of the extra particle are designated by $\vec{r}_0$, and of the other N particles by $\vec{r}_1, \vec{r}_2, \ldots \vec{r}_N$, then $U(\vec{r}^{N+1}) = U(\vec{r}^N) + \Sigma_i^N \; u(\vec{r}_0, \vec{r}_i)$ where $u(\vec{r}_0, \vec{r}_i)$ is the pair potential energy of interaction of the added particle with particle i, and the additivity of pair potentials is assumed. It should be noted, however, that $u(\vec{r}_0, \vec{r}_i)$ depends in fact only on the scalar distance $r = |\vec{r}_i - \vec{r}_0|$, (see Fig. 2.1) and not on the position $\vec{r}_0$. Hence integration over $\vec{r}_0$ may be performed separately from that over $\vec{r}_1, \vec{r}_2, \ldots \vec{r}_N$, to give the volume V, and (1.84) may be transformed into

$$\exp(-\mu/kT) = (q\Lambda^{-3})\rho^{-1} \int \exp(-U(\vec{r}_0, \vec{r}_i)/kT)d(\vec{r}^N) \tag{1.85}$$

where the number density $\rho = N/V$ is introduced for $(N+1)/V$ for large N. The quantity $U(\vec{r}_0, \vec{r}_i) = \Sigma u(\vec{r}_0, \vec{r}_i)$ may be construed as an external field acting on the N particles by the extra particle at r_0.

The process leading to (1.85) may be split into two stages. First, the extra particle is added at a *fixed* position $\vec{r}_0$ so that it is distinguishable from the other particles by not having translational degrees of freedom, and the integration over $\vec{r}_0$ is not performed. The change in Helmholtz energy is

$$\Delta A(\vec{r}_0) = A(T, V, N+1, \vec{r}_0) - A(T,V,N) \tag{1.86a}$$

$$\exp(\Delta A(\vec{r}_0)/kT) = q\int \exp(-U(\vec{r}_0, \vec{r}_i)/kT)d(\vec{r}^N) \tag{1.86b}$$

In the second stage the particle is allowed to move throughout the volume V, so that it gains the communal properties (p. 6) and indistinguishability described by $\rho = N/V$, and the kinetic energy described by Λ^3. A comparison of (1.85) and (1.86) yields

$$\mu = \Delta A(\vec{r}_0) + kT \ln \rho\Lambda^3 \tag{1.87}$$

so that μ may be interpreted as the work done in first placing a new particle at the fixed position $\vec{r}_0$ in a system of N particles of constant T and V, and then removing the constraint of the fixed position.

Processes in the laboratory are commonly carried out at constant pressure rather than at constant volume, so that a system characterized by (T, P, N) and its Gibbs (free) energy G is more useful than the system considered hitherto, characterized by (T, V, N) and by the Helmholtz (free) energy A. The main effect of changing to the new variables is that in (1.74) $\exp(-PV/kT - U/kT)$ replaces $\exp(-U/kT)$. Since the volume is now a function of the pressure, rather than an independent variable, it is an average volume $\langle V \rangle$ which must be factored out on going from

(1.84) to (1.85). The formal form of (1.87) is retained, except that $\Delta G(\vec{r}_0)$ and $\langle\rho\rangle$ replace $\Delta A(\vec{r}_0)$ and ρ. The computation of the average volume, however, is quite complicated in the general case, and results are not conveniently obtained in the (T, P, N) system.

It is generally possible to describe thermodynamically quantities in terms of fluctuations of the volume V, the configurational potential energy $U(\equiv U(\vec{r}^N))$ or the intermolecular virial function $\Phi = -V(\partial U/\partial V)_T$ from their mean values. In general, the mean value of a function X, averaged over all space and all time is given by

$$\langle X\rangle = \frac{\iint X \exp(-H/kT)d(\vec{r}^N)d(\vec{p}^N)}{\iint \exp(-H/kT)d(\vec{r}^N)d(\vec{p}^N)}$$

$$= Q^{-1}\iint X \exp(-H/kT)d(\vec{r}^N)d(\vec{p}^N) \quad (1.88)$$

The time-dependence is of interest in dynamic processes, such as self diffusion, which is connected with the time evolution of the singlet distribution function $n^{(1)}(\vec{r}, t)$ (section 2A(a)), but is outside the scope of this book. Hence, integration over the momenta will be performed, and only the configurational average will be discussed

$$\langle X\rangle_{conf} = Z^{-1}\int X \exp(-U/kT)d(\vec{r}^N) \quad (1.89)$$

The average of the fluctuations of a quantity from its mean value is zero, by definition $\langle X - \langle X\rangle\rangle = 0$, hence it is the mean square of the fluctuation which is of interest

$$\langle(X - \langle X\rangle)^2\rangle = \langle X^2\rangle - \langle X\rangle^2 \quad (1.90)$$

The average potential energy of a canonical ensemble (an N, V, T system) is given by

$$\langle U\rangle = Z^{-1}\int U \exp(-U/kT)d(\vec{r}^N) = kT^2(\partial(\ln Z/\partial T)_V = E_{conf} \quad (1.91)$$

from (1.77). That is the mean value of the intermolecular potential energy is the configurational part of the internal energy. From (1.91) and $C_v = (\partial E/\partial T)_V$ it follows that

$$C_{v\ conf} = (kT^2)^{-1}\langle(U - \langle U\rangle)^2\rangle \quad (1.92)$$

As the system absorbs energy at a given temperature and volume, this goes to increase the mean square fluctuations of the intermolecular potential energy (disregarding the translational and internal degrees of freedom). The mean of the intermolecular virial function is

$$\langle\Phi\rangle = Z^{-1}\int \Phi \exp(-U/kT)d(\vec{r}^N)$$

$$= Z^{-1}\int -V(\partial U/\partial V)_T \exp(-U/kT)d(\vec{r}^N)$$

$$= PV - NkT = NkT(B_2(T)V^{-1} + B_3(T)V^{-2} + \ldots) \quad (1.93)$$

which gives directly the virial expansion of the equation of state (cf. eq. (1.44)). The fluctuations in the intermolecular virial function depend on the second differential of the intermolecular potential energy with respect to distance, which is not directly related to observable quantities but can be calculated if $u(r)$ is a known function, (cf. section 1C)

$$\langle\Phi^2\rangle - \langle\Phi\rangle^2 = kT[PV - V\kappa_T^{-1} + (1/9)\sum^{N} r(du/dr) + r^2(du^2/dr^2)] \qquad (1.94)$$

However, the cross-fluctuations between the potential energy and virial functions again relate directly to macroscopic thermodynamic quantities

$$\langle U\Phi\rangle - \langle U\rangle\langle\Phi\rangle = kT^2 V(\partial P/\partial T)_V - Nk^2 T^2 \qquad (1.95)$$

where the significance of the quantity $(\partial P/\partial T)_V$ has been discussed in section 1B(b).

These relationships concerning fluctuations are quite general, and do *not* depend on an explicit relationship between U and $u(r)$ (cf. eq. 1.38), e.g., on the pair-additivity assumption

$$U(\vec{r}^N) = (1/2)\sum_{ij}^{N} u_{ij}(r) \qquad (1.96)$$

This relationship, or a better approximation in the case of dense fluids far below their critical temperatures where three-body interactions are important, is required in order to relate the thermodynamic quantities and the fluctuation functions to local fluctuations of the density of the fluid (or fluctuations of its volume for an (N, P, T) system), i.e. to its structure. For studying the mean square fluctuation in the number density $\rho = N/V$ in a system of volume V, i.e., the mean square fluctuation in the number of particles N, it is necessary to consider the grand canonical ensemble $(\mu\ V, T)$, and replace the Hamiltonian H in eq. (1.88) not by U but by μ. The result is

$$\langle N^2\rangle - \langle N\rangle^2 = kT(\partial\langle N\rangle/\partial\mu)_{V,T} = \langle N\rangle^2 kTV^{-1}\kappa_T \qquad (1.97)$$

or in terms of the density flucuations

$$\langle\rho^2\rangle - \langle\rho\rangle^2 = \langle\rho\rangle^2 kTV^{-1}\kappa_T \qquad (1.98)$$

The fluctuations in the volume are intimately connected with the radial distribution function, discussed in Chapter 2.

References

1. A. V. Grosse, *J. Inorg. Nucl. Chem.*, **22**, 23 (1961).
2. C. A. Angell, *J. Chem. Educ.*, **47**, 583 (1970); *J. Phys. Chem.*, **70**, 2793 (1966); W. Kauzmann, *Chem. Rev.*, **43**, 219 (1948).
3. T. L. Hill, *Introduction to Statistical Thermodynamics*, Addison-Wesley, Reading, Mass., 1960, p. 290; J. Hirschfelder, D. Stevenson and H. Eyring, *J. Chem. Phys.*, **5**, 726 (1937).
4. P. G. Tait, *Voyage of H.M.S. Challenger, Vol. 2*, Part 4, 1–76, 1889.
5. W. Westwater, H. W. Frantz and J. H. Hildebrand, *Phys. Rev.*, **31**, 135 (1928); J. H.

Hildebrand, *Phys. Rev.*, **34**, 649 (1929); J. H. Hildebrand and J. M. Carter, *J. Am. Chem. Soc.*, **54**, 3592 (1932); B. J. Alder, E. W. Haycock, J. H. Hildebrand and H. Watts, *J. Chem. Phys.*, **22**, 1060 (1954); E. B. Smith and J. H. Hildebrand, *J. Chem. Phys.*, **31**, 145 (1959).

6. C. Antoine, *Compt. Rend.*, **107**, 681, 778 (1888); G. W. Thomson, *Chem, Rev.*, **38**, 1 (1946).
7. J. S. Rowlinson, *Liquids and Liquid Mixtures*, Butterworth, London, 2nd edn., 1969, pp. 84, 97.
8. G. Scatchard, *Chem. Rev.*, **44**, 7 (1949).
9. E. Schadow and R. Steiner, *Z. Phys. Chem. (Frankfurt am Main)*, **66**, 105 (1969); Y. Marcus, *Proc. Symp. Ion Pairs*, Louvain, May 1976; further material to be published.
10. D. C. Grahame, *J. Chem. Phys.*, **21**, 1054 (1953).
11. O. Sinanoglu, in *Intermolecular Forces, Advances in Chemical Physics, Vol. 12*, edited by J. O. Hirschfelder, 1967, p. 283.
12. J. E. Lennard-Jones, *Proc. Roy. Soc., Ser. A*, **106**, 463 (1924); **112**, 214 (1926).
13. T. Kihara, *Rev. Mod. Phys.*, **25**, 831 (1953).
14. F. London, *Z. Phys. Chem., Abt. B*, **11**, 222 (1930); *Trans. Faraday Soc.*, **33**, 8 (1937).
15. J. D. Lambert, G. A. H. Roberts, J. S. Rowlinson and V. J. Wilkinson, *Proc. Roy. Soc., Ser. A*, **196**, 113 (1949).
16. E. R. Lippincot and R. Schroeder, *J. Chem. Phys.*, **23**, 1099 (1955); *J. Phys. Chem.*, **61**, 921 (1957).
17. E. A. Robinson, H. D. Schreibner and J. N. Spencer, *J. Phys. Chem.*, **75**, 2219 (1971); *Spectrochim. Acta, Part A*, **28**, 397 (1972); J. N. Spencer, G. J. Casey, Jr., J. Buckfelder and H. D. Schreibner, *J. Phys. Chem.*, **78**, 1415 (1974).
18. R. S. Mulliken and W. B. Person, *Molecular Complexes*, Wiley–Interscience, New York, 1969; R. S. Mulliken, *J. Am. Chem. Soc.*, **64**, 811 (1952).
19. N. F. Mott, *Proc. Camb. Phil. Soc.*, **32**, 281 (1936).
20. N. H. March, *Liquid Metals*, Pergamon Press, Oxford, 1968; N. H. March and A. M. Murray, *Proc. Roy. Soc., Ser. A*, **261**, 119 (1961).
21. T. L. Hill, *Statistical Mechanics*, McGraw-Hill, New York, 1956; A. Münster, *Statistical Thermodynamics, Vol. 1*, Springer, Berlin, 1969; ref. 3a.
22. A. Ben-Naim, in *Solutions and Solubilities, Part I*, edited by R. J. Dack, Wiley–Interscience, New York, 1975, p. 33.

Chapter 2

The Structure of Liquids

A. Experimental Determination of Structure

An investigation into the structure of liquids may be likened to the task of an archaeological expedition, digging into the remains of the culture of a primitive warlike people who have conquered, destroyed and settled the land of a highly cultured nation. Some vestiges of the early culture, with its hierarchical structure of the society, its temples and public buildings, remain, albeit in ruins. The new people are classless nomads, perhaps organized into families and clans, living in tents or wooden huts of which only the foundations remain for the archaeologist to examine. He is supposed to deduce from his finds the relationships between the persons in the family, clan and tribe, and the way they lived their daily life.

One may go on with this metaphor, and draw a parallel between the tools that the archaeologist brings to his digging site and the experimental methods used to investigate the structure of a liquid. Limiting oneself to the equilibrium properties, the structure of a homogeneous isotropic fluid is completely described by the pair correlation function $g(r)$ introduced in section 2A(a). This should properly be written $g(r; T, \rho)$, since it is a function of the temperature and the density. Several experimental methods have been developed to determine $g(r)$, but all have their limitations and inaccuracies, in particular for the region of low r (where repulsive forces are dominant). The methods include physical experiments on the diffraction of radiation from fluids (mainly liquids; only few highly compressed gases have been examined), computer experiments on systems of up to a few hundreds of particles with a specified pair potential $u(r)$ (cf. section 1C), and the examination of models comprising macroscopic spheres packed in a random fashion. With these tools it is necessary to examine whether any structure at all exists in the fluid, and if so, what its features are. The number of nearest-neighbours and their distance, together with the number of neighbours at greater distances for any given particle in the fluid, and the nature of the interactions between them, are the results of such an investigation. A comparison with the structure of the solid, from which the liquid has arisen on melting may be instructive for the understanding of the liquid structure. The relationship of these considerations and the archaeological metaphor should be obvious.

a. Distribution and correlation functions

In the system of particles characterized by (T, V, N), consider the volume element $d\vec{r}_1$ at the coordinates $\vec{r}_1$ (Fig. 2.1). The probability of finding a particle in

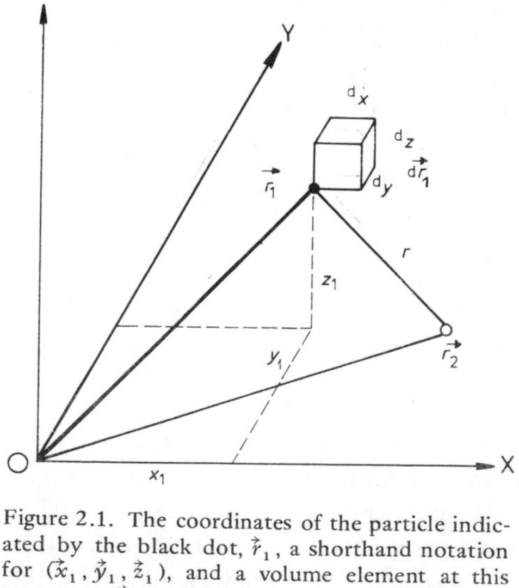

Figure 2.1. The coordinates of the particle indicated by the black dot, $\vec{r}_1$, a shorthand notation for $(\vec{x}_1, \vec{y}_1, \vec{z}_1)$, and a volume element at this location, $d\vec{r}_1$, a shorthand notation for $dV = dx\, dy\, dz\, (\vec{x}_1, \vec{y}_1, \vec{z}_1)$. A second particle (empty circle) is located at r_2, and the scalar distance between them is $r = |\vec{r}_2 - \vec{r}_1|$.

this volume element is $n^{(1)}(\vec{r}_1)$, and the average number of particles found in this volume element is $n^{(1)}(\vec{r}_1)\mathrm{d}(\vec{r}_1)$. The quantity $n^{(1)}(\vec{r}_1)$ is called the *singlet distribution function* and is the average number density of particles at $(\vec{r}_1)$. When a force field exists which affects the particles, $n^{(1)}(\vec{r}_1)$ and the particle density at $(\vec{r}_1)$ depend on this field, but in the absence of such a field, in an isotropic fluid $n^{(1)}(\vec{r}_1)$ is constant. From the normalizing condition that all N particles are to be found in the volume of the system

$$\int n^{(1)}(\vec{r}_1)\mathrm{d}\vec{r}_1 = n^{(1)}(\vec{r}_1)\int \mathrm{d}\vec{r}_1 = n^{(1)}(\vec{r}_1)V = N \tag{2.1}$$

the result is obtained that the singlet distribution function in a homogeneous isotropic fluid equals the bulk number density

$$n^{(1)}(\vec{r}_1) = N/V = \rho \tag{2.2}$$

The probability of finding a particle at volume element $\mathrm{d}\vec{r}_2$, located at $\vec{r}_2$ (Fig. 2.1), *given* a particle already at $\vec{r}_1$, is designated the *pair distribution function*, $n^{(2)}(\vec{r}_1, \vec{r}_2)$. The normalizing condition is

$$\int n^{(2)}(\vec{r}_1, \vec{r}_2)\mathrm{d}\vec{r}_2 = (N-1)n^{(1)}(\vec{r}_1) = N(N-1)/V \tag{2.3}$$

If the particle at $\vec{r}_1$ exerts a spherically symmetrical force field, then $n^{(2)}(\vec{r}_1, \vec{r}_2)$ depends only on the scalar distance $r = |\vec{r}_2 - \vec{r}_1|$. As $r \to \infty$, the probability of occupancy of $\mathrm{d}\vec{r}_2$ becomes independent of the presence of the particle at $\vec{r}_1$, and $n^{(2)}(\vec{r}_1, \vec{r}_2, r \to \infty) = \rho^2$. At small values of r, however, the particle at $\vec{r}_1$ is correlated with that at $\vec{r}_2$ through the force field, and a *pair correlation function* $g(r)$ is defined as

$$n^{(2)}(\vec{r}_1, \vec{r}_2) = g(r)\rho^2 \tag{2.4}$$

Obviously from the above, $g(r \to \infty) \to 1$. Also, for an ideal gas $g(r) = 1$ for any r, but for fluids with volume-occupying particles $g(r < \sigma) \to 0$, as r becomes smaller (for σ see Fig. 1.6).

The number of particles in a spherical shell of thickness $\mathrm{d}r$ at a distance r from a particle at the origin is the *radial distribution function*

$$\mathrm{d}n(r, \mathrm{d}r) = \rho g(r)4\pi r^2\,\mathrm{d}r \tag{2.5}$$

Thus, the integral of (2.5) from zero (in practice from σ) to r is the number of neighbours a particle has up to distance r. Since $\rho 4\pi r^2\,\mathrm{d}r$ is the average number of particles in such a spherical shell of volume $4\pi r^2\,\mathrm{d}r$, the excess number of particles, because of the presence of a particle at the origin, $(g(r) - 1)4\pi r^2\,\mathrm{d}r$, is of interest. The quantity $h(r) = g(r) - 1$ is called the *total* (pair) *correlation function*, and decreases to zero as r increases and the correlation between the particles vanishes. The integral

$$G = \int_0^\infty h(r)4\pi r^2\,\mathrm{d}r \tag{2.6}$$

is the *affinity* between particles in this system.

As the density of a fluid decreases, the pair correlation function $g(r)$ approaches a Boltzmann distribution

$$\lim g(r)(\rho \to 0) = \exp(-u(r)/kT) \tag{2.7}$$

$u(r)$ being the pair potential. Where $u(r)$ has a minimum, at r_0 (for an 'LJ 12-6' potential at $2^{1/6}\sigma$), $g(r)$ has a peak. As the particle density increases, $g(r)$ is no longer given by (2.7), but $U(r)$, the *potential of mean force* between the particles, replaces $u(r)$. The force acting on a particle situated at $\vec{r}_1$ is

$$f(\vec{r}_1) = -\partial U(\vec{r})/\partial\vec{r}_1 = -\partial u(\vec{r}_1, \vec{r}_2)/\partial\vec{r}_1 - \int [n^{(3)}(\vec{r}_1, \vec{r}_2, \vec{r}_3)/ n^{(2)}(\vec{r}_1, \vec{r}_2)] (\partial u(\vec{r}_1, \vec{r}_3)/\partial\vec{r}_1) d\vec{r} \tag{2.8}$$

where $n^{(3)}(\vec{r}_1, \vec{r}_2, \vec{r}_3)$ is the *triplet distribution function*, defined, just as the corresponding $n^{(2)}(\vec{r}_1, \vec{r}_2)$ and $n^{(1)}(\vec{r}_1)$, as the probability of simultaneously finding particles at $\vec{r}_1$, $\vec{r}_2$ and $\vec{r}_3$. The force on a particle at $\vec{r}_1$ is thus taken as the sum of the direct force exerted by the particle at $\vec{r}_2$, $-\partial u(\vec{r}_1, \vec{r}_2)/\partial\vec{r}_1$, and the average force exerted by all the other particles. At such higher densities, three-body interactions come into play. Before pursuing the consequences of (2.8) further, it is instructive to look at the results of some approximate calculations[1], Fig. 2.2, the approximation being necessary since $n^{(3)}(\vec{r}_1, \vec{r}_2, \vec{r}_3)$ *cannot* be evaluated exactly. Instead of the single peak in $g(r)$ that arises at low densities from (2.7), at higher densities further peaks appear at roughly 2σ, 3σ, etc. The height of the peaks becomes more pronounced as the potential-well beomes deeper or as the temperature decreases, i.e. with ϵ/kT. However, it decreases rapidly for successive peaks, and $g(r)$ decays practically to unity for $r > 5\sigma$. This means that in a fluid, as dense as it may be, there exists no correlation between particles located more than five molecular diameters apart. Fluids, thus, show only *short-range order* and lack long-range regularity, in sharp contrast to crystalline solids.

The formal relationship

$$g(r) = \exp(-U(r)/kT) \tag{2.9}$$

requires, to be useful, the connection between $U(r)$ and the pair potential $u(r)$ (cf. eq. (1.38)). For fluids of low density, i.e. for gases or for highly expanded liquids near the critical point, the additivity of pair potentials (eq. (1.96)) is an acceptable approximation. This permits the calculation of observable quantities, such as the internal energy, the pressure, the compressibility and the second virial coefficient for such fluids in terms of $g(r)$ and of $u(r)$.

The internal energy of the system is obtained from (1.77), utilizing (1.74), (1.75) and (1.81) as

$$E = (3/2)NkT + kT^2(\partial(\ln Z)/\partial T)_{N,V} \tag{2.10}$$

If now $\exp(-U(\vec{r}^N)/kT)$ in (1.75) is replaced by $g(r)$, according to (2.9), the result is

$$E = (3/2)NkT + (1/2)N\rho \int u(r)g(r)d\vec{r} \tag{2.11}$$

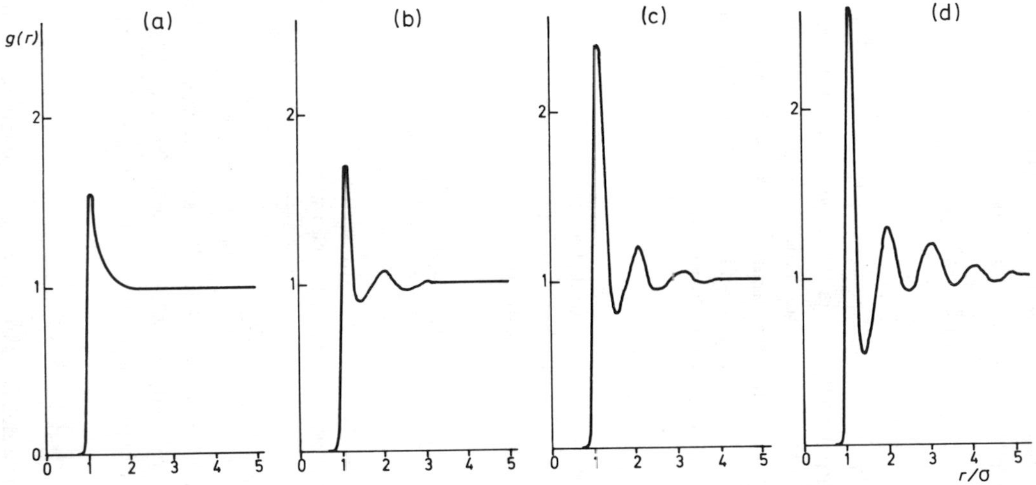

Figure 2.2. The pair correlation function, as calculated by Ben-Naim[1] by a numerical solution of the Percus–Yevick equation (section 2B(b)) for particles with an 'LJ 12-6' potential (section 1C(a)). (a) $\rho = 10^{-3}\ \sigma^{-3}$, $\epsilon = 0.5$ kT; *(b)* $\rho = 0.6\ \sigma^{-3}$, $\epsilon = 0.25$ kT; (c) $\rho = 0.6\ \sigma^{-3}$, $\epsilon = 1.0$ kT; *(d)* $\rho = 1.0\ \sigma^{-3}$, $\epsilon = 0.5$ kT. As the number density increases, the number of peaks and their height increases, the height increasing also with an increase in ϵ.

The first term is the contribution of the kinetic energy to the internal energy, the second is that of the configurational potential energy. This contribution can also be obtained as follows. The product of the number of molecules in a spherical shell from a given molecule, $n(r, \mathrm{d}r)$ according to eq. (2.5), with the pair potential between them and the molecule at the origin, $u(r)$, is integrated over the whole volume, $\int_0^\infty u(r)\rho g(r) 4\pi r^2 \,\mathrm{d}r$. This quantity is then multiplied by N, the number of molecules that serve as the one at the origin, and divided by two, in order not to count each pair of molecules twice over. The result is

$$E_{\mathrm{conf}} = 2\pi N\rho \int_0^\infty u(r)g(r)r^2 \,\mathrm{d}r \tag{2.12}$$

which is seen to be equivalent to the second term in (2.11). Since the first term of (2.11) is the internal (kinetic) energy of an ideal gas, the configurational energy (2.12) equals the energy that must be added to the liquid to vaporize it isothermally, $-U^{\mathrm{V}}$. The pressure in the system is obtained by applying (1.79) and (1.80), and noting that $(\partial U(\vec{r}^N)/\partial V)_{N,T} = (1/3V)\,\Sigma r(\mathrm{d}u(r)/\mathrm{d}r)$, where the sum extends over all the pairs of molecules. As above, this leads to

$$P = \rho kT - (1/6)\rho^2 kT \int r(\mathrm{d}u(r)/\mathrm{d}r)g(r)\mathrm{d}\vec{r} \tag{2.13}$$

A particularly simple relationship is found for the isothermal compressibility $\kappa_T = \rho^{-1}(\partial\rho/\partial P)_T$, the so-called *compressibility equation*, which is valid independently of any relationship between $U(r)$ and $u(r)$

$$\kappa_T = (\rho kT)^{-1} + (kT)^{-1}\int (g(r) - 1)\mathrm{d}\vec{r} = (\rho kT)^{-1}\left(1 + \rho\int_0^\infty h(r)4\pi r^2 \,\mathrm{d}r\right) \tag{2.14}$$

a relationship which will be found useful in section 2A(b). The second virial coefficient, B_2 in (1.44) or (1.93), can be obtained immediately from (2.13)

$$B_2 = -(1/6)NkT \int r(\mathrm{d}u(r)/\mathrm{d}r)g(r)\mathrm{d}\vec{r} \tag{2.15}$$

If the dilute gas value $g(r) = \exp(-u(r)/kT)$, the substitution $4\pi r^2 \,\mathrm{d}r = \mathrm{d}\vec{r}$, and integration by parts are employed, eq. (1.45) results from (2.15). Fluctuations in the particle density can be expressed in terms of (1.89) and (1.90) as

$$\langle(\rho(r) - \rho)^2\rangle = \rho\left(1 + \rho\int (g(r) - 1)\,\mathrm{d}\vec{r}\right) \tag{2.16}$$

since the mean square particle density fluctuation is proportional to the isothermal compressibility (eq. (1.98)). If pressure is applied to a system at constant volume and temperature, these mean fluctuations increase, and a solid with discrete arrangements of particles at certain distances from a given particle results finally, if the temperature is sufficiently low.

For fluids with densities characteristic for liquids, the additivity approximation (1.96) is no longer valid, as already mentioned, so that eqs. (2.10) to (2.13) and

(2.15) cannot be used (although eq. (2.14) or the equivalent eq. (2.16) remain valid). One approach to this problem is to assume, according to Kirkwood, the superposition of probabilities

$$n^{(3)}(\vec{r}_1, \vec{r}_2, \vec{r}_3) \approx \rho^3 g(\vec{r}_1, \vec{r}_2) g(\vec{r}_2, \vec{r}_3) g(\vec{r}_3, \vec{r}_1) \quad (2.17)$$

which relates, through (2.8) and (2.9), $U(\vec{r}^N)$ with $u(r)$. Another approach is to determine $g(r)$ experimentally (section 2A(b)), and use approximate integral equations (section 2B), or numerical calculations by computers (section 2A(c)).

b. Diffraction methods

Electromagnetic radiation (in this context, X-rays), neutrons and electrons, diffracted from fluid samples, yield information on the structures of the fluid through the relationship of the intensity of the diffracted beam, I, with the angle θ by which it has been diffracted. Electromagnetic radiation is scattered by the electrons located in the entire bulk of an atom, neutrons are scattered by the nucleus in the centre of an atom occupying practically no space, and electrons are scattered by the charge, both positive and negative, in the bulk of an atom. Individual atoms are thus the scattering centres, rather than molecules, and chemical bonding can be recognized only through the shortness of certain distances between atoms compared with the sums of their van der Waals' radii. This is a serious limitation, since only monatomic liquids (inert gases and metals) are amenable to convenient examination in this way, with the addition of certain simple spherical molecules or other rotating highly symmetrical molecules (carbon tetrachloride, nitrogen) and of molten salts made up of monatomic ions (alkali halides). The results for other molecular liquids are dominated by intramolecular distances between the bonded atoms and yield no information on liquid structure.

The diffraction of electrons is practical only for fluids of low density and hence will not be discussed further in this book on liquids. The diffraction of neutrons and of X-rays has its respective merits and disadvantages, and the two methods complement each other. Consider Fig. 2.3, which is a schematic representation. The incoming radiation is characterized by its wavelength λ (for neutrons of energy E, the wavelength is hcE^{-1}) and its intensity I_0. Monochromatic radiation is required: crystal monochromators, or a β-filter for isolating the K_α X-radiation together with a pulse height analyser, define the radiation within 5 pm. A typical value for the wave length is 0.1 nm (corresponding to *ca.* 0.08 eV for neutrons); for X-rays the target in the radiation source is selected so that it has a higher atomic number than any of the atoms in the liquid sample, to minimize absorption.

The liquid sample is characterized by the scattering properties of its atoms and by the structure to be determined, at a given temperature T and pressure P (usually atmospheric). The dimensions of the sample are large compared to the wavelength λ (which is commensurate with the average distance between the atoms), but not so large that absorption of the radiation becomes a problem.

The scattering results are presented as the angle-dependent scattering intensities,

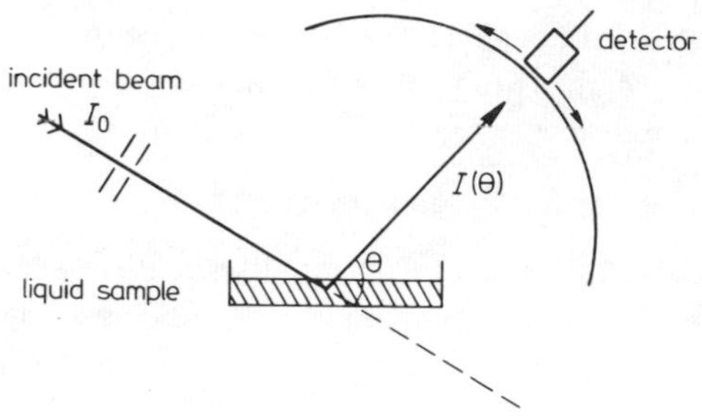

Figure 2.3. Schematic representation of a scattering experiment, defining the scattering angle θ between the incident beam, of intensity I_0, and the diffracted beam, of intensity $I(\theta)$.

$I(\theta)$, where instead of the angle the variable k is used

$$k = \lambda^{-1} 4\pi \sin(\theta/2) \tag{2.18}$$

The properties of the atoms as scattering centres differ for neutron and for X-ray diffraction, occurring from nuclei and from electrons, respectively.

For neutron scattering, the atom is characterized by the scattering length a of its nucleus, or its *bound atom scattering length* $b = a(1 + M_{\text{neutron}}/M_{\text{atom}})$, which is of the order of the nuclear diameter, 0.01 pm, or the corresponding *scattering cross-section* $\sigma = 4\pi b^2$. The atoms of a pure elemental substance may still differ with respect to the isotopes of the nucleus or its spin state, so that an average value, $\langle b \rangle$ must be used. Furthermore, only the coherent scattering contains information on the structure, not the incoherent scattering. The scattered intensity, corrected for absorption, can be written as

$$I(\theta) = \alpha(\theta) \langle b \rangle^2 (S(k) + \Delta) \tag{2.19}$$

Here $\alpha(\vartheta)$ is a proportionality constant converting absolute intensities (differential scattering cross-sections, i.e. the numbers of neutrons scattered per unit time per unit solid angle) to observed intensities $I(\theta)$, and is instrument-dependent but sample-independent. As noted, $\langle b \rangle$ is angle-, θ-, independent. The quantities $S(k)$ and Δ are related to the coherent and incoherent parts of the scattered intensity; the former is called the *structure factor*, and is discussed in detail below. For the present, its limiting dependence on the variable k is to be noted: $S(0) = \rho \mathrm{k} T \kappa_T$ (which is of the order of 0.01) and $S(\infty) = 1$. The intensities scattered at very large angles $I(\infty)$ (i.e. $k \to \infty$) and at very small angles $I(0)$ are used to eliminate $\alpha(\vartheta)$ and Δ from (2.19)

$$S(k) = [I(k) - I(k)S(0) - I(0) + I(\infty)S(0)] / [I(\infty) - I(0)]$$

$$\simeq [I(k) - I(0)] / [I(\infty) - I(0)] \tag{2.20}$$

For X-ray diffraction, the quantity corresponding to $\langle b \rangle^2$ above is the *atomic scattering intensity* for isolated atoms, $a^2(k)$, where

$$a(k) = \int_0^{r_{atom}} \rho_{ia}(r) 4\pi r^2 (\sin kr)(kr)^{-1}\, dr \tag{2.21}$$

where $\rho_{ia}(r)$ is the electron density in the isolated atom. It is seen that in contrast to $\langle b\rangle^2$, a^2 is angle-, θ-, or k-dependent. The structure factor is simply the ratio of the diffracted radiation intensity from the sample normalized for the incident intensity I_0 to that from the N individual atoms of atomic number Z (containing each Z electrons)

$$S(k) = (I(k)/I_0)Z^2 N^{-1}(a(k))^{-2} \tag{2.22}$$

At large values of k, $I(k)/I_0$ tends to $Z^{-2}Na^2(k)$ so that $S(k)$ approaches unity, and this fact is used for normalizing the data at lower k.

The structure factor is related to the radial distribution function[2] of the atoms in the liquid $n(r)$ (eq. (2.5)) or the pair correlation function $g(r)$ (eq. (2.4)), since it reflects the interference of radiation scattered from correlated atoms

$$S(k) = 1 + \rho \int_0^{\infty} 4\pi r^2 (g(r) - 1)(\sin kr)(kr)^{-1}\, dr \tag{2.23}$$

This can be Fourier-transformed to give $g(r)$ explicitly in terms of $S(k)$

$$g(r) = 1 + (2\pi^2 \rho r)^{-1} \int_0^{\infty} (S(k) - 1)k(\sin kr) dk \tag{2.24}$$

An accurate determination of $g(r)$ presupposes an accurate knowledge of $S(k)$ for all values of k, since errors or premature truncation of the integral are seriously magnified by the Fourier-transformation.

There are difficulties with these requirements for both the neutron scattering and the X-ray diffraction methods. Absorption of the radiation in the sample is one such problem, since the path-length of the radiation in the sample is angle-dependent, and hence so is the necessary correction. For X-ray diffraction, a particular problem is incomplete knowledge of $a(k)$ for small k, while $a(k)$ and $I(k)$ become very small for large k and cannot be determined accurately beyond $k = 200\ (\text{nm})^{-1}$. For neutron scattering $\langle b\rangle$ is independent of k, hence $I(k)$ can be determined sufficiently accurately for high k. On the other hand, the incoherent scattering (Δ in eq. (2.19)) may become a large part of the total scattered intensity, leading to poor accuracy. A major source of error is the truncation of the integral (2.23) at both ends of the k scale, and the effect of this on the Fourier-transform. It is difficult to obtain reliable data for $I(k)$ as $k \to 0$ because of the dominance of the non-scattered I_0 at these angles. At the other extreme, k is limited to $4\pi\lambda^{-1}$ because of the geometrical limitation on θ, so $k_0 = k_{max} \cong 120\ (\text{nm})^{-1}$. If the true function $r\rho g(r)$ has a maximum at r_0 spurious minima will appear at $r_0 \pm 45/k_0$ $(\text{nm})^{-1}$, spurious maxima at $r_0 \pm 77/k_0\ (\text{nm})^{-1}$, and so on, because the integration was not carried out to $k = \infty$.

Once the structure factor $S(k)$ is known, it may be used to evaluate other distribution functions besides $g(r)$. For instance, the *direct correlation function*, $f(r)$, which was introduced by Orenstein and Zernicke[3], is obtained from a suitable Fourier-transform as

$$f(r) = (2\pi^2 \rho r)^{-1} \int_0^\infty (1 - 1/S(k))k(\sin kr)\mathrm{d}k \qquad (2.25)$$

If (2.23) is taken to the limit $k = 0$, the result is

$$\lim S(k \rightarrow 0) = 1 + \rho \int (g(r) - 1) 4\pi r^2 \,\mathrm{d}r = \rho \mathrm{k} T \kappa_\mathrm{T} \qquad (2.26)$$

in view of the compressibility equation (2.14), since $\lim(k \rightarrow 0)\ (\sin kr)/kr = 1$. This result was already made use of for the normalization of neutron scattering results. The direct correlation function is related to the pair correlation function $g(r)$ and the total correlation function $h(r) = g(r) - 1$ by defining equation

$$h(\vec{r}_1, \vec{r}_2) = f(\vec{r}_1, \vec{r}_2) + \rho \int f(\vec{r}_1, \vec{r}_3) h(\vec{r}_2, \vec{r}_3) \mathrm{d}\vec{r}_3 \qquad (2.27)$$

remembering that the variable r in $f(r)$, $g(r)$ and $h(r)$ is in fact the scalar distance between the coordinates of two particles in a homogenous isotropic fluid, $r = |\vec{r}_2 - \vec{r}_1|$. The introduction of the coordinates of a third particle at $\vec{r}_3$ shows that equation (2.27) is relevant to the problem of three-body interactions, and this will be discussed further in section 2C. At present it should be noted that (2.27) is consistent with (2.23) and (2.25).

Once the pair correlation function $g(r)$ has been determined from the diffraction data via the structure factor $S(k)$ and eq. (2.24), it may be used to determine several important structural features. The atomic diameter σ, for instance, is approximately the value of r at which $g(r)$ rapidly approaches zero, while r_{max}, the value where $g(r)$ (or alternatively $n(r) = 4\pi r^2 \rho g(r)$) is a maximum is the mean equilibrium separation between neighbouring atoms. An important concept which can be discussed in terms of the peak and valley form of $g(r)$ or of $n(r)$ is the *coordination number* Z of an atom in the monatomic fluid. The area under a curve of $n(r)$ has the dimension of a number of particles and is therefore a convenient measure of Z, the number of nearest-neighbours that a randomly selected particle at the origin has on average. Figure 2.4 shows how Z is obtained from the data by two commonly used methods, both of which have their respective merits and drawbacks. There is no objectively correct method, so that Z is in fact only defined operationally.

$$Z_\mathrm{I} = 2\int_0^{r_{\mathrm{max}}} \rho g(r) 4\pi r^2 \,\mathrm{d}r \qquad (2.28)$$

$$Z_\mathrm{II} = \int_0^{r_{\mathrm{min}}} \rho g(r) 4\pi r^2 \,\mathrm{d}r \qquad (2.29)$$

The method yielding Z_I presumes that $r^2 g(r)$ is symmetric with respect to r_{max}. In

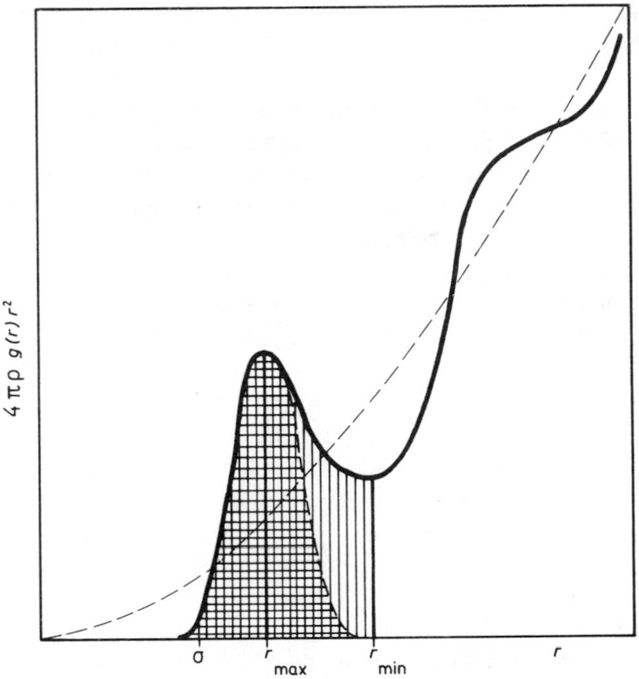

Figure 2.4. The determination of coordination numbers Z from $g(r)$ data. The long-dashed curve is $4\pi\rho r^2$, the continuous curve is $4\pi\rho g(r)r^2$ for a dense fluid at low temperatures, *i.e.* a liquid. The cross-hatched area under the curve is Z_I, and is symmetric with respect to r_{max}, while the vertically hatched area, up to r_{min}, represents Z_{II}. (The curve represents the X-ray diffraction data for liquid sodium at 373 K.)

fact it is not, since atoms in the first coordination shell will tend to remain beyond r_{max} than to be within it, where a high repulsive potential is dominant. On the other hand, r_{max} is clearly defined, as is the leading edge of $g(r)$, i.e., up to r_{max}, spurious peaks discussed above not affecting the results appreciably. The method yielding Z_{II} recognizes the inherent asymmetry of $r^2g(r)$ around r_{max}, but r_{min}, the value of r in the valley beyond the first peak, is not so well defined in the case where the valley is broad or a spurious secondary peak appears in it, as is sometimes the case. Always $Z_I < Z_{II}$, and the results quoted in this book normally are Z_I values.

Quantities applying to liquid structure, obtained by diffraction methods for the various classes of liquids appearing in Table 1.1, are discussed in Chapter 3.

c. Computer experiments

Fast electronic computers have provided the investigator with a powerful tool to carry out experiments on model systems. Two methods have attracted the most attention, and provided most of the results: the *Monte-Carlo* method[4] and the *molecular dynamics* method[5]. The former leads to the equilibrium properties of the fluid, $g(r)$ and the bulk thermodynamic properties, given the intermolecular potential $u(r)$ and the additivity assumption. The latter, on the same basis, leads in addition also to the transport properties. Both methods are limited by the available computer time to the examination of a finite number of configurations of a relatively small number (a few hundred) of particles, but include devices which make these calculations representative of infinite systems. The success of the methods can be measured by comparison with the results of physical experiment, which in the end is a measure of the degree of realism of the potential function selected, $u(r)$. On the other hand physical experiments are neither sufficiently sensitive nor accurate enough to provide a real test of the correspondence of the computed results with the microstructure of the fluid. Another approach is therefore to use simple mathematical functions for the potential, as obviously inadequate as the hard-sphere model, in order to study the theoretical consequences, which sometimes are quite surprising. For instance, condensation and freezing of an ordered crystalline solid is predicted for a hard-sphere fluid devoid of any attractive forces.

The system that the Monte-Carlo and the molecular dynamics methods deal with is represented schematically in two dimensions in Fig. 2.5. In this figure thirty particles are located in a box of edge 8σ so that the reduced (dimensionless) number density is $\rho\sigma^2 = 30/64 = 0.469$. The actual fraction of the area (volume) occupied is $(\pi/4)\rho\sigma^2 = 0.368$. Close-packing corresponds to $\rho_0\sigma^2 = \pi \times 12^{-1/2} = 0.906$, so that the relative density is $\rho/\rho_0 = 0.406$. Indeed, great successes for these methods have been claimed for one-dimensional systems (hard rods) and two-dimensional systems (hard discs) where the equations of motion can be solved analytically, in agreement with the computed results. For three-dimensional systems, however, no analytical results can be obtained, since the basic eq. (2.8) cannot be solved. It is then that the computed results have their predictive value.

In order to avoid boundary problems with the small number of particles that can be handled, a periodic system is considered. The particles inside the square (cubic) box of edge L have 'ghosts' outside the walls that exactly mimic their positions and motions: the prime- and double-prime-numbered particles outside the box. As particle 4 moves outside the box through the left-hand wall, particle $4'$ enters it from the opposite side. In a two-dimensional system, the same idea can be realized if the square is thought to be extended on all sides and then bent to the shape of a torus, hence the appellation *toroidal boundary condition* for what otherwise is the *periodic boundary condition*. The 'ghosts' have a further function: since intermolecular forces with ranges larger than L, the dimension of the box, such as the coulombic forces between ions in a molten salt or the potential field of the free electron gas in metals (section 1C(c)), are difficult to handle, only interactions between particles with $r < L/2$ are considered, and for this purpose, particle $3'$ is used instead of particle 3, when the interaction with particle 1 is computed.

The initial configuration in position space (Monte-Carlo) or position and

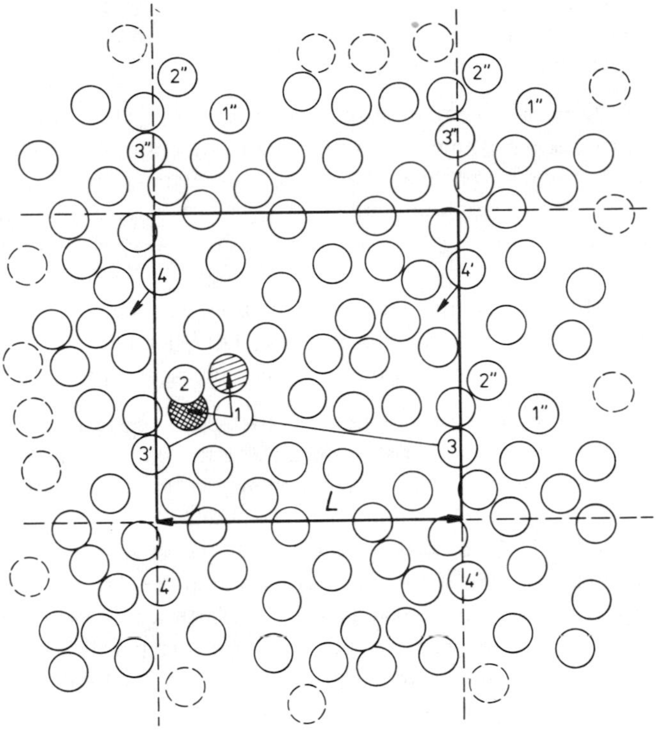

Figure 2.5. A two-dimensional schematic representation of a periodic boundary system of 30 particles of diameter σ (this corresponds to $(30)^{3/2} = 164$ particles in three dimensions). The central square of edge $L = 8\sigma$ is the basic system, the squares around it contain 'ghosts'. The behaviour of the numbered particles (and their primed and double-primed ghosts) is explained in the text.

momentum space (molecular dynamics) is based on a table of random numbers between zero and unity (hence the name Monte-Carlo). Each of the N particles is assigned coordinates in the box of edge L,($L^3 = V$, selected so to give the required number density $\rho = N/V$), so that r_x/L, r_y/L and r_z/L (and the corresponding momenta) are given by these random numbers. If angle-dependent potentials are employed (section 1C(b)), then the corresponding orientations are specified similarly. (In the following, only spherically symmetrical force fields will be discussed.) The configurational energy $U(\vec{r}^N)$ of the system is calculated from the $u(r)$ appropriate for the model under discussion (e.g. one of those shown in Fig. 1.6) with the pair-wise additivity assumption and the given temperature T. Further configurations are then generated by the use of a Markov-chain.

In the Monte-Carlo method, the Markov-chain corresponds to a random walk for each particle in turn, subject to the following rules. Starting from the tth configuration $\vec{r}^N(t)$, a trial configuration $\vec{r}_i'$, $\vec{r}^{N-1}$ is chosen, where only the coordinates of particle i, selected at random, are different, by an amount $\xi\Delta\vec{r}$ (i.e.

$r'_{ix} = r_{ix} \pm \xi_x \Delta r$, $r'_{iy} = r_{iy} \pm \xi_y \Delta r$ and $r'_{iz} = \xi_z \Delta r$, where ξ_x, ξ_y and ξ_z are random numbers between zero and one). The potential energy U' for the trial configuration is now computed, compared with $U(t)$, and

(a) if $U' \leqslant U(t)$, the new configuration is accepted, $r_i(t+1) = r'_i$ and $U(t+1) = U'$, and the process is continued with the random selection of a new particle, j, the coordinates of which are changed as before, to obtain configuration $t + 2$;
(b) if $U' > U(t)$, another random number $0 < \xi_U < 1$ is selected and compared with $\exp(-U'/U(t))$. If now $\exp(-U'/U(t)) \geqslant \xi_U$, the new state is accepted as in (a) above;
(c) if $\exp(-U'/U(t)) < \xi_U$, the trial configuration is rejected, and the previous configuration is taken to be the new one, i.e. $\vec{r}_i(t+1) = \vec{r}_i(t)$ and $U(t+1) = U(t)$. A new particle j is then chosen to obtain configuration $t + 2$.

In Fig. 2.5, the hatched new position of particle 1 corresponds to case (a), $U' < U(t)$, if an attractive potential is operative between particles 1 and 2 at this distance, while the cross-hatched new position may correspond to either case (b) or case (c), because of the repulsive potential. The interaction of particle $i = 1$ with all the other $N - 1$ particles (or their 'ghosts' if $r > L/2$) is considered in computing $U(t)$. The parameter Δr, the maximum allowed step in the Markov-chain, should be chosen so as to produce rapid convergence. Too small a step will produce a great number of accepted configurations, and too large a step will lead to too many rejected configurations, and in both cases the maximal t required to arrive at the average values of the system as defined in (1.89) will be excessive. The procedure is terminated once the equilibrium (average) value of the function, e.g. $\langle U \rangle = (1/t_{max}) \Sigma_t^{t_{max}} U(t)$, has been reached within a predetermined margin. This averaging process is tantamount to weighting equally all the configurations, which, however, are selected as is required with a frequency proportional to $\exp(-U/kT)$, the Boltzmann factor, provided t_{max} is large enough. Tens of thousands of configurations are usually computed. Once $\langle U \rangle$ is known, the thermodynamic properties can be calculated from the configurational partition function Z, eq. (1.75), or from fluctuation theory, eqs. (1.91) to (1.95).

The computation of $g(r)$ presents a difficulty, however, because of the non-spherical symmetry of the periodic boundary system, Fig. 2.5. For each configuration, the function

$$g'(r) = N^{-1} \langle \sum_{\nu} \sum_{i} \sum_{j} \delta(r - |L + r_{ij}|) \rangle \tag{2.30}$$

is computed, where $\delta(x < 0) = 0$ and $\delta(x > 0) = 1$ is the unit step function, ν counts the 'boxes' in the periodic boundary system, and r_{ij} is the distance between any two particles. The averaging is done as before over the t configurations selected by the Markov-chain procedure. The pair correlation function $g(r)$ is then obtained as

$$g(r) = (4\pi\rho r^2)^{-1} \mathrm{d}(g'(r))/\mathrm{d}r \tag{2.31}$$

Further results for the system may then be obtained from eqs. (2.12) to (2.16), etc.

Summarizing, the Monte-Carlo method is capable of calculating average values of such properties as $\langle U \rangle$ or intermolecular distances (leading to $g'(r)$ via (2.30)), and thence the macroscopic thermodynamic properties and the pair correlation function. It requires for this purpose the specification of the pair potential function $u(r)$ (or the orientation-dependent version if required), and of N, ρ (or V, leading to L) and T (leading to $u(r)/\mathrm{k}T$ in the Boltzmann weighting expressions). Only the configuration-dependent equilibrium properties of the system are obtained by this method.

In the molecular dynamics method, the N particles confined in the box with edge L at a temperature T, which defines the total kinetic energy as $(3/2)N\mathrm{k}T$, have positions $\vec{r}^N$ and momenta $\vec{p}^N$ chosen at random (subject to a constant total kinetic energy). The particles are apt to collide with each other and the sequence of occurrence of these collisions can be computed. The first collision in the sequence is then examined and the coordinates and momenta of the two particles involved are computed after a time interval Δt, subject to the laws of motion, i.e. conservation of momentum, and conservation of total energy, kinetic plus potential. It is here that the function $u(r)$ comes in, since a collision does not occur only when $r = |\vec{r}_j - \vec{r}_i| \leqslant \sigma$, but any approach within the range of the intermolecular forces is considered to be a collision. Therefore it is very difficult to compute with $u(r)$ functions other than the hard-sphere potential (Fig. 1.6(a)), where collisions are elastic and occur only for $r < \sigma$. Between collisions the particles move in straight lines with constant velocities. After a collision the original sequence of expected collisions is examined and modified, if necessary, to allow for the results of the collision that has just occurred. Then the positions and momenta for all particles are recorded for the succession of time intervals Δt until the next collision occurs, when the process is repeated. A total of some tens of thousands of collisions is accumulated.

This process permits one to follow the trajectory of a given particle in phase space (section 1D(a)), the singlet correlation function, $n^{(1)}(\vec{r},t)$. It also gives a succession of configurations, over which quantities such as U may be averaged. Transport and relaxation processes may also be examined, but for the present purpose, the quantities $\langle U \rangle$ and $g(r)$ computed as averages over many configurations are of interest. It can be shown that within a term in N^{-1} the results of the molecular dynamics and Monte-Carlo methods are equivalent, so that extrapolation to $N \rightarrow \infty$ at a given ρ would make both methods exact, while for finite N they give complementary results. The methods are often, however, applied for different problems, the molecular dynamics method to the dynamical behaviour of simple particles (hard spheres), the Monte-Carlo method to the equilibrium properties of particles with more 'realistic' intermolecular forces ('LJ 12.6' potentials).

An interesting result obtained for hard spheres by both the Monte-Carlo and the molecular dynamics methods is the appearance of two distinct curves of the reduced equation of state $P/\rho_0\mathrm{k}T$ as a function of the density ρ. At the highest densities, from ρ approaching $\rho_0 = 2^{1/2}\sigma^{-3}$, the density of a close-packed face-centered cubic lattice, down to $\rho = 0.85\rho_0$, only the 'L branch' was observed. At the lowest densities, $\rho < 0.625\rho_0$, only the 'H branch' was observed. At intermediate densities both branches were produced by the computations, while in the region $0.625 < \rho/\rho_0 < 0.635$ wild fluctuations between the two branches were obtained. The existence of the two branches was interpreted as indicating a first-order phase transition, i.e. freezing of the fluid (liquid) as its density is increased at a given temperature and pressure. That a fluid of hard discs devoid of attractive forces condenses to an ordered phase is brought about by the geometric necessities of high densities, and by the shielding of a particle from thermal knocks from other particles by its nearest-neighbours. The geometrical requirements have been studied by physical experiments on actual spheres, steel balls, as described below.

d. Experiments on models

An examination of configurations generated by the Monte-Carlo method after many steps ($t > 10^5$, say) shows them to be truly representative of a random-packed 'model liquid' with the assumed potential law, $u(r)$. In the case of hard spheres, this can be equally simulated by an assembly of steel balls packed into an irregularly shaped rubber balloon, or by a randomly assembled ball-and-spokes model, i.e. by a physical model.[6] These, in turn, can be discussed from a purely geometical standpoint. A useful concept in this context is that of Voronoi polyhedra (Fig. 2.6). The centres of all the particles in the model are connected by straight lines (the spokes in the ball-and-spokes model), which are bisected by perpendicular planes. The smallest polyhedron around a particle formed by these planes is its Voronoi polyhedron, and these polydedra fill all the space that the particles occupy. A physical model of these polyhedra can be attained if plasticine balls are used instead of steel balls and the collection is compressed. Individual 'balls' retrieved from the collection will be in the shape of these polyhedra. The average number of faces and of sides per face can be calculated theoretically, and these come out to be 13.56 and 5.1 respectively. The models, in agreement, have triangular to heptangular faces, with pentagons being the most common, and on average thirteen faces per polyhedron.

A very significant insight obtained from these experiments is the realization of the inherently irregular packing that this geometry entails. Only regular tetrahedra, cubes or octahedra can be packed in an ordered fashion to fill space completely. Close-packed arrays of steel balls have the hexagonal-close-packed or the equivalent face-centered-cubic lattice. The spheres then occupy $2^{1/2}\pi/6 = 0.7405$ of the volume. It is, however, impossible to fill out space completely by other regular polyhedra, such as the pentagon-faced dodecahedron (in two dimensions, it is possible to fill out an area with equilateral triangles, with squares or with regular hexagons, but not with pentagons). This and other regular polyhedra produce a relatively expanded lattice, occupying more volume with empty spaces. If irregular polyhedra are used, a high density *can* be achieved by *random close packing*, where the total volume is 1.56 times the volume of the spheres. This volume is *ca.* 15% more than that for the regular close-packed lattice, where the total volume is $1/0.74 = 1.35$ times that of the spheres. This expansion of 15% from regular to random close packing is observed in the melting of the inert gases. The pentagonal faces of the structures characteristic of liquids can be considered as a barrier to spontaneous crystallization, and as an explanation for the existence of the (metastable) supercooled liquids, since, as stated above, they are not amenable to regular packing.

A count of the number of balls in the ball-and-spokes model which are within a specified distance, r and $r + \Delta r$, from other balls, gives the radial distribution

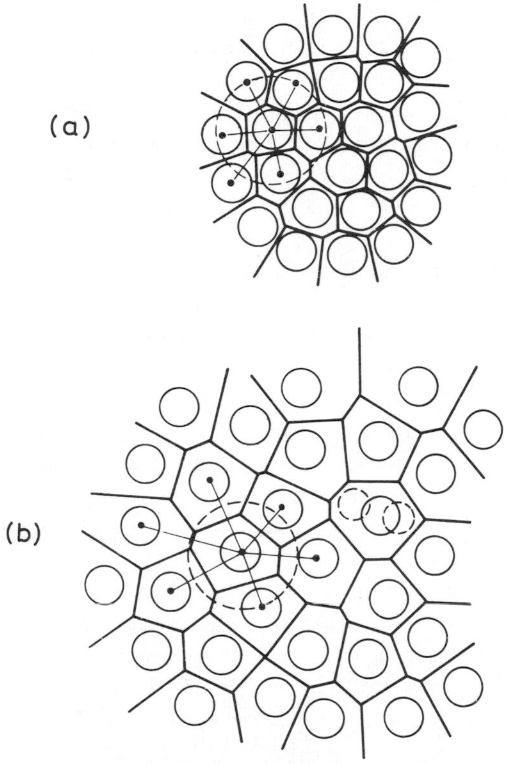

Figure 2.6. Dirichlet polygons (in two-dimensions, representing Voronoi polyhedra in three dimensions) around randomly packed particles[6]. The edges are perpendicular bisectors of the lines joining the centres of the particles. Note that not all particles sharing edges of polygons with a given particle belong to its first coordination shell (here within $2^{1/2}\sigma$). (a) A dense liquid, near the freezing point, $Z = 5$ for the indicated particle, which has a Dirichlet hexagon. (b) An expanded liquid near the critical point, $Z = 1$, although, again, there are six nearest-neighbours for the indicated particle with its Dirichlet hexagon. It is also seen that two particles fit easily into an average polygon that ordinarily 'houses' just one particle.

function $n(r,\Delta r)$, and by the use of (2.5) the pair correlation function $g(r)$. A choice of $\Delta r = 0.2\sigma$ (σ is the diameter of a ball) is convenient, and gives results in qualitative agreement with $g(r)$ of liquid argon. The main discrepancy lies in the low r region, since argon, after all, does not consist of hard spheres.

The coordination number Z (eq. (2.28), Fig. 2.4) in a liquid can also be obtained from the steel-ball model. Paint poured into the pile of balls, left to drain and to dry, remains at contact points between the balls in the form of a ring (in near-contact points in the form of a spot, depending on the capillary forces and surface tension). The average number of rings and spots on each of a large number of balls is the coordination number. Alternatively, a count of the neighbours in the ball-and-spoke model may again be made. The balls whose centres lie within 1.05σ of that of a given ball are considered its *nearest-neighbours*, and their average number is Z. The average number of balls at distances between σ and 2σ is the number of *next-nearest-neighbours*, Z'. The number within $2^{1/2}\sigma$ corresponds to the average number of faces in the Voronoi polyhedra. The experimentally found numbers are $Z = 7.9 \pm 1.2$, $Z' = 35.0 \pm 2.5$, and 13.3 ± 1.3 for the number of faces. Liquid argon has Z around 8.

The effect of *holes* in the liquid structure may be studied with the aid of these models. It is necessary to introduce holes for the liquid to expand from its high-density form near the triple point to its low-density form near the critical point. As mentioned in the introduction (p. 4), the expansion is to 1.5σ in linear dimensions, i.e. to a factor of *ca.* 3.4 in volume. The total volume thus expands from *ca.* 1.56 to *ca.* 6.5 times the volume of a sphere. This cannot occur by rearrangement of the polyhedra alone, but holes of the dimensions of spheres must be introduced and allowed to come into contact as the volume expands. Thus the coordination number falls drastically, but the distance characteristic of the first peak in $g(r)$ hardly changes. A model containing 35% of holes has on average six neighbours for a given ball (within 1.1σ), compared with the nine neighbours in a random close-packed model. A feature which is not demonstrated by a hard-sphere model, but occurs for real liquids with attractive forces (as shown by X-ray diffraction from molten salts, for instance, section 3C(a)), is the shortening of the equilibrium distance r_0 ($= r_{max}$ in $g(r)$, eq. (2.28)), as Z falls in the expanding structure.

B. Theoretical Evaluation of Correlation Functions

a. The Born–Green (BG) approximation

As mentioned on pp. 21 and 46, the main problem in the use of the distribution and correlation functions for fluids at high densities and at temperatures below the critical point, i.e. for liquids, is the relationship of $U(r)$ to $u(r)$. Equation (2.8) defines $U(r)$ as the *potential of mean force*, which is formally related to $g(r)$ by the simple exponential relationship (2.9) or

$$U(r) = -kT \ln (g(r)) \tag{2.32}$$

The assumption of pair-wise additivity of the intermolecular potentials $u(r)$, eq. (1.96), gives one relationship between $U(r)$ and $u(r)$, but this cannot be used to obtain a direct relationship between $g(r)$ and $u(r)$, in view of eq. (2.8), which requires the knowledge of the triplet distribution function, $n^{(3)}(\vec{r}_1, \vec{r}_2, \vec{r}_3)$.

An approximate solution has been proposed and defended in various ways by Kirkwood,[7] Yvon,[8] Born and Green,[8] and Bogoliubov,[8] and is now commonly known as the Born-Green (BG) approximation (the anagrams BGY, BBGYK, etc. are also used). This introduces the additional assumption of the superposition of probabilities, according to Kirkwood,[7] eq. (2.17), which can be restated as

$$U(\vec{r}_1, \vec{r}_2, \vec{r}_3) = u(\vec{r}_1, \vec{r}_2) + u(\vec{r}_2, \vec{r}_3) + u(\vec{r}_3, \vec{r}_1) \tag{2.33}$$

in view of the definitions $n^{(3)} = \rho^3 g$ and (2.32). The superposition of potentials is of course exact, but the approximation (2.33) disregards in U the effects of the fourth and further particles. If (2.17) is introduced into (2.8), an equation relating $g(r)$ to $u(r)$ that involves only pair correlations, is obtained

$$\begin{aligned} \mathrm{k}T\partial(\ln g(\vec{r}_1, \vec{r}_2))/\partial\vec{r}_1 &= -\partial u(\vec{r}_1, \vec{r}_2)/\partial\vec{r}_1 \\ &\quad - \rho^{-3} \int g(\vec{r}_2, \vec{r}_3) g(\vec{r}_1, \vec{r}_3)(\partial u(\vec{r}_1, \vec{r}_3)/\partial\vec{r}_1)\mathrm{d}\vec{r}_3 \end{aligned} \tag{2.34}$$

This is the basic equation in the BG approximation. This integro-differential equation was developed further, and the form convenient for calculation is

$$\begin{aligned} \ln g(r) &= -(1/\mathrm{k}T)u(r) \\ &\quad - (\rho/\mathrm{k}T) \int (g(\vec{r}_1) - 1)\left[\int_{\infty}^{\mathrm{r}} g(r')(\partial u(r')/\partial r')\mathrm{d}r'\right] \mathrm{d}\vec{r}_1 \end{aligned} \tag{2.35}$$

It is immediately seen that at low densities (2.35) approaches the limit already given in (2.7). The second term can be expressed as a density expansion, yielding

$$\ln g(r) = -(1/\mathrm{k}T)u(r) + \ln(1 + \rho(x_1(r)) + \rho^2(x_2(r)) + \cdots) \tag{2.36}$$

The difficulty now rests in giving analytical expressions to the functions $x_1(r)$, $x_2(r)$, etc. The BG theory is essentially a first-order theory, and deals with $x_1(r)$. It uses an iteration procedure, since $x_1(r)$ is a function of $g(r)$, and for densities characteristic for gases the procedure converges rapidly. It yields realistic virial coefficients $B_2(T)$ and $B_3(T)$ (Fig. 2.7), and predicts the existence of two fluid phases (i.e. gas and liquid) at equilibrium below a critical temperature (Table 2.1). However, as the density is increased, it ceases to converge, and it is useless in the region where computer calculations indicate the fluid-solid phase transition, i.e. freezing. It also yields negative pressures when eq. (2.13) is used at high densities with $g(r)$ calculated from the BG approximation. This can be attributed to the incorrect form of the virial coefficients beyond the third, $B_3(T)$ in eq. (1.93).

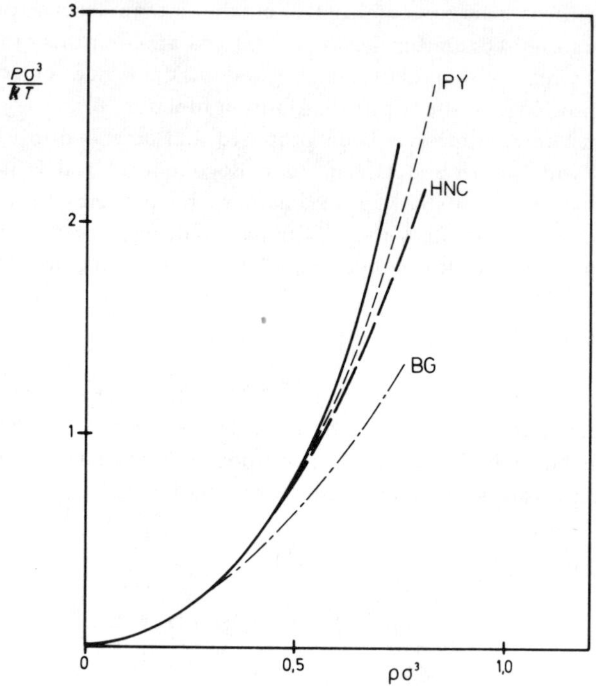

Figure 2.7. Comparison of the equation of state obtained 'experimentally', by molecular dynamics calculations using an 'LJ 12-6' potential at $T = 2.7\epsilon/k$ (continuous curve), with those obtained theoretically (dashed curves) by means of the Born–Green (BG), hypernetted chain (HNC) and Percus–Yevick (PY) approximations.

Table 2.1. Comparison of the critical constants for argon obtained experimentally with those calculated from the BG, the HNC and the PY theories[a].

Source	kT_c/ϵ	$\rho_c\sigma^3$	$P_c/\rho_c kT_c$
Experimental	1.26	0.316	0.297
BG (P)	1.45	0.40	0.44
(κ_T)	1.58	0.40	0.48
HNC (P)	1.25	0.26	0.35
(κ_T)	1.39	0.28	0.38
PY (P)	1.25	0.29	0.30
(κ_T)	1.32	0.28	0.36

[a]The values $\epsilon/k = 119.6$ K and $\sigma = 0.340$ nm were used with an 'LJ 12-6' potential, eq. (1.42), and the critical constants were calculated[9] via the pressure equation (2.13) (P) and the compressibility eq. (2.14) (κ_T).

b. The hypernetted chain (HNC) and Percus– Yevick (PY) approximations

Other approaches to a relationship between $U(r)$ and $u(r)$, not involving the superposition approximation, (2.17) or (2.33), start from the expansion of $\ln g(r)$ in powers of ρ, eq. (2.36), utilizing graph-theory. This is a mathematical device that has been developed for the cluster expansion theory of Mayer (applicable to non-dense fluids), to provide a suitable classification of the integrals covering the interactions of the various particles. When particles at coordinates $\vec{r}_3, \vec{r}_4, \ldots$ affect the correlation between particles at $\vec{r}_1$ and $\vec{r}_2$, then the first of these contributes to integral terms multiplied by ρ^1, the first and second contribute to those multiplied by ρ^2, etc. in the expansion. The integrals are then said, in graph-theory, to have one, two, etc. field points, in addition to the two base points. For the sake of brevity, the coordinates $\vec{r}_1, \vec{r}_2, \vec{r}_3$, etc. will be designated simply 1, 2, 3, . . . in the following. The path between the points i and j involves the interaction term $\phi(i, j) = \exp(-u(i, j)/kT) - 1$ (Fig. 2.8). There are certain topological features to these graphs: articulation points, bridge points and irreducible linkages (Fig. 2.8). An articulation point is a field point through which all paths from one part of the graph to a second part must pass. A bridge point is an articulation point between the two base points, so that all paths have necessarily to pass through it. An irreducible linkage between two field points is a direct path between them, not involving the base points, which furthermore are not directly linked.

It is now possible to prescribe symbolically how the correlation functions $h(r)$, $f(r)$ and the potential of mean force $U(r)$ are to be calculated. To calculate the direct correlation function $f(r) = f(i,j)$ use the set of all the graphs which contain neither articulation points nor bridge points. Symbolically

$$f(i,j) = [\bullet\!\!-\!\!\bullet] + \rho \; \triangle \; + \rho^2 \left[\Box + 2 \boxtimes + \frac{1}{2} \boxtimes + \frac{1}{2} \boxtimes \right] + \cdots \tag{2.37}$$

The set of all the graphs that contain at least one bridge point, $b(i,j)$, given symbolically by

$$b(i,j) = \rho \left[\triangle \right] + \rho^2 \left[\Box + 2 \boxtimes \right] + \cdots \tag{2.38}$$

is added to give the total correlation function $h(i,j)$, which thus involves the set of all the graphs which contain no articulation points

$$h(i,j) = f(i,j) + b(i,j) \tag{2.39}$$

The set of all the graphs that are both (i,j) irreducible and are free from bridge points, $y(i,j)$

$$y(i,j) = \rho^2 \left[\boxtimes \right] + \cdots \tag{2.40}$$

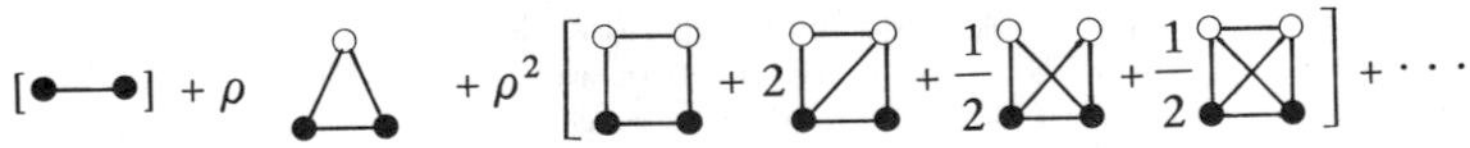

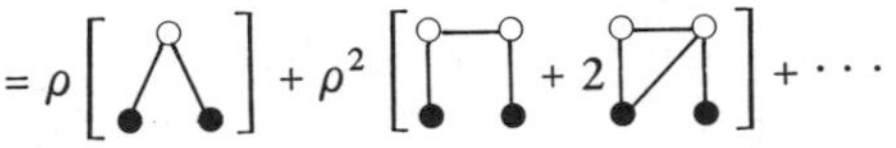

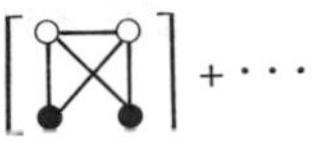

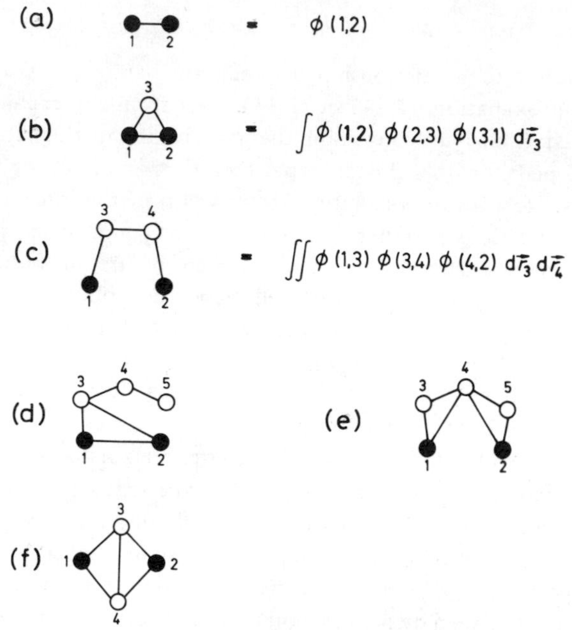

Figure 2.8. A symbolic representation of some interaction graphs, involving base points $i = 1$ and $j = 2$ (filled) and field points 3,4, . . . (empty), and interactions paths between them (connecting lines). (a) Direct interaction between particles at 1 and 2. (b) The addition of the field point at 3 represents an integration over $d\vec{r}_3$ of the product of the ϕ values represented by the connecting lines. (c) A graph involving a double integral, where points 3 and 4 are both articulation and bridge points. (d) Field point 4 is an articulation point, but there is no bridge point corresponding to the direct connection from 1 to 2. (e) Field point 4 is a bridge point, but there are no articulation points. (f) The graph is (1,2)-irreducible, since there is a direct linkage between field points but not between base points.

added to $b(i,j)$ produces the prescription for calculating $U(i,j)$

$$U(i,j) = u(i,j) - kT(b(i,j) + y(i,j)) \tag{2.41}$$

The relationships between the various functions may now be rewritten as, for example

$$h(i,j) - f(i,j) - \ln(1 + h(i,j)) + y(i,j) = u(i,j)/kT \tag{2.42}$$

where, from (2.32) and the definition of $h(r)$, $-\ln(1 + h(i, j)) = U(i,j)/kT$.

The *hypernetted chain theory* (HNC)[10] utilizes this representation and makes the following simplification: the term $y(i,j)$ in (2.42) is dropped, yielding

$$b(r) = h(r) - f(r) = \ln g(r) + u(r)/kT \tag{2.43}$$

This, then, yields the desired connection between $g(r)$ and $u(r)$, involving only the graphs $b(r) = b(i,j)$ defined symbolically in (2.38). The direct correlation function then has the simplified form

$$f(i,j)^{\text{HNC}} = [\bullet\!-\!\!-\!\bullet] + \rho \left[\,\triangle\,\right] + \rho^2 \left[\,\square + 2\,\boxslash + \frac{1}{2}\,\boxtimes\,\right] + \cdots$$

$$= -u(i,j)/kT - (1/2)b^2(i,j) - \cdots \tag{2.44}$$

where the second equality is due to expanding the logarithm in (2.42) and dropping $y(i,j)$

The *Percus–Yevick theory* (PY)[11] has been derived on a completely different basis (see below) but its results can be presented in the above terms. The equation analogous to (2.43) is

$$\ln(1 + b(r)) = \ln(1 + b(r) - f(r)) = \ln g(r) + u(r)/kT \tag{2.45}$$

where, again, the term $y(i,j)$ has disappeared. The symbolic and expanded forms of the direct correlation function are

$$f(i,j)^{\text{PY}} = [\bullet\!-\!\!-\!\bullet] + \rho \left[\,\triangle\,\right] + \rho^2 \left[\,\square + 2\,\boxslash\,\right] + \cdots$$

$$= -u(i,j)/kT - b(i,j)u(i,j)/kT - \cdots \tag{2.46}$$

or, approximately, $f(r) = g(r)(1 - \exp(u(r)/kT))$. The terms missing in (2.46), compared to the full set, are those involving the interaction of the interfering particles with both particles i and j, leaving only those shorter-range interactions interacting with either i or j (i.e. those with non-crossing diagonals in the symbolic representation). A comparison between eq. (2.45) of PY theory and eq. (2.43) of the HNC theory shows agreement when $b(r) - f(r)$ is small so that $\ln(1 + b(r) - f(r)) \approx b(r) - f(r)$. The function $f(r)$ is shorter-ranged than $b(r)$, i.e. it decays to zero for lower values of r, and is therefore advantageous for calculations. For the particular case of hard spheres, it was found that the PY theory, eq. (2.45), can be solved exactly.

The derivation of the PY theory was based originally on the *method of collective variables* combined with fluctuation theory. In a dense fluid, the local density fluctuations due to interactions and thermal motions of the particles can be described in terms of standing waves. For N particles in a cubic box of edge L (number density $\rho = NL^{-3}$) the permitted wavelengths are $\lambda_n = 2Ln^{-1}$ where n can take any integral values from 1 to N. Instead of using $3N$ Cartesian coordinates $\vec{x}_i$, $\vec{y}_i$, $\vec{z}_i$ for the N particles i to describe a configuration (Fig. 2.1), $3N$ collective coordinates $q_n(\vec{x})$ etc.

$$q_n(\vec{x}) = \sum_{j=1}^{N} \sin(2\pi\vec{x}_j/\lambda_n) \tag{2.47}$$

are used to describe the collective effect of all the postions of the particles in the

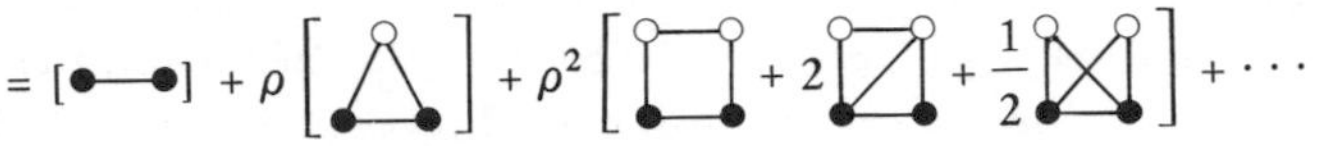

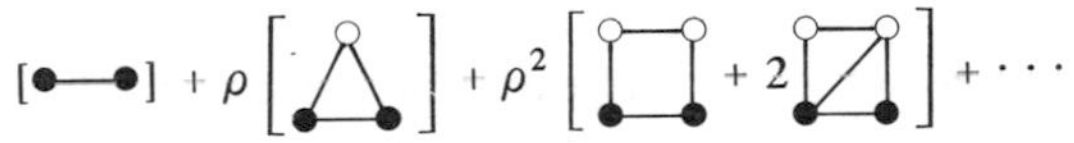

x-direction, etc. on the standing waves for given wavelengths λ_n. The configurational potential energy part of the Hamiltonian H in (1.71) is then expressed in terms of the new coordinate system by means of complicated mathematical transformations, resulting in a sum of terms, each depending on only one of the collective variables. This facilitates the computation of the integrals and thus of the correlation functions. The main difficulty, however, is the fact that the smallest wavelength permissible, $\lambda_N = 2(N/\rho)^{-1/3}$, is of the order of σ, the diameter of the particles or the interparticle distance in a dense fluid. The final result is the eqs. (2.45) and (2.46) given above.

The hypernetted chain and the Percus–Yevick theories have been used to calculate pair correlation functions for given intermolecular potential functions of various forms. Conversely, to calculate the latter from experimental structure factors $S(k)$ (section 2A(b)), the necessary equations are readily obtained from eqs. (2.43) and (2.45) utilizing the total and the direct correlation functions obtained via Fourier-transformations from $S(k)$

$$u(r)^{\mathrm{HNC}} = \mathrm{k}T[h(r) - \ln(1 + h(r) - f(r))] \tag{2.48a}$$

$$u(r)^{\mathrm{PY}} = \mathrm{k}T \ln[1 - f(r)/(1 + h(r))] \tag{2.48b}$$

The potentials obtained in this way have the correct qualitative features as discussed in section 1C for those liquids for which structure factors can be obtained experimentally: the inert gases and metals. However, the calculated results from both the HNC and the PY theories do not reproduce the $u(r)$ values calculated from scattering experiments or free-electron theory adequately. Once both $g(r)$ and $u(r)$ are known, the equation of state and thermodynamic properties follow (section 2A(a)). It was found that, in general, the compressibility equation, (2.14), gives better results than the pressure equation, (2.13), for dense fluids, i.e. liquids well below the critical point. For the critical data, Table 2.1 gives the impression that the opposite is true. It is seen that even for this purpose, the PY equation gives better results than the HNC equation, although the premise of the density-fluctuation calculation of the PY theory requires a close correlation between the particles, which is possible only at densities higher than that at the critical point. At densities typical of liquids nearer the freezing point, the PY theory is indeed superior, as Fig. 2.7 shows for a model liquid.

For hard spheres, the PY theory yields exact results, in the sense that the equations can be solved analytically, as stated above. The pressure equation, (2.13), yields for the equation of state the expression

$$PV/N\mathrm{k}T = (1 + 2\eta + 3\eta^2)/(1 - \eta)^2 \tag{2.49a}$$

while the compressibility equation, (2.14), yields

$$PV/N\mathrm{k}T = (1 + \eta + \eta^2)/(1 - \eta)^3 \tag{2.49b}$$

where $\eta = (\pi/6)\rho\sigma^3$. The discrepancy is real, showing an inadequacy of the PY treatment. Best agreement with the Monte-Carlo data for hard spheres is obtained by a combination of the two equations

$$PV/NkT = (1 + \eta + \eta^2 - \eta^3)/(1 - \eta)^3 \tag{2.50}$$

Other thermodynamic functions may be obtained from

$$A_{\mathrm{conf}}/NkT = -1 + \ln \rho + (4\eta - 3\eta^2)/(1 - \eta)^2 \tag{2.51}$$

which is obtained by integration of the pressure over the density. The agreement attained between the results from eqs. (2.50) and (2.51) with the computed results for hard spheres is impressive. Thus, although the HNC theory is more complete in that it involves more terms in the estimation of $f(r)$ than does the PY theory (compare eqs. (2.44) and (2.46)), it overestimates the 'tail' of the many-body correlations, and yields results agreeing less well with computer experiment data than do those of the PY theory.

c. *Scaled particle theory*

If all that is needed from a theory for liquids is the equation of state and a prediction of the thermodynamic properties, the detailed structure is irrelevant, and the scaled particle theory of Reiss *et al.*[12] has many merits. It is not a theory of the structure of the liquid since it does not yield the pair correlation function $g(r)$ for all r, although it does use the number of particles at one particular distance, namely $g(\sigma)$, as an important parameter. It also estimates $g(r)$ for $r \leqslant (3/2^{1/2})\sigma$. The significance of $g(\sigma)$ is seen when it is realized that the equation of state may be exactly stated as

$$PV/NkT = 1 + (2/3)\pi\rho\sigma^3 g(\sigma) \tag{2.52}$$

for hard spheres.

Consider a grand canonical ensemble (p. 38) of N particles in a volume V at temperature T. If another particle is added, this addition may be considered to occur gradually, through a coupling parameter ξ changing from 0 to 1. Assuming pair-wise additivity of the potentials, the configurational potential energy is

$$U(\vec{r}^{N+1}, \xi) = \sum_1^N \xi u_{0j} + \sum_1^N u_{ij} \tag{2.53}$$

where the new particle has suffix zero. The chemical potential of a particle in the system is (cf. eqs. (1.83), (1.85))

$$\mu = kT \ln \rho\Lambda^3 + 4\pi\rho \int_0^1 \mathrm{d}\xi \int_0^\infty (\partial u/\partial \xi) g(r, \xi) r^2 \,\mathrm{d}r \tag{2.54}$$

Instead of coupling the interactions of the N particles with the new particle through its 'appearance' parameter ξ, they may be coupled through its size, via the *scaling parameter* λ, varying from zero to σ. The analogous expression to (2.54) is then

$$\mu = \mathrm{k}T \ln \rho\Lambda^3 + 4\pi\rho \int_0^\sigma \mathrm{d}\lambda \int_0^\infty (\partial u/\partial\lambda) g(r, \lambda) r^2 \,\mathrm{d}r$$

$$= \mathrm{k}T \ln \rho\Lambda^3 + 4\pi \mathrm{k}T\rho \int_0^\infty g(\lambda, \lambda)\lambda^2 \,\mathrm{d}\lambda \qquad (2.55)$$

The pressure is obtained through the thermodynamic identity

$$P = \int_0^\rho \rho' (\partial\mu/\partial\rho')_{N,T} \mathrm{d}\rho'$$

so that (2.52) results. The quantity $g(\lambda,\lambda)$, is obtained as follows. Consider the formation in a hard sphere fluid of a *cavity* of radius r. Let $\rho G(r,\rho)$ be the concentration of the centres of the spheres on the surface of the cavity. The cavity plays exactly the role of another hard sphere of diameter $2r - \sigma$, since it excludes the centres of the other particles from this spherical region. The cavity can thus be regarded as a *scaled particle*, and, in the case where $r = \sigma$, it is an ordinary particle, hence $G(\sigma) = g(\sigma)$. The function $G(r)$ may be calculated for $r < \sigma/2$ through probability considerations: $4\pi r^2 \rho G(r)\mathrm{d}r$ is the probability that a particle is found in the spherical shell of thickness $\mathrm{d}r$ at distance r from the centre of the cavity. The probability that a cavity of radius $r < \sigma/2$ is empty is $1 - (4/3)\pi r^3 \rho$, since one particle, at most, may be located there. Combining the two expressions leads to

$$G(r < \sigma/2) = (1 - (4/3)\pi r^3 \rho)^{-1} \qquad (2.56)$$

The value of G for cavities of larger radius cannot be expressed analytically, unless the limit of macroscopic cavities ($r \to \infty$) is approached, where the surface tension determines the work that has to be done on the system to create the cavity. The reversible work $W(r)$ gives the probability of the cavity being empty as $\exp(-W(r)/\mathrm{k}T)$ and is connected with $G(r)$ by

$$W(\mathrm{r}) = 4\pi \mathrm{k}T\rho \int_0^r G(r') r'^2 \,\mathrm{d}r' \qquad (2.57)$$

Noting that G is a function of the density (or the reduced density $\eta = (\pi/6)\rho\sigma^3$) in addition to the radius, and utilizing the scaling factor $\lambda = r'/\sigma$, this expression may be written as

$$W(r) = 4\pi \mathrm{k}T\rho\sigma^3 \int_0^{r/\sigma} G(\lambda, \eta)\lambda^2 \,\mathrm{d}\lambda \qquad (2.58)$$

If it is assumed that the relationship between $W(r)$ and $G(\lambda, \eta)$ is valid at small r, namely $\sigma/2 < r \leqslant \sigma$, in the same functional form as for $r \to \infty$, where it is given by the surface tension, $G(\sigma,\eta)$ may be found. The expression may be written as an infinite series or in a closed form

$$G(\sigma, \eta) = 1 + 2.50\eta + 0.765\eta^2 + 0.204\eta^3 + \ldots$$
$$= (4 - 2\eta + \eta^2)/4(1 - \eta)^3 \qquad (2.59)$$

Once this quantity is known, and since $g(\sigma) = G(\sigma)$, eqs. (2.52) and (2.55) can be solved, and the equation of state and thermodynamic properties of a hard-sphere fluid calculated.

The application to real fluids can be made by letting σ be temperature- and density-dependent. The scaled particle theory can also be used to give the results for the hard-core reference of perturbation theories described in the next section. Its merits are the relative conceptual and mathematical simplicity, and the information provided about the surface tension of a fluid. However, the intractability of the function $G(r)$ beyond $r = \sigma/2$, which is due to the probability of two or more particles entering the cavity, hence bringing back the problem of multiple correlation functions, detracts from the usefulness of the theory, and explains the long time interval between its initial proposal and recent developments[12] which may overcome some of the difficulties.

d. Perturbation theories

Returning to the problem of structure, one should note that this is determined mainly by the repulsive forces, while the attractive forces merely provide a background potential through which the molecules move. Accepting this view, it is natural to utilize the perturbation approach[13], in which the steep form of the repulsion potential (Fig. 1.6) is specified for a 'reference fluid', and other fluids are dealt with by applying the attractive potential as a perturbation. Hard spheres have been dealt with so extensively by liquid state theories, and many detailed computer experiments have provided a great deal of data concerning the hard-sphere fluid, that it is natural to use the repulsion potential of a hard sphere $u(r < d) = \infty$, $u(r \geqslant d) = 0$, as the reference potential. Note that the free parameter d is used rather than a molecular diameter σ, since at this stage the appropriate choice of d cannot yet be made. The potential function will be written

$$u(r) = u_0(r) + \gamma u_1(r) \qquad (2.60)$$

where γ is a coupling parameter (similar to ξ in (2.53)), 'turning on' the perturbation potential $u_1(r)$, $u_0(r)$ being the reference potential. For the present, consider only the hard sphere u_0 and the turned-on perturbation, $\gamma = 1$. It is possible to write the Helmholtz energy of the fluid as a series in reciprocal temperatures

$$A = A_0 + (\epsilon/kT)A_1 + (\epsilon/kT)^2 A_2 + \ldots \qquad (2.61)$$

where A_0 is the Helmholtz energy of the reference (hard-sphere) fluid, ϵ is the depth of the potential-well in u_1, and

$$A_1 = (NkT\rho/2\epsilon)\int_0^\infty g_0(r)u_1(r)d\vec{r} \qquad (2.62)$$

$g_0(r)$ being the pair correlation function for hard spheres. However the terms onwards from A_2 contain at least the three- and four-body correlation functions $g_0^{(3)}$ and $g_0^{(4)}$, which even for hard spheres are not well known. An alternative to truncating the series after the first term is to write the exact relationship

$$A = A_0 + 2\pi N\rho \int_d^\infty g_0(r)u(r)r^2\,dr \tag{2.63}$$

where the difficulty, however, has been transferred to the unknown lower limit of the integration, which is both temperature- and density-dependent.

Progress has been made along two lines. The *Barker–Henderson* (BH)[14] theory replaces (2.60) by a more complicated division into reference and perturbation potentials

$$\exp(-u'(r)/kT) = (1 - \delta(d' - \sigma))\exp(-u(d')/kT) + \delta(d' - \sigma)$$
$$+ \delta(r - \sigma)[\exp(-\gamma u(r)/kT) - 1] \tag{2.64}$$

where $\delta(x < 0) = 0$ and $\delta(x > 0) = 1$ is the Heaviside step function, $d' = d + (r - d)/\alpha$, in which d is, again, the as-yet-undetermined hard-core diameter, and α and γ are coupling parameters. For $\alpha = \gamma = 0$, the hard-sphere potential is retrieved for u', while for $\alpha = \gamma = 1$, $u' = u(r)$. The parameter d is obtained from

$$d = \int_0^\sigma [1 - \exp(-u(r)/kT)]\,dr \tag{2.65}$$

which is temperature-, but not pressure-dependent. If the Helmholtz energy is expanded as a Taylor series in α and γ, the choice (2.65) makes the term to order α zero at all temperatures and densities. The first two terms in the expansion are identical to the right-hand side of (2.63), with σ replacing d as the lower limit of integration. The third term is

$$A_2 = (N\rho/kT)\int_d^\infty g_0(r)(u_1(r))^2\,d\vec{r} \tag{2.66}$$

and, to the second order, the only further term that has to be calculated is that in γ^2, since those in α^2 and in $\alpha\gamma$ are negligibly small or zero for liquid densities. The term in γ^2 does contain the three- and four-body correlation functions. It can be evaluated if the Kirkwood superposition approximation, eq. (2.17), is assumed to hold.

Calculations with the square-well potential, eq. (1.40), with R chosen as 3/2, which defines $u_1(\sigma < r < (3/2)\sigma) = -\epsilon$ and zero beyond that range in (2.60), showed reasonable agreement with computer experimental results for such properties as the equation of state, densities of coexisting gaseous and liquid phases, saturated vapour pressures, and the critical constants ($P_c V_c/NkT_c = 0.393$ compared with the computer value 0.306, $kT_c/\epsilon = 1.39$ compared with 1.28). If the 'LJ 12-6' potential is used as the perturbing potential u_1, the parameter d is found to vary considerably with the temperature: from 0.975σ at $T = \epsilon/\mathrm{k}$ to 0.900σ at $T = 10\epsilon/\mathrm{k}$. Again second-order calculations (terms in γ and in γ^2, using $g_0(r)$ for hard spheres from PY theory) of the equation of state and thermodynamic properties show reasonable agreement with computer experimental results, as well as with those for argon, although agreement becomes worse as the temperature decreases or the density increases.

Some of the discrepancy between the behaviour of dense, 'cool' liquids and that predicted by the BH perturbation theory can be removed by, again, modifying the potential function.[15] Starting from the premise that it is the repulsive forces which determine the structure, the reference potential is written

$$u_0(r \leqslant r_0) = u(r) + \epsilon; \quad u_0(r > r_0) = 0 \tag{2.67a}$$

while the perturbing potential is

$$u_1(r \leqslant r_0) = -\epsilon; \quad u_1(r > r_0) = u(r) \tag{2.67b}$$

where $u(r)$ is an arbitrary function, which could be the 'LJ 12-6' potential. This separation is justified by noting that the repulsive *force* between the molecules, $-(\partial u/\partial r)_{\rho,T}$ is continuous at the point $r = \sigma$, so that the molecules are unaffected by the fact that at this point $u(r) = 0$. On the other hand, the force vanishes at $r = r_0$, the minimum in the potential curve. Instead of using the hard-sphere pair correlation function for the reference fluid as in the BH theory, eqs. (2.62) to (2.66), the function (cf. eq. (2.45))

$$g_0(r) = \exp(-u_0(r)/\mathrm{k}T)(1 + b(r)) \tag{2.68}$$

is used, where only the residual $b(r)$ is approximated by $b_{\mathrm{HS}}(r)$ the hard-sphere function, giving the (small) correlation that exists beyond the range of interaction of the reference potential $u_0(r)$. It is advantageous to use the structure factor (section 2A(b)), $S(k)$, rather than its Fourier-transform, $g(r)$, in the development of the argument. It is postulated that even for $k < \pi/\sigma$, $S(k) = S_0(k)$ quite effectively, provided the density is sufficiently high, $\rho\sigma^3 > 0.4$ (for large k this is always a good approximation).

The approximate hard-sphere diameter d is now selected so as to obey the relationship

$$\int_0^d r^2(1 + b_{\mathrm{HS}})[\exp(-u_{\mathrm{HS}}(r)/\mathrm{k}T) - 1]\,\mathrm{d}r$$

$$= \int_0^{r_0} r^2(1 + b_{\mathrm{HS}})[\exp(-u_0(r)/\mathrm{k}T) - 1]\,\mathrm{d}r \tag{2.69}$$

This makes it, contrary to eq. (2.65) of the BH treatment, not only temperature-, but also density-dependent, decreasing with increasing ρ or T, when the other variable is given. A typical value is $d = 1.025\sigma$ for $\rho\sigma^3 = 0.50$ and $T = 1.36\epsilon/\mathrm{k}$. The Helmholtz energy is then calculated from the expression

$$A = A_0 + 2\pi N\rho \int_d^\infty u_1(r)(1 + b_{\mathrm{HS}})(r)\exp(-u_0(r)/\mathrm{k}T)r^2\,\mathrm{d}r \tag{2.70}$$

Table 2.2. Some properties of argon observed experimentally and calculated by various theories of the liquid state.

	$\frac{kT_c}{\epsilon}$	$\frac{P_c\sigma^3}{\epsilon}$	$\frac{v_c}{N\sigma^3}$	$\frac{kT_m}{\epsilon}$	$\frac{v^l(m)}{N\sigma^3}$	$\alpha_p T_m$	$\frac{s^E}{R}$ *	$\frac{u^E}{RT_m}$ *
Solid (expt., T_m)				0.701	1.035		−5.53	−7.14
Liquid (expt., T_m)	1.277	0.119	3.15		1.178	0.368	−3.64	−5.96
Van der Waals'	1.027	0.074	5.20	0.564		0.356		
PY (press eq.)[a]	1.326	0.147	3.22	0.725		0.651		
BH (perturbation)[b]	1.41	0.170	3.25					
Scaled particles[c]	1.305	0.140	3.36	0.715		0.619		
LJD[d]	1.30	0.434	1.77		1.037		−5.51	−7.32
Hole (CE)[e]	2.75	0.469	2.00					
Hole (H)[f]	1.41	0.139	3.39					
Tunnel (B)[g]	1.07	0.37	1.8					
Significant structure[h]	1.306	0.141	3.36	0.711	1.159		−3.89	

*Excess over ideal gas under same conditions.
[a]From eqs. (2.13) and (2.45); [b]from eq. (2.66); [c]from eq. (2.52); [d]from eqs. (2.82), (2.87) and (2.88); [e]according to refs. 23 and 25; [f]according to refs. 24 and 25; [g]according to J. A. Barker, *Aust. J. Chem.*, **13**, 187 (1960); *Proc. Roy. Soc., Ser. A*, **259**, 442 (1961) and ref. 19; [h]according to ref. 25.

which may be compared with eq. (2.63). Good agreement with computer experimental data for 'LJ 12-6' potential fluids under conditions of high density and low temperatures, as well as for the critical state, i.e. over the entire liquid range, is attained for the derivatives of (2.70), the equation of state and the thermodynamic properties (Table 2.2).

e. Many-body potentials

Many-body interactions are a corner-stone of liquid state theories. Even for imperfect gases beyond the very dilute state, many-body interactions come in as the third and higher virial coefficients. For liquids, when the pair correlation function is related to the pair potential through the potential of mean force, eq. (2.9), the force equation, (2.8), brings in the triplet correlation function. The pressure-derivatives of $g(r)$, required for the calculation of, for example, the thermal expansivity, involves the triplet correlation function

$$kT(\partial\rho^2 g(r)/\partial P)_T = \rho^2 \int [g^{(3)}(\vec{r}_1, \vec{r}_2, \vec{r}_3) - g^{(2)}(\vec{r}_1, \vec{r}_2)]\, d\vec{r}_3 + 2\rho g(r) \quad (2.71)$$

The temperature-derivative, required for the calculation of the heat capacity, involves in addition the quadruplet correlation function. Integrals of these derivatives are available from the corresponding derivatives of the structure factor $S(k)$, obtainable in principle experimentally, e.g. for the pressure-derivative

$$\int (\partial g(r)/\partial P)_T (kr)^{-1} \sin kr dr = \rho^{-1}(\partial S(k)/\partial P)_T + (2\pi)^{-3}\rho^{-2}\int (\partial S(k)/\partial P)_T dk \quad (2.72)$$

The derivative itself cannot be obtained thus. The theories discussed in sections 2B(a) to (d) prescribe methods to circumvent the difficulty of the unavailability of the higher correlation functions by direct experiment. However, all these methods assumed, to the order of refinement presented, the pair-wise additivity of the potentials. Computer experiments (section 2A(c)) also assume pair-wise additivity of potentials when applied to models of fluids (hard-sphere, square-well, 'LJ 12-6', etc.), and disregard many-body potentials in the calculation of many-body correlations.

That *triplet potentials* are, indeed, important can be seen by the fact that the configurational internal energy of argon at 100 K and at a molar volume of 2.7×10^{-5} m^3 has a 5% contribution from triplet potentials, while the pressure has as much as a 50% contribution. The apparent success of the 'LJ 12-6' potential in describing the properties of argon ($\epsilon/k = 119.8$ K and $\sigma = 0.3405$ nm from second virial coefficients, $\epsilon/k = 117.7$ K and $\sigma = 0.3504$ nm from a wider range of temperatures, but $\epsilon/k = 142.1$ K and $\sigma = 0.3361$ nm from fitting gas, liquid and solid properties as well as spectroscopic data[14,16]), contrasting with the fact that argon crystallizes in a face-centered-cubic lattice rather than the predicted close-packed-hexagonal lattice, is a further good indication for the importance of many-body, notably triplet, potentials.

Attempts to deal with real fluids, in particular at densities and temperatures characteristic of liquids, thus require the inclusion of many-body potentials, as defined in eq. (1.37). Even the inclusion of the triplet potential u_{ijk} alone (the terms shown in eq. (1.38)) would be a great help. A rather simple triplet potential due to dispersion forces[17]

$$u_{ijk} = (3/4)^2 \alpha^3\, h\nu (r_{ij} r_{jk} r_{ki})^{-3} (3 \cos\theta_i \theta_j \theta_k + 1) \tag{2.73}$$

is the triple dipole interaction, where θ_i is the angle between r_{ij} and r_{ik}, etc. One effect of this is to add a term in $g^{(3)} u_{ijk}$ in those calculations where only $g^{(2)} u_{ij}$, i.e. $g(r)u(r)$ was used hitherto. The same applies to the force equation, that is the r-derivative of u_{ijk} which is needed, for example, in the pressure equation (2.13).

Within the framework of the density expansion, employing the graph-theory which is the basis of the HNC and PY theories, simple prescriptions can be added to eqs. (2.37), (2.38), (2.40), etc. to take into account the triplet potential. The symbolical notation is[18] a shaded triangle between three points, representing the linkage $\phi(i,j,k) = \exp(-u(i,j,k)/kT) - 1$. These graphs all belong to the set $y(i,j)$. Some of the new graphs, to first order in ρ, are

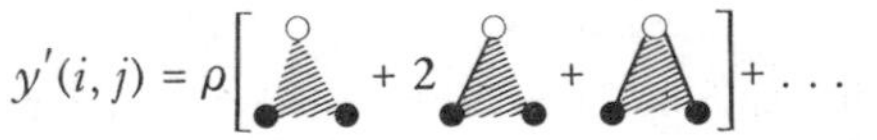

$$y'(i, j) = \rho \left[\triangle + 2 \triangle + \triangle \right] + \ldots \tag{2.74}$$

For the HNC theory, only the subset $y(i, j) - y'(i, j)$ is discarded[18] when going from (2.42) to (2.43). Thus a new effective potential $u^{\text{eff}}(r)$ should replace the pair potential $u(r)$ in (2.43) and (2.44), where

$$u(i, j)^{\text{eff}} = u(i, j) - kT\rho \int \phi(i, j, k)(\phi(i, k) + 1)(\phi(j, k) + 1)\mathrm{d}\vec{r}_k$$

$$\simeq u(i, j) + \rho \int u(i, j, k)(\phi(i, k) + 1)(\phi(j, k) + 1)\mathrm{d}\vec{r}_k \tag{2.75}$$

The second equation is valid for small triplet potentials, on linearizing $\phi(i,j,k)$. For the PY theory, the additional set $d(i,j)$ of all connected graphs free of bridge points and direct i–j linkages is defined (which is the one discarded on going from (2.42) to (2.45)), and the triplet potential is taken into account[18] by not discarding $y'(i,j)$, which is a subset of $d(i,j)$. The same effective potential, (2.75), then serves for (2.45) and (2.46).

New expressions are obtained for the internal energy and for the pressure, replacing (2.11) and (2.13) which consider only pair-wise potentials

$$E = (3/2)N\mathrm{k}T + (1/2)N\rho \int u(i,j)g(i,j)\mathrm{d}\vec{r} + (1/6)N\rho^2 \int u(i,j,k)g(i,j,k)\mathrm{d}\vec{r}_j\mathrm{d}\vec{r}_k \tag{2.76}$$

$$P = \rho\mathrm{k}T - (1/6)\rho^2 \int r(\partial u(r)/\partial r)g(r)\mathrm{d}\vec{r} - (1/6)\rho^3 \int r(\partial/\partial r)u(i,j,k)g(i,j,k)\mathrm{d}\vec{r}_j\mathrm{d}\vec{r}_k \tag{2.77}$$

while the compressibility equation, (2.14), remains unchanged. Lack of knowledge of $g(i,j,k)$ makes direct evaluation of E and of P impossible, and recourse must be taken to approximations, such as the HNC or PY theories, discussed above.

C. Liquid Models

The theories that attempt to relate the radial distribution function or pair correlation function to the intermolecular potential energy consider the fluids at increasing densities essentially as imperfect gases under compression. They can predict the thermodynamic properties of *simple* (monatomic) fluids, such as argon, and also their first-order transition to an ordered phase. Since they show sigmoid-shaped pressure–volume curves (cf. Fig. 1.4), these theories lead to the recognition of a critical point and permit the calculation of the quantities applying to it. Their practical application is based on an expansion of $\ln g(r) + u(r)/\mathrm{k}T$ in terms of the density, eq. (2.40), so that their adequacy decreases as the density increases. These theories are therefore least useful at low temperatures and high densities, i.e. near the freezing point. Indeed, the transition noted in some cases to an ordered structure is not the freezing of a liquid but the condensation of a compressed gas.

Chemical reactions, however, are carried out in solvents that have a far more complicated potential function than the simple fluids considered above, and as a rule at temperatures much nearer the freezing point than the critical point. It is this fact that led to alternaive treatments that regard the liquid as a disordered solid rather than as a compresed gas. These approaches[19] start with a lattice on which the molecules vibrate around ordered lattice points, and introduce some measure of disorder. Instead of stressing the pair correlation function $g(r)$, they emphasize the *partition function* Q in their derivations. Several models of these disordered lattices have been proposed, with more or less success, measured by their ability to account for the properties of liquids found experimentally. Although their development has not been so vigorous in recent years, compared with the advances of the treatments based on correlation functions, such lattice models offer advantages when mixtures are considered (Chapter 4), and merit close examination.

a. The simple cell model

A *dense* assembly of N molecules occupying a volume V at a temperature T is considered, in which each molecule is constrained by its neighbours to occupy a limited region in space – a 'cell'. In fact, molecular dynamics calculations[20] (section 2A(b)) indicated that in a fluid, just as in a solid at comparable densities ($\rho\sigma^3 = 0.49$) hard-sphere particles spend a large time in the vicinity of their original positions, before jumping, in a fluid, to a more remote position. The simplest cell model makes then the following assumptions:

(i) the total volume may be subdivided into cells;
(ii) the centres of the cells form a regular lattice;
(iii) the cells are all identical but may be distinguished by their positions in the lattice (the molecules are indistinguishable);
(iv) each cell contains one molecule, hence there are N cells and the volume per cell equals the volume per molecule, V/N;
(v) each molecule moves independently in its cell;
(vi) the effect of the molecules in the cells neighbouring a given cell on the molecule in that cell is 'smeared out' uniformly over a sphere of radius R.

The validity of these assumptions is discussed further below, but if accepted, they permit the calculation of the configurational partition function $Q_{\text{conf}} = Z$.

For an assembly of hard spheres of diameter σ, it is possible to define a *free volume* by means of the equation $Z(N, V, T) = v_f^N$. The physical picture behind this is that the free volume is the volume available in a cell for the centre of a molecule to move in, Fig. 2.9. For an ordered solid the free volume vanishes but a factor of $(V/N)^N$ enters the partition function because of the confinement of the molecules in the distinguishable cells. For the corresponding low density gas, each molecule may have its centre anywhere in the entire volume, so $v_f = V$ and there is no longer any distinguishability between the molecules, and hence the corresponding factor is $V^N(N!)^{-1}$. The ratio of these factors can be easily calculated for large N using Stirling's approximation, $N! = \exp(N \ln N - N)$, yielding $V^N(N!)^{-1}/(V/N)^N = e^N$. This is the source of the *communal entropy* that fluids have in excess of crystalline solids (see section 1A, eq. (1.3)). For liquids with $v_f \ll V$ (and at high densities, even $v_f < V/N$), only a part of this factor is operative, but its estimation is difficult, so that this is a source of error in any cell theory. The simple cell model applied to hard spheres then calculates the free volume from geometrical reasonings, invoking assumptions (ii) and (vi). The volume of the spherical cavity, Fig. 2.9(c), in which the molecule may move is

$$v_f = (4\pi/3)(R - \sigma/2)^3 \tag{2.78}$$

since each centre may approach up to $R - \sigma/2$ to the surface of this cavity. If N

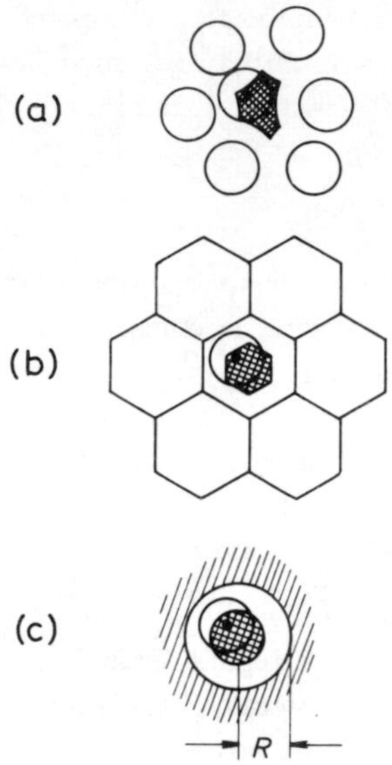

Figure 2.9. A two-dimensional representation of cells and the free volume. (a) With random packing, the centre of the molecule indicated can be anywhere in the cross-hatched area, which is the *free area* for this molecule. (b) In an expanded regular lattice, the free area (cross hatched) has a regular shape. With close packing this area vanishes. (c) When the effect of the molecules in neighbouring cells is smeared out, the free area becomes circular in shape. The extension to three dimensions is obvious.

molecules, of volume $v_m = (4/3)\pi(\sigma/2)^3$ each, are close-packed (hexagonal or face-centered-cubic lattices), the total volume V_0 is $2^{1/2}$ times larger than the actual volume of the molecules (i.e., for a close-packed solid $\rho\sigma^3 = 6/\pi\sqrt{2} = 1.35$). Thus

$$V_0 = (\pi\sqrt{2}/6)N\sigma^3 = 0.7405\ N\sigma^3 \tag{2.79}$$

According to assumptions (ii) and (vi), the cells are spheres of radius R on a close-packed lattice, hence the diameter

$$2R = [(6/\pi\sqrt{2})V/N]^{1/3} \tag{2.80}$$

of the cell can be obtained from the actual volume V. Insertion into eq. (2.78) yields then

$$v_f = 2^{-1/2} N^{-1} (V^{1/3} - V_0{}^{1/3})^3 \qquad (2.81)$$

If this value of the free volume is inserted into its definition $v_f = Z^{1/N}$ and $A_{conf} = -\mathrm{k}T \ln Z$ is used, the thermodynamic properties of this assembly can be calculated. It is evident, however, that the properties of an expanded solid, rather than of a liquid, are obtained by this procedure.

An alternative is to define the free volume operationally, for a system having both attractive and repulsive forces. The existence of attractive forces may be taken as a necessary requirement for the existence of a liquid state, since hard spheres show (section 2A(c)) only gas-like and crystal-like properties. Let U_0 be the potential energy when all the molecules are at the centres of their cells. The free volume is then defined by

$$Z^0(N,V,T) = v_f^N \exp(-U_0/\mathrm{k}T) \qquad (2.82)$$

If the potential energy of the actual liquid is approximated by

$$U = U_0 + \Sigma[u(\vec{r}_i) - u(0)] \qquad (2.83)$$

where the summation extends over all the molecules i, and $\vec{r}_i$ is the displacement of the ith molecule from the centre of its cell, then the free volume is obtained from (1.96) and the last two equations as

$$v_f = \int_{cell} \exp(-[u(\vec{r}_i) - u(0)]/\mathrm{k}T)\mathrm{d}\vec{r}_i \qquad (2.84)$$

If, now, assumption (v) is invoked, the exponent may be estimated and hence the properties of the system.

b. The Lennard-Jones and Devonshire (LJD) model

The cell theory developed by Lennard-Jones and Devonshire (LJD)[21] starts from eq. (2.84). The geometrical estimate of the volume of the cell derived from (2.80), $v_{cell} = V/2^{1/2}N$, is now replaced simply by $v_{cell} = V/N$, and the cells form a face-centered-cubic lattice. Each cell thus has twelve neighbours, so that the cells may be equated with the appropriate Voronoi polyhedra (section 2A(d)), in this case regular dodecahedra (Fig. 2.6). As for the simple model, the cells thus grow as the density decreases and the volume expands on heating, in view of assumption (iv), which prescribes one cell for each molecule. The potential energy of molecule i in its cell, $u(\vec{r}_i)$, is determined by the field of its neighbours. If assumption (vi) is again invoked, a spherically symmetrical field results, and the vector $\vec{r}_i$ may be replaced by the scalar displacement r of the molecule from the cell centre. Consider the concentric shells of cells around that containing molecule i. The distance a_n between the centre of a cell in the nth shell and that of the central cell is $n^{1/2}$ times the distance a between the centres of neighbouring cells for the type of lattice under discussion. To obtain spherical symmetry, integration is carried out over all the angles θ between the radius through molecule i and the radii to the centres of all the cells. If m_n is the number of cells in the nth shell ($m_1 = 12$, $m_2 = 6$, $m_3 = 24, \ldots$) then

$$u(r) - u(0) = \sum_n m_n (1/2) \int_0^{\pi} [u(\{a_n^2 + r^2 - 2ra_n \cos\theta\}^{1/2}) - u(a_n)] \sin\theta \, d\theta \tag{2.85}$$

For molecules obeying the 'LJ 12-6' potential, the result is

$$u(r) - u(0) = 12\epsilon[(N\sigma^3/V)^4 L(x) - 2(N\sigma^3/V)^2 M(x)] \tag{2.86}$$

where $x = (r/a)^2$, and the functions $L(x)$ and $M(x)$ are known, rather complicated functions of x. The value of the integrand in (2.85) becomes effectively zero for $r > 0.5a$, so that the error introduced by replacing the Voronoi dodecahedron by a sphere of radius $0.5a$ is negligible. A sphere of radius $(3 \cdot 2^{1/2}/8\pi)^{1/3}a = 0.553a$, which has the same volume as v_{cell}, may be a better choice. The integration (2.84) over the cell then yields

$$v_f = 2 \cdot 2^{1/2}(V/N) \int_0^{0.5532} x^{1/2} \exp(-[u(r) - u(0)]/kT) dx \tag{2.87}$$

with eq. (2.86) used in the exponent. This has been solved numerically. The partition function is finally calculated from v_f and U_0, eq. (2.82), where U_0 is now given by

$$U_0 = (1/2)N \sum_n m_n u(a_n) = 6N\epsilon[1.011(N\sigma^3/V)^4 - 2.409(N\sigma^3/V)^2] \tag{2.88}$$

where the second equality holds for 'LJ 12-6' potentials. It should be noted that not only cells in the first shell ($n = 1$) contribute to v_f or to U_0 but also cells from other shells, at least for the 'LJ 12-6' potential.

Having calculated the partition function according to (2.82) with the aid of (2.87) and (2.88), the configurational thermodynamic properties follow. The pressure-volume curves have sigmoidal shapes at low temperatures, indicating equilibrium between a dense and an expanded phase, and a critical point is predicted. Substances which conform to the 'LJ 12-6' potential have $T_c/\epsilon k^{-1} = 1.30$, $P_c/\epsilon\sigma^{-3} = 0.434$ and $v_c/\mathbf{N}\sigma^3 = 1.768$, i.e. $P_c v_c/\mathbf{R}T_c = 0.591$. These figures may be compared with those for argon, namely 1.26, 0.116, 3.16 and 0.292 respectively, so that serious discrepancies are evident. In fact, at a reduced temperature $T/\epsilon k^{-1} = 0.70$, corresponding to the triple point of argon, the data calculated by the LJD theory for the experimental liquid volume, i.e. $V = (N/\mathbf{N})v^l$, are closer to those of solid argon than to those of liquid argon: the reduced molar volume is $v/\mathbf{N}\sigma^3 = 1.037$ compared with 1.035 for the solid and 1.186 for the liquid, and the reduced molar entropy is $s/\mathbf{R} = -5.51$ compares with -5.33 for the solid and -3.64 for the liquid. Thus the LJD theory describes a model for an expanded crystalline solid rather than for a liquid.

It is necessary to modify some or all of the assumptions (i) and (vi) made for the cell model in order for the LJD model to be a liquid model. Assumption (i) is easy to defend since if none of the other assumptions is made it involves only a conceptual picture of the system, and the walls of the cells have no influence on its behaviour. Turning to the last assumption, (vi), it was found that it introduces negligible errors at high densities, and, at lower densities, may be corrected for[19]. The error in v_f was found to be *ca.* −13% at $\rho\sigma^3 = 0.50$ and a maximal error of −30% was found at $\rho\sigma^3 = 0.31$, the error decreasing at lower densities. Thus the smearing of the effect of the molecules in neighbouring cells over spherically symmetrical shells is a good approximation.

Assumption (v) is more troublesome, since it precludes correlated movements of neighbouring molecules. To the order of the validity of the pair-wise additivity assumption $U = (1/2)\Sigma u(ij)$, the expression

$$U = (1/2)u(IJ) + \sum_i \sum_j [u(iJ) - u(IJ)] + (1/2)\Delta_{ij} \tag{2.89}$$

can be written. Here, the distance between molecule i and the cell-centre of molecule j is denoted by iJ, the distance between the cell-centres themselves by IJ, and so forth, and the identity

$$u(ij) \equiv u(IJ) + [u(iJ) - u(IJ)] + [u(Ij) - u(IJ)] + \Delta_{ij}$$

$$\Delta_{ij} = u(ij) - u(iJ) - u(Ij) + u(IJ) \tag{2.90}$$

is employed. In eq. (2.89), the first term can be identified as U_0 of eq. (2.83) and eq. (2.88), and the second term includes the sum over all j of the potentials for a given molecule i, that is the cell-field potentials without the smearing assumption, so that

$$U = U_0 + \Sigma_i[u(\vec{r}_i) - u(0)] + (1/2)\Sigma\Delta_{ij} \tag{2.91}$$

The LJD model simply drops the last term, in Δ_{ij}, defined by eq. (2.90). An approximate evaluation of Δ_{ij} will thus approximately correct for the correlation effect, neglected by assumption (v). If only pairs of molecules in nearest-neighbour cells are considered as moving in a partly correlated fashion, then the right-hand side of eq. (2.82) should be multiplied by a factor $\exp(-NZ\Delta_{ij}/2kT)$, where Z is the *coordination number of the lattice* (i.e. 12, and not the distance-dependent quantity, eq. (2.28)), to give a corrected partition function. For an 'LJ 12-6' potential, at reduced temperatures in the normal liquid range of argon $T/\epsilon k^{-1} = 0.75$ to 0.90 and densities $\rho\sigma^3 = 1.00$ down to 0.63, the quantity $\exp(-\Delta_{ij}/kT)$ was found[11] to be in the range 0.024 to 0.051, so that the correlation correction is quite small. At a higher reduced temperature $T/\epsilon k^{-1} = 1.2$ and lower densities, $\rho\sigma^3 = 0.50$ down to 0.25, that is near the critical point, the correlation corrections are also quite small, $\exp(-\Delta_{ij}/kT)$ ranging from 0.05 to −0.07.

Assumption (iv), the single occupancy of the cells, has been dispensed with in modifications of the LJD model, in an attempt to deal with the problem of the communal entropy. At high densities, where $v_{cell} \sim (4\pi/3)(\sigma/2)^3$, multiple occupancy is precluded, but at low densities it is permissible, with the simultaneous existence of empty cells, provided that division into N cells is maintained. Consider the case when double occupancy is allowed. The partition function is then $Z = (N!)^{-1} \Sigma_i Z(m_i)(N!/\Pi m_i!)$, where m_i is the number of molecules in the ith cell, 0, 1 or 2, and $Z(m_i)$ is a restricted configurational integral allowing a given m_i molecules in each particular cell i. The LJD model is thus restricted to the term $Z(1) = Z(m_i \equiv 1)$. A parameter β is defined by

$$\beta^N Z(1) = \Sigma_i Z(m_i)/\Pi m_i! \qquad (2.92)$$

Then the configurational Helmholtz energy is $A_{conf} = -NkT \ln Z(1) - NkT \ln \beta$, and the communal entropy, obtained from the second term, is

$$S_{comm} = Nk[\ln \beta + T(\partial(\ln \beta)/\partial T)_{N,V}] \qquad (2.93)$$

For crystalline solids, single occupancy only is possible, so that from (2.92) $\beta = 1$ and $S_{comm} = 0$. For a liquid $\beta > 1$, and its determination is extremely difficult unless multiple occupancy is restricted to a double one. in this case

$$\beta = \omega_1 + 2(\omega_0 \omega_2/2)^{1/2} \qquad (2.94)$$

where ω_{m_i} is a constant factor which multiplies $Z(m_i)$. The values $\omega_0 = \omega_1 = 1$ can be assumed, and the value of ω_2 has been calculated[11], being in the range of 0.44 to 0.66 for an 'LJ 12-6' fluid at the reduced temperature $T/\epsilon k^{-1} = 1.2$ and at reduced densities of 0.60 down to 0.35, yielding values of β somewhat higher than 2. This correction for communal entropy, or multiple occupancy of the cells, together with those smaller ones discussed above for the potential smearing to spherical symmetry and the correlation of movement, bring the calculated Helmholtz energy and the entropy calculated for argon near its critical point, at reduced densities below ca. $\rho\sigma^3 = 0.50$, very close to the experimental data. Clearly, this treatment cannot overcome the difficulties at high densities, $\rho\sigma^3 > 0.7$, characteristic of liquids not too far from the freezing point.

c. *Disorder and hole models*

The main difficulty with the LJD model, or its modifications discussed above, for dense liquids is seen to rest with assumptions (ii) and (iii). Unless these are dispensed with, the model will continue to correspond to an expanded crystalline solid rather than to a liquid, because too much 'order' is built into the model by its insistence on a regular lattice and division into N cells with $v_{cell} = V/N$.

The disorder theory of melting of Lennard-Jones and Devonshire[22] attempts to deal with this situation, by relaxing the requirement of regularity of the lattice (assumption (ii)). Suppose that the lattice for the N molecules has $2N$ sites, on two interpenetrating sublattice, α and β, having N sites each. The sublattice sites are equivalent and each α-site has six neighbouring β-sites and twelve next-nearest-neighbour α-sites, the combined lattice being simple cubic, such as in sodium chloride. For an *ordered* arrangement, all the molecules occupy sites on one of the sublattices, and the configurational partition function is given by eq. (2.82) $Z_0 = v_f^N \exp(-U_0/kT)$. In the liquid randomness is introduced by letting the N molecules occupy any of the $2N$ sites in a *disordered* manner, so that

$$Z = Z_0(N!)^{-1} \Sigma \exp[-(U - U_0)/kT] \tag{2.95}$$

where the summation extends over all the configurations and $(N!)^{-1}$ is introduced since the molecules themselves are indistinguishable. According to the *Bragg and Williams approximation* only nearest-neighbour and next-nearest-neighbour interactions are taken into account, and their average values, $\epsilon_{\alpha\beta} = \epsilon_{\beta\alpha}$ and $\epsilon_{\alpha\alpha} = \epsilon_{\beta\beta}$, rather than their instantaneous distance-dependent values, are considered. Then the total configurational potential energy of the disordered arrangement is

$$U = (1/2)N_\alpha(6x_\beta\epsilon_{\alpha\beta} + 12x_\alpha\epsilon_{\alpha\alpha}) + (1/2)N_\beta(6x_\alpha\epsilon_{\alpha\beta} + 12x_\beta\epsilon_{\alpha\alpha}) \tag{2.96}$$

where $x_\alpha = 1 - x_\beta = N_\alpha/N$ and N_α is the number of molecules occupying α-sites $(N_\alpha + N_\beta = N)$. Setting $x_\alpha = 1$ (or $x_\beta = 1$) yields the value for the ordered phase: $U_0 = (1/2)N \cdot 12\epsilon_{\alpha\alpha}$. Therefore the difference

$$U - U_0 = 6Nwx(1 - x) \tag{2.97}$$

where $w = \epsilon_{\alpha\beta} - 2\epsilon_{\alpha\alpha}$, and $x = x_\alpha$. The sum over all the configurations can be replaced as a sum over N_α exponents of $-(U - U_0)/kT$ weighted by the combinative weight $N!(N!/N_\alpha!(N - N_\alpha)!)$, and the latter sum by its maximal term. This is found by differentiation, giving

$$3w(2x - 1)/kT = \ln[x/(1 - x)] \tag{2.98}$$

There is one solution for this equation for values $w < kT/3$, $x_0 = 0.5$, giving a maximum, while for $w > kT/3$ there are two maxima x_0 symmetrical around $x = 0.5$ (where there is a minimum), i.e. the system is partly ordered ($x_0 = 1$ and $x_0 = 0$ correspond to complete order). The Helmholz energy of the system is

$$A = -kT \ln Z_0 - 6Nwx_0(1 - x_0) - 2NkT(x_0 \ln x_0 + (1 - x_0)\ln(1 - x_0)) \tag{2.99}$$

the last two terms being the contribution of the disorder. It is likely that w depends on the relative distance between neighbours and next-nearest-neighbours, hence on the density, and if repulsive forces are dominant for 'LJ 12-6' potential molecules then $w = w_0(\rho\sigma^3)^4$ can be written. This permits differentiation of A with respect to volume $V = N\rho$ (remembering that x_0 depends on w) to give the pressure. Certain pressures correspond to two different values of the density, indicating the existence of two condensed phases at equilibrium, the denser one with partial order ($x_0 \neq 0.5$) and the other completely disordered ($x_0 = 0.5$). The parameter w_0 can be found by adjusting the melting curve $T_m(P)$ to one experimental point, e.g. for zero pressure. For argon this gives $w_0 = 0.928\epsilon$, where ϵ is the energy parameter for the 'LJ 12-6' potential. From this follow nearly correct values of other quantities, e.g. $\Delta v^F/v = 0.135$, $\Delta s^F/\mathbf{R} = 1.70$, $\alpha_P = 0.0040\ \mathrm{K}^{-1}$ (for the liquid), P_σ (at

$T = 90.3$ K) = 0.286 Pa, compared with the experimental values 0.12, 1.66, 0.0045 and 0.291. However, this treatment also predicts that the two densities associated with each pressure approach each other as the temperature increases, leading to a critical point for the melting process (at $T = 1.1\epsilon/k$ for substances conformal with argon), which has never been observed, even at much higher pressures (section 1A), and does not appear in Monte-Carlo (section 2A(c)) calculations even at $T = 2.7\epsilon/k$. Faults may be found with the use of a simple cubic lattice for the combined α and β sublattices (it has too large a volume for a given average nearest interatomic distance r_0 or σ), or of a regular lattice at all. However, this order-disorder model is a workable model for melting and for low-temperature liquids. Its semiempirical nature (the adjustment for w_0) can in principle be removed if some more sophisticated treatment of $w = \epsilon_{\alpha\beta} - 2\epsilon_{\alpha\alpha}$ is developed. Other lattices, and perhaps more kinds of sites to be occupied, may remove some of the discrepancies, so that this model should be refined.

Another way of circumventing the necessity of introducing assumptions (ii) and (iii), and the use of cells of equal shape and of a size $v_{\text{cell}} = V/N$ increasing as the density decreases, is to introduce *holes* into the system[23]. These holes are not unoccupied regular cells, as encountered in the modified LJD theory (cf. text around eq. (2.92) to eq. (2.94)), but N_0 cells which are empty, additional to the N singly occupied cells. The volume per cell is now $v_{\text{cell}} = V/(N + N_0)$ and need not increase as the density decreases; in fact it may be equated with the cell volume in the solid, so that N_0 is allowed to carry the burden of the expansion. The results of X-ray and neutron diffraction studies (section 2A(b)) for liquids not too far above T_m show that r_0 does not increase very fast with increasing temperatures and decreasing densities, but that Z, below that for the solid already at T_m, falls further (Table 3.2). This confirms the presence of molecule-sized holes, and is contrary to the assumptions on which the LJD model is based. The random distribution of the N_0 holes on the $N + N_0$ sites is a modification of the order-disorder model discussed above, and may be treated on analogous lines. Only interactions between nearest-neighbour molecules will be considered, with an energy parameter ϵ, when the molecules are at the centres of adjacent cells, the displacement from the centre being taken care of by the free volume v_f, as in eq. (2.82). The overall average number of occupied neighbouring cells is $(1/2)Z_0N \cdot N/(N + N_0)$, where Z_0 is the *lattice*-coordination number, which does not vary with the density or temperature. There is a total of $(N_0 + N)!/N_0!$ different arrangements, so that

$$Q_{\text{conf}} = (N_0 + N)!(N_0!N!)^{-1} v_f^N \exp[-Z_0 N^2 \epsilon/2kT(N_0 + N)] \quad (2.100)$$

If v_f is assumed to depend on the temperature only but not on the volume, then the pressure is

$$P = kT(\partial(\ln Q_{\text{conf}})/\partial V)_T = -kT(v_{\text{cell}})^{-1} [\ln(1 - \rho v_{\text{cell}}) + (Z_0\epsilon/2kT)(\rho v_{\text{cell}})^2] \quad (2.101)$$

Sigmoid-shaped curves are obtained at low temperatures, indicating an equilibrium between a condensed and a gas phase. At higher temperatures monotonic curves are obtained, extrapolating to the perfect gas law $P = kT\rho$ when the second term in the square bracket is dropped and the logarithm is expanded. A critical temperature $T_c = (Z_0/4)\epsilon/k$ is predicted, which, however, is much too high for liquids conformal with argon, when the value $Z_0 = 12$ valid for their solid lattices is used. A more refined treatment, allowing v_f to depend on the density and taking into account interactions with next-nearest-neighbours, actually brings the hole model near to the order-disorder model. This is brought about by setting $N_0 = N$, but requires then smaller cells since there are not that many molecule-sized holes in dense liquids near T_m : $\Delta v^F/v_s$ is only about 0.15. Another way to make v_f density-dependent is to regard it as a combination of v_f^0, the LJD value (which is temperature-dependent) and v_{cell}, e.g. $v_f = (Z/Z_0)v_f^0 + (1 - Z/Z_0)v_{cell}$, where $Z = (1/2)Z_0N^2/(N + N_0)$ (ref. 24) is the average coordination number, as given already above. This gives a considerable improvement to calculated values of the critical constants compared with the simple hole theory (Table 2.2).

d. Significant structure theory

A significant improvement over the simple hole model was brought about by specifying in a definite way the number of holes, the properties of the holes, and the degrees of freedom, hence the partition functions, of the holes and molecules: the significant structure theory[25]. It should be realized that holes of molecular sizes will be favoured in the liquid. Smaller holes deny access to molecules and hence lower the entropy of the liquid, while larger holes, while not contributing more to the entropy, raise the potential energy of the liquid. The holes may be regarded as *fluidized vacancies* (Fig. 2.10) that move around in the volume of the liquid. A molecule in the liquid ordinarily has solid-like degrees of freedom, that is, it vibrates around its position. Molecules that are adjacent to a vacancy acquire translational degrees of freedom. In as far as the vacancies can be looked at as moving just like gas molecules, only in the opposite direction (Fig. 2.10), these vacancies confer on adjacent molecules gas-like degrees of freedom. This should not be construed as stating that there are gas-like molecules in the liquid, a notion that has given rise to confusion and to criticism.

If one compares the volume V of a liquid to that of a close-packed solid V_s, containing the same number N of molecules, then $V - V_s$ is the excess volume due to the fluidized vacancies, and the mole fraction of this volume is $x = (V - V_s)/V_s$. Only those vacancies completely surrounded by molecules confer gas-like degrees of freedom on them (a vacancy surrounded entirely by vacancies does not contribute to the properties). For random distribution of vacancies, a fraction V_s/V of sites around a vacancy is occupied by molecules, hence the fractional contribution of gas-like degrees of freedom is $[(V - V_s)/V_s](V_s/V) = (V - V_s)/V = x/(1 + x)$. The remaining fraction, $1/(1 + x)$ of the degrees of freedom is solid-like. The contribution to the heat capacity at constant volume of monatomic liquids is k for a solid-like, k/2 for a gase-like degree of freedom, hence the expected value for liquid argon is

$$c_v = 3Nk[1/(1 + x) + x/2(1 + x)] \tag{2.102}$$

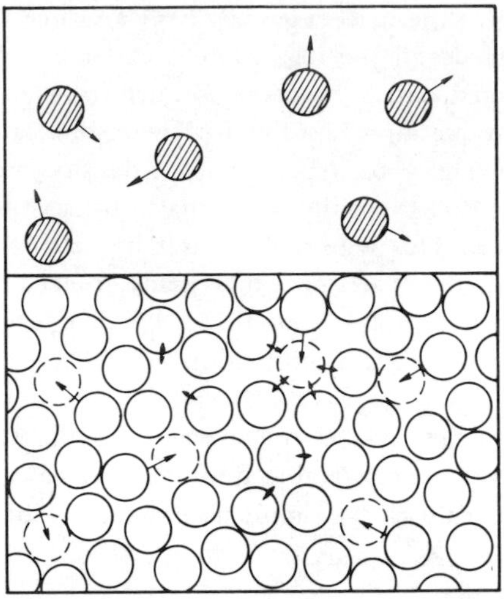

Figure 2.10. A schematic two-dimensional representation of the fluidized vacancies (broken circles) moving in a liquid as if they were mirroring gas molecules (upper portion of figure, hatched circles). Solid-like molecules have normally only vibrational degrees of freedom (symbolized by double arrows), except when they have sufficient kinetic energy to outcompete the other neighbours of a vacancy into which they can move, the vacancy then moving in the opposite direction.

This relationship was claimed to conform to the experimental results[25], but this has been challenged[26].

The properties of the liquid can now be derived from the partition function, which is factored to the contributions from the two kinds of behaviour

$$Q = Q_s^{N/(1+x)} Q_g^{Nx/(1+x)} \tag{2.103}$$

Once the values of Q_s for the solid-like and Q_g for the gas-like behaviour are specified, thermodynamic data can be obtained as functions of T and $\rho = N/V = N/V_s(1 + x)$, requiring only the proportionality factor V_s. This can be taken as either the volume of the actual solid extrapolated to $T = 0$ K, or as the close-packed value $N\sigma^3/\sqrt{2}$, with σ as the parameter.

In order to calculate Q_s it must be realized that several particles compete for a given vacancy but only one of sufficient kinetic energy has the chance of denying it to other particles and occupying it. Therefore a positional degeneracy factor, the number of available vacant sites plus one (the original site), must be applied to such a particle. The number of available additional sites is the product of the total number of vacancies around a solid-like molecule, n_h, and the probability that the molecule has the required energy ϵ_h, i.e. $\exp(-\epsilon_h/kT)$. The value of n_h is obtained from the liquid volume at the melting point, $V_{l(m)}$ and the coordination number of the solid, Z_0

$$n_h = xZ = Z_0(V - V_s)/V_{l(m)} \tag{2.104}$$

taking $Z = V_sZ_0/V_{l(m)}$. The average kinetic energy of each of the $Z - 1$ neighbours of the vacancy, competing with that of energy ϵ_h, is $(2Z_0)^{-1}(3/2)kT$, since half the time it moves in the direction of the vacancy, and has Z_0 directions to move to. This multiplied by $(Z - 1)$ is then the minimum energy, ϵ_h, that a particle requires to be able to outcompete the others. At the melting point, $T_m = \Delta u^F/(3/2)R$, since for monatomic fluids the entropy of fusion $\Delta s^F \simeq (3/2)R$, while $\Delta u^F \simeq \Delta u^S(V_{l(m)} - V_s)/V_{l(m)}$, where Δu^S is the molar energy of sublimation, the proportionality factor being the fraction of holes introduced by melting, at a constant kinetic energy. Thus

$$\begin{aligned}\epsilon_h &= (Z - 1)(2Z_0)^{-1}(3/2)k\Delta u^S(V_{l(m)} - V_s)/V_{l(m)}(3/2)R \\ &= (Z - 1)(2Z_0)^{-1}(\Delta u^S/N)(V_{l(m)} - V_s)/V_{l(m)}\end{aligned} \tag{2.105}$$

The positional degeneracy factor is therefore $(1 + n_h\exp(-\epsilon_h/kT))$, with n_h and ϵ_h given by eq. (2.104) and eq. (2.105), respectively. The Einstein oscillator can be taken to represent the vibrational degrees of freedom of a solid-like molecule for a monatomic fluid, hence

$$Q_s = \exp(\Delta u^S/RT)(1 - \exp(-\theta_E/T))^{-3}(1 + n_h \exp(-\epsilon_h)kT)) \tag{2.106}$$

where θ_E is the Einstein characteristic temperature.

The partition function for the gas-like degrees of freedom is calculated for $Nx/(1 + x)$ indistinguishable indeal-gas particles moving in the excess volume $V - V_s$

$$Q_g = \Lambda^{-3}(V - V_s)\{[(Nx/(1 + x)]!\}^{-Nx/(1+x)} \tag{2.107}$$

Introduction of Q_s from (2.106) and Q_g from (2.107) into (2.103) yields then the partition function for the liquid.

This model may be improved in several ways. The Einstein oscillator approximation for the solid-like degrees of freedom may be replaced by the LJD approximation (eqs. (2.82) and (2.87)), and the 'LJ 12-6' potential may be used in these equations for $u(r)$. The result is

$$Q = \Lambda^{-3N}[\exp(-u(0)/2kT)(1 + Z_0(V - V_s)/V_{1(m)} \exp(-\epsilon_h/kT)v_f]^{Nx} \cdot (eV/N)^{Nx/(1+x)} \tag{2.108}$$

where $u(0) = Z_0\epsilon[(V_s/N\sigma^3)^4 - (12/5)(V_s/N\sigma^3)^2]$, v_f is given by (2.87), and $e^{Nx/(1+x)}$ arises from using the Stirling approximation with the factorial in (2.107). The calculation is not free from empirical parameters: in addition to ϵ and σ, the values of Z_0, Δu^S, V_s and $V_{1(m)}$ are required in order to obtain $Q_{conf}(T, V, N) = Q\Lambda^{3N}$. For hard spheres, the values $\epsilon_h = 0$ (obviating a need for Δu^S), $Z_0 = 12$, $V_{1(m)} = V_s = 3N\sigma^3/8^{1/2}$ were used, giving an equation of state and an excess entropy (excess over the ideal gas at the same (T, V, N)) in good agreement with computer calculated values[27]. For 'LJ 12-6' potentials, the configurational energy U_0, given by eq. (2.88) (where $Z_0 = 12$ was selected), may replace $\Delta u^S/\mathbf{N}$. The final numerical result for simple fluids (inert gases, nitrogen, methane, etc.) is

$$Q_{\mathrm{conf}}(N,V,T) = (\mathrm{e}V/N)^N (\mathrm{e}V/N\sigma^3)^{-1} \exp(8.388\mathrm{k}T/\epsilon)(\mathrm{k}T/35.01)^{3/2}$$
$$[1 + 10.7((V/N\sigma^3) - 1)] \exp(-0.0436\mathrm{k}T(V - N\sigma^3)/\epsilon N\sigma^3)^{N^2\sigma^3/V} \tag{2.109}$$

where the only remaining parameters are ϵ and σ. This equation gives very good results (Table 2.2) for the equation of state, vapour pressure, density, excess entropy and internal energy of such simple fluids. In their papers and book, Eyring et al.[25] show also how the significant structure theory can be applied to much more complicated liquids (organic solvents, polar and hydrogen-bonded liquids, molten salts, liquid metals) and other properties (surface tension, dielectric constant), by using the liquid property at one given value of temperature and density, and then calculating the whole curve.

References

1. A. Ben-Naim, in *Solutions and Solubilities, Part I*, edited by R. J. Dack, Wiley–Interscience, New York, 1975.
2. F. Zernike and J. A. Prins, *Z. Phys.*, **41**, 184 (1927).
3. L. S. Orenstein and F. Zernike, *Proc. Acad. Sci. Amsterdam*, **17**, 793 (1914).
4. N. A. Metropolis, A. W. Rosenbluth, M. N. Rosenbluth, A. H. Teller and E. Teller, *J. Chem. Phys.*, **21**, 1087 (1953).
5. B. J. Alder and T. W. Wainwright, *J. Chem. Phys*, **27**, 1208 (1957); **31**, 459 (1959); W. W. Wood and J. D. Jacobson, *J. Chem. Phys.*, **27**, 1207 (1957).
6. J. D. Bernal, *Nature (London)*, **183**, 141 (1959); **185**, 68 (1960); G. D. Scott, *Nature (London)*, **188**, 908 (1960); **194**, 956 (1962); **201**, 382 (1964).
7. J. G. Kirkwood, *J. Chem. Phys.*, **3**, 300 (1935).
8. J. Yvon, 'La Theorie Statistique des Fluides et l'Equation d'Etat', *Actualités Sci. Ind.*, **203** (1935); M. Born and H. S. Green, *Proc. Roy. Soc., Ser. A*, **188**, 10 (1946); N. N. Bogoliubov, *J. Phys. USSR*, **10**, 257, 265 (1946).
9. D. Levesque, *Physica (Utrecht)*, **32**, 1965 (1966).
10. J. M. J. van Leeuwen and J. de Boer, *Physica (Utrecht)*, **25**, 792 (1959); M. S. Green, *J. Chem. Phys.*, **33**, 1403 (1960).
11. J. K. Percus and G. J. Yevick, *Phys. Rev.*, **110**, 1 (1958).
12. H. Reiss, H. L. Frisch and J. L. Lebowitz, *J. Chem. Phys.*, **31**, 369 (1959); D. M. Tully-Smith and H. Reiss, *J. Chem. Phys.*, **53**, 4015 (1970); R. M. Gibbons, *Mol. Phys.*, **17**, 81 (1969); **18**, 809 (1970).
13. R. W. Zwanzig, *J. Chem. Phys.*, **22**, 1420 (1954); J. A. Pople, *Proc. Roy. Soc., Ser. A*, **221**, 498, 508 (1954).
14. J. A. Barker and D. Henderson, *J. Chem. Phys.*, **47**, 2856, 4714 (1967); *Ann. Rev. Phys. Chem.*, **23**, 439 (1972).
15. J. D. Weeks, D. Chandler and H. C. Andersen, *J. Chem. Phys.*, **54**, 5237 (1971); **55**, 5422 (1971).
16. P. A. Egelstaff, *Ann. Rev. Phys. Chem.*, **24**, 159 (1973).
17. B. M. Axilrod and E. Teller, *J. Chem. Phys.*, **11**, 299 (1943).
18. G. S. Rushbrooke and M. Silbert, *Mol. Phys.*, **12**, 505 (1967); J. S. Rowlinson, *Mol. Phys.*, **12**, 513 (1967).
19. J. A. Barker, *Lattice Theories of the Liquid State*, Pergamon Press, Oxford, 1963.

20. T. W. Wainwright and B. J. Alder, *Nuovo Cimento,* **9**, Suppl. 1, 116 (1958).
21. J. E. Lennard-Jones and A. F. Devonshire, *Proc. Roy. Soc., Ser. A,* **163**, 53 (1937).
22. J. E. Lennard-Jones and A. F. Devonshire, *Proc. Roy. Soc., Ser. A,* **169**, 317 (1939); **170**, 464 (1939).
23. H. Eyring, *J. Chem. Phys.,* **4**, 283 (1936); F. Cernuschi and H. Eyring, *J. Chem. Phys.,* **7**, 547 (1939).
24. D. Henderson, *J. Chem. Phys.,* **37**, 631 (1962).
25. H. Eyring, T. Ree and H. Hirai, *Proc. Nat. Acad. Sci. U.S.A.,* **44**, 683 (1958); H. Eyring and M. K. Jhon, *Significant Liquid Structures*, Wiley, New York, 1969.
26. J. S. Rowlinson, *Trans. Faraday Soc.,* **67**, 576 (1971).
27. D. Henderson, *J. Chem. Phys.,* **39**, 1857 (1963).

Chapter 3

The Properties of Liquids

When the properties and the structure of individual liquids are to be reviewed, it is useful to classify the liquids according to some broad criteria. Such a classification has been attempted in Chapter 1, according to the intermolecular forces, and it was shown (Table 1.2) that some, but not all, the bulk properties of the liquids follow this order of classification. The classes recognized were simple fluids, non-polar molecular liquids, polar molecular liquids, hydrogen-bonded molecular liquids, ionic liquids, and liquid metals. For some purposes it is convenient to employ a somewhat different classification, for instance it may be useful to distinguish between atomic liquids (inert gases and molten metals), monatomic-ionic liquids (a certain group of molten salts) and molecular and polyatomic-ionic liquids. The distinction is that the former group can be studied by means of diffraction experiments, and structural conclusions concerning the liquid may be derived from the data (section 2A(b)), while in the latter group, information mainly on intramolecular (ionic) interactions and structure is obtained from diffraction experiments, and little is learned about the liquid itself – with obvious exceptions (e.g. water).

As far as the intermolecular forces are concerned, there is little difference between the simple fluids (i.e. liquified inert gases), which are monatomic, and spherically symmetrical (or practically so) non-polar molecular fluids, such as methane, niopentane (2,2′-dimethylpropane), nitrogen or carbon dioxide. These liquids differ from other non-polar liquids with non-spherical molecules (e.g. chain-like, such as hexane, or plate-like such as benzene) or large spherical molecules which are highly polarizable (such as carbon tetrachloride), where the intermolecular forces are not central. Because of the short range of the forces, interactions occur mainly between adjacent segments of neighbouring molecules, rather than between the molecules as entities.

Strong intermolecular forces that lead to self-association (section 1A(c)) of liquids may be of several kinds. Dipole–dipole interactions may lead to pair-wise association, but rarely to higher aggregates. Hydrogen bonding may lead to dimers primarily (e.g. carboxylic acids), to chain-like oligomers (e.g. liquid hydrogen fluoride, alcohols, *N*-alkylamides) or to three-dimensional networks (e.g. water, formamide). Donor–acceptor interactions may operate between the ions of an ionic melt, and lead to autocomplexation into discrete complex ions (e.g. mercury bromide) or to extensive networks (e.g. boron oxide).

The classification used in this chapter mainly follows that of the intermolecular forces, with allowance for the molecular complexity (atomic *versus* molecular), shape (central *versus* non-central forces) and extent of interaction (single particles *versus* associated species).

The properties that will be reviewed in this chapter are mainly the bulk thermodynamic and mechanical properties of the liquids, and the static optical and electrical properties (Section 1B), as well as the relevant molecular properties (size, shape, charge, polarizability, dipole moment). The structural features that can be discussed are mainly the distances to the nearest- and next-nearest-neighbours and the appropriate coordination numbers. These properties and features depend on the conditions under which the liquid is studied: the temperature and the pressure, which, in turn, determine the density. Wherever possible, liquids will be reviewed at comparable reduced temperature and pressures, i.e. at corresponding states.

A. Atomic Liquids

a. The noble gases

Liquified noble gases are hardly of interest to chemists, except for their extremely important role as model systems[1], against which theories of the liquid state must be tested, and which serve as starting points for the deviations which more complex fluids show. Liquid helium is of great use to chemists as a cryogenic liquid but since it is the lowest boiling liquid (T_b = 4.215 K) and because of the large role that quantum-effects have on its properties, it is an extremely interesting fluid for liquid state physicists, but not to liquid state chemists, since it is *not* a good model for general liquid behaviour. Liquid radon is hardly of interest either, since the high specific radioactivity of even its longest-lived isotope, ^{222}Rn, makes its properties too dependent on the ever present radiation and growing-in daughters to

Table 3.1. Some properties of liquified noble gases[a].

Property	Ne	Ar	Kr	Xe
Bulk properties				
M (kg mol^{-1})	0.020183	0.039948	0.08380	0.13130
T_c (K)	44.40	150.85	209.35	289.74
P_c (MPa)	2.722	4.894	5.502	5.840
v_c (10^{-6} m^3 mol^{-1})	41.74	74.52	92.24	119.36
T_{tr} (K)	24.55	83.78	115.95	161.30
$v^l(T_{tr})$ (10^{-6} m^3 mol^{-1})	16.12	28.15	34.13	42.69
$\Delta v^F/v^l(T_{tr})^a$	0.133	0.126	0.131	0.131
Δs^F (J K^{-1} mol^{-1})	13.65	14.02	14.10	14.23
T_b (K)	27.07	87.27	119.80	165.05
$v^l(T_b)$ (10^{-6} m^3 mol^{-1})	16.72	28.54	34.73	42.95
$\Delta s^V(T_b)$ (J K^{-1} mol^{-1})	63.99	74.67	78.86	76.56
$\Delta u^V/v^l(T_b)$ (MJ m^{-3})	90.1	202.9	245.9	262.2
Atomic properties				
I (10^{-18} J)[b]	3.440	2.512	2.232	1.936
α (10^{-30} m^3)[c]		1.63	2.48	4.01
σ(LJ) (nm)[c]	0.252[d]	0.3504	0.3827	0.4099
ϵ/k(LJ) (K)[c]	35.8[d]	117.7	164.0	222.3

[a] V. G. Fastovskii, A. E. Rovinskii and Yu. V. Petrovskii, *Inert Gases*, Atomizdat, Moscow, 1964;
[b] *Handbook of Chemistry and Physics*, Chemical Rubber Co., Dayton;
[c] A. E. Sherwood and J. M. Prauswitz, *J. Chem. Phys.*, **41**, 429 (1964);
[d] P. E. Siska, J. M. Parson, T. P. Schäfer and Y. T. Lee, *J. Chem. Phys.*, **55**, 5762 (1971).

be usefully measurable in the present context. This leaves the four noble gases neon, argon, krypton and xenon, of which the first has properties which are not completely free from quantum-effects (owing to its low mass).

Table 3.1 gives some of the properties of the liquified heavier noble gases. Some other properties can be derived from the listed ones by standard thermodynamic formulas, e.g. $h^V = T_b S^V(T_b)$, etc. The narrow existence limits of these substances as liquids under atmospheric pressures, $T_b - T_{tr}$, should be noted (from 2.5 K for neon to 3.8 K for krypton and xenon) but under higher pressure the liquid range is extended considerably. However, the fraction that the liquid temperature range takes of the total existence of condensed phases, $(T_c - T_{tr})/T_c = 0.4456 \pm 0.0016$, is smallest for the noble gases of almost all substances, compared with > 0.5 for most other substances (boron trifluoride is a notable exception, with the fraction being 0.438, cf. footnote, p. 10). The critical pressure does not vary much among the noble gases (except for neon), and the critical volume is proportional to the cube of the molecular diameter, the ratio l_c/σ, eq. (1.1), being 1.414 ± 0.010 (again, except for neon, where it is 1.63). The quotient $Z_c P_c V_c / \mathbf{R} T_c = 0.2906 \pm 0.0008$ (for neon it is 0.3071, for helium 0.3030). The three heavier noble gases are thus conformal substances, cf. eqs. (1.25), (1.26) and (1.43).

The process of freezing of the liquid is accompanied by a volume reduction of 13%. The volume reduction on freezing plays an important role in the significant structure theory of the liquid state, section 2C(d). Some of the properties of liquid argon at the triple point are compared with theoretical calculations in Table 2.2. Freezing involves also a decrease of entropy of $(1.68 \pm 0.02)\mathbf{R}$, of which $\mathbf{R}$ is the communal entropy lost on freezing, eq. (1.3). The remaining $0.68\mathbf{R}$ is rather large compared with the change of entropy on freezing of other simple fluids and no satisfactory explanation for this amount has been put forward. On the other hand, the entropy-increase on boiling at the normal boiling point is rather smaller (−10%) than the average for normal liquids, Trouton's rule (1.9), and indeed, the normal boiling points, T_b, are not corresponding temperatures for the noble gases. T_b/T_c decreases from 0.608 for neon to 0.570 for xenon. Corresponding temperatures are those where the vapour volumes are equal (Hildebrand's rule, p. 9), and the best comparison is at a given fraction of the critical pressure, for example[2] at $P_c/50$, $\Delta s^V = (9.02 \pm 0.03)\mathbf{R}$.

As mentioned above, the heavier noble gases are conformal fluids and the triple point is a corresponding temperature $T_{tr}/T_c = 0.5544$. The properties of liquid krypton and xenon may therefore be derived for that temperature from those of liquid argon, shown in Table 1.1. For instance the vapour pressures are 1/72.5 times the critical pressures, and also the cohesive energy densities, $\Delta u^V/V$ (p. 10), are proportional to them.

The molecular interactions in the liquid are confined to dispersion forces and to repulsion at close distances. The leading term of the attractive energy is given by (1.49), the coefficient C (in units of 10^{-78} J m^6) being 5.00 for argon, 10.30 for krypton, and 23.35 for xenon. The depth of the potential-well is given by ϵ according to the 'Lennard-Jones 12.6' potential (1.42). It is known, however, that this potential is not an adequate representation of the potential function and many alternative expressions have been proposed.[3] Of these, only the Kihara potential (1.47) is reported here, with the parameters[4] (ϵ/k, σ, δ) being (147.2 K, 0.3314 nm, 0.0368 nm) for argon, (215.6 K, 0.3521 nm, 0.0459 nm) for krypton, and (298.8 K, 0.3878 nm, 0.0797 nm) for xenon, the hard-core diameter increasing relative to the atomic diameter in this series from 11 to 13 to 20%. Even this potential can be fitted to the data with different parameters[5], with δ/σ about constant at 9%. Because of the non-unique state of these expressions, it is impossible to obtain a definite molecular collision diameter σ, or equilibrium distance between nearest-neighbours $r_0 = 2^{1/6}\sigma$ from the data leading to the potential functions (virial coefficients B_2, gas viscosities, etc).

The intermolecular distances have, however, been obtained from diffraction studies[6]. At a reduced temperature of $T/T_c = 0.559 \pm 0.004$, the distance to the first peak in the $n(r) = \rho g(r) 4\pi r^2$ curve is at $r_{max} = 0.318$ nm for neon, 0.378 nm for argon, 0.402 nm for krypton, and 0.443 nm for xenon (the latter three r_{max} are 96% of r_0 from B_2). The values do depend on the temperature and on the number density but not to a great extent: for argon in the vicinity of the critical point r_{max} (in units of nm) $= 0.346 + 2.6 \times 10^{-3}(T)$(in K) $- 4 \times 10^{-4}\rho$(in nm^{-3}).

The coordination number Z_I (eq. (2.28)), however, depends more on the temperature and much more on the number density of particles in the fluid. At the above reduced temperature and at $\rho = 37$ nm^{-3}, $Z = 8.4$ for neon from X-ray diffraction and at $\rho = 32$ nm^{-3}, $Z = 8.5$ for argon from neutron diffraction. At the same temperature but at $\rho = 21\ \mathrm{nm}^{-3}$, $Z = 11.7$ for argon from X-ray diffraction (near the coordination number of the crystalline solid, $Z = 12$, while a lower Z would have been expected at the lower density). Nearer the critical point[7], at $T/T_c = 0.949$, Z increases from 5.4 at $\rho = 13.7\ \mathrm{nm}^{-3}$ to 5.8 at $\rho = 14.7\ \mathrm{nm}^{-3}$, and at $T/T_c = 0.982$ from 4.9 at $\rho = 11.7\ \mathrm{nm}^{-3}$ to 6.2 at $14.7\ \mathrm{nm}^{-3}$. The higher sensitivity of the coordination number to density compared to temperature is apparent. Three peaks are discernable in the radial distribution curve $n(r)$ for the liquid: for argon the second peak occurs at $1.92r_{max}$, the third at $2.82r_{max}$. These further peaks are absent at similar temperatures in the gas. The coordination number of liquid krypton[7] is obtained by neutron diffraction: as the reduced temperature increases from $T/T_c = 0.559$ to 0.874, the number density falls from $\rho = 17.6\ \mathrm{nm}^{-3}$ to $13.0\ \mathrm{nm}^{-3}$ and the coordination number falls from 8.5 to 6.5. For liquid xenon[6], X-ray diffraction results show Z falling from 9.0 at $T/T_c = 0.563$, $\rho = 14.2\ \mathrm{nm}^{-3}$ to 8.7 at $T/T_c = 0.598$, $\rho = 13.5\ \mathrm{nm}^{-3}$. These experimental results, and in more detail, the structure factors $S(k)$, which are the primary data of the diffraction studies (section 2A(b)) have been used to obtain the pair correlation function $g(r)$, the radial distribution function $n(r)$, and the direct correlation function $f(r)$ by suitable Fourier-transforms (eq. (2.25)). These, in turn, have been used for intensive testing of the various models and theories of liquids, discussed in sections 2B and 2C.

b. Liquid metals

Metals form liquids which have very complex interatomic (ion–ion, ion–electron and interelectron) interactions. The only justification for discussing them here, just after the noble gases, is that their structures too can be readily determined by diffraction methods, and a considerable amount of information has been accumulated. Table 3.2 gives the values of r_{max} and Z for the first peak at the lowest and the highest temperatures for which data could be found. The general features observable in the Table is that, apart from the alkali metals, the metals have $0.27 < r_{max}(\mathrm{nm}) < 0.34$, i.e. a narrow variability of the atomic diameter among the metals, and a coordination number Z that falls somewhat with increasing temperatures.

It is interesting to compare distances and coordination numbers for the liquid metals and their crystalline solids near the melting point. Contrary to the non-metallic monatomic substances, i.e. the noble gases, the crystals of which have close packed structures with $Z = 12$ and the liquids have invariably $Z^l < Z^s$, the metals can be divided into two groups, according to whether $Z^l < Z^s$ or $Z^l > Z^s$ for the nearest-neighbours. In the former group, the metal atoms have metallic bonds and the crystals have closed-packed structures. The liquids also have close-packed structures but because of expansion on melting the coordination number falls, while the interatomic distances becomes somewhat smaller. To this group belong Ag, Al, Au, Cd, Hg, Pb, Tl and Zn. In the second group, homopolar bonding occurs between the metal atoms, and some atoms in the crystal are therefore nearer each other than for close packing, while others are further away. In the liquids, thermal

Table 3.2. Nearest-neighbours in liquid metals.

Metal	T(K)	T/T_c	Method	r_{max} (nm)	Z	Ref.
Ag	~1273		X-ray	0.283	10.0	a
Al	~935	0.120	X-ray	0.288	10.0	b
	1123	0.145	X-ray	0.305		c
Au	1373		X-ray	0.286	11.0	d
Bi	573	0.124	Neutron[t]	0.335	7.7	e
	823	0.178	Neutron	0.335	7.5	e
Cd	623	0.175	X-ray	0.306	8.3	f
Cs	303	0.147	Neutron	0.531	9.0	g
	573	0.278	Neutron	0.515		c
Cu	1363		X-ray	0.257	11.5	h
Ga	273	0.036	X-ray[t]	0.283	9.2	i
	423	0.056	Neutron	0.284	6.8	j
Ge	1273	0.158	X-ray	0.270	8.0	d
Hg	237	0.137	X-ray	0.303	7.5	k
	291	0.168	X-ray	0.300[l]	6.0	d
	423	0.244	X-ray	0.310	7.7	m
In	438	0.062	X-ray[t]	0.317[l]	8.0	d
	663	0.094	X-ray	0.330	8.5	f
K	338	0.150	Neutron	0.464	9.0	g
	618	0.289	Neutron	0.480		c
Li	453	0.122	Neutron	0.315	9.5	g
Mg	948	0.269	X-ray	0.320	10.0	n
Na	373	0.155	Neutron	0.382	9.0	g
	476	0.198	Neutron	0.380		c
Pb	603	0.112	Neutron	0.330	11.6	p
	648	0.120	X-ray	0.340[l]	8.0	d
	1103	0.204	Neutron	0.340	7.7	p
Rb	313	0.149	Neutron	0.497	9.5	g
	513	0.244	Neutron	0.505		c
Sb	938		X-ray	0.312[l]	6.1	q
	1073		Neutron	0.333	8.6	u
Sn	505	0.056	X-ray[t]	0.327	7.6	r
	1023	0.114	X-ray	0.313	7.2	s
Tl	648		X-ray	0.330[l]	8.0	d
Zn	733	0.252	X-ray	0.294	10.8	f

[a]O. Pfannenschmid, *Z. Naturforsch. A.* **15**, 603 (1960); [b]R. Hezel and S. Steeb, *Phys. Kondens. Mater.*, **14**, 314 (1972); [c]M. D. Johnson, P. Hutchinson and N. H. March, *Proc. Roy. Soc., Ser. A*, **282**, 283 (1964); [d]H. Hendus, *Z. Naturforsch. A*, **2**, 505 (1947); **3**, 416 (1948); [e]P. C. Sharrah and G. P. Smith, *J. Chem. Phys.*, **21**, 228 (1953); [f]H. Richter, G. Breitling and F. Herre, *Z. Naturforsch. A*, **12**, 896 (1957); [g]N. S. Gingrich and L. Heaton, *J. Chem. Phys.*, **34**, 873 (1961); [h]B. R. Orton, B. A. Shaw and G. I. Williams, *Acta Met.*, **8**, 177 (1960); [i]S. E. Rodriguez and C. J. Pings, *J. Chem. Phys.*, **42**, 2435 (1965); [j]P. Ascarelli, *Phys. Rev.*, **143**, 36 (1966); [k]R. F. Kruh, G. T. Clayton, C. Head and G. Sandlin, *Phys. Rev.*, **129**, 1479 (1963); [l]a pronounced second peak was observed with (r_{max} (nm), Z) being for Hg (0.347, 4.0), for In (0.388, 4.0), for Pb (0.437, 4.0), for Sb (0.440, 11.7) and for Sn (0.422, 4.0); [m]B. R. Orton and R. L. T. Street, *J. Phys. C*, **5**, 2089 (1972); [n]S. Woerner, S. Steeb and R. Hezel, *Z. Metallk.*, **56**, 682 (1965); [p]B. I. Khrushchew, A. M. Bogomolov and L. S. Sharipova, *Fiz. Met. Metaloved.*, **22**, 279 (1966); [q]H. K. F. Müller and H. Hendus, *Z. Naturforsch. A*, **12**, 102 (1957); [r]K. Furukawa, B. R. Orton, J. Hamor and G. I. Williams, *Phil. Mag.*, **8**, 141 (1963); [s]H. Richter and D. Handtman, *Z. Phys.*, **181**, 206 (1964); [t]electron diffraction data from R. Leonhardt, H. Richter and W. Rosstentscher, *Z. Phys.*, **165**, 121 (1961) generally confirm the X-ray and neutron diffraction results; [u]Y. Waseda and K. Suzuki, *Phys. Status Solidi B*, **47**, 581 (1971).

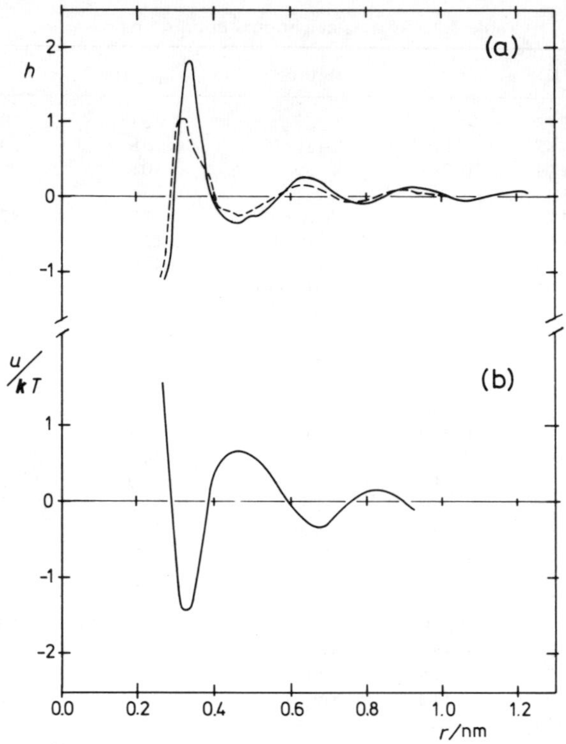

Figure 3.1. Results of diffraction experiments on molten lead; (a) the total correlation function $h(r)$, and (b) the pair potential $u(r)/kT$, (———) at 613 K, (– – – –) at 1373 K (after J. E. Enderby, *Adv. Struct. Res. by Diffraction Methods,* 4, 65 (1972); N. H. March, '*Liquid Metals*', Pergamon Press, Oxford, 1968, p. 42).

agitation has destroyed the favoured geometry of the few nearest-neighbours, so that the liquids are also close-packed, while the nearest-neighbour distance is smaller, but the number of these neighbours is larger than in the solid. To this group belong Bi, Ga, Ge, In, Sb and Sn.

Some of the metals which have $Z < 8$ at the lowest temperature have a pronounced adjacent secondary peak, which, added to the first, brings the combined Z to > 10. To this group belong Hg, In, Pb, Sb and Tl. This additional structure in the liquid metal may be enhanced at a higher temperature, as Fig. 3.1 shows for lead. Figure 3.1 also shows that the short-range order holds to at least three molecular diameters, and the maxima move slightly towards shorter distances as the temperature is increased. Information on the interactions that lead to the observed behaviour is obtained when the structure factor $S(k)$ is analysed, since it shows more clearly the difference between metals and non-metals. The low angle limit for metals is $S(0) \sim 0.01$ compared with a value of ~ 0.06 for the noble gases. The function $(1 - 1/S(k))/(1 - 1/S(0))$ for the noble gases is strongly oscillating, while for liquid metals it is much more localized, having few contributions from $k > 20$ nm^{-1}, and most of the effect is at the lowest angles $k < 4$ nm^{-1}. If attention is focused on the direct correlation function $f(r)$, which is the Fourier-transform of $1 - 1/S(k)$ according to (2.25), it is found that $f(r)$ becomes positive at longer distances for the metals than for the noble gases and that it has a negative portion for the metals at some range of distances still farther away from the origin, contrary to the noble gases, for which $f(r)$ remains positive. This negative portion corresponds to a repulsive interaction, and is reflected in a corresponding positive portion of the pair potential $u(r)$ (as a rough approximation $f(r) = -u(r)/kT$ for larger r, eqs. (2.44), (2.46) and (2.48)). The Born–Green theory (section 2B(a)) was found[8,9] to be most appropriate for liquid metals, and Fig. 3.1 shows the pair potential for liquid lead calculated from this theory. Detailed potential functions have by now been obtained[8,9] for all the liquid metals in Table 3.2, except Ag, Au, Cu and Ge, with similar general features, which may be compared to the theoretical expectations (pp. 35 to 37).

Two concepts have been of importance in the development of the knowledge of interatomic bonding in metals, which have been less emphasized in connection with liquid metals than in connection with the solid phases and with intermetallic alloys and compounds. These are the valency and electronegativity, which still play a role in the understanding of the behaviour of liquid metal mixtures (Chapter 8). The valency is determined by the number of electrons outside the inert gas core, the orbitals available to them without excessive promotion energies, and the number of electrons in the free state (or metallic orbitals). It is generally agreed that the metals in the first six (A) columns of the Periodic Table have valencies according to the column numbers, e.g. 1 for potassium, 2 for calcium, and so on up to 6 for chromium. The average molar energy of vaporization of these liquid metals at the freezing point is u^V (kJ mol^{-1}) = $(100 \pm 20)n$, where n is the valency, and thus a measure of the strength of bonding per bond in the melt. Beyond the sixth group, the valency does not increase with the number of electrons. According to Pauling[10], the valency is $n = 6$ for the Mn, Fe, Co, and Ni groups, and $n = 6.56 - c$, where c is the B-group column number (i.e. 1 for Cu up to 5 for As). The fractional number comes from the participation of some of the orbitals and electrons in the common free-electron cloud, and the crowding of the increasing number of electrons into the remaining orbitals so that fewer single electrons can participate in bonding. These valencies, however, do not conform to the assignments necessary for the Hume-Rothery[11] ratios of electrons to atoms in crystalline intermetallic compounds, where the elements of the groups Cu to As, i.e. B-subgroups, must be assigned valencies $n = c$, while the Fe-, Co- and Ni-subgroups must be assigned $n = 0$. Energies of vaporization from the melts will conform to the same value $(100 \pm 20)n$, as found for the A-group metals, only if the following values of n are assigned: 6 for the Fe-group, 5 for the Co-group, 4 for the Ni-group, 3 for the Cu-group, 1 for the Zn-group, 2 for the Ga-group, 3 for the Ge-group, and 2 for the As-group. Since the ratio u^V (in kJ mol^{-1})/100 = n is completely empirical, and considerable variations in the proportionality factor occur within the subgroups, these values of n should not be taken too seriously. The electronegativities that are of interest in the formation of (liquid) alloys are listed in Table 8.3, and their use discussed in sections 8A(b) and 8A(c).

Table 3.3. Bulk properties of liquid metals freezing below 1500 K[a].

Metal	T_m(K)	T_b (K)	T_c (K)[b]	$v^l(T_m)$ (cm^3 mol^{-1})	$\Delta v^F/v^s$	$\Delta s^F/R$	$\Delta u^V(T_m)$ (kJ mol^{-1})	σ (N m^{-1})[c]
Li	454	1590	3720	13.66	0.0165	0.79	153.7	0.389
Na	371	1163	2400	24.75	0.025	0.84	102.4	0.191
K	336	1039	2140	47.2	0.0255	0.86	84.4	0.101
Rb	312	974	2100	58.0	0.025	0.89	80.8	0.078
Cs	302	958	2060	72.2	0.026	0.88	73.9	0.068
Mg	923	1380	3850	15.47	0.0412	1.17	127.2	0.570
Ca	1123	1765	4590	28.1		0.88	164.2	0.360
Sr	1043	1657	3810	35.5		1.15	151	0.303
Ba	983	1910	3920	32.4		1.17	165	0.277
Al	932	2740	7740	11.38	0.065	1.39	290	0.914
La	1193	3740		23.3		0.54	414	0.70
Ce	1077	3740		21.0	−0.008	0.58	374	0.70
U	1405	4090	12500	13.3	0.022	1.10	527	1.550
Pu	913	3510		14.5	−0.025	0.51	314	0.550
Cu	1357	2868	8900	7.71	0.042	1.16	309.1	1.350
Ag	1234	2485	7500	11.60	0.038	1.16	244.7	0.921
Au	1336	3239		11.5	0.051	1.12	347.8	1.130
Zn	693	1180	2910	9.82	0.047	1.28	116.0	0.810
Cd	594	1038	3560	14.02	0.040	1.24	98.9	0.560
Hg	234	630	1763	14.8	0.037	1.18	57.0	0.480
Ga	303	2510	7620	9.95	−0.032	2.22	270.1	0.735
In	429	2270	6680	16.32	0.020	0.93	243	0.560
Tl	577	1730		18.1	0.022	0.85	169.3	0.490
Ge	1210	3100	8050	13.2	−0.051	3.20	334.9	0.620
Sn	505	2960	8720	17.06	0.023	1.67	289.2	0.575
Pb	601	2024	5400	19.55	0.035	0.96	184.2	0.480
Sb	903	1653		18.73	0.008	2.61	233	0.384
Bi	545	1832	4620	20.78	−0.0335	2.41	190.1	0.376

[a]Data from J. R. Wilson, *Met. Rev.*, **10**, 381 (1965) with some additions; [b]data from ref. 12; and A. V. Grosse and A. D. Kirschenbaum, *J. Inorg. Nucl. Chem.*, **24**, 739 (1962); J. A. Cahill and A. D. Kirschenbaum, *J. Phys. Chem.*, **66**, 1080 (1962); [c]data from J. R. Wilson, *Met. Rev.*, **10**, 381 (1965) and from J. Bohdansky, H. E. J. Schins and E. J. Hubert, *J. Inorg. Nucl. Chem.*, **29**, 2173 (1967); **30**, 2331, 3362 (1968).

The bulk properties of liquid metals are summarized in Table 3.3 for those metals with $T_m < 1500$ K. The entropy of vaporization at the melting point* is $(\Delta u^V(T_m)/T_m + R = \Delta s^V(T_m) = (33 \pm 6)R$, i.e., it is much less invariant than the Trouton constant $(10.2 \pm 1.2)R$, eq. (1.9), which applies at the normal boiling point but is still essentially constant. Hence, $T_m < 1500$ K implies $\Delta u^V <$ 400 kJ mol^{-1} or $n < 4$. In fact, the metals with $n \geqslant 4$ on any valency scheme have high melting points. Those with relatively low melting points, which are the ones dealt with almost exclusively when liquid metals are discussed, are thus confined to the first three A-groups, and to the B-groups of the Periodic Table.

The liquid range of metals is quite extensive, T_b is two to three times higher than T_m, except for the alkaline-earth metals, where it is only 1.5 times as high, and for the elements Ga, In and Sn, where it is 8.3, 5.3 and 5.9 times as high, respectively. In fact, gallium has the longest relative liquid range of all metals. The critical temperature[12] is not known accurately for most metals: direct measurments exist only for mercury, where $T_c = 1763 \pm 15$ K was found. For the other metals recourse was taken to the law of rectilinear diameters (1.29) to extrapolate vapour and liquid densities, or to the observation that Δs^V is a unique function of the reduced temperature T/T_c. In view of the values listed in Table 3.3, the actual liquid range of metals is seen to be generally very large, the fractional existence of metals as liquids from their total existence as condensed phases, $(T_c - T_m)/T_c$, being 0.8 (except for the alkaline-earth metals, where it is lowest for strontium, 0.73), and becomes as large as 0.96 for gallium. The correlation[13] between the critical temperature and the depth of the pair interaction potential-well can be applied to the potential data of Waseda and Ohtani[9] to estimate critical temperatures, the best correlation being T_c (K) $= 8.3\epsilon/k$. However, gallium, indium and tin are outside the correlation, with *ca.* twice as high critical temperatures than predicted. The correlation leads to $T_c = 6440$ K for Sb and $T_c = 4110$ K for Tl, missing in Table 3.3.

The liquid volumes of the metals show the typical effects of the atomic sizes and play an important role in the understanding of the behaviour of mixtures and liquid alloys. Within ± 7%, $v^l = 1.53 \times N(\pi/6)r_{max}^3$, where r_{max} is the distance to the first peak in $g(r)$, equal to the mean atomic diameter in the melt, and 1.53 is a packing factor. Zinc has apparently a lower packing factor, and Bi, Ge and Sb a higher one. The volume change on melting, Δv^F, is found[14] to correlate well with the entropy of melting, so that $\Delta s^F(v^s/\Delta v^F) = 260 \pm 30$ J K^{-1} mol^{-1} for most metals (Al, Au, In, Li, Sb and Sn are exceptions, besides those that have negative Δv^F, i.e., Bi, Ga, Ge and Pu). The exceptions (apart from Au and Li) are on the border line between those elements which are metallic conductors and those which are semi-conductors in the crystalline state. The entropy of melting is within 20% of **R**, the communal entropy (p. 6), the glaring exceptions being again Bi, Ga, Ge and Sb where it is more than twice **R**, and the rare-earth metals La and Ce, where it is only half as large. Both Δv^F and Δs^F depend, of course, on the crystal structure of the solid as well as on the liquid structure.

*At a non-standard and variable pressure, the equilibrium vapour pressure.

The thermal expansivity α_P for the solid metals is known to correlate well with the melting point: $\alpha_P^s(293\ \mathrm{K}) = 0.06/T_m$. A good correlation exists[15] also for the expansivities of the liquid metals: $\alpha_P^l(T_m) \sim 0.23/(T_b - 0.23T_m)$, except for the metals of groups VIII and IIB of the Periodic Table. A further correlation exists[14] between the heat capacities and expansivities: $c_p^l/T_m\alpha_P^l = 305 \pm 30\ \mathrm{J\ K^{-1}\ mol^{-1}}$, with the usual exceptions of Bi, Ga and Sn, and also Hg and Pb, which conform to the other correlations. Compressibility data[16] for the liquid metals are scarce. The alkali metals have large compressibilities $10^{10}\kappa_T$ varying from $2.10\ \mathrm{Pa^{-1}}$ for sodium to $6.73\ \mathrm{Pa^{-1}}$ for caesium, while other low melting metals have much lower compressibilities, $10^{10}\kappa_T$ being in the range of 0.24 to $0.43\ \mathrm{Pa^{-1}}$ for Bi, Cd, Ga, In, Pb, Sn, Tl and Zn.

The vapour pressures of several liquid metals have been measured but the enthalpy of vaporization is best obtained from the enthalpy of sublimation of the solid metal, corrected for the enthalpy of fusion (which is $\mathbf{R}T_m$ times the $\Delta s^F/\mathbf{R}$ values listed in Table 3.3) and for the heat capacities of the solid, liquid and vapour. The energy of vaporization at the melting point, $\Delta u^V(T_m) = h^V(T_m) - \mathbf{R}T_m$, listed in Table 3.3, is a measure of the cohesive energy of the liquid metal at this temperature. This parameter is connected to the pair potential, $u(r)$, and the pair correlation function, $g(r)$, by means of eq. (2.12). Because of the long range and oscillating nature of the functions $u(r)$ and $g(r)$, the integration of the data has not been attempted. The ratio of Δu^V and the molar volume of the liquid v^l is the cohesive energy density, and the square root of this quantity, δ, is the solubility parameter (p. 10). This plays an important role in understanding the properties of mixtures of liquid metals, and a comprehensive list of values of δ is given in Table 8.3.

The heat capacity of the liquid metals has been measured both for constant volume and for constant pressure. The former is lower than the high-temperature limiting value for the solid, $3\mathbf{R}$, the decrease on melting being[17] $0.15\mathbf{R}$ on average, and it falls beyond that, so that $c_V^l = (3 - 0.15T_m^{-1}T)\mathbf{R}$. The heat capacity at constant pressure at the melting point is[14] $c_p^l = (3.72 \pm 0.22)\mathbf{R}$, i.e. $0.87\mathbf{R}$ larger than c_V, instead of the difference $\mathbf{R}$ expected if the liquid metal behaved as an ideal gas. There is no clear trend in c_p on melting – for some metals it increases, for others it decreases; also its temperature coefficient is irregular, usually negative just beyond the melting point, but for some metals it turns positive at some higher temperature[17]. The conclusion is that there is little change with melting of either the vibrational modes of the atoms, of the valence electron configuration, but that structural changes do occur at higher temperatures.

The surface tension of liquid metals (Table 3.3) has been determined quite accurately. This property correlates with the cohesive energy density, the low melting metals have $\sigma < 1.4\ \mathrm{Nm^{-1}}$, while for the refractory metals $\sigma > 1.4\ \mathrm{Nm^{-1}}$ (the value for zirconium), the highest value observed being $2.7\ \mathrm{Nm^{-1}}$ (for rhenium). For all the metals, however, the surface tension is higher than for molecular liquids (Tables 3.5, 3.6) or for molten salts (Table 3.10). This can be attributed to the small surface area per atom; in fact the product $\sigma(v^l)^{2/3}$ for liquid metals does not deviate from that of molecular liquids. The surface tension decreases with rising temperatures almost linearly and should vanish at the critical temperature. However, on extrapolation, the observed linear functions tend to meet at a common temperature of 3550 K for the alkali metals and of 5080 K for the alkaline-earth metals at a somewhat negative value of σ. The intersection temperatures with the $\sigma = 0$ line do not coincide with the critical temperatures estimated from other data, but for some metals (Na, K, Ca, Bi) the agreement is reasonable. Estimates based on the vanishing of σ on linear extrapolation of high temperature data are $T_c = 6530$ K for silver, and 4500 K for thallium, values not otherwise available from Table 3.3.

B. Molecular Liquids

It is simple to distinguish molecular liquids from atomic ones by the requirement that the substance in the liquid state will exhibit per mole of substance at least one mole of essentially covalent bonds. These are absent in the liquified noble gases and the metallic elements considered hitherto (elements such as silicon or arsenic were therefore excluded from the latter group). It is less easy to distinguish molecular from ionic liquids, since borderline cases abound, examples being molten mercury bromide and boron oxide, where the nature of the bonds is intermediate. The definition used above does not exclude, however, types of bonding additional to the covalent bonds, e.g. dipolar association or hydrogen bonding. If the intermolecular forces are to be used as a basis for classification it is convenient to distinguish between the following classes: fluids (liquified gases) with small molecules, where the intermolecular forces are essentially central, and mainly of the dispersion type; liquids with non-polar molecules which are sufficiently large that the forces are between neighbouring segments of two molecules, rather than between the molecules as separate entities; liquids with polar molecules, where the forces are dominated by dipole interactions; and liquids with hydrogen-bonded molecules. These classes will now be discussed in turn.

a. Fluids with small or spherical molecules

It is quite arbitrary to assign molecular liquids to the class having 'small' molecules. Some examples of such substances are included in Table 3.4, from which it is evident that most are liquified gases rather than ordinary liquids ($T_b < 273$ K) and have small molar volumes as liquids ($v^l < 50$ cm^3 mol^{-1}). Most of these fluids are conformal in the thermodynamic sense, in that the quotient $P_c v_c / RT_c$ is within $\pm$ 0.007 of the mean value 0.282 (cf. 0.2906 $\pm$ 0.0008 for the three heaviest noble gases p. 90). Exceptions are oxygen ($P_c v_c / RT_c = 0.332$), bromine (0.375), nitrogen oxide (0.250) and hydrogen chloride (0.248).

Since the vaporization entropies of most molecular liquids are in agreement with Trouton's rule (1.9), $\Delta s^V / R = 10.2 \pm 1.2$, the low boiling points indicate also low enthalpies of vaporization and low cohesive energies, which in turn indicate fairly weak intermolecular forces. Some of the fluids are polar: liquid Cl_2 has $\mu = 1.07$ Debye (1 Debye = 3.33564×10^{-30} C · m), liquid SO_2 has $\mu = 1.61$ Debye, liquid

Table 3.4 Some properties of fluids which consist of small molecules.

Fluid	T_m (K)	T_b (K)	T_c (K)	$v^l(T_m)$ (cm³ mol⁻¹)	$\Delta s^F/R$	$\Delta s^V/R$
N_2	63.1	77.3	126.1	32.3	1.36	8.7
O_2	54.3	90.1	154.8	24.4	1.01	9.1
Cl_2	172.2	238.6	417.	46[a]	4.47	10.3
Br_2	266.0	331.9	584	54.6[a]	4.78	10.9
HCl	158.9	188.3	324.6	36.4	1.51	10.3
HBr	184.7	206	363.2	29.2[a]	1.56	10.3
HI	222.3	238.1	423	45.7[a]	1.56	10.0
H_2S	187.6	212.5	373.6	35.4[a]	1.51	10.6
PH_3	139.4	185.8	324.5	45.6[a]	0.97	9.4
CH_4	90.7	111.7	190.7	35.3	1.24	8.8
C_2H_6	89.9	184.6	305.5	55[a]	3.83	9.6
C_3H_8	185.5	231.1	370.0	76[a]	2.29	9.8
CO	68.1	81.7	133	35.3	1.46	8.9
N_2O	182.3	184.7	309.6	35.8	4.31	10.8
NO	109.6	121.4	309.7	22.5	7.18	16.4
SO_2	200.5	263	430.7	44.7	4.44	11.4
BF_3	146.5	173.3	260.9	38.3	3.48	13.1
CF_4	84.5	145	225.9	45[a]	1.01	10.2

[a]At T_b rather than T_m.

H_2S has $\mu = 0.94$ Debye, the other fluids in Table 3.4 have lower or zero dipole moments. However, orientation forces play a minor role in the interactions, as seen in Table 1.3, though their importance increases with decreasing temperature. The main interaction forces are dispersion forces, section 1C(b), which act centrally, between the molecules as separate entities, because of their small dimensions. The pair potential can be described by an 'LJ 12-6' potential, eq. (1.42), with the parameters ϵ/k and σ, describing the depth of the potential-well and the collision diameter, obtainable from viscosity and second virial coefficient measurements on the gases. Since these liquids can be described by means of this two-parameter potential function, the liquids should indeed be conformal fluids, describable by means of the critical temperatures and pressures and two scaling factors. The depth of the potential-well, in particular, should be proportional to the critical temperature, and indeed for many of the fluids listed in Table 3.4, and for other similar ones beside, $\epsilon/kT_c = 1.36 \pm 0.16$. There are, however, many exceptions: bromine, hydrogen chloride, hydrogen bromide and nitrogen oxide are not conformal with the others. Furthermore, estimates of ϵ vary widely[18]; recent estimates for oxygen, nitrogen and carbon monoxide are only a third as large as previous values. In order to attain better agreement with the corresponding states principle, it was proposed to use a third parameter, which should describe the deviation from the centric forces typical for the monatomic noble gases. This is the *acentric factor*[19]

$$\omega = -\log(10p_\sigma(T = 0.7T_c)/P_c) \qquad (3.1)$$

based on the fact that at $T/T_c = 0.7$, $p_\sigma/P_c \simeq 1/10$ for the noble gases. The magnitude of ω is illustrated by the values 0.013 for the tetrahedrally symmetrical methane, 0.105 for ethane and 0.152 for propane, also 0.040 for nitrogen and 0.100 for hydrogen sulphide. Unfortunately, values of the acentric factor are not available for those fluids which deviate (see above) from conformality with the majority of the other fluids discussed here. The depth of the potential-well, ϵ/k, is related to the ionization potential and to the polarizability. The ionization potentials do not vary much for most of the fluids of Table 3.1 and similar ones, $I = 11.6 \pm 1.3$ V, with notable exceptions, having appreciably higher values (near 17), being nitrogen, boron trifluoride and carbon tetrafluoride. The polarizabilities, α, on the other hand, do vary considerably, from less than 2×10^{-30} m^3 for compact molecules such as oxygen, nitrogen and carbon monoxide, to more than 4×10^{-30} m^3 for polyatomic molecules such as ethane, propane or carbon tetrafluoride, or those containing very heavy atoms such as hydrogen iodide.

Not very much is learned about the structure of these liquids from diffraction studies. The liquified non-metallic elements oxygen, nitrogen, chlorine and bromine have been so studied, as well as methane[6]. The interatomic distance corresponding to the first maximum in the $g(r)$ or $\rho r^2 g(r)$ curve corresponds to the *intra*molecular distance, as well as to the nearest intermolecular distance, and for the diatomic molecules, the atomic coordination number Z_{atoms} is $1 + Z_{molecules}$. A near approximation to close packing is exhibited by some of the fluids: $Z_{molecules}$ is 10.0 for nitrogen, 10.3 for oxygen, 11.4 for methane and 12.0 for bromine. Only chlorine shows a more complicated structure in the liquid: an atom has six nearest-neighbours at 0.41 nm, and ten somewhat further neighbours at 0.52 nm. Chlorine, it must be remembered, although a homopolar diatomic molecule, is polar in the liquid state ($\mu = 1.07$ Debye), and this must be due to the influence of the, apparently unsymmetrically placed, six nearest-neighbours. The coordination numbers, though not the distances, for all the liquids studied depend on the number density of molecules (and on the temperature) to an extent similar to that shown by the noble gases (p. 91). The above-mentioned Z values are valid at densities ρ of 12 to 24 nm^{-3}, typical for these liquids between the melting and the normal boling points. The r_{max} values from diffraction studies, divided by $2^{1/6}$ (cf. eq. 1.42), agree very well with σ obtained from viscosity or virial coefficient studies for nitrogen, oxygen and methane, but not so for bromine and chlorine, where they are distinctly smaller (e.g. $r_{max}/2^{1/6} = 0.374$ nm and $\sigma = 0.427$ nm). The distance r_{max} is apparently affected overmuch by the shorter intramolecular distance.

b. Liquids with non-spherical, non-polar molecules

The class of liquids to be dealt with now is of more interest to chemists than the foregoing ones, since it comprises liquids which are commonly used as solvents. Among the solvents, those liquids which consists of non-polar molecules have often only slight interactions with the solutes, hence they are called *inert solvents*, and are useful in that they permit the solute to show most of its own properties, unencumbered by interactions with the solvent. The price that has to be paid for

Table 3.5. Some properties of

Property	$C(CH_3)_4$	cyclo-C_6H_{12}	n-C_6H_{14}	n-C_8H_{18}	n-$C_{10}H_{22}$	n-$C_{12}H_{26}$
Bulk properties[a]						
T_m(K)	256.6	279.7	177.8	216.4	243.5	263.6
T_b (K)	282.7	353.9	341.9	398.8	447.3	489.4
T_c (K)	433.8	553.	507.3	568.7	617.6	658.3
v^l (cm^3 mol^{-1} at 298 K)	123.3	108.7	131.6	163.5	195.9	228.6
α_P (10^{-3} K^{-1} at 298 K)		1.217	1.361	1.150	1.034	0.965
κ_T (10^{-10} Pa^{-1} at 298 K)		11.4	15.9	12.1		
$\Delta s^F/\mathbf{R}$	1.48	1.15	8.85	11.53	14.19	16.81
$\Delta v^F/v^S$		0.0516	0.1177	0.0612	0.0737	0.103
$c_p/\mathbf{R}$ (at 298 K)		19.1	24.0	31.2	38.7	46.2
$\Delta s^V/\mathbf{R}$	9.68	10.18	10.15	10.38	10.56	10.72
δ ($J^{1/2}$ $cm^{-3/2}$ at 298 K)	12.8	16.8	15.0	15.4	16.1	15.9
σ (10^{-2} N m^{-1} at 298 K)	1.152	2.438	1.791	2.114	2.337	2.491
n_D (at 298 K)	1.342[b]	1.4235	1.3723	1.3951	1.4097	1.4195
ϵ (at 293 K)	1.801	2.023	1.880	1.948	1.991	2.002[c]
Molecular properties[e]						
ϵk^{-1} (K) (LJ 12-6)	233	313	423	333		
σ (nm) (LJ 12-6)	0.745	0.614	0.592	0.741		
ω	0.195	0.209	0.301	0.398		
ϵk^{-1} (K) (Kihara)	554					
σ (nm) (Kihara)	0.579	0.570	0.604			
δ/σ (Kihara)	0.35	0.44	0.47			
α (10^{-30} m^3)		10.87	11.78	15.44	19.10	22.75
I (V)		9.50	10.43	10.24	10.19	

[a]Data mainly from J. A. Riddick and W. B. Bunger, *Organic Solvents*, Wiley—Interscience, New

this is the often low solubility exhibited by the solutes, especially the more polar ones.

These solvents are all liquid at room temperature, and have a fairly wide normal liquid range, say 100 to 200 K, and a total liquid range $(T_c - T_m)/T_c > 0.5$. Some examples of such liquids are shown in Table 3.5, where it is apparent that neopentane (tetramethylmethane, or 2,2-dimethylopropane) is an obvious exception, having only a narrow liquid range. The liquids are mainly hydrocarbons or halogenated hydrocarbons, and it is difficult to find other examples mainly because the inorganic liquids which physically fit the classification (e.g. germanium tetrachloride or borazole $B_3N_3H_6$) are chemically reactive, usually hydrolyse in moist air and have not been sufficiently well characterized as liquids in contrast to their well-known properties as chemicals. The liquids listed, and others of this class, have molar volumes exceeding *ca.* 100 cm^3 mol^{-1} (carbon disulphide is a notable exception, benzene and carbon tetrachloride are only slightly low). The liquids listed are conformal thermodynamically, in that $P_c v_c/\mathbf{R}T_c$ is constant at 0.264 to better than 5%, and they obey Trouton's rule $\Delta s^V/\mathbf{R} = 10.4$ to within 4%.

The fusion properties depend upon the packing in the solid and do not reflect important liquid-state characteristics. The more spherical molecules (globular molecules such as neopentane or carbon tetrachloride) can rotate to some extent in typical non-polar ('inert') liquids.

C_6H_6	$C_6H_5CH_3$	1,4-C_6H_4-$(CH_3)_2$	1,3,5-$C_6H_3(CH_3)_3$	C_6F_6	CCl_4	CS_2
278.7	178.2	286.4	228.4	278.3	250.2	161.6
353.3	383.8	411.5	437.7	353.4	349.9	319.4
562.1	591.7	616.2	637.3	516.7	556.3	552.
89.9	106.9	123.9	139.6	115.8	97.1	60.3
1.229	1.060	1.020	0.95	1.412	1.220	0.618
9.7	9.2		6.8		11.1	
4.26	4.48	7.19	5.01	5.01	1.17	3.27
0.1332	0.113	0.221	0.034		0.0530	0.033
16.7	18.6	21.7[c]	26.2	26.7	16.4	9.3
10.47	10.40	10.52	10.73	10.78	10.30	10.07
18.8	18.3	18.1	18.1	17.4	17.6	20.4
2.818	2.792	2.776	2.831	2.26	2.615	3.152
1.4979	1.4941	1.4933	1.4968	1.3781	1.4574	1.6241
2.275[d]	2.379[d]	2.270	2.279		2.238	2.641
335					327	488
0.563					0.588	0.444
0.215	0.241				0.255	0.115
832						
0.503					0.550	0.460
0.43					0.43	0.28
10.32	12.26	14.92			10.5	8.74
9.24	8.92				11.46	10.03

York, 3rd edn., 1970; [b]at 293 K; [c]at 303 K; [d]at 298 K; [e]data mainly from refs. 4, 5 and 18.

the solid just below the melting point, and hence have small entropies of fusion, Δs^F, which nearly equal the communal entropy of the liquid. Low entropies of fusion are also connected with low volume changes on fusion, Δv^F. Flexible chain molecules, such as the aliphatic hydrocarbons, or elongated molecules, like *p*-xylene, are constrained from rotation and configurational changes in the solid, so their entropy and volume change on fusion are considerable. A quantitative interpretation of the observed behavior move: for example, the low T_m of toluene (by *ca* 100 K) compared with benzene and *p*-xylene, which have similar Δs^F and Δv^F values, and also the high Δv^F value of n-hexane, in view of its very low T_m, is difficult, however.

In the liquid state, these solvents behave in a similar way, as already mentioned. They absorb energy according to their flexibility and possible vibrational modes, and the observed c_p values are as expected. The surface energies are low, reflecting the low cohesive energy densities (in fact δ^2/σ is nearly constant at 1.2×10^{10} m^{-1}, within ± 7%). These latter (δ^2) are lowest for the aliphatic hydrocarbons*, and increase within their series as the ratio of surface to volume increases,

*Fluorinated hydrocarbons have even lower values of δ^2, e.g. 146 J cm^{-3} for perfluoro-n-hexane, C_6F_{14}, compared with 223 J cm^{-3} for n-hexane, C_6H_{14}. For values of δ see Table 5.2.

signifying that the interactions are between neighbouring segments of the molecules rather than between entire molecules. This idea was developed by Huggins[20], who described the vaporization energy (i.e., $\Delta u^{V} = v\delta^2$) in terms of the product of an interaction energy per unit contact area of a given type, ϵ_σ, the average intermolecular contacting area per mole of segments of this type, σ^0, and half the number of such segments per molecule $n_\sigma/2$; $\Delta u^{V} = -\epsilon_\sigma \sigma^0 n_\sigma/2$. For carbon tetrachloride, for instance, the 'segment' is a bonded chlorine atom and $n_\sigma = 4$, for benzene or cyclohexane, the segments are $-CH=$ and $-CH_2-$ groups, respectively, and $n_\sigma = 6$. It was impossible[20] to give individual values of ϵ_σ and σ^0, but the ratios of the value for carbon tetrachloride, tin tetrachloride and cyclohexane to that for benzene could be evaluated, being 1.080, 1.112 and 1.393 respectively. If there are two types of segments in a molecule (methyl and methylene groups in an aliphatic hydrocarbon, methyl and phenyl groups in toluene, etc.) the relationships become more complicated. This concept is valuable for dealing with mixtures, see section 5A(e). The essential point in this approach is that the interaction is only between segments in contact, hence the energy is a step function and not related to the molecular separation r in a continuous way.

The ordinary approaches relate Δu^{V} to $u(r)$ and $g(r)$, and the former is made up of a repulsive and an attractive contribution. These can be obtained from the properties of the liquids and the attractive contribution will be discussed first. Since the dielectric constant is almost exactly equal to the square of the refractive index n_D^2, no contribution from orientation forces or dipole–induced dipole forces must be added to that from dispersion forces for the attractive term of the intermolecular potential. The only exception is toluene, which in fact is the only liquid listed in Table 3.5 which has a permanent dipole moment, $\mu = 0.31$ Debye $= 1.03 \times 10^{-30}$ C · m. The dispersion forces can be evaluated according to eq. (1.49), noting that the ionization potentials I do not vary much from one liquid to another, but that the polarizabilities α do, reflecting the molecular size and complexity. 'Lennard-Jones 12-6' potentials, eq. (1.42), have been fitted to the data with the values of ϵk^{-1} and σ shown in Table 3.5, but it is found that several sets of the two parameters can fit the second virial coefficient and gas viscosity data, so that the values given are not unique.

In fact, for these relatively large and non-spherical molecules two-parameter interaction functions are not adequate. The Kihara potential with its hard core (or mean diameter δ), eq. (1.49), is a useful[4,5,21] three-parameter equation, see Table 3.5. Another approach uses the critical data and the experimental acentric factor ω (eq. (3.1), values listed in Table 3.5) in a three-parameter Morse-type equation[22]

$$u(r) = 4\epsilon(y^2 - y)$$

$$y = \exp[c(1 - r/\sigma)]$$

$$c = 5 + 11\omega \tag{3.2}$$

with the parameters ϵ, σ and ω. The depth of the potential-well is related to the critical temperature and ω by

$$\epsilon k^{-1} = 1.25T_c[1 - 2\omega/(1 - 2\omega)] \quad (3.3)$$

and the collision diameter σ is related to P_c in a more complicated way.

The molecular pair correlation function $g(r)$ cannot, unfortunately, be derived for these liquids from diffraction studies. Most of these merely confirm that the interatomic distances within a molecule of the liquid are the same as in the gas, while the number of nearest-neighbours of a given atom is not obtained with great precision. Narten *et al.*[23] have attempted to obtain the intermolecular structure characteristics of carbon tetrachloride and of benzene. For benzene they found an average arrangement in the liquid similar to that in the solid (with 12 nearest-neighbours), while for carbon tetrachloride they propose a model where each molecule has 13 nearest-neighbours (6 at 0.577 nm between the carbon centres, one at 0.613 nm and 6 at 0.677 nm), while each chlorine atom has 12 nearest-neighbours (at 0.380 ± 0.002 nm between the chlorine centres). An important feature of this model is that it excludes free rotation of the molecules in the liquid. However, these structures are much too solid-like, and do not take the greater volume and entropy available to the molecules into account. Orientational disorder[24] is produced in these substances often at a certain temperature below the melting point, but this does not mean that the molecules are free to rotate since the available volume may limit rotation to one of the axes only. With increasing temperatures and volumes these restrictions are relaxed, and a detailed structural investigation of these liquids over a series of temperatures (and pressures) should be of great value.

c. Dipolar aprotic liquids

Polar liquids which contain at least one proton bound to an electronegative atom are called *protic solvents* (liquids), and may enter into hydrogen-bond formation, and also dissociate ionically as Brønsted acids. Polar liquids which do not contain such protons are called *aprotic dipolar solvents*, and do not show these complications. Some liquids, such as dimethylformamide or chloroform may be borderline cases, since the carbon atom to which the lone hydrogen atom is bound is sufficiently positive to be able to share this hydrogen atom with a strongly electronegative atom which has a free pair of electrons. Pure liquids, however, suffer little from this complication. Thus the only consideration, additional to those discussed for the liquids dealt with in the previous sections, is the presence of strong orientation forces which operate between the molecules of the present class of liquids, as a result of their large dipole moments, typically μ (Debye) = μ (3.33564×10^{-30} C · m) > 1. If the dipole is embedded deep within the molecule, the liquid shows only weak signs of its presence, while in contrast, an exposed dipole causes the molecules to associate. This, in turn, affects the ability of the solvent to interact with a solute in mixtures, hence its solvent properties.

The class of dipolar aprotic liquids comprises a large number of members, and a selection of common organic solvents which belong to this class is listed in Tables

Table 3.6. Some bulk properties of selected dipolar aprotic liquids[a].

Liquid	T_m (K)	T_b (K)	v [b] (cm^3 mol^{-1})	$\Delta_S F/R$	$10^2 \sigma$ [b] ($N\ m^{-1}$)	ϵ[f]	g[b]
1-Chlorobutane	150.1	351.6	105.1		2.34	7.39	0.97
Chlorobenzene	227.6	404.8	102.2	5.05	3.27	5.62[b]	0.74
1,1-Dichloroethane	176.2	330.4	84.7	5.35	2.42	10.0	1.02
1,2-Dichloroethane	273.5	356.6	79.4	3.89	3.15	10.36	1.08
Chloroform	209.6	334.3	80.7	5.47	2.65	4.81	0.99
1,1,2,2-Tetrachloroethane	229.4	419.4	105.8		3.54	8.20	1.16
Diethyl ether	156.9	307.7	104.7	5.59	1.65	4.34	1.36
2,2′-Bis-chloroethyl ether	226.4	451.9	117.9	4.60	3.70	21.2	1.80
Tetrahydrofurane	164.7	339.2	81.6		2.64	7.58	0.91
1,4-Dioxane	285.0	374.5	85.7	5.26	3.30	2.21[b]	–
Acetone	178.5	329.4	74.0	3.83	2.27	20.70	1.19
Cyclohexanone	241.1	428.8	104.2		3.45	18.3	1.01
Ethyl acetate	189.2	350.3	98.5	6.67	2.31	6.02[b]	0.82
γ-Butyrolactone	229.6	477.	96.2			39.	1.13
Ethylene carbonate	309.6	511.	66.7[c]	3.90		89.6[c]	1.43
Propylene carbonate	224.2	514.9	85.2			66.1	1.19
Triethylamine	158.5	362.7	140.0		2.01	2.42[b]	0.41
Pyridine	231.6	388.4	80.9	3.85	3.63	12.4	0.74
Acetonitrile	229.3	354.8	52.9	4.29		37.5	0.99
Succinonitrile	331.0	540.	81.2[d]	1.34	4.68[d]	56.5[d]	2.02
N,N′-Dimethylformamide	212.7	426.2	77.4	9.13	3.52	36.71[b]	1.00
N,N′-Dimethylacetamide	253.	439.3	93.0	4.95	3.32	37.78	1.30
Tetramethylurea	272.0	448.4	120.3			23.06	1.13
N-Methylpyrrolidone	248.8	475.	96.4			32.0[b]	0.90
Nitromethane	244.6	374.4	54.0	4.77	3.65	35.87	0.85
Nitrobenzene	278.9	484.0	102.7	5.02	4.28	34.82	0.96
Dimethylsulphoxide	254.6	462.2	71.3	6.58	4.29	46.68	1.06
Sulpholane	301.6	560.5	95.3[e]	0.57	3.55[e]	43.3[e]	0.88
Tri-n-butyl phosphate	<193.	562.	273.8		2.72	7.96[e]	1.02
Hexamethyl phosphora-mide	280.4	506.	174.5	7.27[f]	3.38[f]	30.54	0.83

[a]Data from J. A. Riddick and W. B. Bunger, *Organic Solvents*, Wiley–Interscience, New York, 3rd edn., 1970; [b]at 298 K; [c]at 313 K; [d]at 333 K; [e]at 303 K; [f]at 293 K.

3.6 and 3.7, along with some of their bulk- and molecular-properties, respectively. Chemically, these liquids are halogenated hydrocarbons, organic oxygen compounds (ethers, ketones, esters, lactones), organic nitrogen compounds (tertiary amines, nitriles, *N,N′*-dialkylamides), nitro- and other aprotic mixed oxygen–nitrogen compounds, thio-analogues of the oxygen compounds, sulphoxides, sulphones, sulphite, sulphate, phosphate esters, etc. Some typical inorganic liquids such as thionyl chloride and phosphorus oxychloride also belong to this class. All these liquids have a reasonably wide liquid range, $(T_c - T_m)/T_c > 0.6$ for most of them, although for many the critical temperature is unknown and may be too high for thermal stability of the molecules. Two exceptions among the liquids in Tables 3.6 and 3.7 with $(T_c - T_m)/T_c = 0.514$, only, are 1,2 dichloroethane and 1,4-dioxane, the liquid range of which is narrow ($\leqslant 90$ K); this is also so by the criterion $T_b - T_m$, which for the other liquids in the Tables is > 120 K. Most, again, are liquids at room temperature, notable exceptions being ethylene carbonate (m.p. 36.4 °C), succinonitrile (m.p. 57.8 °C) and sulpholane (m.p. 28.4 °C) among the liquids listed.

In spite of their relatively large dipole moments (Table 3.7), many of the liquids are not associated sufficiently at their boiling points for this to affect the Trouton constant. The ratio $\Delta s^V/R$ is within the limits 10.2 ± 1.2 (mainly in the upper half of the range) for the liquids in the Tables, except for ethylene carbonate (11.80), and possibly also propylene carbonate, for which no Δh^V could be found, dimethylacetamide (11.86), bis-2-chloroethyl ether (12.04), tetramethylurea (12.23), tributyl phosphate (13.14), butyrolactone (13.17), hexamethyl phosphoramide (13.45) and sulpholane (13.5). Naturally, many other exceptions can be found among liquids not listed in the Tables. As the temperature is lowered, association may be expected to increase, and should be appreciable also for the liquids which do obey Trouton's rule at the boiling point. This is shown by the angular correlation parameter *g* in Table 3.6, which is a measure of the deviation of the dielectric constant from the value it would have in a non-associated liquid of

Table 3.7. Some molecular properties of selected dipolar aprotic liquids.

Liquid	μ (D)	α (10^{-30} m^3)	*DN*	*AN*	ϵk^{-1} (K)	σ (nm)
1-Chlorobutane	1.90					
Chlorobenzene	1.54	12.25		5.7[a]		
1,1-Dichloroethane	1.98			7.9[a]		
1,2-Dichloroethane	1.86		0.1	10.8[a]		
Chloroform	1.15	8.23		23.1	327	0.543
1,1,2,2-Tetrachloroethane	1.71					
Diethyl ether	1.15	8.73	19.2	3.9	351	0.544
2,2′-Bis-chloroethyl ether	2.58					
Tetrahydrofurane	1.75		20.0	8.0		
1,4-Dioxane	0.45	9.44		10.8		
Acetone	2.69	6.33	17.0	12.5	519	0.467
Cyclohexanone	3.01			9.5[a]		
Ethyl acetate	1.88	8.82	17.1	6.9[a]	531	0.517
γ-Butyrolactone	4.12			18.6[a]		
Ethylene carbonate	4.87		16.4			
Propylene carbonate	4.98	8.56	15.1	18.3		
Triethylamine	0.87			1.2[a]		
Pyridine	2.37	9.5	33.1	14.2		
Acetonitrile	3.44	4.45	14.1	18.9		
Succinonitrile	3.68					
N,N′-Dimethylformamide	3.86	7.91	26.6	16.0		
N,N′-Dimethylacetamide	3.72		27.8	13.6		
Tetramethylurea	3.47		29.6	9.8[a]		
N-Methylpyrrolidone	4.09		27.3	11.2[a]		
Nitromethane	3.56		2.7	20.5		
Nitrobenzene	4.03	12.92	4.4	14.8		
Dimethylsulphoxide	3.96	7.97	29.8	19.3		
Sulpholane	4.81		14.8	13.3[a]		
Tri-n-butyl phosphate	3.07		23.7			
Hexamethyl phosphoramide	5.54	18.8	38.8	10.6		

[a]From relationship between *AN* and transition energy of betaines.[27]

the same dipole moment and refractive index (where $g \equiv 1$). The values were calculated from the modified[25] Kirkwood equation

$$g = (9k\epsilon_0/4\pi N)\, vT\mu^{-2}(\epsilon - 1.1n_D^2)(2\epsilon + 1.1n_D^2)/\epsilon(1.1n_D^2 + 2)^2 \quad (3.4)$$

where $1.1n_D^2$ takes the place of the dielectric constant at infinite frequency (cf. eq. 1.35). There is, however, no correlation between the deviation of $\Delta s^V/\mathbf{R}$ from 10.2 ± 1.2 and the deviation of g from unity. Parallel alignment, i.e. head-to-tail, of the associating dipoles leads to $g > 1$, while antiparallel alignment leads to $g < 1$, and it is difficult to interpret any given g value. The association does, however, interfere with the orientation of the molecules in an electric field, so that the net dielectric constant is not proportional to the dipole moment. The sensitivity of the dielectric constant to the structure and interactions of these liquids extends also to its temperature- and pressure-derivative, and a detailed study of these functions could yield very valuable information, but insufficient data are, as yet, available.

The behaviour of these liquids on freezing is rather complicated: for those liquids where Δs^F is known it is usually $\geqslant 4\mathbf{R}$, corresponding to the release of the communal entropy and several degrees of freedom of rotation and vibration. Since free rotation of the molecules is hindered to some extent by orientation of the dipoles, the individuality of the various liquids becomes evident. Densities of the corresponding solids at the melting point are generally not available, so that Δv^F cannot be computed and discussed. The molar volumes of the liquids vary widely, but many representative liquids have rather small volumes, say 80 to 100 cm^3 mol^{-1}. In view of the high boiling points, $T_b > 350$ K for most of the liquids listed, and since $\Delta s^V \geqslant 10.2\mathbf{R}$, the cohesive energy densities $\Delta u^V/v$ are seen to be quite large. These quantities can be described as made up from contributions of the dispersion forces, λ^2, and from contributions of the orientation forces, τ^2. These quantities play an important role in the description of the behaviour of mixtures and are discussed in section 5A(f), with a list of values displayed in Table 5.4. The ratio of the cohesive energy density to the surface tension, $(\lambda^2 + \tau^2)/\sigma$, which is nearly constant at 1.2×10^{10} m^{-1} for non-polar liquids (where $\tau = 0$ and $\lambda = \delta$), is usually larger for the aprotic polar liquids. This is probably due to the less efficient attraction of the oriented dipoles at the surface than in the bulk of the liquid. The surface tension is therefore not much larger for the aprotic dipolar liquids (Table 3.6) than for non-polar liquids (Table 3.5).

Apart from the dipole moment, the molecular properties of the aprotic polar liquids are not known extensively, since these properties are determined on the gases rather than on the liquids. The collision diameter σ, for those cases where it is known (Table 3.7), is fairly large but the energy scaling factor ϵk^{-1} is not. The values given, which are consistent with second virial and visocosity data for the vapours, are not unique, however, and furthermore, several potential functions, such as the Stockmayer–Keesom, eq. (1.56), or the Kihara, eq. (1.47), with or without a dipole–dipole interaction term, can yield parameters consistent with the data[26]. Structural data from diffraction are not available for the polar aprotic liquids in general, so that radial distribution functions, pair correlation functions

$g(r)$, and theoretically derived potential functions $u(r)$ cannot be obtained, to compare with the parameters obtained from the gas-phase data.

There are two other properties listed in Table 3.7, the donor and acceptor numbers[27], *DN* and *AN*. The first is a measure of the donor properties of the molecules and is $(4.184 \times 10^3)N$ times the bond energy (J molecule^{-1}) between the donor molecule and a standard acceptor molecule, antimony pentachloride, in an ideally inert solvent. In practice, the number is set equal to $-\Delta h^0$ (kcal mol^{-1}) (1 cal = 4.184 J) for the reaction of adduct formation between dilute solutions of the liquid and antimony pentachloride in a solvent such as 1,2 dichloroethane. This measure does include contributions from dipole interactions but is mainly determined by the binding effect arising from the availability of the free electron pair in the aprotic dipolar molecules. A similar concept which has been applied qualitatively to donor liquids is that of hardness and softness. These are related to the relative abilities to be polarized (softness) and to undergo electrostatic interactions (hardness), as well as to engage in covalent bonding. Some combination of the quantities α, μ and *DN* should therefore be a measure of the softness of the solvent molecules, but contrary to the case of ions (section 7B(b)), no quantitative softness parameters have been proposed for these uncharged molecules.

The acceptor number *AN* of a liquid is a measure of its readiness to accept electrons from donors, and describes its capability to solvate anions. A quantitative measure has been devised[27] by measuring the chemical shift of the NMR signal of ^{31}P in dilute solutions of triethyl phosphate, $(C_2H_5O)_3P{=}O$, in the examined solvent, relative to hexane, corrected for bulk susceptibility, and normalized to that in 1,2-dichloroethane containing the standard acceptor antimony pentachloride, which is assigned the value $AN = 100$. Extrapolation of the measured chemical shifts to infinite dilution of the triethyl phosphate in the liquid ensures absence of solute–solute interactions, so that the chemical shift becomes dependent only on the electron-accepting properties of the liquid, from the strong donor phosphoryl oxygen of the triethyl phosphate (estimated $DN \sim 40$). The acceptor number is a molecular property in so far as the NMR signal is sensitive to the individual electron-accepting properties of the molecule interacting with the donor, but this is affected by the field exerted by the neighbouring (acceptor) solvent molecules, and therefore describes the properties of the acceptor in bulk. Some acceptors, notably acids not listed in Table 3.7, have very large *AN* values (e.g. 126.3 for methylsulphonic acid), reflecting their readiness to dissociate a proton.

d. Hydrogen-bonded liquids

Along with the aprotic dipolar solvents, the protic ones, which are ordinarily intermolecularly hydrogen-bonded, are useful solvents. This class of liquids has, therefore, been extensively studied. It includes water, which is discussed separately in the next section, and various oxygen- and nitrogen-containing molecules, with hydrogen atoms bonded to these atoms forming the hydrogen bonds. Aliphatic and aromatic alcohols and amines, carboxylic acids and amides are typical representatives of this class of liquids, including also those which have several of these

Table 3.8. Some properties of selected hydrogen-bonded liquids[a].

Liquid	T_m (K)	T_b (K)	v (cm^3 mol^{-1})[b]	$\Delta v^F/v^s$	$\Delta s^F/\mathbf{R}$	$10^2\sigma$ (N m^{-1})[b]	ϵ[b]	μ (D)	g[b]	AN
Methanol	175.5	337.9	40.7	0.085	2.2	2.21	32.70	2.87	0.99	41.3
Ethanol	159.1	351.4	58.7	0.111	3.8	2.19	24.55	1.66	3.01	37.1
n-Propanol	147.0	370.4	75.1		4.4	2.23	20.35	3.09	0.88	31.0[f]
Isopropanol	185.2	355.4	76.9		3.5	2.12	19.90	1.66	3.08	33.5
n-Butanol	184.5	390.8	92.0	0.125	6.1	2.32	17.50	1.75	2.79	30.0[f]
n-Hexanol	228.6	430.2	125.2	0.128	8.1	2.41	13.3	1.55	3.47	
Benzyl alcohol	257.9	478.6	103.9		8.0	3.95	13.1[e]	1.66	2.01[e]	31.2[f]
Phenol	314.1	455.0	89.2[c]	0.109	4.4	3.82[c]	9.78[c]	1.45	1.72[c]	
m-Cresol	285.4	475.4	105.0	0.098	4.5	3.70	11.8	1.54	2.08	
Ethylene glycol	260	470.5	55.9		5.4	4.60	37.7	2.28	2.13	42.8[f]
Glycerol	291.3	563.2	73.2		7.5	6.33	42.5			44.3[f]
Ethoxyethanol	<183	408.8	97.4			2.82	29.6	2.08	3.60	
Acetic acid	289.8	391.1	57.5	0.199	4.9	2.69	6.15[e]	1.68	0.57[e]	52.9
Butyric acid	268.0	436.4	92.4	0.121	4.7	2.62	2.97[e]	1.65	0.22[e]	
n-Butylamine	224.1	350.6	99.6			2.35	4.88[e]	1.37	1.00[e]	
Diethylamine	223.4	328.6	104.2			1.94	3.58[e]	1.11	0.93[e]	
Ethylenediamine	284.5	390.4	67.5		8.2	4.01	12.9	1.90	1.13	12.6[f]
Aniline	267.2	457.6	91.5	0.108		4.28	6.89[e]	1.51	0.86[e]	17.5[f]
m-Toluidine	242.8	476.6	108.5		1.9	3.76	5.95[e]	1.45	1.29[e]	
2-Ethanolamine	283.7	444.1	60.4		8.7	4.83	37.72	2.27	4.13	33.3[f]
Formamide	275.7	483.7	39.9		2.9	5.79	111.0[e]	3.37	2.01[e]	39.8
N-Methylformamide	269.4	455	59.1			3.87	182.4	3.86	3.90	32.1
N-Methylacetamide	303.7	479	77.0[d]		3.3	3.37[d]	191.3[d]	4.39	4.23[d]	33.7[f]

[a]J. A. Riddick and W. B. Bunger, *Organic Solvents*, Wiley–Interscience, New York, 3rd edn., 1970; [b]at 298 K; [c]at 318 K; [d]at 303 K; [e]at 293 K; [f]From relationship with transition energy of betaines, Ch. Reichardt and K. Dimroth, *Fortschr. Chem. Forsch.*, **11**, 1 (1968), and reference 27.

functional groups. Selected liquids of this class are listed in Table 3.8, along with some of their properties.

The liquids of this class are naturally (that is to say by definition) strongly associated. The entropy of vaporization at the boiling point is $\Delta s^V/R = 13.3 \pm 0.8$, that is *ca.* 30% higher than for 'normal', i.e. non-associated, liquids. Notable exceptions are the carboxylic acids, which are associated also in the vapour phase (acetic acid has therefore $\Delta s^V/R = 7.5$ only), and the amines, which have normal Trouton's constants, since they are only weakly associated. A comparison of the amines with the corresponding hydroxy-compounds is instructive: compare the butyl-, phenyl-, *m*-tolyl- and ethylene-compounds in Table 3.8. The freezing properties (T_m, $\Delta s^F/R$) do not show regularities since they depend more on the packing of the solids than on the properties of the liquids. The critical temperatures (T_c is 563 and 561 K for the butyl-, 694 and 699 K for the phenyl-, and 706 and 709 K for the *m*-tolyl-compounds) are almost the same, since in the critical state there is no longer any extensive hydrogen bonding, but Δs^V, T_b, ϵ and g, which are typical properties of the liquids, are lower for the amine-compounds than for the corresponding hydroxy-compounds. Contrary to expectation, the surface tension does not show this regularity, being similar for the two groups of liquids, with some amines having higher, some lower values of σ than the hydroxy-compounds. These results can be explained structurally by noting that the oxygen atoms in the hydroxy-compounds have two free pairs of electrons and one hydrogen atom available for hydrogen bonding, and the amines only one pair of electrons and two hydrogen atoms. The former arrangement is more suitable for chain-like association (Fig. 3.2(a)), the two hydrogen atoms in the latter probably interfering with each other. To this should be added the fact that the dipole moment of the aliphatic amines is only *ca.* 80% as strong as in the corresponding alcohols (it is similar in the aromatic examples).

The alcohols show regularly changing liquid properties as the aliphatic chain length is increased or as the number of hydroxy-groups is increased. The freezing properties and the dipole moments (also g), however, do not show such a regularity, while the entropy change on freezing does show a regular increase with increasing size and flexibility of the molecules. The fractional volume change on freezing, $\Delta v^F/v^s$, however, shows alternative low (odd number of carbon atoms) and high (even number) values: 8 to 9% for the former, 11 to 13% for the latter. These clearly reflect packing efficiencies in the solid, rather than liquid-phase properties. The dielectric constant drops as the chainlength increases in the alcohols, but increases with an increasing number of hydroxy-groups. Since however the dipole moment varies irregularly, so does the angular correlation parameter g. The carboxylic acids have similar dipole moments to the alcohols, but much lower dielectric constants and g parameters. This is due to the effective pair-wise association by two hydrogen bonds in a cyclic manner, so that the molecular dipoles of the partners cancel each other (Fig. 3.2(b)). If the vapour were not associated, these acids would have normal Trouton's constants, as is the case, indeed, with propionic acid, butyric acid (both isomers) and valeric acid. Formic and acetic acids have $\Delta s^V/R = 7.5$, signifying associated vapours. Carboxylic acids

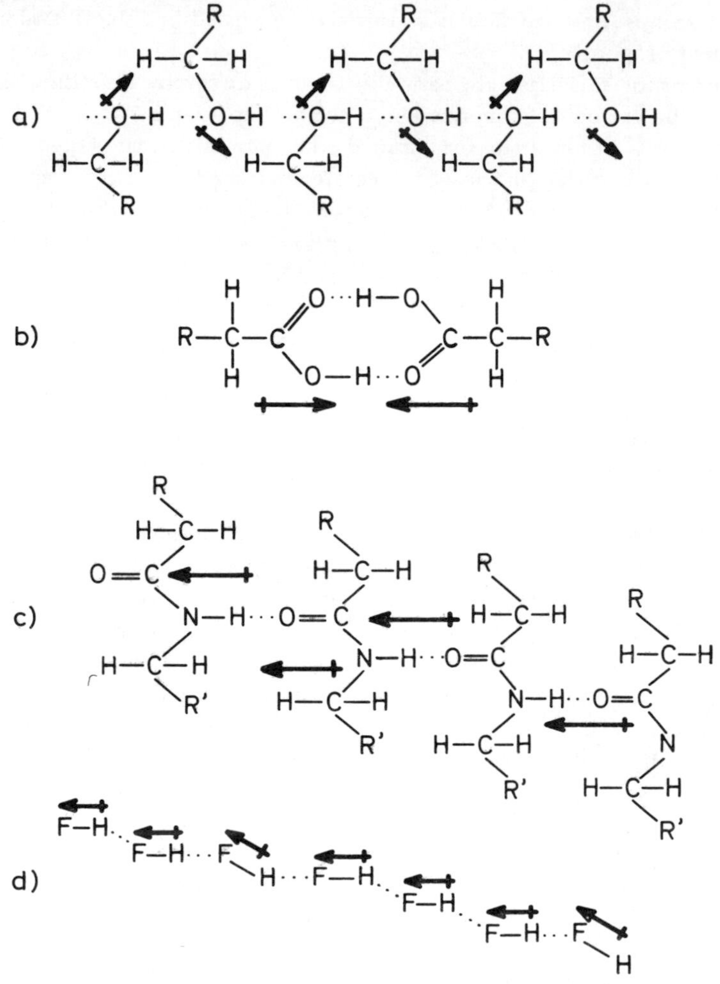

Figure 3.2. Schematic representations of the hydrogen bonding in some protic liquids. The arrows represent the directions of the electrical dipole vectors of the individual molecules. (a) alcohols, R = H or alkyl; (b) carboxylic acids, R = H or alkyl; (c) amides, R and R′ = H or alkyl; (d) hydrogen fluoride.

with six or more carbon atoms have high $\Delta s^{V}/\mathbf{R}$, 13.5, i.e. similar to the other hydrogen-bonded liquids, so that at the boiling points the pair-wise association has probably yielded to less specific but more extensive association. The contraction of the first two carboxylic acids on freezing is very large: $\Delta v^{F}/v^{S}$ is 18.96% for formic acid and 19.93% for acetic acid, while from propionic acid on it is smaller but still considerable, *ca.* 12%, and again shows fluctuations with the odd and even number of carbon atoms, but not as regularly as for the alcohols. The melting points, however, show the odd–even fluctuation very regularly.

The amides are a particularly interesting group of associated liquids since in the monoalkyl-substituted amides the cooperative effect of parallelly aligned dipoles has attained a very impressive magnitude[28]. The dielectric constants are extremely large (*N*-methylpropionamide at 233 K has the largest value found for any liquid, 348), and so are the correlation factors *g*, signifying a chain-like association with a *trans*- configuration of the · · · O=C—N—H · · · amide-group (Fig. 3.2(c)). The mean chain length obtained from an analysis of the *g* values is *ca.* six, and per hydrogen bond formed *ca.* 15 kJ mol^{-1} are released ($-\Delta h^0$) with a change of entropy (Δs^0) of *ca.* 25 J K^{-1} mol^{-1}. Chain-like association is characteristic also for the association of the aliphatic alcohols, and these too have appreciable *g* values (the lower values for methanol and n-propanol are connected with their higher dipole moments rather than with a different mode of association). When more extensive networks of hydrogen bonds are favoured, as with unsubstituted amides (e.g. formamide in Table 3.8) or with water (see next section), the dielectric constant is still high, but *g* becomes somewhat lower.

The association of these liquids by hydrogen bonding leads to conflicting properties as solvents. On the one hand, the cooperatively enhanced orientation of the dipoles leads to high dielectric constants, and hence to electrostatic effects favourable to electrolytic solution and dissociation. The hydrogen-bonding capacity also leads to effective solvation of small anions, in preference to large anions. On the other hand, while aprotic dipolar solvents can be characterized quantitatively by their donor numbers[27] *DN*, the lone pair of electrons in the protic, i.e. hydrogen-bonded, solvents is not as easily available, so that the solvation of electron acceptors (cations) is not readily correlated. Donor numbers of 33.8 for methanol, 30.4 for ethanol, 29.0 for butanol, 31.0 for n-propylamine, 39.1 for formamide and 42.3 for water have been assigned by comparison with dipolar aprotic solvents (on this basis, benzene is assigned *DN* = 4.9), for their effect on the shift of the spectral bands of oxovanadium(IV)bis(acetylacetonate). The donor numbers determined by direct interaction with the standard acceptor antimony pentachloride are lower however (15.6 for ethanol, 18.0 for water), but these were determined under conditions where the donor molecules were mutually associated rather than at infinite dilution, so that the spectroscopic values are preferred when the donating ability of the individual molecules is discussed. The acceptor properties of the liquids, expressed by *AN* as given in Table 3.8, are quite marked, since the self-hydrogen-bonding ability corresponds to an ability to hydrogen-bond with other molecules, and thus to accept electron pairs from donors. Not listed in the Table are several strong acids, such as trifluoroacetic acid (*AN* = 105.3) or methylsulphonic acid (*AN* = 126.3), which have decidedly better electron-accepting abilities than the alcohols or the amides or even of water (Table 3.9).

e. *Water-like liquids*

Of all the liquids, water has naturally received the most attention. In it the cooperative effect of hydrogen bonding is manifested to the greatest extent, owing to the presence of two hydrogen atoms and two free electron pairs on each

Table 3.9. Some properties of hydrogen fluoride, water and ammonia.

Property	HF	H_2O	NH_3
Bulk properties			
T_m (K)	189.78	273.15	195.46
$\Delta s^F/R$	2.86	2.65	3.48
$\Delta c_p^F/R$		4.48	2.70
$\Delta v^F/v^s$	0.360	−0.0825	0.113
T_b (K)	292.66	373.15	239.78
$\Delta s^V/R$	3.54	13.10	11.71
$\Delta c_p^V/R$	−2.62	−4.99	−5.28
T_c (K)	461.	647.35	405.6
P_c (10^6 Pa)	6.49	22.12	11.38
v_c (10^{-6} m^3 mol^{-1})	69.	55.2	72.1
p_σ (10^3 Pa) (273 K)	48.50	0.611	390
(T_r)	11.25	12.34	8.66
v (10^{-6} m^3 mol^{-1}) (273 K)	20.00	18.02	26.67
(T_r)	18.53	18.23	23.39
α_p (10^{-3} K) (273 K)	2.26	−0.068	2.16
(T_r)	2.29	0.458	1.54
κ_T (10^{-12} Pa^{-1}) (273 K)	48.18	509.8	12.5
(T_r)	29.3	441.7	4.94
c_V (J K^{-1} mol^{-1}) (273 K)	42.8	76.0	48.1
(T_r)	37.0	72.5	50.8
c_p (J K^{-1} mol^{-1}) (273 K)	48.7	76.0	78.9
(T_r)	45.1	75.3	73.6
σ (10^{-2} N m^{-1}) (273 K)	1.026	7.563	2.341
$(\partial\sigma/\partial T)_p$ (10^{-5} N m^{-1} K^{-1}) (273 K)	−9.1	−15.6	−33.7
n_D (298 K)	1.1574	1.3325	1.325[a]
$(\partial n_D/\partial T)_p$ (10^{-3} K^{-1}) (298 K)		−0.105	
ϵ (273 K)	83.6	87.74	19.50
(T_r)	120.2	69.91	26.3
$(\partial(\ln\epsilon)/\partial T)_p$ (10^{-3} K^{-1}) (273 K)	−11.0	−4.56	−6.1
$(\partial(\ln\epsilon)/\partial P)_T$ (10^{-12} Pa^{-1}) (273 K)		451	
η (10^{-3} Pa · s) (273 K)	0.256	1.770	0.166
(T_r)	0.41	0.547	
κ_0 (Ω^{-1} m^{-1})	1.0×10^{-4}	1.38×10^{-6}	2.5×10^{-9}
Molecular properties			
μ (Debye)	1.48	1.834	1.82
α (10^{-30} m^3)	2.46	1.456	2.26
I (V)	16.04	12.62	10.2
σ (nm)		0.274	0.302
ϵk^{-1} (K)		732	528
$s^0(g)/R$ (298 K)	24.03	22.696	23.15
DN		18 (33)	59
AN		54.8	

[a] At 289 K.

molecule, which permit the formation of four hydrogen bonds. There are several other substances where very extensive hydrogen bonding is possible, including both organic (glycols, glycerol, etc.) and inorganic (sulphuric acid, hydrogen peroxide, etc.) compounds. In this section only water will be discussed in detail, and compared with the related isoelectronic substances ammonia and hydrogen fluoride. Even so, the treatment here must be very incomplete in view of the wealth of available information and the many theories proposed for its explanation.

It has often been stated that water is an exceptional liquid, but it is important to examine in what respects it is actually exceptional. If compared with hydrogen fluoride and with ammonia, Table 3.9*, it is seen that for most properties it is not really exceptional (although it is, when compared with CH_4 or H_2S, with respect to melting point, boiling point, heat capacity, dielectric constant, etc.). Even the expansion on freezing is observed for other substances, see Tables 3.3 and 3.10. Exceptional are the negative sign of the isobaric expansibility, α_p, up to 4 K above the melting point, Figure 3.3(a), (i.e. the contraction that solid water (ice) undergoes as it melts continues in the liquid, Fig. 3.3(b)) and the negative sign of the temperature coefficient of the isothermal compressibility, $d^2(\ln d)/dT\,dP$, up to 46 K above the melting point. These anomalies exist at moderate pressures (10^5 Pa), but vanish at high pressures (3×10^8 Pa). Along with hydrogen fluoride and ammonia, however, water is definitely not a 'normal' liquid, and the properties of all three are strongly affected by extensive hydrogen bonding, and to some extent by ionic dissociation.

Whereas a water molecule can donate two hydrogen atoms and two electron pairs to form four hydrogen bonds (tetrahedrally arranged), an ammonia molecule can donate three protons but only one electron pair, and a hydrogen fluoride molecule one proton and three electron pairs, to form only two hydrogen bonds each. A three-dimensional network of hydrogen bonds therefore characterizes water, while chain-like (or ring) associates characterize the other two substances. In solid water the three-dimensional arrangement of the oxygen atoms is regular, each being surrounded tetrahedrally by four others, 0.276 nm away, in a manner isomorphous with the tridymite form of silica. Although the cohesion is strong (T_m = 273.15 K, compared with 90.7 K for CH_4 and 187.6 for H_2S), the structure is very open and the molar volume large. There is thus a contraction of 8.25% in volume on melting. In hydrogen fluoride, on the other hand, the chain-like associates (Fig. 3.2(d)) are packed very efficiently in the solid, and a very large expansion of 36.0% is observed on melting. The dipole forces between the chains, however, are not so large, and the melting point is relatively low (T_m = 189.78 K) as for the isoelectronic NH_3 (T_m = 195.46). It is of interest to note that the changes of entropy on melting are not large for all three substances, about three times the communal entropy gained in the liquid relative to the solid, but the substances differ markedly in the heat capacity change on melting. Hydrogen fluoride, like covalently bonded chain molecules such as normal paraffins beyond octane, exhibits a premelting increase in heat capacity[29], so that at $T_m - 3.7$ K, $c_p = 55.19$ J K^{-1} mol^{-1} for the solid, while for the liquid at $T_m + 8.1$ K, $c_p = 43.01$ J K^{-1} mol^{-1} only, so that the apparent change on melting is negative[29]. Water, on the other hand, exhibits a large positive change from c_p = 38.70 J K mol^{-1} for the solid, which is mainly vibrational in origin, to

*For the comparison of temperature-dependent properties, values are given at two temperatures: at the fixed temperature 273 K (which is within the normal liquid range of HF, the lower liquid limit of H_2O, and at almost four times the atmospheric pressure for NH_3), and at a reduced temperature $T_r \sim 1/2 T_c$ (actually 240 K for HF, 323 K for H_2O and 200 K for NH_3), although these substances are not really conformal. Properties of water at 298 K are listed in Table 1.1, and at several other temperatures in Table 6.1.

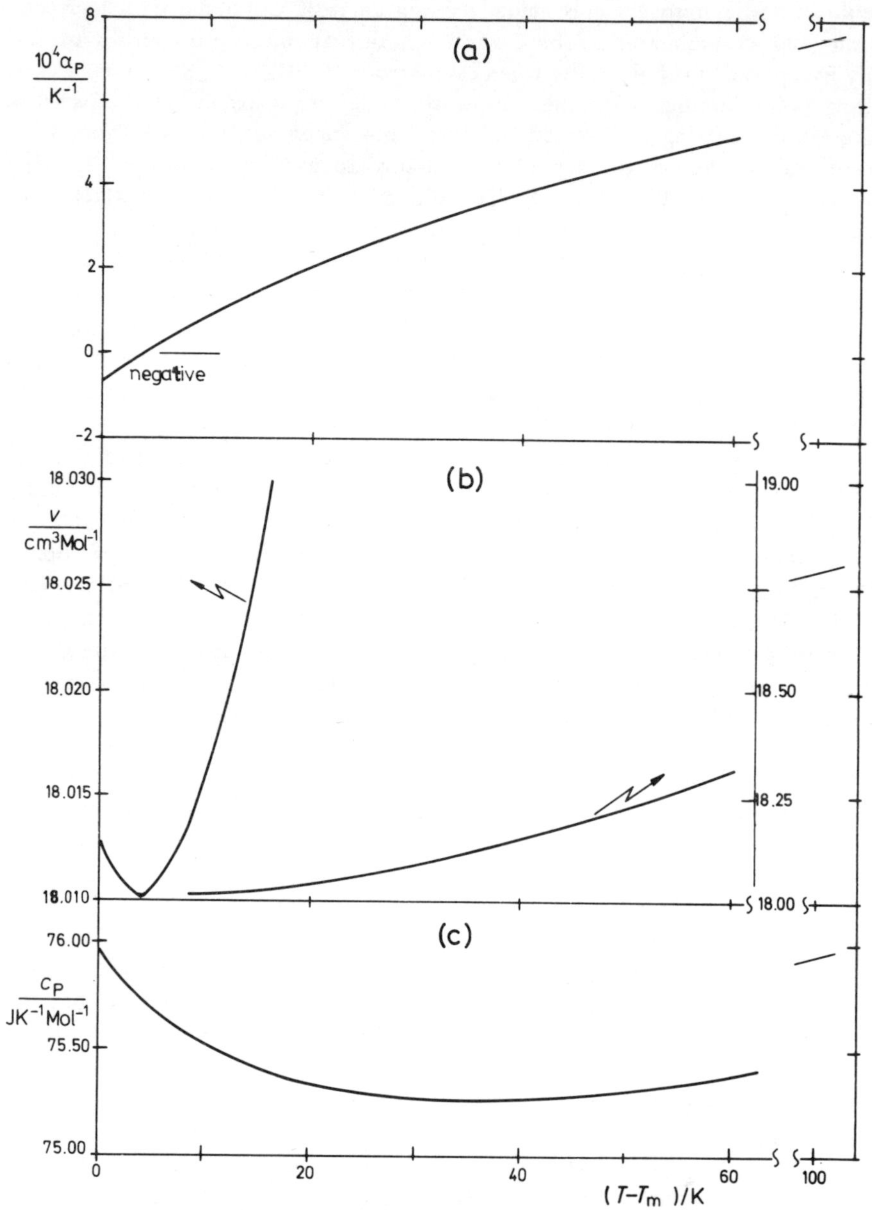

Figure 3.3. Some properties of water and their temperature-dependence. (a) The isobaric thermal expansibility; note the negative portion below 4 °C. (b) The molar volume; note the minimum corresponding to maximum density at 3.98 °C (= 277.13 K). (c) The molar heat capacity at constant pressure; note the shallow minimum and the regaining at the boiling point of the value at the melting point.

75.99 J K mol^{-1} for the liquid (which exhibits a shallow minimum near 35 °C, Fig. 3.3(c)), the difference being the configurational heat capacity that liquid water possesses. The change in heat capacity of ammonia is impressive too, though not as large.

At their normal boiling points, water and ammonia behave as expected for associated liquids (see previous section): compare T_b for H_2O, 373.15 K, and for H_2S, 212.5 K; for NH_3, 239.78 K and for PH_3, 185.8 K; and the Trouton constants 13.10 for water and 11.71 for ammonia with the normal $\Delta s^V/R$ = 10.2 ± 1.2. Hydrogen fluoride is exceptional, and has a uniquely low Trouton constant of only 3.54, in spite of a high boiling point: T_b = 292.66 K for HF compared with 188.3 K for HCl. The low entropy (and heat capacity) change on vaporization of hydrogen fluoride is due to an anomaly of the vapour rather than of the associated liquid: gaseous hydrogen fluoride is also strongly associated, containing aggregates similar in size to those in the liquid, but remote from each other. Thus only relatively weak dipole associations between the aggregates in the liquid are broken on vaporization, not the strong hydrogen bonds. On further heating and at low pressures these hydrogen-bonded aggregates eventually dissociate too, but even at the critical point they persist, and the compressibility factor $P_c v_c/RT_c$ is very low: 0.117 only compared with 0.227 for water, 0.243 for ammonia, 0.264 for normal liquids, and 0.291 for the noble gases. The heat of vaporization of hydrogen fluoride is correspondingly low, since it does not include the energy of separating the molecules completely from one another. The mean association number at the boiling point is *ca.* four, so that there are three hydrogen bonds to break, or 3/2 bonds per molecule of HF, requiring each *ca.* 28 kJ mol^{-1}. Therefore the observed latent heat of boiling Δh^V must be increased by *ca.* 42 kJ mol^{-1}, to give the corrected value of *ca.* 51 kJ mol^{-1} for complete separation of the molecules. The 'corrected' nominal solubility parameter δ = $(\Delta u^V/v)^{1/2}$ is thus *ca.* 49 (J $cm^{-3})^{1/2}$, comparable to that of water, 49.0 (J $cm^{-3})^{1/2}$, and much larger than that of ammonia, 26.8 (J $cm^{-3})^{1/2}$, and is now a reasonable measure of the cohesive energy density in this liquid.

The interaction of two separated molecules in the gas phase, when they approach each other, can be described in terms of dispersion and orientation forces for ammonia and for water (because of the hydrogen bonding in the gas, the second virial coefficient and gas viscosity of hydrogen fluoride cannot be analysed in these terms). Reasonable fits can be obtained with the two parameter 'Lennard-Jones 12-6' potential, as well as with three parameter Kihara hard-core or Stockmayer-Keesom polar potentials, eqs. (1.42), (1.47) and (1.56), respectively. The core size for the Kihara potential is, however, very small, $\delta/\sigma = 0.02$ for ammonia and only 0.003 for water, and as good a fit is obtained without inclusion of the polar term[26] as with it. The dipole moments of these molecules are relatively small, and without hydrogen bonds, no unusual interactions take place in the gas phase. The donor number of water*, DN = 18, is also not impressively large, that of

*This is the value obtained from the heat of interaction with antimony pentachloride. The value 42.3, obtained from the spectral shift of oxovanadium(IV)bis(acetylacetonate) is probably more correct, see p. 113.

ammonia being much larger. The acceptor number of water, $AN = 54.8$, is however among the largest, and water has therefore outstanding capabilities of solvating anions, which are electron donors, as well as cations, which are electron acceptors. The success of water as a solvating ligand for cations is partly due to its small size and the ability of a cation to accommodate several water molecules around it. A cooperative process sets in, so that liquid water behaves as if it had the much larger $DN = 33$ rather than 18.

The two unique thermodynamic properties of water, i.e.

$$\alpha_p < 0 \quad \text{for} \quad T < (T_m + 4\ \text{K})$$

and

$$(\partial \kappa_T / \partial T)_p < 0 \quad \text{for} \quad T < (T_m + 46\ \text{K})$$

have already been mentioned. The thermal expansibility of water is relatively small (Most liquids have α_p (10^{-3} K) near unity of larger), but in contrast, its compressibility is very large, and most liquids show values of κ_T (10^{12} Pa^{-1}) = 100, compared with several hundred for water. The surface tension is also some two to three times higher than for normal liquids or for most hydrogen-bonded ones (with the obvious exceptions of glycerol and formamide among those listed in Table 3.8), and the high dielectric constant is indicative of a cooperative alignment of the dipoles. Indeed, the value of g is only 0.86 in ammonia at 237 K, and not much larger at lower temperatures, signifying an antiparallel orientation of the dipoles in the hydrogen-bonded structure of this liquid. In water and in hydrogen fluoride g is respectively 2.67 and 7.82. The extremely large value in the latter liquid is due to the stability of linear dimers HF . . . HF and chain oligomers (Fig. 3.2(d)) $H(F \ldots H)_nF$ (along with cyclic hexamers) with nearly parallel, hence cooperatively enhancing, dipoles. In water the situation is more complicated, since the tetrahedral environment of each oxygen atom leads to a complicated angular correlation of the dipoles.

Information on the structure of liquid water has been obtained from diffraction and spectroscopic measurements[30]. The radial distribution function is dominated by scattering from O—O pairs, with only minor contributions from O—H and H—H scattering. The first peak in $g(r)$ is located at $r_{max} = 0.282$ nm and the coordination number is $Z = 4.4$ at $T = (T_m + 4\ \text{K})$. The second peak is located at 0.45 nm, which corresponds to 1.63σ, where σ is the O—O distance in ice (0.276 nm) or in the gas phase (Table 3.9). The distance 1.63σ corresponds almost exactly to the requirement imposed by the tetrahedral nature of the hydrogen bonds, $2\sigma \sin(\theta/2)$, where θ is the tetrahedral angle 109.46°, and a coordination number of four characterizes the crystal structure of ice. The immediate environment of a water molecule in liquid water, that is nearest- and second-nearest-neighbours, is thus very ice-like. This environment does not, however, persist beyond 2σ from a given water molecule. As the temperature is increased, the value of r_{max} gradually increases (to 0.294 nm at 470 K), the coordination number rises slightly, and the second peak diminishes and all but vanishes by 373 K.

Theories of the structure of water, which lead to predictions of structural, thermodynamic, spectroscopic and other properties have been reviewed recently[30] and will not be discussed here extensively. These reviews differ in their details and basic assumptions: proposals of major significance are the continuous model of Pople[31] with bent hydrogen bonds, the clathrate model of Pauling[10] and Frank and Quist[31], the interstitial model of Samoilov[31] and of Mikhailov[31], the flickering cluster model of Frank and Wen[31] and of Némethy and Scheraga[31], and other two-state mixture models. However computer calculations by the Monte-Carlo and the molecular dynamics techniques (section 2A(c)), and theoretical evaluation by the Percus–Yevick and similar integral equations (section 2B(b)) have led recently to fair agreement of calculated with experimental properties[32]. One of the properties of water resulting from these theories is its 'structuredness'. Many controversies arose from the difficulty in defining this quantity. In a qualitative manner, a model such as that of the flickering clusters of Frank and Wen[31] involves a large number of water molecules which are momentarily all hydrogen-bonded with each other, with from one to four hydrogen bonds per molecule, and a few non-bonded molecules existing too. A water molecule participating in one bond is likely to participate in another, since the bonding is cooperative, and conversely, if owing to thermal agitation one bond is broken, several more are likely to be broken in the vicinity, so that the clusters 'flicker'. It is possible to calculate the average number of hydrogen bonds formed by a molecule in the cluster, by constructing the partition function and minimizing the Helmholz free energy of the system at a given (N, V, T). A quantity S, which is the fraction of the number of hydrogen bonds that is left intact in a given sample of water under given conditions, out of those occurring in the same quantity of ice, is thus a measure of its 'structuredness'

$$S = \sum_{0}^{4} nN(n)/4N \qquad (3.5)$$

where $N(n)$ is the number of water molecules forming n hydrogen bonds, N being the total number of molecules. The difficulty remains, however, of defining those bonds which can be considered as remaining intact: how much distortion or bending is allowed before a hydrogen bond is considered broken? Since this is arbitrary[33], the controversy has little point. However changes in S can be followed for any arbitrary but operational definition of intact hydrogen bonds. Thus a concept such as a structural temperature T_s, for an aqueous solution at temperature T that has the same amount of structure as pure water at temperature T_s is valid, if the amount of structure is operationally defined (see Section 6B(a)), see frontispiece.

A final feature of hydrogen fluoride, water and ammonia that affects their properties is their ionic dissociation. This is of less importance in the latter two molecules than in the former, in their pure states, but it is of paramount importance in solutions of Brønsted acids or bases in these liquids as solvents. The schematic reaction

$$2QH(QH)_p = [QH_2(QH)_m]^+ + [Q(QH)_n]^- \qquad (3.6)$$

where Q^- stands for F^-, OH^- or NH_2^-, lies far to the left in all three liquids. The sum of the indices $m + n$ is not necessarily equal to p, since it is likely that the cation and the anion are more extensively solvated than the neurtral molecule. In liquid ammonia, at its boiling point, the ion concentration product is 3×10^{-21} M^2. The specific conductivity is accordingly only 1×10^{-9} Ω^{-1} m^{-1}, in spite of the low viscosity. In water at room temperature, with almost four times the viscosity, the specific conductivity is 6×10^{-6} Ω^{-1} m^{-1}, i.e. 6000 times larger, because the ion concentration product is 1.008×10^{-14} M^2 (at 298 K). For hydrogen fluoride the viscosity is as low as for ammonia, but the specific conductivity is another 15 times higher than in water, because of the still larger ion concentration product, 2.6×10^{-12} M^2 (at 273 K). Conductance in all three liquids is by the bond flipping mechanism, rather than by migration. Liquid hydrogen fluoride has strongly acid properties: on the H_0 acidity function scale it has a value of −11, comparable with that of sulphuric acid, while the value for water is +7, and it behaves as a typical amphiprotic solvent.

C. Ionic Liquids

Ionic liquids range over several types of substance, from those which are typical molten salts, completely dissociated to monatomic ions, such as potassium chloride (properties of which are listed in Table 1.2), or to polyatomic ions, such as potassium nitrate or tetra-*n*-pentylammonium bromide, to those which are only partially dissociated to ions, such as molten mercury bromide (which has a specific conductivity of only 0.0233 Ω^{-1} m^{-1} compared with several hundreds for typical ionic melts), and to those which form networks of highly associated ions in the melt, such as beryllium fluoride or boron oxide. The structural units are in all cases charged ions, interacting with each other by coulombic forces, although in strongly associated ionic liquids, the forces can become directional rather than isotropic, having a considerable covalent character, as in the boron oxide melt. The internal structure of the molecular ions is of little interest and such units as the nitrate anion interact unchanged with other ions, and move unchanged by diffusion or under the influence of shear or electrical forces.

a. Dissociated molten salts with monatomic ions

Completely dissociated molten salts have common properties, irrespective of whether they dissociate to monatomic or to polyatomic ions. However, the structure of the former can be studied by diffraction methods, and therefore this group of ionic liquids merits separate consideration, see Table 3.10.

Both X-ray and neutron diffraction studies have been made, mainly of the molten alkali halides[34]. The derived radial distribution data for molten lithium chloride are shown in Fig. 3.4. The prominent peak at 0.385 nm is due to diffraction from pairs of like-charged ions – mainly from closely situated pairs of chloride anions with only a minor contribution from pairs of lithium cations. The X-ray data show a small peak at smaller distances, and since the scattering amplitude of ^7Li for neutrons is negative, a valley is seen for the neutron scattering data, at 0.245 nm. This is due to scattering from a pair of nearest-neighbours, which

Table 3.10. Nearest- and next-nearest neighbours in alkali halide melts obtained by X-ray diffraction or neutron scattering[34].

Salt	Method	r_1 (nm)	Z_1	$\Delta r_1/r_1$ [a]	r_2 (nm)	Z_2	$\Delta r_2/r_2$ [a]
LiF	X-ray	0.195	3.7	−0.077	0.30	8	+0.010
LiCl	Neutron	0.245	3.5	−0.086	0.380	8.3	+0.011
	X-ray	0.247	4.0	−0.077	0.385	12.0	+0.023
LiBr	X-ray	0.268	5.2	−0.063	0.412	12.8	+0.022
LiI	X-ray	0.285	5.6	−0.095	0.445	11.3	+0.009
NaF	X-ray	0.230	4.1	−0.043	0.344	9.	+0.015
NaCl	X-ray	0.280	4.7	−0.054	0.42	9.	+0.007
NaI	X-ray	0.315	4.0	−0.063	0.480	8.9	+0.013
KF	X-ray	0.27	4.9	−0.037	0.386	9.	−0.026
KCl	Neutron	0.310	3.5	−0.052	0.47[c]		
	X-ray	0.310	3.7	−0.052		>12	
RbCl	X-ray	0.330	4.2	−0.033		>12	
CSCl	X-ray	0.353	4.6	−0.011	0.487	7.1	−0.037
CsBr	Neutron	0.355	4.7	−0.048[b]	0.52	7.9	−0.012[b]
	X-ray	0.355	4.6	−0.048[b]	0.54	8.3	+0.026[b]
CsI	X-ray	0.385	4.5	−0.023[b]	0.55	7.2	−0.013[b]

[a] $\Delta r/r = (r^{l} - r^{s})/r^{s}$ is the contraction or expansion in the liquid relative to the distance in the solid at the melting point. For the solid $Z_1 = 6$ and $Z_2 = 12$.
[b] Values for metastable forms of the solid, with the same Z values as for the other salts.
[c] Ref. 36.

is thus a cation–anion pair, as expected from the coulombic forces operating in the liquid. Further peaks due, mainly, to diffraction from Cl–Cl pairs, are seen at 0.70 and 1.02 nm, but the details do not persist much beyond 1.0 nm. Similar results were obtained for the other alkali halides, and are summarized in Table 3.10, which shows the following features. The coordination number of nearest-neighbours, Z_1, is always smaller than in the solid (where it is 6 for the NaCl structure, and 8 for the CsCl structure, which applies for CsBr and CsI just below their melting points). Concomitant with this decrease in Z_1 there is a decrease also in r_1 relative to the solid, averaging *ca.* − 5%. The coordination number Z_2 for second-nearest-neighbours, that is of like ions, is usually, but not in all cases, lower in the liquid than in the solid (where it is 12), and there is a concomitant increase in distance of *ca.* + 2% on average, although in certain cases (KF and CsF with certainty) there is a small decrease in r_2 too. The interpenetrating quasi-lattice microstructure of alternating cationic and anionic sites is thus quite well manifested and persists in the generally random and isotropic macrostructure (non-structure) over three to five ionic diameters before the order is destroyed. Since the peaks in the radial distribution function are rather broad, however, it is only over average quantitites that these statements hold. The said microstructure is, however, consistent also with small clusters of regular crystalline structure separated by disordered surfaces[35]. These crystallites rotate, although they need not be spherical, and thus permit a range of distances and coordinations.

Much less is known about the structures of other molten salts dissociating to monatomic ions. Calcium fluoride and barium chloride have been studied[35], and (r_1, Z_1) values of (0.235 nm, 6.8) and (0.32 nm, 5.5) respectively have been found,

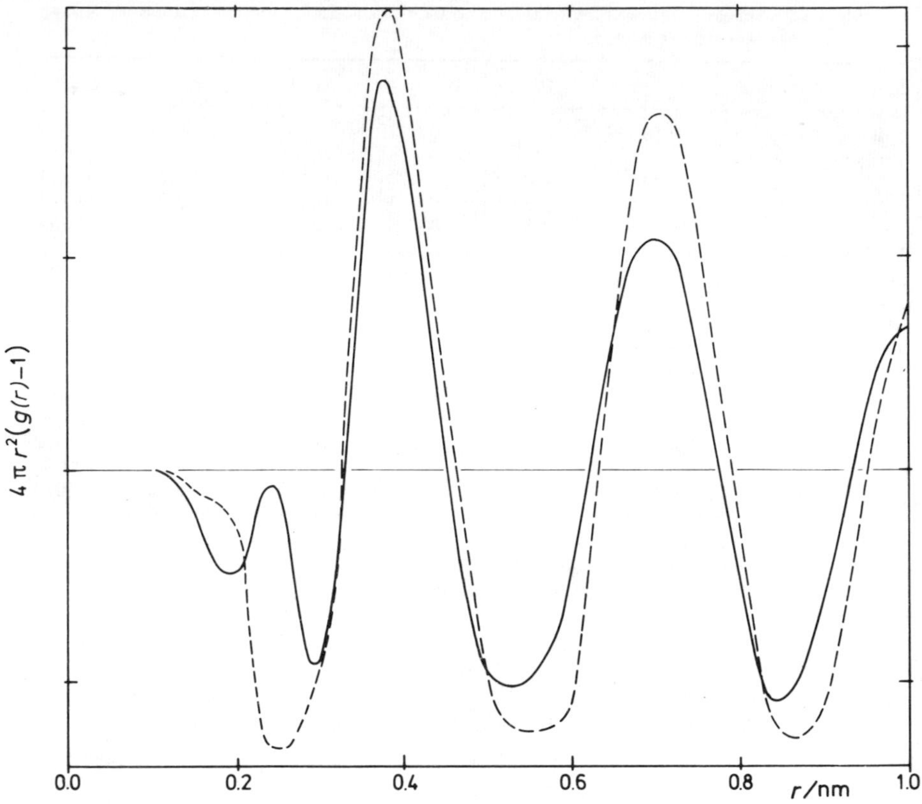

Figure 3.4. The excess radial distribution function, $(n^{(2)}(r)/n^{(1)}) - 1 = 4\pi r^2(g(r) - 1)$, with an arbitrary ordinate scale, of lithium chloride melts at 900 K obtained by X-ray diffraction (———), and by neutron diffraction from ^{7}LiCl (— — — —).

again r_1 being somewhat shorter than in the expanded solid near the melting point, and Z_1 smaller than the value for the solid, $Z_1 = 8$. For a zinc chloride melt[36], cold neutron scattering gave a Cl–Cl distance of 0.379 nm, i.e. close to the value for lithium chloride melts, but much shorter than for sodium or potassium chloride melts, in obvious correlation with cation size.

Interionic distances, from which ionic radii are derived by the additivity rule, are important parameters for correlating and predicting the properties of ionic liquids. So also are the coordination numbers for nearest- and next-nearest-neighbours, particularly for the properties of mixtures of ionic liquids (Chapter 7). It is therefore unfortunate that not much more information is available concerning these quantities: even their temperature-dependencies for the alkali halide melts have not been determined experimentally with any precision. That they are not temperature-invariant is obvious[37], but not often realized in discussions of the properties.

Progress has recently been made by applying computer simulation methods, e.g. the Monte-Carlo and molecular dynamics methods discussed in section 2A(c), to ionic liquids. A formidable difficulty, arising from the long range of the coulombic forces, had first to be overcome before the methods that proved so successful for systems with shorter-range forces, such as dispersion or core repulsion, could be applied[38]. It is necessary, of course, to select an appropriate potential function $u(r)$ to describe the pair interactions between the ions. The widely used potential function of Tosi and Fumi[39] has been challenged recently[40], but in all cases a core repulsion, coulombic attraction and repulsion according to the sign of the charge, and induced dipole interactions are taken into account. The calculations permit the derivation of pair correlation functions $g_{ij}(r)$, for the different pairs of ions: ii, ij and jj, and of thermodynamic properties that can be checked by experiment. The molecular dynamics method also yields correlation times, i.e. the times for which ions are within a specified distance of each other, and transport properties, which again can be compared with experimental results. Such calculations have been applied by now[38,40,41] to liquid LiI, NaCl, KCl, CsBr and generally to the liquid alkali halides at several temperatures. The results are satisfactory in the sense that they give confidence in the potential functions employed, so that they may now be used to study systems for which experimental data are lacking.

The ionic liquids with monatomic ions include sulphides, oxides, hydrides and halides, but only the latter have been extensively studied. Table 3.11 shows some of the bulk properties of the alkali halides, of the chlorides of many other metals, and of the bromides and iodides of some heavy metals. The melting points of all are known, but some melt only under pressure ($AlCl_3$, $InCl_3$, $ZrCl_4$). The boiling points of practically all are also known, though some of the tri- and higher-valent salts have very narrow liquid ranges under atmospheric pressure (e.g. 27 K for $HgCl_2$ or 23 K for $TaCl_5$). Other salts, mainly the monovalent and divalent ones, have rather long liquid ranges, of a few hundred K. Critical temperatures are known mainly for the low-melting high-valency salts, and the liquid fraction out of the total range of existence of condensed phases, $(T_c - T_m)/T_c$, is not very large for these compounds, *ca.* 0.35 on average. The quantity $P_c v_c/RT_c = 0.274 \pm 0.014$ is fairly constant, so that these polyvalent halides are nearly conformal fluids. For the alkali halides, on the other hand, no critical data have been measured, but there exist estimates, based on extrapolation to high temperatures, of the liquid and vapour densities according to the law of rectilinear diameters, eq. (1.29), or of the surface tension to vanishing point.

The molten halides freeze with a release of entropy amounting to *ca.* $\Delta s^F = (2.94 \pm 0.21)\mathbf{R}$ for the alkali halides (except CsCl), and generally to 1.6**R** per ion (i.e. $s^F = 3.15\mathbf{R}$ for NaCl, 5.26**R** for $MgCl_2$, 5.80**R** for $LaCl_3$, 9.28**R** for $NbCl_5$), which is close to the value for freezing liquified noble gases, p. 93. This is a good indication of complete dissociation to ions in the liquid near the freezing point. There are however many exceptions: either because of rearrangements in the solid near the melting point, as for instance for barium chloride where $\Delta s^F = 1.63\mathbf{R}$ only, but another 1.73**R** is released in the solid state transition, or because of extensive association in the liquid, as in zinc chloride, where $\Delta s^F = 2.10\mathbf{R}$ only, compared with the 4.8**R** expected in both cases. The volume shrinkage on freezing is somewhat higher for the molten alkali halides, as the values in Table 3.11 show,

Table 3.11. Properties of selected molten halide salts[a].

Salt	T_m (K)	T_b (K)	T_c [b] (K)	$\Delta v^F/v^s$	$10^{-6}\ v^l$ [d] ($m^3\ mol^{-1}$)	$10^4\ \alpha_p$ [f] (K^{-1})	$10^{10}\ \kappa_T$ [e] (Pa^{-1})	$10^2\ \sigma$ [f] ($N\ m^{-1}$)	κ [f] ($\Omega^{-1}\ m^{-1}$)
LiF	1121	1966	4140	0.294	14.34	2.78		22.76	907.3
LiCl	883	1656	3080	0.262	28.23	2.96	2.47	12.54	616.3
LiBr	823	1562	3020	0.243	34.34	2.63	2.94		513.8
LiI	742	1449	3250		43.05	2.99	2.34		424.0
NaF	1269	1983	4270	0.274	21.32	2.97		17.88	528.2
NaCl	1074	1738	3400	0.250	37.56	3.63	2.87	10.82	390.2
NaBr	1020	1666	3200	0.224	43.95	3.27	3.36	9.81	320.8
NaI	933	1577	3160	0.186	54.68	3.56	4.73	8.80	253.3
KF	1130	1783	3460	0.172	30.42	3.58		13.65	390.0
KCl	1043	1710	3200	0.173	48.80	3.98	3.84	9.52	241.2
KBr	1007	1671	3170	0.166	55.95	4.03	4.38	8.40	186.1
KI	954	1618	2980	0.159	67.83	4.13	5.99	7.53	148.0
RbF	1048	1663	3280					12.10	
RbCl	988	1654	3140	0.143	53.79	4.09		9.26	176.0
RbBr	953	1625	3130	0.135	60.86	4.10		8.41	131.6
RBI	913	1577	3035		73.09	4.08		6.52	101.4
CsF	976	1504	2915		41.63	3.61		10.20	281.4
CsCl	918	1597	3040	0.105	60.31	3.95	5.12	8.62	138.5
CsBr	908	1573	3045	0.268	67.92	4.05	6.71	7.95	102.7
CsI	894	1553	3020	0.285	81.63	3.85		5.65	82.2

	T_m (K)	T_b (K)	T_c [c] (K)	$\Delta v^F/v^s$	$10^6\ v^l$ [d] ($m^3\ mol^{-1}$)	$10^4\ \alpha_p$ [f] (K^{-1})	$\Delta s^F/R$	$10^2\ \sigma$ [f] ($N\ m^{-1}$)	κ [f] ($\Omega^{-1}\ m^{-1}$)
$BeCl_2$	688	828			53.58	7.62	1.51		0.76
$MgCl_2$	987	1691			56.76		5.25	6.61	120.6
$CaCl_2$	1045	2273		0.009	53.35	2.07	3.28	14.13	250.4
$SrCl_2$	1146	2273		0.042	58.16	2.17	1.67	16.38	243.1
$BaCl_2$	1235	2103		0.035	65.62	2.20	1.63	15.80	256.7
$AlCl_3$	465.8[g]		628	0.839	111.14		9.12	0.83	5.6×10^{-5}
$ScCl_3$	1233	1240			95.				83.7
YCl_3	994	1783			77.47	2.01			40.
$LaCl_3$	1128	2000			76.63	2.50	5.80		130.

$TiCl_4$	249.1	409	638	0.124	105.22				
$ZrCl_4$	710[g]		778						
$HfCl_4$	705[g]		723						
$ThCl_4$	1043	1193			113.				73.2
$NbCl_5$	477	523	807		130.28		9.28		2.2×10^{-5}
$TaCl_5$	490	513	767		133.44		9.14		3×10^{-5}
$MnCl_2$	923	1504			53.48		4.91		157.7
$FeCl_2$	950	1280					5.45		
$CoCl_2$	1013	1297					7.01		
$NiCl_2$	1303						7.15		
CuCl	703	1485			26.77	2.09	1.75	9.2	365.2
AgCl	728	1845		0.089	29.42	1.81	2.14	17.62	410.3
$ZnCl_2$	591	1005			53.70	1.80	2.09	5.38	1.29
$CdCl_2$	841	1234			54.10	2.47	4.31	8.36	376.1
$HgCl_2$	550	577	973		62.21	6.80	4.25	5.61	0.0052
$GaCl_3$	350.7	473			85.73	10.52		2.64	
$InCl_3$	859[g]				103.32	10.72			42.6
TlCl	702	1089			42.61	3.27	2.72		134.5
$SnCl_2$	520	925			56.32	3.80	2.95	10.15	389.9
$SnCl_4$	239	388	592		109.74		4.63		~0
$PbCl_2$	768	1226			56.12	3.10	3.74	13.19	180.2
$SbCl_3$	346	493	794				4.51		8.5×10^{-5}
$BiCl_3$	653	712	1179		80.62	6.72		6.85	58.1
AgBr	703	1835			33.52	1.88	1.57	15.27	301.5
AgI	831	1778			42.11	1.84	1.36		248.7
$HgBr_2$	514	592	1012		70.54	6.54	4.19	6.45	0.0233
HgI_2	523	627	1078		86.91	6.36	4.33	6.70	2.52
TlBr	733	1098			47.18	3.27	2.68		98.5
TlI	713	1118					2.47		64.9
$BiBr_3$	492	735	1219		95.86	5.71		6.77	27.3
BiI_3	681								30.8

[a]Data mainly from G. J. Janz, *Molten Salts Handbook*, Academic Press, New York, 1967.
[b]A. D. Kirshenbaum, J. A. Cahill, P. J. McGonigal and A. V. Grosse, *J. Inorg. Nucl. Chem.*, **24**, 1287 (1962).
[c]J. F. Mathews, *Chem. Rev.*, **72**, 71 (1972).
[d]At T_m.
[e]At 1073 K.
[f]At $T = 1.1T_m$.
[g]Melting under pressure.

than for liquified noble gases, where it is 13%, but the few results available for higher-valent salts show a shrinkage of only a few percent, with the notable exception of liquid aluminium chloride, which undergoes extremely large shrinkage, of *ca.* 46%, as it solidifies. This, however, is far from a typical molten salt, as examination of its conductivity will show.

The molar volumes of the molten halides vary over a considerable range, from the highly compact copper (I) chloride and sodium and lithium fluorides, to the relatively bulky polyvalent halides. The equivalent volumes of these, however, are quite regular: for the chlorides the values are 27.9 ± 2.0 cm^3 mol^{-1}, i.e. not larger than for the monovalent copper or silver chlorides, and only barium, indium and mercury are outstandingly larger. Thus the volume is dicated mainly by the anion for the polyvalent molten halides (*ca.* 38 cm^3 mol^{-1} for the bromides and 43 cm^3 mol^{-1} for the iodides), but for the alkali halides the volume depends on both cation and anion. This is an important structural indicator, signifying that for the polyvalent halides the anionic quasi-lattice is dominant, and the relatively small cations take up positions in the *vacancies* left between the spherical anions which are almost in contact. The alkali cations are too large and too numerous to permit such liquid structures. The thermal expansibilities of the molten halides are relatively small, typically 3×10^{-4} K^{-1}, about a third as high as for typical organic liquids, or a tenth as high as for the liquified noble gases. The isothermal compressibility is also about only one third as high as for organic liquids. This behaviour is consistent with the small molar volumes and the strong cohesive forces in the liquid. These are shown also by the surface tension, which is very appreciable, amounting to some five times higher than for typical organic liquids. It is, however, noteworthy that as the ionic liquid halides are heated to their boiling points, they behave as very ordinary liquids as far as their entropies of vaporization are concerned. The mean value for the Trouton constant for the alkali halides is $(11.8 \pm 0.8)\mathbf{R}$ i.e. almost in the range of ordinary, non-ionized liquids, eq. (1.9). This is because the state of the vapour is similar to that of the liquid, i.e. separate ions. For the other halides in Table 3.11, the value is not very different: $\Delta s^{\mathrm{V}} = (12.1 \pm 1.4)\mathbf{R}$. Some of the salts contributing to the somewhat higher average are known to be strongly associated in the liquid state near the melting point, and presumably also near the boiling point, e.g. zinc chloride, for which $\Delta s^{\mathrm{V}} = 14.27\mathbf{R}$.

Some of the properties of molten salts are strongly temperature-dependent, and hence should be compared at corresponding temperatures, because there is no standard temperature at which all are liquid and can be studied. Table 3.11 brings some of the properties of the ionic liquid halides at a 'corresponding' temperature of $T = 1.1T_{\mathrm{m}}$ (except for the compressibility, where the data are known only at a few discrete temperatures, and the molar volumes, given at T_{m}, but these may readily be recalculated for $T = 1.1T_{\mathrm{m}}$ with the aid of α_{P}). The selection has been made according to arguments presented by Reiss, *et al.*[42], according to which the vapour pressures of the alkali halides at multiples of T_{m}, say $1.1T_{\mathrm{m}}$, bear constant ratios to those at T_{m}. This is a consequence of the potential function describing the interactions, eq. (1.41), rewritten as

$$u(r) = \infty \quad \text{for} \quad r < a$$
$$= Kz_a z_b e^2 r^{-1} \quad \text{for} \quad r \geqslant a \tag{3.7}$$

where a is taken as the interionic separation, and of the fact that for the alkali halides $aT_m = 3 \times 10^{-7}$ K m, i.e. a constant. A reduced temperature $T_r = K^{-1}z^{-2}aT$ is defined, but because K, the reciprocal of the effective dielectric constant, is unknown, $T_{r,m} = K^{-1}z^{-2}aT_m$ is assumed to be a constant, so that $T_r = (T_{r,m}/T_m)T$. Setting $T_r/T_{r,m} = 1.1$ therefore produces a set of corresponding temperatures $T = 1.1T_m$. (Another ratio $T_{r,m}/T_r$ could have been selected.) Many properties of molten salts have, therefore, been reported[43] at $T = 1.1T_m$, based on the acceptance of (3.7) for the potential function, on the invariability of K with z, a and T, and on the constancy of aT_m (which breaks down for the lithium halides or the alkaline-earth oxides, which should conform[42]). For the alkali halides, the reduced temperature $1.1T_m$ is indeed a reasonably constant fraction, 0.333 ± 0.027, of the estimated critical temperature T_c, and hence a corresponding temperature also in that sense. Rather than emphasize the melting point T_m, which is a property of the ionic crystal rather than of the ionic liquid, a different approach has been proposed[44], which utilizes a liquid-phase property to define reduced temperatures. This is the isobaric thermal expanisivity α_p, so that $T_r = 2.5\alpha_p T$, the coefficient 2.5 being arbitrary. The temperatures $T/T_r = 0.4/\alpha_p$ are therefore corresponding. For the alkali halides (except the lithium salts), these latter temperatures are not very different from the former ones, $1.1T_m$, and hence are again similar fractions of T_c, 0.35 ± 0.05, but they show greater variation. Since, however, the critical temperatures are only uncertain estimates, it is pointless to define corresponding temperatures with reference to T_c.

The potential function (3.7) has a hard-sphere repulsion term, and therefore scaled particle theory, section 2B(c), has been applied[45] to it, yielding the equation of state (cf. eq. (2.49b))

$$Pv/\mathbf{R}T = (1 + \eta + \eta^2)/(1 - \eta)^3 \tag{3.8}$$

the isobaric thermal expansibility

$$\alpha_P = (1 - \eta^3)/(1 + 2\eta)^2 T \tag{3.9}$$

the isothermal compressibility

$$\kappa_T = v(1 - \eta)^4/\mathbf{R}T(1 + 2\eta)^2 \tag{3.10}$$

and the surface tension

$$\sigma = a\mathbf{R}T(2 + \eta)/4v(1 - \eta)^2 \tag{3.11}$$

where $\eta = (\pi/6)\,\rho a^3$ is the reduced number density, and a, as above, is the interionic distance. The product of compressibility and surface tension is

$$\sigma\kappa_T = a(2 - 3\eta + \eta^3)/4(1 + 2\eta)^2 \tag{3.12}$$

and permits the one to be calculated from the other, provided the molar volume $v = N/\rho$ and the distance parameter a are known. Agreement with experimental surface tension values is obtained if a is selected not as the sum of the ionic radii in the solid but as the equilibrium bond distance in the molecular salt, measured in the gas by electron diffraction or microwave techniques. These can be approximated[46] by taking 0.820 times the sum of the Pauling crystal radii r_c, or better by

$$a = 0.820[r_{c+} + r_{c-} - (\alpha_+ + \alpha_-)/2a^3] + 0.175[(\chi_{F^-} - \chi_{Cs^+}) - (\chi_- - \chi_+)] \quad (3.13)$$

where α is the polarizability of the ions (this is a small correction term, calculated iteratively) and χ is their electronegativity. Generally, however, (3.13) or its approximation $a = 0.820(r_{c+} + r_{c-})$ is ignored, and a is set equal to $(r_{c+} + r_{c-})$ in applications of scaled particle theory to molten salts, and, in particular, molten salt mixtures (Chapter 7).

The last property shown in Table 3.11 is the specific conductance κ, again given for the 'corresponding' temperature, $1.1T_m$. This book would not ordinarily deal with transport properties such as the conductance, but an exception will be made in the case of molten salts since this is their outstanding characteristic property as ionic liquids. The equivalent conductance Λ is readily calculated by dividing κ into z_+ times the molar volume $v = v_m\ (1 + 0.1\alpha_p T_m)$ at $T = 1.1T_m$, where v_m is the volume at T_m, listed in Table 3.11. The specific conductivities are typically a few hundreds $\Omega^{-1}\ m^{-1}$, and for those cases where they are less than *ca.* $1\ \Omega^{-1}\ m^{-1}$, the occurrence of a considerable amount of association or of covalent bonding in the liquid is invoked as an explanation. Extensive association to a network of ions is invoked in the cases of beryllium and zinc chloride among the salts shown in Table 3.11, whilst covalent bonding is invoked in the cases of aluminium, niobium, tantalum, mercury, tin (IV) and antimony chlorides, and mercury bromide and iodide; see section 3C(c).

b. Molten salts with polyatomic ions

The main difference between the present group of ionic liquids and those with monatomic ions is that the former may be thermally unstable at elevated temperatures, and therefore such properties as boiling points and critical points are, as a rule, unknown. Only for sodium hydroxide and cyanide, and for thallium nitrate from among those listed in Table 3.12, have boiling points been determined, namely 1663 K, 1803 K and 706 K, respectively. While nitrates of the alkali metals can be heated about 100 K above their melting points before extensive decomposition sets in, those of the alkaline earths and other polyvalent metals decompose on melting. On the other hand, alkali sulphates and carbonates may be heated several hundred degrees above their melting points.

Another important difference between molten salts with monatomic and polyatomic ions is that only for the former can diffraction methods be applied usefully in order to gain some knowledge about their structures. For the latter, the

Table 3.12. Properties of selected molten salts with polyatomic ions[43].

Salt	T_m (K)	$\Delta s^F/R$	$\Delta v^F/v^s$	$10^6\ v^l$ ($m^3\ mol^{-1}$)	$10^4\ \alpha_p$ (K^{-1})	$10^2\ \sigma$ ($N\ m^{-1}$)	κ ($\Omega^{-1}\ m^{-1}$)
NaOH	593	1.29		22.42	2.72		264
NaCN	835	1.27					
NaSCN	583	4.98					104.2
$NaNO_2$	558			38.13	4.22	11.97	160.2
Na_2CO_3	1123	4.77		53.65	2.33	20.76	327.0
Na_2SO_4	1157	3.59		68.66	2.40	18.90	262.6
Na_2MoO_4	960	2.40		73.47	2.29	20.80	69.7
Na_2WO_4	971	2.95		76.24	2.11	19.94	60.3
$LiNO_3$	525	5.87	0.214	38.71	3.12	11.45	108.0
$NaNO_3$	579	3.03	0.107	44.60	3.84	11.80	122.1
KNO_3	610	2.31	0.033	54.06	3.99	10.87	81.3
$RbNO_3$	583	0.96	−0.0023	59.42	4.01	10.54	55.0
$CsNO_3$	687	2.47	0.121	69.13	4.26	8.84	66.5
$AgNO_3$	483	2.88		42.89	2.61	14.83	87.2
$TlNO_3$	480	2.06		54.31	3.89	9.31	47.2
n-$(C_4II)_4$NBr	395	4.72	0.082	319.1	7.04		239.3
n-$(C_5H_{11})_4$NCl	295	0.53					
n-$(C_5H_{11})_4$NBr	376	13.26	0.090	412.7			
n-$(C_5H_{11})_4$NI	412	11.48	0.107	404.2		2.8	
n-$(C_5H_{11})_4$NSCN	323	7.32	0.045	398.6			31.1
n-$(C_5H_{11})_4$NClO$_4$	391					3.0	
n-$(C_6H_{13})_4$NBr	377	5.08	0.031	698			
n-$(C_7H_{15})_4$NBr	369	11.74					

studies[34] merely confirm that the ionic units remain intact, e.g. as CO_3^{2-} or SO_4^{2-} in the melts.

In their bulk properties, molten salts with polyvalent ions resemble on the whole those with monatomic ions. Some melt at fairly high temperatures (sulphates, carbonates), most of the others at fairly low temperatures, so that molten salts with organic cations (tetraalkylammonium) or anions (carboxylate) can also be studied. Table 3.12 shows some properties of selected molten sodium salts, molten nitrates, and molten tetraalkylammonium salts. The entropies released on freezing are not as regular as for the typical monatomic salts, $\Delta s^F = 3.2R$ for uni-univalent and $4.8R$ for uni-divalent salts. Free rotation of the bulky anions and configurational randomization for long chain parts of cations can occur in the solids at temperatures below the melting point, so that only a fraction of the entropy change is obtained at the melting point proper, while hindered rotation may occur in the liquid even at temperatures appreciably higher than the melting point. Internal degrees of freedom may also be locked-in in the process of freezing, as for the tetraalkylammonium salts shown, leading to a high value of the entropy change. The volume changes on freezing also show great variation, and the increase in volume (negative $\Delta v^F/v^s$) as rubidium nitrate freezes should be noted. Lithium nitrate on the other hand, contracts strongly, as also do sodium and caesium nitrates. These properties depend on the packing in the solid as has been mentioned several times above, and not solely on the structure of the liquid. The other properties shown, the molar volume of the liquid at T_m and the isobaric thermal expansibility, the surface tension and the specific conductance at $T = 1.1T_m$, are all comparable in magnitude to those shown by salts with monatomic ions. The tetraalkylammonium salts, however, tend to have large molar volumes and correspondingly low surface tension, because of the concomitant low cohesive energy. Table 3.12 does not show the isothermal compressibilities, which are known for some molten nitrates; at 673 K they are κ_T (10^{-10} Pa^{-1}) = 2.34 for $LiNO_3$, 2.16 for $NaNO_3$ and 2.34 for KNO_3, comparable in magnitude with the values for the corresponding halides.

Molten nitrates have received most attention since they are readily available, are low melting and are thermally stable, all relatively speaking. The relative configurations of alkali cations and nitrate anions are shown in Fig. 7.3; it should be noted that no particular position is favoured in the liquid state but that lithium ions have better opportunities to approach the anion very closely. It is also apparent that a concept such as ionic radius has very little applicability in molten nitrates, although much use has been made of it in studies of molten salt mixtures, Chapter 7. Another concept which is useful in this connection is the hardness and softness of the ions, so that molten salts can be described in terms of the charges, the sizes and the softness parameter (section 7A(b), Table 7.2) of its ions. The softness parameters bear a relationship to the polarizabilities of the ions, α, and these, in turn, to the molar refractivities, $\mathbf{R} = (4\pi/3)N\alpha$, which should be additive in the ionic values. The molar refractivities are related to the experimentally measured refractive indices and molar volumes by eq. (1.30). The refractive index n_D is, of course, temperature-dependent, and so is the molar volume, but these dependencies cancel out each other to a certain extent, so that R becomes nearly temperature-independent. Values of R are shown in Table 3.13.

The dielectric constant of ionic liquids is an intriguing quantity[47], but hardly any data exist with which theories may be compared, owing to the high conductivities of these liquids. Various estimates in the range from n_D^2 up to about 6 have been made[43], but most theoretical approaches to the properties of molten salts or their mixture circumvent the need to use a dielectric constant, and discuss the electrical interaction of nearest- and next-nearest-neighbours only, assuming in this case the (relative) dielectric constant of vacuum, $\epsilon = 1$, and its permittivity, $\epsilon_0 = 8.8542 \times 10^{-12}$ C^2 J^{-1} m^{-1}, to hold.

Table 3.13. Molar refractivities of ions in molten salts, obtained from measured refractive indices[39] by the additivity rule.

Cations	Li^+	Na^+	K^+	Rb^+	Cs^+	Ag^+	Tl^+	Zn^{2+}	Cd^{2+}	Hg^{2+}	Pb^{2+}
R (cm^3)	0.11	0.64	2.98	5.01	8.65	6.08	12.36	0.73	4.15	5.86	12.09

Anions	F^-	Cl^-	Br^-	NO_2^-	NO_3^-	SCN^-	SO_4^{2-}	WO_4^{2-}
R (cm^3)	2.67	8.67	12.22	8.84	10.26	16.67	14.96	23.30

c. *Associated molten salts*

Perusal of Table 3.11 shows, as already mentioned, several cases of salts with low conductivity, signifying association. Such cases are rarer among the fluorides than among the chlorides exhibited in the Table, but beryllium fluoride is a well known example. This compound belongs, along with the chlorides of beryllium and zinc and the oxides of boron, silicon and germanium to the group of network-formers among the associated molten salts. Here, the transport properties, and in particular the viscosity, the electrical conductivity and associated temperature-dependencies, are of prime importance for understanding the properties. Table 3.14 shows some of the properties of these salts, mainly the transport properties, at three temperatures: the melting point, the (purportedly) corresponding temperature of $1.1T_m$, and a somewhat higher temperature, $T_m + 200$ K. Remarkable are the very high activation energies, corresponding to the Arrhenius plots

$$\kappa(\text{or } \eta^{-1}) = A_{\kappa(\text{or } \eta)} \exp(-E_{\kappa(\text{or } \eta)}/RT) \tag{3.14}$$

i.e. the E_κ or the E_η values, approach, or in some cases exceed, the vaporization enthalpies (given in the Table at the boiling points). This means that in order to move in the melt, an entity in the liquid must break a number of its strong bonds, comparable with that required for complete escape from the melt. In the case of ordinary molten salts, activation energies of only approximately 10 kJ mol^{-1} are required, at least an order of magnitude lower than for these network-forming ionic

Table 3.14. Some properties of network-forming molten salts.

	BeF_2	$BeCl_2$	$ZnCl_2$	B_2O_3	SiO_2	GeO_2
T_m (K)	828	688	591	723	2001	1388
T_b (K)	1600	828	1005	1520	3220	
Δh^V (kJ mol^{-1})	224	108	119			
At T_m						
κ (Ω^{-1} m^{-1})	1×10^{-6}	8×10^{-2}	1.7×10^{-1}	4×10^{-7}		10^{-3}
E_κ (kJ mol^{-1})	>420	235	134	109		
η (Pa s^{-1})	$>10^3$		3.6×10^0	2.0×10^3	2.9×10^5	1×10^6
E_η (kJ mol^{-1})	420		159	63	640	750
σ (10^{-2} N m^{-1})						25.0
At 1.1 T_m						
T (K)	911	757	650	795	2200	1523
κ (Ω^{-1} m^{-1})	3.9×10^{-5}	7.6×10^{-2}	1.3×10^0	1.9×10^{-6}		4.1×10^{-3}
E_κ (kJ mol^{-1})	290	100	50	109		195
η (Pa s^{-1})	1.5×10^2		5.5×10^{-1}	7.0×10^2	5.6×10^3	
E_η (kJ mol^{-1})	360		75	63	640	
σ (10^{-2} N m^{-1})			5.4		30.7	
α_p (10^{-4} K^{-1})	0.35	7.62	1.80		<0.1	
At T_m + 200 K						
T (K)	1028		791	923		1588
κ (Ω^{-1} m^{-1})	2.8×10^{-3}		9.9×10^0	1.9×10^{-5}		1×10^{-2}
E_κ (kJ mol^{-1})	290			109		
η (Pa s^{-1})	7.7×10^0			1.4×10^2		1×10^4
E_η (kJ mol^{-1})	290			63		670
σ (10^{-2} N m^{-1})			5.4	8.3		
α_p (10^{-4} K^{-1})			2.82	1.06		0.12

liquids. Furthermore, E_κ and E_η for ordinary ionic liquids are independent of the temperature over a wide range, that is, the Arrhenius equation is strictly adhered to. For the network-forming liquids, on the other hand, the Arrhenius equation is not obeyed over a wide temperature range at the lower temperatures. Rather, a modified equation is followed, having $T - T_0$ in the denominator of the exponent rather than T, where T_0 is the ideal glass transition temperature, where all configurational entropy and the free volume are lost, see p. 6. This behaviour causes the apparent E_κ or E_η, calculated according to (3.14), to fall at increasing temperatures, see for example E_η for beryllium fluoride or E_κ for beryllium chloride in Table 3.14. The decreasing energies of activation mean that fewer bonds have to be broken to effect movement, or in other words the liquid is less extensively polymerized into a network at the higher temperature. For boron oxide, apparently, the temperatures shown are not yet sufficiently high to affect the polymerization and E_κ and E_η do not show a dependence on temperature. The fact that changes take place at a higher temperature is evident from the behaviour of α_P, which falls from the already low value of $1.06 \times 10^{-4}\ K^{-1}$ at $T_m + 200$ K to the even lower value of $0.60 \times 10^{-4}\ K^{-1}$ at $T_m + 400$ K. As is well known, silica, SiO_2, has an extremely low expansibility in the glassy state, and the same is true of liquid. Extensive information on surface tension is not available but the values shown for liquid silica and germania are quite high, reflecting the strong bonding. The viscosity is correspondingly high while the electrical conductivity is extremely low. For the chlorides the polymerization is not so extensive and the bonding not so strong, therefore the viscosity is not as high and the conductivity not as low.

Association, however, by no means leads inevitably to polymerization. The covalent nature of the bonding, expecially with the halides, is easily saturated with rather small molecular species, monomers or at most dimers. Aluminium chloride and bromide are well known dimeric species even in the gas phase or in dilute solutions. Diffraction studies with X-rays[48] indeed confirm that in liquid aluminium chloride at 493 K the entities are distorted edge-sharing double tetrahedra. In indium iodide melts at 483 K the same double tetrahedra are present, but they are much less rigid and some monomeric polyhedra are also present, since the mean coordination number of iodide ions round an indium ion is 4.7. In cadmium iodide at 723 K, the Cd–I bond length is decidedly shorter than in the solid, and the I–Cd–I bond angle is 110° rather than 90°, so that the cadmium ions have two nearer-neighbouring iodide ions and four more distant ones, rather than the regular CdI_6^{4-} octahedra present in the crystal, but the total coordination number of iodide ions round cadmium ions is 6.1. Real monomeric units are found in tin(IV) iodide at 438 K, where the radial distribution function is unequivocally explained by SnI_4 tetrahedra, with 2.2 tin atoms on average near each tin atom, 9 iodine atoms near each iodine (it is 12 in the solid), but exactly 4.0 iodine atoms near each tin atom.

These associated salts have usually low melting points and a narrow liquid range, some subliming without melting at atmospheric pressure (cf. aluminium, zirconium, hafnium and indium chlorides in Table 3.11). The conductivities are again quite low, as are the viscosities, so that the barrier to movement under the influence of an electric field is not the tight network of the surrounding neighbours, but rather the scarcity of ions. The dissociation constants for the mercury (II) halides according to

$$2HgX_2 \rightleftharpoons HgX^+ + HgX_3^- \qquad (3.15)$$

are 1.275×10^{-9} for X = Cl, 5.3×10^{-8} for X = Br, and 6.1×10^{-3} for X = I, at the respective melting points, and the corresponding degrees of ionic dissociation are 0.0035%, 0.023% and 7.8%, respectively[49]. The specific conductivities at 20 K above the freezing point are 0.0075, 0.0173 and 2.79 Ω^{-1} m^{-1}, respectively, and the activation energies, according to (3.14) are 25.7, 25.9 and −12.6 kJ mol^{-1}, respectively. Thus, although the conductivities are of the same order of magnitude as those of beryllium and zinc chlorides, Table 3.14, the activation energies are an order of magnitude lower for the mercury halides (for the iodide E_K is negative, signifying electronic rather than ionic conduction). The activation energies for viscosity are 14.4, 14.9 and 19.2 kJ mol^{-1} for mercury chloride, bromide and iodide, respectively, even less than for the conductivity, since these latter contain also the dissociation enthalpy for reaction (3.15). These activation energies are only one half (conductance) or one third (viscosity) as high as the vaporization enthalpies at the boiling point, while for the network-forming ionic liquids they are comparable (Table 3.14). The other associated ionic liquids listed in Table 3.11 (and the corresponding bromides and iodides) behave in general like the mercury halides.

References

1. B. L. Smith, *The Inert Gases, Model Systems for Science*, Wykeham Publications, London, 1971.
2. J. S. Rowlinson, *Liquids and Liquid Mixtures*, Butterworth, London, 2nd edn. 1969, p. 268.
3. J. A. Barker and D. Henderson, *Ann. Rev. Phys. Chem.*, **23**, 439 (1972).
4. A. E. Sherwood and J. M. Prausnitz, *J. Chem. Phys.*, **41**, 429 (1964).
5. O. Sinanoglu, in *Intermolecular Forces, Advances in Chemical Physics*, Vol. 12, edited by J. O. Hirschfelder, Wiley, New York, 1967 p. 283.
6. P. W. Schmidt and C. W. Tompson, in *Simple Dense Fluids*, edited by H. L. Frisch and Z. W. Salsburg, Academic Press, New York, 1968, p. 31.
7. P. G. Mikolaj and C. J. Pings, *J. Chem. Phys.*, **46**, 1401 (1967); G. T. Clayton and L. Heaton, *Phys. Rev.*, **121**, 649 (1961).
8. N. H. March, *Liquid Metals*, Pergamon Press, Oxford, 1968, Ch. 5.
9. J. E. Enderby and N. H. March, *Adv. Phys.*, **14**, 453 (1965); Y. Waseda and K. Suzuki, *Acta Met.*, **21**, 1065 (1973); Y. Waseda and M. Ohtani, *Sci. Rep. Res. Inst. Tohoku Univ.*, **24A**, 218 (1973), see also recent discussion in *The Properties of Liquid Metals*, edited by S. Takeuchi, Taylor and Francis, London, 1973, especially papers by J. E. Enderby, P. A. Egelstaff, W. S. Howells, L. E. Ballentine and D. Schiff, where the merits of the HNC equation are expounded, in contrast to the failure of the BG equation to give temperature-independent potentials, and the failure of the superposition approximation (2.17) to conform to experimental results for pressure-derivatives of the structure factor $S(k)$ (2.72).
10. L. Pauling, *Nature of the Chemical Bond*, Cornell University Press, Ithaca, N.Y. 3rd ed., 1960.
11. W. Hume-Rothery, *J. Inst. Metals*, **35**, 295 (1926).
12. A. V. Grosse, *J. Inorg. Nucl. Chem.*, **22**, 23 (1962); 28, 2125 (1966); A. V. Grosse and P. J. McGonigal, *J. Phys. Chem.*, **68**, 414 (1964); I. Z. Kopp, *Zh. Fiz. Khim.*, **39**, 360 (1965);

41, 1474 (1967); L. D. Volyak, *Zh. Fiz. Khim.*, **42**, 501 (1968); E. U. Franck and F. Hensel, *Phys. Rev.*, **147**, 109 (1966); *Rev. Mod. Phys.*, **40**, 697 (1968).
13. H. D. Jones, *Phys. Rev.*, **A, 8**, 3215 (1973).
14. O. Kubachevski, *Trans. Faraday Soc.*, **45**, 931 (1949).
15. D. J. Steinberg, *Met. Trans.*, **5**, 1341 (1974).
16. O. J. Kleppa, *J. Chem. Phys.*, **18**, 1331 (1950).
17. R. Grover, *J. Chem. Phys.*, **55**, 3435 (1971).
18. J. O. Hirschfelder, C. F. Curtiss and R. B. Bird, *Molecular Theory of Gases and Liquids*, Wiley, New York, 1964 (corrected printing).
19. K. S. Pitzer, D. Z. Lippmann, R. F. Curl, Jr., C. M. Huggins and D. E. Petersen, *J. Am. Chem. Soc.*, **77** 3433 (1955).
20. M. L. Huggins, *J. Phys. Chem.*, **74**, 371 (1970).
21. L. S. Tee, S. Gotoh and W. E. Stewart, *Ind. Eng. Chem., Fundam.*, **5**, 363 (1966).
22. D. D. Konowalow, *J. Chem. Phys.*, **46**, 818 (1967); D. D. Konowalow and S. L. Guberman, *Ind. Eng. Chem., Fundam.*, **7**, 622 (1968).
23. A. H. Narten, M. D. Danford and H. A. Levy, *J. Chem. Phys.*, **46**, 4875 (1967); A. H. Narten, *J. Chem., Phys.*, **48**, 1630 (1968).
24. A. R. Ubbelohde, *Melting and Crystal Structure*, Clarendon Press, Oxford, 1965, pp. 106, 113, 230.
25. R. H. Cole, *J. Chem. Phys.*, **27**, 33 (1957).
26. A. Das Gupta and T. S. Storvick, *J. Chem. Phys.*, **76**, 1470 (1972).
27. V. Gutmann and E. Wychera, *Inorg. Nucl. Chem. Lett.*, **2**, 257 (1966); V. Gutmann, *Coordination Chemistry in Non-Aqueous Solutions*, Springer, Vienna, 1968, p. 19; U. Mayer, V. Gutmann and W. Gerger, *Monatsh. Chem.*, **106**, 1235(1975).
28. R. Y. Lin and W. Dannhauser, *J. Phys. Chem.*, **67**, 1805 (1963); S. J. Bass, W. I. Nathan, R. M. Meighan and R. H. Cole, *J. Chem. Phys.*, **68**, 509 (1964).
29. J. H. Hu, D. White and H. L. Johnston, *J. Am. Chem. Soc.*, **75**, 1232 (1953).
30. For recent reviews see A. Ben-Naim, *Water and Aqueous Solutions*, Plenum Press, New York, 1974; *Water*, Vol. 1, edited by F. Franks, Plenum Press, New York, 1972; D. Eisenberg and W. Kauzmann, *The Structure and Properties of Water*, Oxford University Press, Oxford, 1969.
31. J. A. Pople, *Proc. Roy. Soc. Ser.*, A, **205**, 163 (1951); H. S. Frank and A. S. Quist, *J. Chem. Phys.*, **34**, 604 (1961); O. Ya. Samoilov, *Zh. Fiz. Khim.*, **20**, 1411 (1946); V. A. Mikhailov, *Zh. Strukt. Khim.*, **9**, 397 (1968); H. S. Frank and W. Y. Wen, *Discuss Faraday Soc.*, **24**, 133 (1957); G. Némethy and H. A. Scheraga, *J. Chem. Phys.*, **36**, 3382, 3401 (1962).
32. J. A. Barker and R. O. Watts, *Chem. Phys. Lett.*, **3**, 144 (1969); H. Popkie, H. Kistenmacher and E. Clements, *J. Chem. Phys.*, **59**, 1325 (1973); A. Rahman and F. H. Stillinger, *J. Chem. Phys.*, **55**, 3336 (1971); *J. Am. Chem. Soc.*, **95**, 7943 (1973); A. Ben-Naim, *J. Chem. Phys.*, **52**, 5531 (1970).
33. A. Ben-Naim, in *Water and Aqueous Solutions*, edited by R. A. Horne, Wiley, New York, 1972; cf. also A. Ben-Naim in ref. 30 above.
34. H. A. Levy and M. D. Danford in *Molten Salt Chemistry*, edited by M. Blander, Wiley, New York, 1964, Ch. 2.
35. G. Zarzycki, *J. Phys. Radium*, **18A**, 65 (1957); **19A**, 13 (1958).
36. J. K. Wilmhurst and J. M. Bracker, in *Molten Salts*, edited by G. Mamantov, Dekker, New York, 1969, pp. 303–304.
37. M. E. Melnichak and O. J. Kleppa, *Rev. Chim. Miner.*, **9**, 63 (1972).
38. L. V. Woodstock and K. Singer, *Trans. Faraday Soc.*, **67**, 12 (1971).
39. F. G. Fumi and M. P. Tosi, *J. Phys. Chem. Solids*, **25**, 31 (1964); M. P. Tosi and F. G. Fumi, *J. Phys. Chem. Solids*, **25**, 45 (1964).
40. J. Michelsen, P. Woerlee, F. van der Graaf and J. A. A. Ketelaar, *J. Chem. Soc. Faraday Trans. II*, **71**, 1730 (1975).
41. D. J. Adams and I. R. McDonald, *Solid State Phys.*, **7**, 2761 (1974); F. Lantelme, P. Turq, B. Quentree and J. W. E. Lewis, *Mol. Phys.*, **28**, 1537 (1974); C. Margheritis and C. Sinistri, *Z. Naturforsch. A.*, **30**, 83 (1975).
42. H. Reiss, S. W. Mayer and J. L. Katz, *J. Chem. Phys.*, **35**, 820 (1961).
43. G. J. Janz, *Molten Salt Handbook*, Academic Press, New York, 1967.
44. J. A. A. Ketelaar, *Z. Phys. Chem. (Frankfurt am Main)*, **98**, 453 (1975).

45. H. Reiss and S. W. Mayer, *J. Chem. Phys.*, **34**, 2001 (1961); S. W. Mayer, *J. Phys. Chem.*, **67**, 2160 (1963).
46. A. Honig, M. Mandel, M. L. Stitch and C. H. Townes, *Phys. Rev.*, **96**, 629 (1954).
47. F. H. Stillinger, in *Molten Salt Chemistry*, edited by M. Blander, Wiley, New York, 1964, Ch. 1.
48. R. L. Harris, R. E. Wood and H. L. Ritter, *J. Am. Chem. Soc.*, **73**, 3151 (1951); R. E. Wood and H. L. Ritter, *J. Am. Chem. Soc.*, **74**, 1760, 1763, (1952); **75**, 471 (1953).
49. G. Janz and J. McIntyre, *J. Electrochem. Soc.*, **109**, 842 (1962); M. S. Jhon, G. Clemena and E. R. Van Artsdalen, *J. Phys. Chem.*, **72**, 4155 (1968).

Chapter 4

A General Discussion of Liquid Mixtures

A. Thermodynamics of Mixing

a. Mixtures

In general, when n_A moles of component A, n_B moles of component B, n_C moles of component C, etc. are mixed, forming a homogeneous phase, the result is a mixture containing $n = n_A + n_B + n_C + \ldots$ moles of substances, providing there is no chemical reaction among them. If there are only two components, A and B, this will be a *binary mixture*, in which case the components will sometimes also be called 1 and 2. If one of the components, say B, changes its state (from gaseous or solid to liquid) on forming the mixture, the mixture is designated a *solution* of the solute B (and if $n_B \ll n^A$, it will be a *dilute solution*) in the solvent A.

The composition of the mixture can be stated in several ways. The most straightforward is in terms of the masses of the components, W_j (kg), defining the *mass fraction* w_i of component i as

$$w_i = W_i/(W_A + W_B + \ldots) = W_i/\Sigma_j W_j \tag{4.1}$$

If molar masses M_j are assigned to the components, then the *mole fraction* x_i is given by

$$x_i = n_i/(n_A + n_B + \ldots) = n_i/\Sigma_j n_j = n_i/n \tag{4.2}$$

$$= W_i/\Sigma_j W_j(M_i/M_j) \tag{4.3}$$

If one of the components, say A, is designated as the solvent, the number of moles of any other component i per unit mass of A is its *molality*

$$m_i = n_i/W_A = W_i/M_iW_A \tag{4.4}$$

There are two distinct ways of expressing the volume concentration. The *concentration* c_i is the number of moles of i per unit volume of the mixture, the total volume of which is V_m, or the number density of particles of species i, divided by Avogadro's number

$$c_i = n_i/V_m = \rho_i/\mathrm{N} \tag{4.5}$$

This can be converted to the *volume fraction* ϕ_i by multiplication with the partial molar volume v_i of the appropriate component at the temperature of mixing

$$\phi_i = n_iv_i/V_m = c_iv_i \tag{4.6}$$

The total volume of the mixture is given by

$$V_m = \Sigma_i n_i v_i = \Sigma_i n_i v_i^* - nv^M \tag{4.7}$$

where v_i^* is the molar volume of the pure component and v^M is the molar volume change of mixing, which is not directly determinable from the properties of the pure components. The other method of expressing volume concentration is by defining volume fractions φ_i according to

$$\varphi_i = n_iv_i^*/\Sigma_j n_j v_j^* = x_i/\Sigma_j x_j(v_j^*/v_i^*) \tag{4.8}$$

In this mode only the properties of the pure components are used to describe the mixture (Fig. 4.1). Both representations of volume composition suffer from being temperature-dependent.

In the case where there are chemical reactions between the components of the mixture to yield new chemical species, it is necessary to distinguish between the *nominal composition* and the *actual composition* of the mixture. The former is given in terms of the components that can be independently added to form the mixture, i.e. the components in the thermodynamic sense. From these nominal components, other chemical species may be formed by association, dissociation, solvation, etc., yielding interdependent actual species. If there are n nominal components, and c external constraints (such as the electroneutrality condition), and s actual species, there are $r = s - n - c$ independent chemical reactions required among these species. Consider an associated mixture of chemical species A (nominal component 1) and B (nominal component 2) forming general species A_iB_j ($i = 0,1,2, \ldots; j = 0,1,2, \ldots$). The number of moles of component 1 is given by

$$n_1 = n_A + \Sigma_{ij} i n_{A_iB_j} \tag{4.9}$$

and similarly for component 2. The number of moles of the unassociated

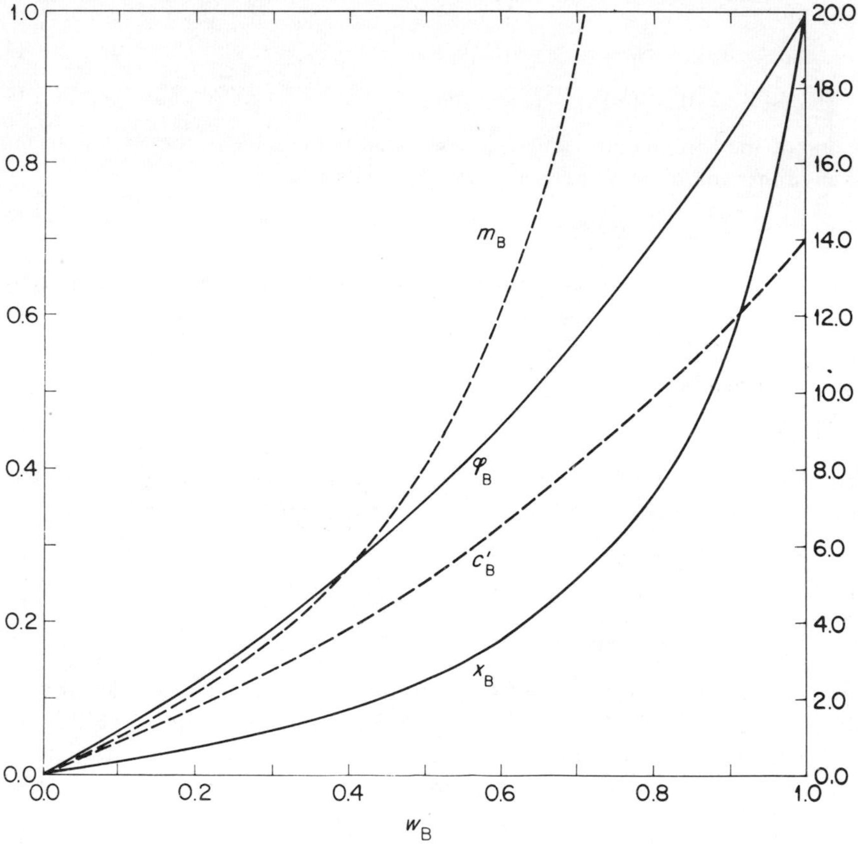

Figure 4.1 . The mole fraction x_B and volume fraction φ_B (left-hand ordinate), and the molality m_B (mol kg^{-1}) and approximate molarity c'_B.(mol dm^{-3}) (right-hand ordinate) of 2-bromoethanol (B) in water (A) plotted against the weight fraction w_B. (Since v^M is unknown, it is assumed equal to zero, and $c'_B = n_B/(n_A v_A^* + n_B v_B^*)$ is taken to represent n_B/V_m). The temperature $t = 20\,°C$, $d_A^* = 0.9982$ kg dm^{-3}, and $d_B^* = 1.7629$ kg dm^{-3}.

monomeric actual species A is therefore

$$n_A = n_1 - \Sigma_{ij} i n_{A_i B_j} \tag{4.10}$$

and again, similarly, for species B. Compositions and thermodynamic functions can then be stated either in terms of n_1 and n_2 or in terms of n_A and n_B, according to requirements.

When a mixture is formed from its components, any extensive thermodynamic property Y (e.g., G, S, V, etc.) can be expressed as the sum

$$Y_m = \Sigma_i n_i y_i \tag{4.11}$$

where Y_m refers to the mixture, and y_i is the *partial molar thermodynamic property* of i in the mixture. The quantity y_i equals the change of Y_m when one mole of this component is added to an infinitely large quantity of the mixture specified by the composition n_i, i = A, B, ... Alternatively it equals **N** times the change in Y_m when one particle of i is added to a macroscopic sample of the specified composition. The partial molar property is obtained from the partial differentiation with respect to the number of moles of the component.

$$y_i = (\partial Y_m/\partial n_i)_{n_{j\neq i},T,P} \tag{4.12}$$

For a binary mixture the partial molar property is obtainable from

$$y_i(x_i) = y_m(x_i) + (1 - x_i)(\partial y_m/\partial x_i)_{T,P} \tag{4.13}$$

where $y_m = Y_m/n$, through a well known geometric construction (Fig. 4.2).

One of the more important partial molar properties is the chemical potential, which is the partial molar Gibbs energy

$$\mu_i = g_i = (\partial G_m/\partial n_i)_{n_{j\neq i},T,P} \tag{4.14}$$

When an infinitesimal change in the composition is made, the condition of equilibrium requires that $\Delta G_m = 0$, hence the *Gibbs–Duhem relationship* arises

$$\Sigma n_i d\mu_i = 0 \text{ (at constant } T,P) \tag{4.15a}$$

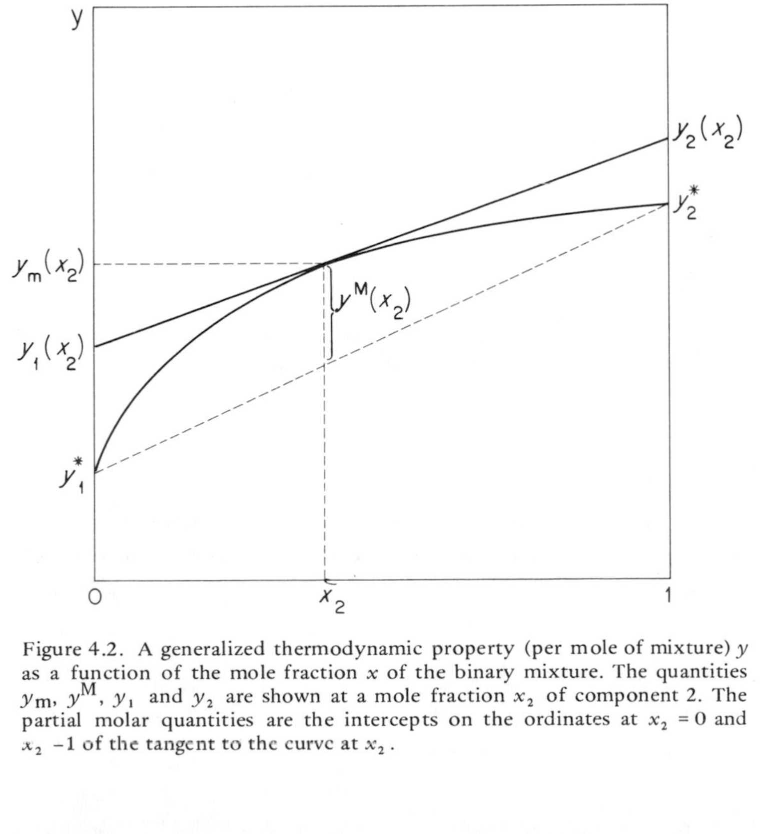

Figure 4.2. A generalized thermodynamic property (per mole of mixture) y as a function of the mole fraction x of the binary mixture. The quantities y_m, y^M, y_1 and y_2 are shown at a mole fraction x_2 of component 2. The partial molar quantities are the intercepts on the ordinates at $x_2 = 0$ and $x_2 = 1$ of the tangent to the curve at x_2.

or for a binary mixture

$$n_A d\mu_A + n_B d\mu_B = x_A d\mu_A + x_B d\mu_B = 0 \tag{4.15b}$$

from which follows

$$d\mu_A / d(\ln x_A) = d\mu_B / d(\ln x_B) \text{ (at constant } T,P) \tag{4.16}$$

since $dx_A = -dx_B$.

The chemical potential of each component is related to its *absolute activity*[1] in the mixture, λ_i, by

$$\mu_i = \mathbf{R}T \ln \lambda_i \tag{4.17a}$$

or

$$\lambda_i = \exp(\mu_i / \mathbf{R}T) \tag{4.17b}$$

while the absolute activity is related to the composition in several ways. To anticipate the following, for a perfect mixture the simple relationship

$$\lambda_i \text{ (perfect mixture)} = x_i \lambda_i^* \tag{4.18}$$

holds, where $\lambda_i^* = \exp(g_i^* / \mathbf{R}T)$ is a constant independent of composition at constant T,P. Equation (4.18) may be applied as a definition for such a mixture.

b. *Ideal mixing*

Since the liquid state is still relatively little understood, it is best to describe the mixing of liquids in terms of a perturbation of an ideal mixing process. The thermodynamic functions for the mixture will then be sums of three terms: one for the pure components in their standard state, one for the ideal process of mixing, and one for the excess function for the real mixture. For example, for the Gibbs energy the expression

$$G_m = \Sigma_i n_i g_i^* + G^M = \Sigma_i n_i g_i^* + G^{M\,id} + G^E \tag{4.19}$$

is written. Setting $\mathbf{R}T\ \Sigma_i n_i \ln x_i$ for the ideal Gibbs energy of mixing $G^{M\,id}$ is one way of expressing this idea, but the ideal mixing process can be described in several ways, leading to different expressions. For the sake of simplicity, binary mixtures of stable species A and B will be considered, but the expressions can be easily modified for multicomponent and reactive mixtures.

(i) The *statistical mechanical approach*[2] starts with an assembly of N_A molecules of species A and N_B molecules of species B ($N_A + N_B = N$) in a total volume V_m and at a temperature T. The partition function for this assembly is given by (see Appendix of Chapter 1)

$$Q(T, V_m, N_A, N_B) = \Sigma_r \exp(-E_r / \mathrm{k}T) \tag{4.20}$$

where the summation is over all the energy states r of the system. For a *perfect mixture* the following conditions hold.

(a) It is possible to factorize out of the partition function the contributions from the translational and the internal energy states of the molecules (electronic, vibrational, rotational), and these are the same for each molecule in the mixture as in the pure liquids.

(b) The energy E_r is the same for all the mutual configurations of the molecules, and is independent of the composition.

The external, configurational, part of the partition function is Q_{conf}, and the common energy of all the states is E, so that if all the N molecules were indistinguishable, the sum in (4.20) could be written as

$$Q_{conf,\ indist} = N!\ \exp(-E/\mathbf{k}T) \tag{4.21}$$

there being $N!$ arrangements of the molecules. Since, in fact, the molecules are distinguishable, it is necessary to divide this expression by $N_A!\,N_B!$

$$Q_{conf} = Z = N!(N_A!N_B!)^{-1}\ \exp(-E/\mathbf{k}T) \tag{4.22}$$

The mole fraction of species A, $x_A = N_A/N$, will be called simply x, so that $x_B = 1 - x$. The configurational partition function for the mixture is now factorized to contributions from the pure substances A and B, weighted according to their mole fractions, and to a contribution from the mixing process

$$\ln Z_m = x \ln Z_A + (1 - x) \ln Z_B + \ln Z^M \tag{4.23}$$

For the pure substances, (4.22) gives $Z_A = Z_B = \exp(-E/\mathbf{k}T)$ on setting N in turn equal to N_A and to N_B, hence

$$\ln Z^M = \ln[N!(xN)!^{-1}((1 - x)N)!^{-1}] \simeq -N[x \ln x + (1 - x) \ln(1 - x)] \tag{4.24}$$

if Stirling's approximation $\ln(N!) \simeq N \ln N - N$ is used. The Helmholz energy of mixing is then given as

$$A^M = -\mathbf{k}T \ln Z^M = N\mathbf{k}T[x \ln x + (1 - x) \ln(1 - x)] \tag{4.25}$$

For one mole of mixture $N = \mathbf{N}$, Avogadro's number, so that $N\mathbf{k} = \mathbf{R}$. For a perfect mixture also the following condition holds.

(c) There is no volume change on mixing, i.e. $V^M = V_m - \Sigma n_i v_i^* = 0$. Hence, for the Gibbs energy of mixing

$$G^M = A^M + PV^M = n\mathbf{R}T[x \ln x + (1 - x) \ln(1 - x)] \tag{4.26}$$

Equation (4.26) is thus the statistical mechanical–thermodynamic definition of a perfect mixture, based on the fulfilment of conditions (a), (b) and (c) listed above.

(ii) The *thermodynamic approach* requires the following conditions for the mixing of n_A moles of component A and n_B moles of component B to give a perfect mixture[3].

(a) There is no heat of mixing, i.e.

$$H^M = H_m - \Sigma_i n_i h_i^* = 0 \tag{4.27}$$

(b) There is no volume change of mixing, i.e.

$$V^{M} = V_{m} - \Sigma_{i} n_{i} v_{i}^{*} = 0 \tag{4.28}$$

(c) The entropy of mixing is a function of the composition only, given specifically by the expression

$$S^{M} = -n\mathbf{R}[x \ln x + (1-x) \ln(1-x)] \tag{4.29}$$

This describes then a mixing process which is perfect in terms of mole fractions. An alternative formulation, which describes mixtures which are perfect in terms of volume fractions, is

$$S^{M}(\text{perfect vol.}) = -n\mathbf{R}[x \ln \varphi + (1-x) \ln(1-\varphi)] \tag{4.30}$$

where $\varphi = \phi$ (cf. eqs. (4.6) and (4.8)) since $V^{M} = 0$. Usually, however, a mixture which is perfect in terms of (4.30) rather than of (4.29) is said to deviate from perfect mixing (see section 5A(b) on athermal mixtures).

From (4.27) and (4.29) the Gibbs energy change for perfect mixing is

$$G^{M} = H^{M} - TS^{M} = n\mathbf{R}T(x \ln x + (1-x)\ln(1-x)] \tag{4.31}$$

identical with (4.26). Also, from (4.27) and (4.28) the thermodynamic functions of mixing which do not contain the entropy, as well as derivatives of the entropy with respect to volume, pressure and temperature, are all zero.

(iii) The *operational approach*[4] defines a perfect mixture by requiring the vapour pressures of the components of the mixture to obey *Raoult's law*

$$p_{i} = x_{i} p_{i}^{*} \tag{4.32}$$

where p_{i}^{*} is the vapour pressure of the component i. However, equilibrium between the liquid and the vapour phases requires that $\mu_{i}(\text{liq}) = \mu_{i}(\text{vap})$, but

(a) the vapour is not a perfect gas (otherwise it would not liquefy!), and contributes a term $B_{i}(p_{m} - p_{i}^{*})$ to the chemical potential (where B_{i} is the second virial coefficient, higher terms being neglected), for the non-ideal work of expansion from the pressure of the pure component, p_{i}^{*}, to the pressure of the mixture, p_{m};

(b) since the liquid is not under the same pressure in the pure states and in the mixture, a compression term $v_{i}^{*}(p_{i}^{*} - p_{m})$ is added to the chemical potential. The requirement that $\mu_{i}(\text{liq}) = \mu_{i}(\text{vap})$ therefore leads to

$$\ln p_{i} = \ln x_{i} p_{i}^{*} - [B_{i}(p_{m} - p_{i}^{*})/\mathbf{R}T] - [v_{i}^{*}(p_{i}^{*} - p_{m})\mathbf{R}T] \tag{4.33}$$

The validity of Raoult's law (4.32) then requires that in (4.33) the last two terms on the right-hand side exactly cancel each other, i.e. $B_{i} = v_{i}^{*}$. Since v_{i}^{*} is positive, if the vapours are perfect, the liquid mixture cannot be perfect in terms of Raoult's law. Usually $|B_{i}| \gg v_{i}^{*}$, so that a perfect mixture in terms of $H^{M} = 0$ and $V^{M} = 0$ will not obey Raoult's law (and *vice versa*), although the deviations are quite small.

Often, the terms 'perfect mixture' and 'ideal mixture' are used interchangeably. It is, however, convenient to apply the term 'ideal mixtures', or *ideal solutions* to mixtures in which the solvent A obeys Raoult's law

$$p_A = x_A p_A^* \tag{4.34}$$

over a composition range where for the solute B, *Henry's law*

$$p_B = K_{B(A)} x_B \tag{4.35}$$

holds, where $K_{B(A)}$ is a constant characteristic of both B and A, and is different from p_B^*. If the solute B is a gas, then $K_{B(A)}^{-1}$ is its Bunsen solubility coefficient. Solutions are usually not ideal over the whole composition range – which may often not be realizable because of solubility limitations – but approach ideality as they become dilute in B. The *ideal dilute solution* is thus the limiting case (Fig. 4.3)

$$\lim_{x_B \to 0} \mathrm{d}p_B/\mathrm{d}x_B = K_{B(A)} = const. \tag{4.36}$$

$$\lim_{x_A \to 1} \mathrm{d}p_A/\mathrm{d}x_A = p_A^* \tag{4.37}$$

Equation (4.19) implies that for perfect mixtures $G^E = 0$. Introduction of (4.34)

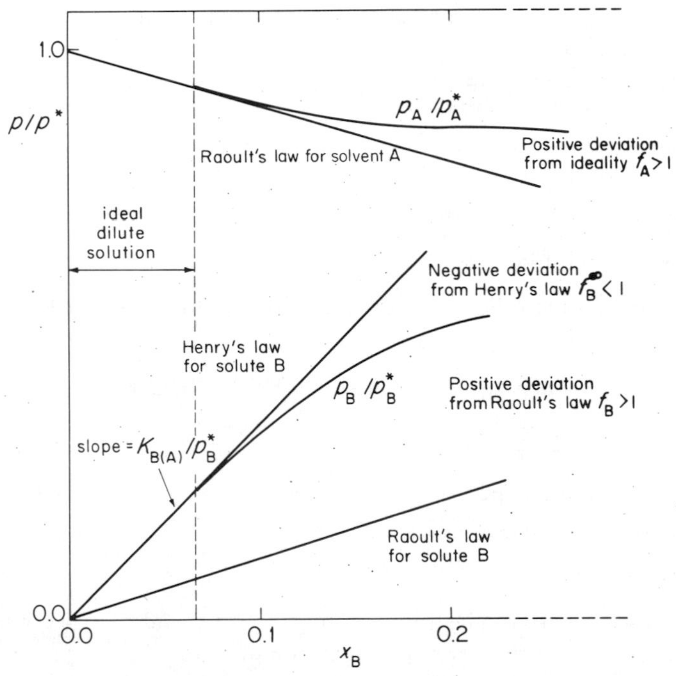

Figure 4.3. The vapour pressure diagram of a dilute solution of the solute B in the solvent A. The region of ideal dilute solutions, where Henry's and Raoult's laws are obeyed by the solute and solvent respectively, is marked off, as are the deviations from ideality at higher concentrations

and (4.26) (or 4.31) into (4.14) yields for the perfect mixture

$$\mu_i = g_i^* + (\partial G^M/\partial n_i)_{n_{i\neq j},T,P} = \mu_i^* + \mathbf{R}T \ln x_i \qquad (4.38)$$

which leads directly to (4.18). For non-perfect mixtures, eq. (4.38) does not hold, and an excess chemical potential μ_i^E must be added on the right-hand side. The deviations from ideal (perfect) mixing behaviour will now be discussed.

c. *Deviations from ideality*

Real systems do not, as a rule, conform to the behaviour of perfect mixtures expressed in (4.31) or (4.32). This deviation is expressed in terms of the thermodynamic excess functions. For the Gibbs energy

$$G^E = G^M - \mathbf{R}T\Sigma_i n_i \ln x_i \qquad (4.39)$$

and similar expressions can be written for S^E (= S^M + $\mathbf{R}\Sigma n_i \ln x_i$), H^E(= H^M since $H^{M\,id}$ = 0), V^E(= V^M), etc. The corresponding molar quantities are obtained by dividing by n, thus

$$g^E = g^M - g^{M\,id} = g_m - \Sigma_i x_i g_i^* - \Sigma_i x_i \ln x_i \qquad (4.40)$$

The partial molar quantities can be obtained by differentiating the extensive excess functions with respect to n_i, thus

$$\mu_i^E = g_i^E = (\partial n g^E/\partial n_i)_{n_{j\neq i},T,P} = \mu_i - \mu_i^* - \mathbf{R}T \ln x_i \qquad (4.41)$$

The deviations from ideality are designated as positive or negative according to the sign of the excess functions, and in general according to the sign of g^E. Deviations for ideality also imply deviations from Raolut's law, such that (Fig. 4.3)

$$p_i/p_i^* \gtreqless x_i \qquad (\text{for } \mu_i^E \gtreqless 0) \qquad (4.42)$$

From (4.17) and (4.41) follows the definition of the *(rational) activity coefficient* f_i

$$\ln f_i = \mu_i^E/\mathbf{R}T = (\mu_i - \mu_i^*)/\mathbf{R}T - \ln x_i = \ln \lambda_i/\lambda_i^* - \ln x_i$$

$$= \ln p_i/p_i^* - \ln x_i \qquad (4.43)$$

Alternatively, f_i can be defined in terms of the activities, introducing the *(relative) activity* a_i

$$a_i = \lambda_i/\lambda_i^* = p_i/p_i^* = x_i f_i \qquad (4.44)$$

Positive deviations from ideality thus imply $f_i > 1$ and negative ones $f_i < 1$. However g^E is the sum of the deviations of all components

$$g^E = \Sigma_i x_i \ln f_i \qquad (4.45)$$

so that $g^E > 0$ does not prevent some $f_i < 1$ over a part of the composition range. The Gibbs–Duhem relationship (4.16), however, can be written in an integral form, involving the excess chemical potentials

$$\int_0^1 \mu_A^E \mathrm{d}x_A = \int_0^1 \mu_B^E \mathrm{d}x_B \quad \text{(at constant } T,P) \tag{4.46}$$

which can be written as

$$\int_0^1 (\mu_A^E - \mu_B^E)\mathrm{d}x_A = 0 \tag{4.47}$$

$$\int_0^1 \ln(f_A/f_B)\mathrm{d}x_A = 0 \tag{4.48}$$

Therefore if $f_A/f_B > 1$ over some part of the composition range, it must be < 1 over some other part, in a compensating way, so that (4.48) is obeyed strictly[5]. The Gibbs–Duhem relationship also leads to another useful result for the activity coefficients

$$x_A(\mathrm{d}(\ln f_A/\mathrm{d}x_A) = x_B(\mathrm{d}(\ln f_B)/\mathrm{d}x_B)) \tag{4.49}$$

From this the one can be determined if the behaviour of the other as a function of the composition is known

$$\ln f_A = \int_{x_B}^1 (x_B/x_A)\mathrm{d}(\ln f_B) \tag{4.50}$$

When dilute solutions are discussed, designated by superscript ∞, it is convenient to introduce a new standard state, that of the infinitely dilute solution of all solutes B, C, . . . in the solvent A. In this situation, the standard state for the solvent is still the pure solvent

$$a_A^{\infty 0} = \lim_{x_A \to 1} a_A^{\infty} = a_A^* = 1 \tag{4.51}$$

$$f_A^{\infty 0} = \lim_{x_A \to 1} f_A^{\infty} = f_A^* = 1 \tag{4.52}$$

For a solute B, however, the activity coefficient f_B^{∞} is defined by (4.43) with the reference chemical potential $\mu_B^{\infty 0}$ arbitrarily fixed in such a way that (Fig. 4.3)

$$\lim_{x_B \to 0} \ln f_B^{\infty} = \ln f_B^{\infty 0} = \lim_{x_B \to 0} [(\mu_B^{\infty} - \mu_B^{\infty 0})/RT - \ln x_B] = 0 \tag{4.53}$$

or that $f_B^{\infty 0} = 1$. This choice of the standard state applies to each of the solutes B, C . . . individually. It is this asymmetric choice of standard states that may be used to separate semantically solutions from mixtures. Solutes that obey Henry's law (4.36) over a certain concentration range in dilute solutions have $f_B^{\infty} = 1$ over this range. The constant $K_{B(A)}$ is related to $\mu_B^{\infty 0}$ simply as

$$K_{B(A)} = p_B^* \exp[(\mu_B^{\infty 0} - \mu_B^*)/RT] = p_B^* \exp[\Delta g_{B(A)}^{tr}/RT] \tag{4.54}$$

The molar Gibbs energy change for transferring B from the standard state of its infinitely dilute solution in solvent A to the standard state of the pure liquid B, $\Delta g^{tr}_{B(A)}$, is thus defined as the difference between the two standard chemical potentials $\mu_B^{\infty 0}$ and μ_B^*. It must be remembered that often the standard state of a pure liquid at the temperature at which $\mu_B^{\infty 0}$ applies cannot be realized, and is only hypothetical. The relationship between f_B^∞ and f_B is obtained similarly from $\Delta g^{tr}_{B(A)}$ as

$$f_B^\infty = f_B \exp(-\Delta g^{tr}_{B(A)}/RT) \tag{4.55}$$

In the cases where the composition of the solution is expressed in terms other than mole fractions, namely in terms of molalities (4.4) or concentrations (4.5), it is convenient to define the respective activity coefficients in these scales. The standard chemical potential is always defined according to the appropriate scale, and the superscript ∞ will be omitted, since the infinitely dilute standard state is implied. For the *molal activity coefficient*

$$\ln \gamma_i = (\mu_i - \mu^0_{(m)i})/RT - \ln m_i \tag{4.56}$$

and for the *molar activity coefficient*

$$\ln y_i = (\mu_i - \mu^0_{(c)i})/RT - \ln c_i \tag{4.57}$$

Since the infinitely dilute standard states are chosen so that

$$\lim_{m_i \to 0} \gamma_i = \lim_{c_i \to 0} y_i = \lim_{x_i \to 0} f_i^\infty = 1 \tag{4.58}$$

the relationships between the activity coefficients and the concentration scales may be readily derived. For example

$$\ln f_B^\infty = \ln \gamma_B + \ln(1 + m_B M_A)$$
$$= \ln y_B - \ln d_A^* + \ln(d - c_B(M_B - M_A)) \tag{4.59}$$

where d is the density of the solution, $\Sigma_i n_i M_i / V_m$, and d_A^* is the density of the solvent M_A/v_A^*. Consequently, the relationship between the molal and molar activity coefficients is

$$\ln y_B = \ln \gamma_B - \ln d_A^* - \ln[(1 + m_B M_B)/d] \tag{4.60}$$

The derivatives of the activity coefficients with respect to the temperature and the pressure are of interest

$$(\partial(\ln f_i/\partial T)_{P,x_i} = -h_i^E/RT^2 = (h_i^* - h_i)/RT^2 \tag{4.61}$$

$$(\partial(\ln f_i^\infty)/\partial T)_{P,x_i} = (h_i^\infty - h_i)/RT^2 \tag{4.62}$$

$$(\partial(\ln f_i)/\partial P)_{T,x_i} = v^E/RT = (v_i - v_i^*)/RT \tag{4.63}$$

$$(\partial(\ln f_i^\infty)/\partial P)_{T,x_i} = (v_i - v_i^\infty)/RT \tag{4.64}$$

where h_i^∞ and v_i^∞ are, respectively, the partial molar enthalpy and volume of component i at infinite dilution. Expressions similar to (4.62) and (4.64) can be written for γ_i, since the interconnection of f_i^∞ and γ_i is independent of both temperature and pressure, but not for y_i, since the densities are pressure- and temperature-dependent.

d. Phenomenological expressions

In order to gain more insight into the behaviour of liquid mixtures and solutions, it is important to examine the dependence of the excess functions, or of the activity coefficients, on the composition. Several empirical expressions have been suggested, and according to the degree of simplification permitted in definite cases, these can be ordered into several classes of mixtures.

Since all the thermodynamic functions can be obtained from the Gibbs energy by appropriate differentiation, it suffices to write for a binary mixture the Redlich–Kister equation[6]

$$g^E = x_1 x_2 \Sigma_j b_j(T,P)(x_1 - x_2)^{j-1} \quad (1 \leqslant j \leqslant k) \tag{4.65}$$

From the expression $\mathbf{R}T \ln f_i = (\partial(ng^E)/\partial n_i))_{T,P,n_j}$ the k-suffix Margules expressions for the activity coefficients are obtained

$$\mathbf{R}T \ln f_1 = b_1' x_2^2 + b_2' x_2^3 + \ldots b_k' x_2^{k+1} \tag{4.66}$$

$$\mathbf{R}T \ln f_2 = b_1'' x_1^3 + b_2'' x_1^3 + \ldots b_k'' x_1^{k+1} \tag{4.67}$$

where

$$b_1' = b_1 + 3b_2 + \ldots (2k-1)b_k \qquad b_1'' = b_1 - 3b_2 + \ldots -(2k-1)b_k$$

$$b_2' = -4(b_2 + 4b_3 + \ldots (k-1)^2 b_k) \qquad b_2'' = 4(b_2 + 4b_3 + \ldots (k-1)^2 b_k)$$

$$b_3' = 12(b_3 + 5b_4) \qquad b_3'' = 12(b_3 - 5b_4)$$

$$b_4' = -32b_4 \qquad b_4'' = 32b_4 \tag{4.68}$$

and terms higher than b_4 (i.e., $k > 4$) are very seldom used. It should be noted that the leading term in $\ln f_i$ is the one in $(1 - x_i)^2$, since the first-power terms cancel out because of the Gibbs–Duhem equation. The terms with odd j in (4.65) have even powers of $(x_1 - x_2)$ or $(2x_1 - 1)$, and hence, are symmetric with regard to the composition, while those with even j contribute to the asymmetry of the curve $g^E(x_1)$. It should also be noted that in the general case the coefficients b_j are functions of the temperature and the pressure. The temperature-dependence can be written as[7]

$$b_j = \alpha_j + \beta_j T + \gamma_j T \ln T \quad \text{(at constant } P\text{)} \tag{4.69}$$

Whenever this expression holds, the excess heat capacity of the binary mixture is given by

$$c_p^E = -x_1 x_2 \Sigma_j \gamma_j (x_1 - x_2)^{j-1} \quad (j \leqslant k) \tag{4.70}$$

and will be independent of the temperature, and if $k = 1$, it will be a simple parabolic function of the composition. Similarly, if (4.69) holds

$$s^E = -x_1 x_2 \Sigma_j(\beta_j + \gamma_j(1 + \ln T))(x_1 - x_2)^{j-1} \quad \text{(at constant } P, j \leqslant k) \qquad (4.71)$$

which again is considerably simplified if $k = 1$ and $\gamma_1 = 0$, since then $s^E = -\beta x_1 x_2$.

For cases where 1-suffix Margules expressions are adequate, and if also $\beta = 0$, and only $\alpha \neq 0$, that is if b is independent of the temperature, the excess entropy vanishes, and the mixture is called a *regular solution*[8,9], where $h^E = g^E = bx_1x_2$ (for $k = 1$). The excess enthalpy is obtained in the general case as

$$h^E = x_1 x_2 \Sigma_j(-\alpha_j + \gamma_j T)(x_1 - x_2)^{j-1} \quad \text{(at constant } P, j \leqslant k) \qquad (4.72)$$

The special case where $b/T = \beta = constant$ ($\alpha = \gamma = 0$, for $k = 1$) leads to *athermal solutions*, where $h^E = 0$ (Fig. 4.4)[10].

The series expansion (4.65) yields in a simple manner the activity coefficient of one component if that of the other is known as an explicit function of the composition. When (4.66) holds for component 1, then from the Gibbs–Duhem relationship

$$\ln f_2 = \Sigma_j b_j' x_2^{j-1} - \Sigma_j b_j'(j-2)^{-1}[(j-1)x_2^{j-2} - 1] \quad (j \leqslant k) \qquad (4.73)$$

for component 2.

There are binary mixtures for which (4.65) is not the most convenient

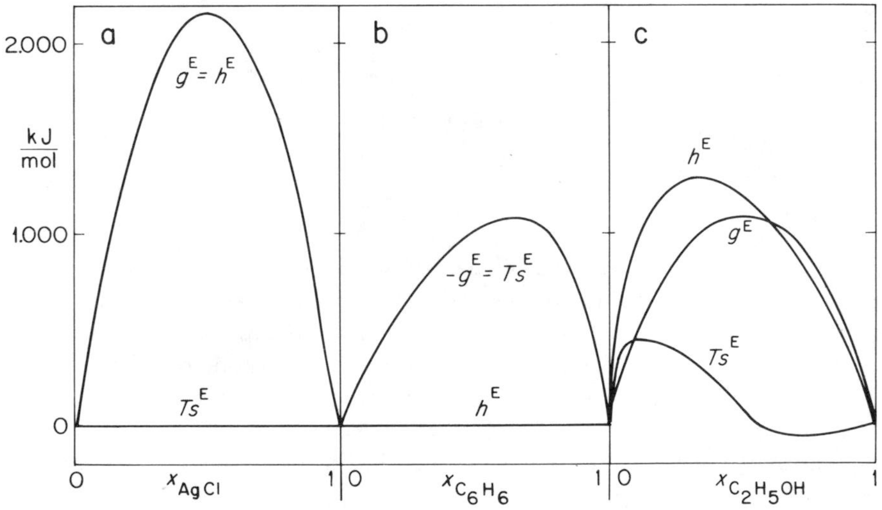

Figure 4.4. Excess functions in binary liquid mixtures. (a) Regular solution: molten silver chloride and lithium chloride at 873 K, $s^E = 0$; data from E. J. Salstrom, T. J. Kew and T. M. Powell, *J. Am. Chem. Soc.*, **58**, 1848 (1936). (b) Athermal solution: low molecular weight polystyrene $H[-(C_6H_5)CHCH_2-]_{10}H$ in benzene at 298 K; note that g^E is negative in this system; after R. F. Blanks and J. M. Prausnitz, *Ind. Eng. Chem., Fundam.*, **3**, 1(1964). (c) A highly complex non-ideal system: ethanol and iso-octane at 298 K, after C. G. Savini, D. R. Winterhalter and H. C. Van Ness, *J. Chem. Eng. Data*, **10**, 168, 171 (1965).

expression for g^E, but which can be better described by means of z_i fractions instead of the mole fractions x_i. The fraction z_i is defined as[11]

$$z_1 = x_1/(x_1 + rx_2); \quad \text{and} \quad z_2 = rx_2/(x_1 + rx_2) \tag{4.74}$$

with r an arbitrary parameter (e.g. the ratio of the free volumes). In the special case, where $r = v_2^*/v_1^*$, the ratio of the molar volumes of the components, then $z_i = \varphi_i$, the volume fraction. The expression for the excess Gibbs energy becomes

$$g^E/\mathbf{R}T(x_1 + rx_2) = b_{12}z_1z_2 + b_{112}z_1^2z_2 + b_{122}z_1z_2^2 + \ldots \tag{4.75}$$

the Wohl expression, where r can be treated as another, arbitrary parameter. The activity coefficients can again be obtained from the derivative with respect to n_i. For the special case where all the b values above b_{12} vanish, then

$$g^E/\mathbf{R}T = b_{12}rx_1x_2/(x_1 + rx_2) \tag{4.76}$$

and for the activity coefficients the van Laar equations[12]

$$\mathbf{R}T \ln f_1 = b_{12}(1 + x_1/rx_2)^{-2} \tag{4.77}$$

$$\mathbf{R}T \ln f_2 = b_{12}r(1 + rx_2/x_1)^{-2} \tag{4.78}$$

are obtained, in two arbitrary parameters, b_{12} and r.

e. *Demixing*[13]

Mixtures for which at every composition, temperature and pressure

$$(\partial^2 g^M/\partial x_i^2)_{T,P,n_{j\neq i}} > 0 \tag{4.79}$$

are stable as a single phase. However, mixtures which at a given pressure over a temperature range have a region for which

$$(\partial^2 g^M/\partial x_i^2) < 0 \tag{4.80}$$

or equivalently,

$$(\partial^2 g^E/\partial x_i^2) + \mathbf{R}T/x_i(1 - x_i) < 0 \tag{4.81}$$

or

$$(\partial \mu_i/\partial x_i) < 0 \tag{4.82}$$

or

$$(\partial a_i/\partial x_i) < 0 \tag{4.83}$$

or

$$(\partial f_i/\partial x_i) < 0 \tag{4.84}$$

are in this region unstable, will demix and split into two phases (Fig. 4.5). The set of compositions and temperatures for which $(\partial^2 g^M/\partial x_i^2) = 0$ is called the *spinodal*

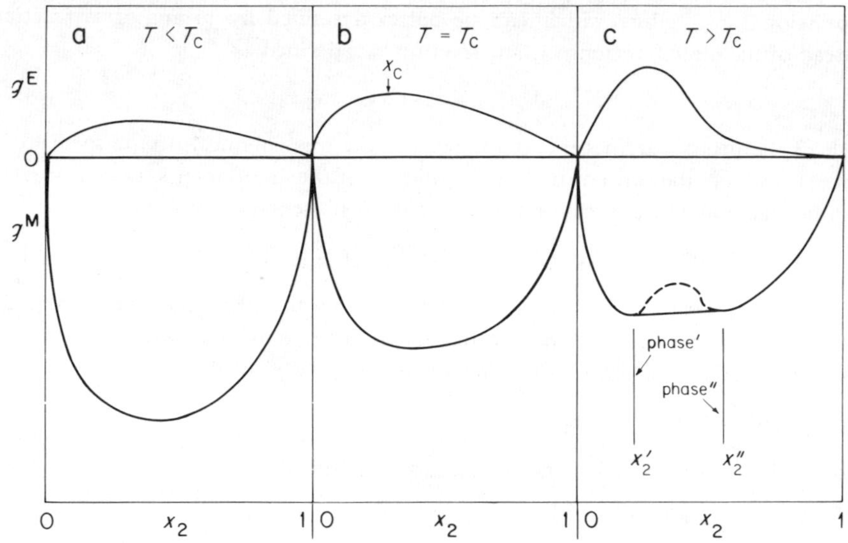

Figure 4.5. A system with a lower critical solution point at (x_c, T_c). (a) At $T < T_c$, a single phase is stable, (4-79) holds. (b) At $T = T_c$, (4-87) holds. (c) At $T > T_c$ the system splits into two phases, of compositions x'_2 and x''_2 and (4-80) to (4.84) hold for compositions in between.

curve, while the *coexistence curve* for a binary mixture requires at any temperature

$$\mu'_i = \mu''_i \tag{4.85}$$

$$a'_i = a''_i \tag{4.86}$$

where the prime and double-prime represent the two coexisting phases at equilibrium with each other. It is clear that $f'_i \neq f''_i$, since otherwise condition (4.86) will require also $x'_i = x''_i$, that is, both phases have the same composition, i.e. are identical and represent one stable phase. The points x_c and T_c where (Fig. 4.5)

$$(\partial^2 g^M/\partial x^2) = 0 \quad \text{and} \quad (\partial^3 g^M/\partial x^3) = 0 \tag{4.87}$$

are the *critical solution points.* A system may have none, one or two critical solution points. In the first case there is partial miscibility over the whole temperature range where the mixture is liquid, if the condition of instability applies, or complete miscibility, if (4.79) applies throughout. In the second case there may be a *lower critical solution temperature* (LCST) if on raising the temperature the system goes from complete miscibility to a two-phase system (the rarer case), or an *upper critical solution temperature* (UCST) if this transition occurs on lowering the temperature (the more common case). The former situation requires $(\partial^2 h^M/\partial x^2) > 0$ and $T_c(\partial^2 s^E/\partial x^2) > (\partial^2 h^E/\partial x^2)$ (this usually involves $h^E < 0$) and the latter situation requires the reverse relationships. In the third case, there may be a closed solubility loop, with both a UCST and an LCST, or the system may have at a low temperature a UCST, and at a higher temperature an

LCST, with a complete miscibility region in between – a very rare case indeed (Fig. 4.6).

Consider now the special case[7] where

$$g^E = z_1 z_2 (\alpha + \beta T + \gamma T \ln T) = z_1 z_2 b(T) \quad (4.88)$$

is obeyed. This expression describes a type of regular solution where a correction for $r \neq 1$ has been made, such that, at a given temperature it has activity coefficients $\mathbf{R}T \ln f_i = b(1 - z_i)^2$. Such a solution will have an instability domain

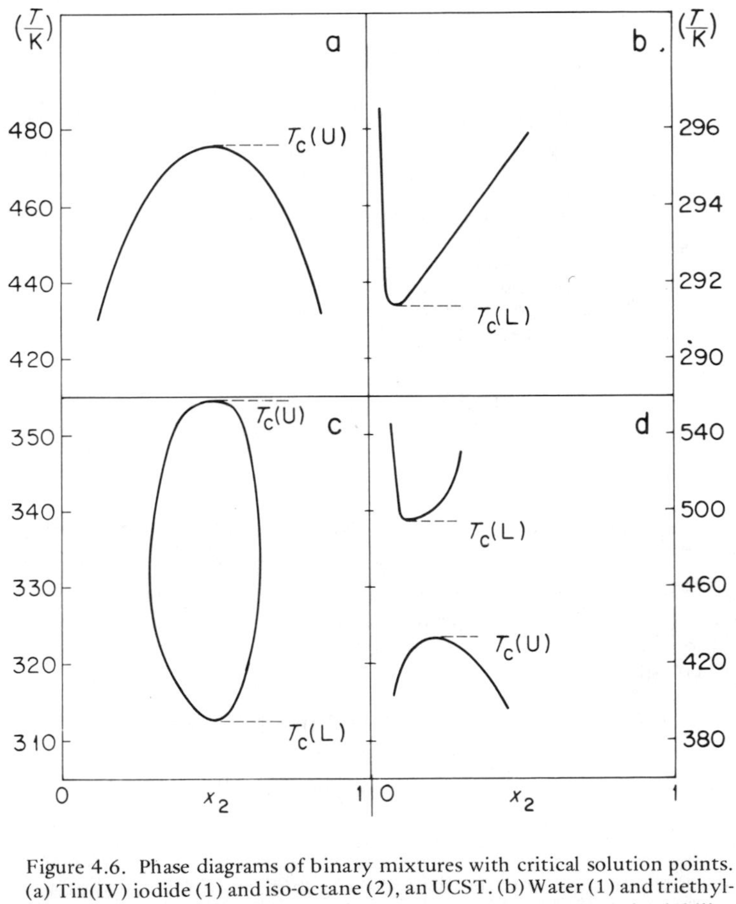

Figure 4.6. Phase diagrams of binary mixtures with critical solution points. (a) Tin(IV) iodide (1) and iso-octane (2), an UCST. (b) Water (1) and triethylamine (2), an LCST. (c) Guaiacol (1) and glycerol (2), a closed solubility loop with a lower LCST and an upper UCST. (d) Sulphur (1) and benzene (2), a system with two two-phase regions with an intermediate region of complete miscibility, a lower UCST and an upper LCST. Data from: (a) M. E. Dice and J. H. Hildebrand, *J. Am. Chem. Soc.*, **50**, 3023 (1928); (b) B. J. Hales, G. L. Bertrand and L. G. Hepler, *J. Phys. Chem.*, **70**, 3970 (1966); (c) H. Kehiaian, *Bull. Acad. Polon. Sci., Ser. Sci. Chim.*, **10** 585 (1962); (d) W. D. Groves and J. S. Forsyth, *Proc. Intern. Solvent Extraction Conf., ISEC 71*, Soc. Chem. Ind., London, 1971, p. 1221.

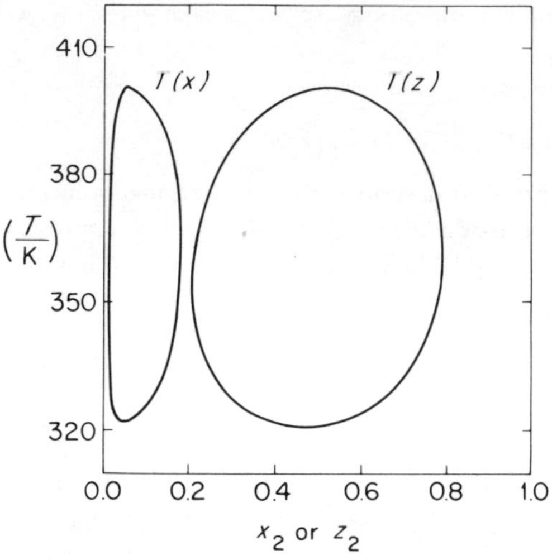

Figure 4.7. Phase diagram of water (1) and 2-butoxyethan-l-ol (2) mixtures, plotted against mole fractions, $T(x)$, and corrected for asymmetry with $r = 0.058$ (whereas $v_1^*/v_2^* = 0.137$), for $T(z)$. Data from H. Kehiaian, *Bull. Acad. Polon. Sci., Ser. Sci. Chim.*, **10**, 585 (1962).

determined by

$$b(T) > \mathbf{R}T/2z_i(1 - z_i) \tag{4.89}$$

which will be asymmetric with respect to x for $r \neq 1$, but symmetric with respect to z (Fig. 4.7). In fact, it is possible to select an arbitrary r value (not necessarily (v_2^*/v_1^*)), to convert from an asymmetric (T,x) curve to a symmetric (T,z) curve. If, in (4.88), $\gamma = 0$, only one critical solution point

$$T_c = \alpha/(2\mathbf{R} - \beta) \tag{4.90}$$

is possible, a UCST for $\alpha > 0$ and $\beta < 2\mathbf{R}$, and an LCST for $\alpha < 0$ and $\beta > 2\mathbf{R}$. Only if $\gamma \neq 0$ is a closed solubility loop possible, if $\gamma < 0$, $\alpha < 0$ and $\beta > 2\mathbf{R} - \gamma(1 + \ln \alpha/\gamma)$. The coexistence curve is given for all cases by

$$b(T) = \mathbf{R}T(1 - 2z)^{-1} \ln(1 - z)/z \tag{4.91}$$

for all the values of z, except that it is given by $b(T) = 2\mathbf{R}T(T = T_c)$ at $z = 1/2 = z_c$.

B. Molecular Aspects of Mixing

The previous discussion of mixing was wholly from the thermodynamic standpoint dealing with the macro-properties of bulk phases, with complete disregard of the molecular nature of the liquids (except in the statistical derivation of G^M for perfect mixtures) and of the forces between the molecules. Of greater predictive value is the molecular approach, which yields an insight into the relationships of the particles in the mixture. However, the concept of excess functions permits the molecular description of the pure component liquids to be taken as granted, and focuses attention on the *changes* which are caused by the process of mixing.

As for pure liquids, any molecular theory of mixtures considers the interactions of pairs of molecules in a volume V, containing N_A molecules of species A and N_B molecules of species B, as a function of their coordinates. It is assumed that the internal and translational contributions to the partition function (section 1D) can be factorized out, leaving the configurational contribution

$$Z = Q_{conf}(T, V, N, x) \tag{4.92}$$

It is instructive to consider the simplified situation, where the former contributions cancel out in the excess thermodynamic functions, and only the latter, the configurational partition function, remains. In an isotropic liquid, the pair mutual interaction energy becomes a function of the distance r, rather than of the coordinates, and may be described as

$$\epsilon_{ij}(r) = \epsilon_{ij}^* \psi(r/r_{ij}^*) \tag{4.93}$$

in terms of the two scaling parameters ϵ_{ij}^* and r_{ij}^* and a function ψ, i and j taking on the designations AA, BB and AB for the three kinds of mutual pair interactions.

The various molecular theories of mixing differ according to the restrictions they put on the function ψ, and according to the averaging processes that are used in evaluating Z. At the outset there are two different ways of carrying out the averaging process. If results for constant volume thermodynamic functions are desired as a primary goal, the treatment in terms of pair correlation functions $g_{ij}(r)$ is convenient, but a change from a (T, V, N, x) system to a (T, P, N, x) system is rather cumbersome. An alternative method is to expand Z in a Taylor series with the aid of perturbation parameters, around the properties of a reference substance. These approaches will now be briefly examined individually.

a. *The pair correlation function method*[14]

As in section 1C, dealing with pure liquids, the additivity of pair mutual potential energies is assumed, so that the potential energy of the pure liquid i will be taken as (see p. 48)

$$U_i^* = 2\pi n_i \mathrm{N}^2 (v_i^*)^{-1} \int \epsilon_i(r) g_i(r) r^2 \, dr \tag{4.94}$$

for n_i moles of liquid. In a binary mixture, the probability of finding at a distance r from a molecule of component 1 another molecule of component 1 will be $g_{11}(r)$, while that of finding there a molecule of component 2 will be $g_{12}(r)$, and the respective pair interaction energies are $\epsilon_{11}(r)$ and $\epsilon_{12}(r)$. Similarly, the functions $g_{21}(r)$, $g_{22}(r)$, $\epsilon_{21}(r)$ and $\epsilon_{22}(r)$ are now defined. For a constant volume system the total potential energy is

$$U_m = 2\pi N^2 \{(n_1/v_1^*)\int \epsilon_{11}(r)g_{11}(r)r^2\,dr + (n_1/v_2^*)\int \epsilon_{12}(r)g_{12}(r)r^2\,dr$$

$$+ (n_2/v_1^*)\int \epsilon_{21}(r)g_{21}(r)r^2\,dr + (n_2/v_2^*)\int \epsilon_{22}(r)g_{22}(r)r^2\,dr\} \qquad (4.95)$$

where $\epsilon_{21}(r) = \epsilon_{12}(r)$. The limiting value of $g_{11}(r)$ and of $g_{21}(r)$ is simply the probability of finding anywhere a molecule of 1, and this equals the volume fraction φ_1. Similarly, $g_{12}(r)$ and $g_{22}(r)$ approach φ_2 at large r, hence $\varphi_2 g_{21}(r) = \varphi_1 g_{12}(r)$. The excess potential energy of mixing, $U^E = U_m - U_1^* - U_2^*$ is then given by

$$U^E = 2\pi N^2(n_1 v_1^* + n_2 v_2^*)\{(\varphi_1/v_1^*)^2 \int \epsilon_{11}(r)[g_{11}(r)/\varphi_1 - g_{11}^*(r)]\,r^2\,dr$$

$$+ (\varphi_2/v_2^*)^2 \int \epsilon_{22}(r)[g_{22}(r)/\varphi_2 - g_{22}^*(r)]\,r^2\,dr$$

$$+ (\varphi_1\varphi_2/v_1^* v_2^*)[2\int \epsilon_{12}(r)g_{12}(r)\varphi_2^{-1}\,r^2\,dr$$

$$- (v_2^*/v_1^*)\int \epsilon_{11}(r)g_{11}^*(r)r^2\,dr - (v_1^*/v_2^*)\int \epsilon_{22}(r)g_{22}^*(r)r^2\,dr]\} \qquad (4.96)$$

There is only one cross-term, containing ϵ_{12} and g_{12}, which characterizes the mixture, but it is the problem of relating ϵ_{12} to ϵ_{11} and ϵ_{22}, and g_{12} to g_{11} and g_{22}, that is crucial for this approach.

As a zeroth approximation, a system comprising spherical molecules with central forces and equal sizes will be considered. For these the assumption of *random mixing* holds, and

$$g_{11} = g_{11}^* = g_{22} = g_{22}^* = g_{12}/\varphi_2 = g \qquad (4.97)$$

Also, $v_1^* = v_2^* = v^*$ and $v^E = 0$. For such a mixture (4.96) can be readily simplified to give the excess potential energy per mole.

$$u^E = 2\pi N^2 v^{*-1} x_1 x_2 \int (2\epsilon_{12}(r) - \epsilon_{11}(r) - \epsilon_{22}(r))g(r)r^2\,dr \qquad (4.98)$$

In the zeroth approximation, ψ of (4.93) is the same for all ij, and $r_{11}^* = r_{22}^* = r_{12}^* = r^*$ (the equal size requirement), so that the integral

$$\int \psi(r/r^*)g(r)r^2\,dr = \psi^* \qquad (4.99)$$

can be factorized out. The product $2\pi N^2 \psi^* \epsilon_{ij}^* = C_{ij}$ is now introduced into (4.98) giving the simple equation

$$u^E = x_1 x_2(2C_{12} - C_{11} - C_{22})/v^* \qquad (4.100)$$

Because of the random mixing stipulation, there is no excess entropy, $s^E = 0$, and therefore $g^E = h^E = a^E = u^E$ are all given by (4.100).

For a first approximation the random mixing restriction, (4.97), is retained but the restrictions of equal sizes and $v^E = 0$ are relaxed. In this case (4.96) reduces to

$$u^E = 2\pi N^2 (x_1 v_1^* + x_2 v_2^*)\varphi_1\varphi_2 [(2/v_1^* v_2^*) \int \epsilon_{12}(r)g(r)r^2 \,dr$$

$$- (1/v_1^*)^2 \int \epsilon_{11}(r)g(r)r^2 \,dr - (1/v_2^*)^2 \int \epsilon_{22}(r)g(r)r^2 \,dr] \qquad (4.101)$$

Now, if the function ψ is of the Lennard-Jones type (section 1C) and the arithmetic mean (Lorenz) rule holds for the scaling parameter $r_{12}^* = (r_{11}^* + r_{22}^*)/2$, it is again possible to factorize out the integral, but this time $C_{ij}(r_{ij}^{*3})$ will have to be used, since the size-dependence cannot in the general case be factorized out, giving

$$u^E = (x_1 v_1^* + x_2 v_2^*)\varphi_1\varphi_2 [(2/v_1^* v_2^*)C_{12}(r_{12}^{*3}) - (1/v_1^*)^2 C_{11}(r_{11}^{*3})$$

$$- (1/v_2^*)^2 C_{22}(r_{22}^{*3})] \qquad (4.102)$$

where the C_{ij} values are integral functions of the molecular sizes. Again $a^E = u^E$, but since $v^E \neq 0, g^E$ cannot be readily derived.

b. Pertubation methods

The perturbation methods are based on the theorem of corresponding states, on the concept of *conformal liquids* (section 1B) and on that of *conformal mixtures* introduced by Longuet-Higgins. The properties of the components and of the mixture are related to those of a reference liquid, which may be chosen either to give some convenient values for the *perturbation parameters*, or else be one of the pure components themselves. There are four perturbation parameters, obtained by applying (4.93) to the interactions ii, jj and ij. If component 1 is selected as the reference liquid, the parameters are[15]

$$\delta = (\epsilon_{22}^* - \epsilon_{11}^*)/\epsilon_{11}^* = (\epsilon_{22}^*/\epsilon_{11}^*) - 1 \qquad (4.103)$$

$$\rho = (r_{22}^* - r_{11}^*)/r_{11}^* = (r_{22}^*/r_{11}^*) - 1 \qquad (4.104)$$

for the pure liquids, and

$$\theta = (\epsilon_{12}^* - \tfrac{1}{2}\epsilon_{11}^* - \tfrac{1}{2}\epsilon_{22}^*)/\epsilon_{11}^* \qquad (4.105)$$

$$\sigma = (r_{12}^* - \tfrac{1}{2}r_{11}^* - \tfrac{1}{2}r_{22}^*)/r_{11}^* \qquad (4.106)$$

for the mixture (Fig. 4.8).

The first-order conformal mixture theory[16] applies the random mixture restriction, and specifies that δ and ρ should be small compared with unity, and $\sigma = 0$, so that the arithmetic mean rule for r_{12}^* applies. The configurational partition function for a pure liquid is written as

$$Z^*(T,V) = r^{*3N} Z^\circ(T/\epsilon^*, V/r^{*3}) \qquad (4.107)$$

The Helmholz energy $A = -kT \ln Z$ can then be written as

$$A^*(T,V) - \epsilon^* A^\circ(T/\epsilon^*, V/r^{*3}) - 3NkT \ln r^* \qquad (4.108)$$

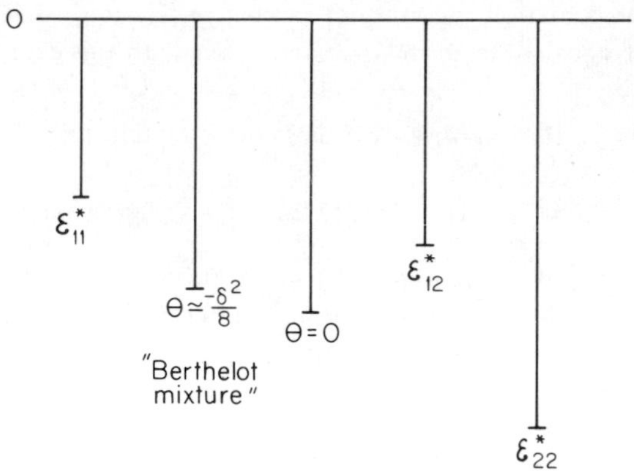

Figure 4.8. Potential energy minima for pair interactions in pure liquids (ϵ^*_{11} and ϵ^*_{22}) and for mixtures: athermal, $\theta = 0$; Berthelot, or geometric means, $\epsilon^*_{12} = (\epsilon^*_{11}\ \epsilon^*_{22})^{1/2}$ or $\theta \simeq -\delta^2/8$; and general, where the case $|\epsilon^*_{12}| <$ |geometric mean| is shown.

which may be expanded as a Tayor series, truncated after the first term for the first-order approximation

$$\begin{aligned} A^*(T,V) - A^\circ(T,V) &= \{A^\circ - T(\partial A^\circ/\partial T)_V(\epsilon^* - 1) \\ &\quad + 3(NkT + V(\partial A^\circ/\partial V)_T)(r^* - 1)\} \\ &= U^\circ(\epsilon^* - 1) + 3(NkT - p^\circ V)(r^* - 1) \end{aligned} \tag{4.109}$$

The last equation applies, since $T(\partial A^\circ/\partial T)_V = TS^\circ_V$ and $A^\circ - TS^\circ_V = U^\circ$, and $(\partial A^\circ/\partial V)_T = -p^\circ$, the equilibrium (vapour) pressure. (If * applies to component 2 and $^\circ$ to component 1, then $(\epsilon^* - 1) = \delta$ and $(r^* - 1) = \rho$.) Now the liquid marked * is taken as an *equivalent liquid* to the mixture, having the same average potential energy between its molecules as in the mixture. Then, dropping the *

$$A(T,V) - A^\circ(T,V) = \Sigma\Sigma x_1 x_2 \{U^\circ(\epsilon^*_{ij} - 1) + 3(NkT - p^\circ V)(r^*_{ij} - 1)\} \tag{4.110}$$

The first-order dependence of $A(T,V)$ on ϵ^* and r^{*3} equals that of $G(T,P)$. (This is not true for entropies and energy—enthalpy relationships: to the first order in v^E, $s^E_V - s^E_P = -(\alpha_m/\kappa_m)v^E = (u^E_V - u^E_P)/T$, where the suffixes V and P denote constant volume and pressure respectively, and α_m and κ_m are the coefficients of thermal expansion and of isothermal compressibility of the mixture, taken as x-averages of those of the components.) Furthermore, the double sum over $(r^*_{ij} - 1)$ yields $2r^*_{12} - r^*_{11} - r^*_{22} = 0$ if $\sigma = 0$, hence the pV term cancels out, so that, per mole of mixture

$$g^E = x_1 x_2 u^\circ[2\epsilon^*_{12} - \epsilon^*_{11} - \epsilon^*_{22}] = x_1 x_2 [2\epsilon^*_{11} u^\circ \theta] \tag{4.111}$$

which is a one-(arbitrary) parameter (θ) equation, since u° and ϵ^*_{11} are known properties of the reference liquid, component 1.

A second-order conformal mixture theory has been derived[17], but it was shown that it contains terms involving averaging processes which cannot be related directly to the thermodynamic properties of the pure components. This second-order theory can be simplified considerably if the random mixing approximation and the Lennard-Jones type of potential energy are assumed to apply. It then provides for asymmetry in g^E, in terms of x_i, and for more realistic temperature- and pressure-dependencies of g^E. In particular, if a *Lorentz* ($\sigma = 0$) – *Berthelot* ($\epsilon_{12}^* = (\epsilon_{11}^* \epsilon_{22}^*)^{1/2}$ or $\theta = -\delta^2/8$) *random mixture case* is considered, the second-order results are

$$g^E = x_1 x_2 [(-\tfrac{1}{4}u^\circ + \tfrac{1}{2}Tc_p^\circ)\delta^2 - \tfrac{3}{2}\mathbf{R}T\delta\rho - \tfrac{1}{4}(nmu^\circ + 3(n + m + 1)\mathbf{R}T\rho^2] \tag{4.112}$$

where n and m are the negative powers of the distance-dependence of the attractive and repulsive terms in the Lennard-Jones equation (e.g. 6 and 12), respectively. Again, g^E is symmetric in x, since the Berthelot (geometric mean) energy approximation has been made. This case has also been called the *single-liquid approximation*, and for it

$$g^E > h^E > 0 > Ts^E \qquad 0 > v^E \tag{4.113}$$

(whereas the first-order conformal mixture theory has all the excess functions of the same sign, either positive or negative).

c. *The average potential model*[18]

In order to get away from the random mixing restriction, a brief consideration will be given to the *average potential model*, which is one type of a *two-liquid approximation.* If the differences θ and σ are large, they produce the dominant terms of the excess functions, since then the first-order conformal mixture theory, which is independent of the random mixture assumption, applies. Only for Lorentz–Berthelot type mixtures is it relatively easy to introduce corrections for ordering effects, as a perturbation of the random assembly. The two-liquid treatment considers one equivalent pure liquid, which has the same average potential as the mixture around a molecule of component 1, surrounded at random with molecules of either species, and another equivalent pure liquid with the average potential of the mixture around a molecule of component 2. At this point it is best to avoid expanding the Helmholz energy, since the series converges slowly, but to define average potentials

$$\langle \epsilon_1(r) \rangle = x_1 \epsilon_{11}(r) + x_2 \epsilon_{12}(r) = \langle \epsilon_1^* \rangle \psi(r/\langle r_1^* \rangle) \tag{4.114}$$

and similarly for $\langle \epsilon_2(r) \rangle$. The volume of the system is

$$V = N_1 \langle r_1^* \rangle^3 \langle v_1^* \rangle + N_2 \langle r_2^* \rangle^3 \langle v_2^* \rangle \tag{4.115}$$

where the average reduced volume, $\langle v_i^* \rangle$, has to be determined by minimizing $A(V,T,x)$, assuming unequal sharing of the available volume. This allows for unequal sizes and non-random mixing. The reduced temperature $\tilde{T} = T/c^*$ and pressure $\widetilde{P} = Pr^{*3}/\epsilon^*$, and the functions

$$\eta(\widetilde{T},\widetilde{P}) = -Z(\widetilde{T},\widetilde{P}) - 3 \ln r^* \quad (4.116)$$

$$\omega(\widetilde{T}) = \widetilde{T}(\partial\eta(\widetilde{T},\widetilde{P})/\partial\widetilde{P})_{P=0} \quad (4.117)$$

$$\eta^\circ(\widetilde{T}) = \eta(\widetilde{T},0) \quad (4.118)$$

are now used to give the excess functions

$$v^E = \langle r^* \rangle^3 \omega(\langle \widetilde{T} \rangle) - x_1 r_{11}^{*3} \omega(\widetilde{T}_{11}) - x_2 r_{22}^{*3} \omega(\widetilde{T}_{22}) \quad (4.119)$$

$$g^E = \mathbf{k}T\{x_1(\eta^\circ(\langle \widetilde{T}_1 \rangle) - \eta^\circ(\langle \widetilde{T}_{11} \rangle)] + x_2[\eta^\circ(\langle \widetilde{T}_2 \rangle) - \eta^\circ(\langle \widetilde{T}_{22} \rangle)]$$
$$- 3[x_1 \ln(\langle r_1^* \rangle / r_{11}^*) + x_2 \ln(\langle r_2^* \rangle / r_{22}^*)]\} \quad (4.120)$$

The functions $\omega(\widetilde{T})$ and $\eta^\circ(\widetilde{T})$ have been obtained from the carefully measured critical data of krypton, as power series in $\widetilde{T}$. The values of δ and ρ are known only to ±20 %, even if $\epsilon_{11}^*/\epsilon_{22}^* = T_{11}^c/T_{22}^c$ is known to ±2 % for several liquids, so θ and σ are best taken as adjustable parameters, rather than calculated with preconceived combination rules (such as the Lorentz–Berthelot rules), to give agreement with experiment. The average potential treatment described here predicts $g^E > 0$, $h^E > 0$, but s^E and v^E may be either positive or negative, and the excess functions are not symmetrical in x, which agrees with the behaviour of many real mixtures.

d. The quasi-chemical approach[19]

Another way to get around the restriction of random mixing is to apply the quasi-chemical approach. This is done at the expense of introducing other restrictions, namely the assumption of a quasi-lattice with a constant coordination number Z, and the disregard of interactions other than those of nearest-neighbours. Originally this approach was developed for solid solutions (superlattices of metal alloys) where indeed a lattice exists, and where long-range order is manifested[20]. If only a short-range order is considered, the same approach may still be applied for a liquid mixture, the average coordination number Z being assumed to remain independent of the temperature and the composition.

Consider the two liquid components, 1 and 2. In the first there are only contacts between 1–1 particles (ions, metal atoms, molecules), each particle having Z such pair-wise interactions of potential energy ϵ_{11}. This energy is independent of r, since only the energy of nearest-neighbour interactions is considered to contribute to the configurational energy of the system. The same applies to liquid 2, with the same Z (since Z is assumed to be independent of the composition), and an energy ϵ_{22}. If now a particle from component 1 is exchanged with one of component 2, Z 1–1 and Z 2–2 interactions disappear, and $2Z$ 1–2 interactions, of energy ϵ_{12} each, appear in their stead. The potential energy change for one such interchange is

$$w = Z(2\epsilon_{12} - \epsilon_{11} - \epsilon_{22}) \quad (4.121)$$

It is further assumed that the interaction energy of any pair of particles is independent of the nature of the other partners that these particles have.

Consider now one mole of mixtures, containing $x\mathbf{N}$ particles of 1 and $(1-x)\mathbf{N}$ particles of 2. If the distribution were completely random, each particle would be surrounded by xZ particles of 1 and $(1-x)Z$ particles of 2. Since, however, there is an energy bias, a certain degreee of order is established in the mixture. The degree of order may be expressed in different ways, for example as the total number N_{12} of 1–2 contacts, or as the probability p that the neighbour of a 1-particle is a 2-particle, rather than another 1-particle. Again, the condition of equlibrium may be expressed in different ways, utilizing for example the Boltzman distribution law, or equating the rates of transfer of particles of a given type to and from positions adjacent to a specified particle, or from calculating the combinatory values of the entropy of the equilibrium distribution and equating it with T^{-1} times its energy. Explicit expressions relating $N_{12}(x, T)$ or $p(x, T)$ to w have been derived. The degree of order in the mixtures, σ may be defined as

$$\sigma = (p - (1-x)Z)/(1-(1-x)Z) \tag{4.122}$$

Complete order, $\sigma = 1$, corresponds to $p = 1$, a certainty that each 1-particle is surrounded by Z 2-particles, and *vice versa*; complete randomness, $\sigma = 0$, arises when $p = (1-x)Z$.

The configurational potential energy of one mole of pure component 1, according to the assumption that only interactions of pairs of neighbours will be considered, is $\frac{1}{2}\mathbf{N}Z\epsilon_{11}$, and similarly for pure component 2, $\frac{1}{2}\mathbf{N}Z\epsilon_{22}$. The configurational potential energy of the mixture is then

$$\begin{aligned} u_{\text{conf}} &= \tfrac{1}{2}(x\mathbf{N}Z - N_{12})\epsilon_{11} + \tfrac{1}{2}((1-x)\mathbf{N}Z - N_{12})\epsilon_{22} + N_{12}\epsilon_{12} \\ &= \tfrac{1}{2}x\mathbf{N}Z\epsilon_{11} + \tfrac{1}{2}(1-x)\mathbf{N}Z\epsilon_{22} + \tfrac{1}{2}N_{12}w/Z \end{aligned} \tag{4.123}$$

It is now necessary to find the average, equilibrium, value $\langle N_{12}\rangle$ of the number of 1–2 contacts in the mixture. The quasi-chemical approach consists of considering the equilibrium constant of the exchange 'reaction'

$$[1\text{–}1] + [2\text{–}2] = 2[1\text{–}2] \tag{4-124}$$

the energy of one such interchange having already been given by (4.121). The mass law expression (giving rise to the 'quasi-chemical' appellation)

$$\langle N_{12}\rangle^2 = (xZ\mathbf{N} - \langle N_{12}\rangle)((1-x)Z\mathbf{N} - \langle N_{12}\rangle)\exp(-w/Z\mathbf{k}T) \tag{4.125}$$

was first suggested on an intuitive basis, but later rationalized on mechanical–statistical grounds from calculations of partition functions or combinatory expressions. The special case of $w = 0$ leads to random mixing and $\langle N_{12}\rangle = x(1-x)Z\mathbf{N}$, which is sometimes called the *Bragg–Williams approximation*[20]. The number of 1–2 pairs divided by $x\mathbf{N}$, the number of 1-particles, equals the probability p defined above, which for random mixing is $(1-x)Z$, leading to zero order, $\sigma = 0$, according to (4.122). However, generally, for non-zero w values, $\langle N_{12}\rangle$ is obtained by solving (4.125)

$$\langle N_{12}\rangle = \tfrac{1}{2}Z\mathbf{N}\{[1 + 4x(1-x)(e^{w/Z\mathbf{k}T} - 1)]^{1/2} - 1\}/(e^{w/Z\mathbf{k}T} - 1) \tag{4.126}$$

Expansion of the square root and of the exponent leads to the approximate expression

$$\langle N_{12}\rangle = x(1-x)Z\mathbf{N}[1 - x1 - x)w/Z\mathrm{k}T + \ldots] \tag{4.127}$$

If the quasi-lattice with constant coordination number also entails a contant volume, independent of compostion, i.e. $v^{\mathrm{E}} = 0$, then the excess molar enthalpy of mixing equals the excess energy per mole of mixture and is obtained from the configurational energy (4.123)

$$h^{\mathrm{E}} = u^{\mathrm{E}} = \langle N_{12}\rangle w/2Z = x(1-x)(\tfrac{1}{2}\mathbf{N}w)[1 - x(1-x)w/Z\mathrm{k}T + \ldots] \tag{4.128}$$

The last equality becomes approximately, for very small w relative to $Z\mathrm{k}T$

$$h^{\mathrm{E}} \simeq x_1 x_2(\mathbf{N}w/2) \tag{4.129}$$

which has the familiar composition-dependence (4.100). Whatever the number of terms from the expansion (4.127) that are used in (4.128), the resulting expression for h^{E} remains symmetric with respect to the composition. Asymmetry must be introduced as a bias, so that the probability of a 1-particle being surrounded by 2-particles does not equal the reverse situation.

Since $h^{\mathrm{E}} = g^{\mathrm{E}} - T(\partial g^{\mathrm{E}}/\partial T)_x$, and $\langle N_{12}\rangle$ is an explicit function of the temperature, and if w and Z are assumed to be temperature-independent, g^{E} should be obtainable by integration, except for the unknown integration constant (unknown with respect to composition-dependence). It turns out that a calculation of this constant, the configurational entropy of the mixture, is in general quite complicated, involving a combinatory calculation of the number of distinguishable configurations, weighted according to their energy factors. The excess entropy comes out to be

$$s^{\mathrm{E}} = -x^2(1-x)^2(w/2\mathrm{k}T)^2/Z - \ldots \tag{4.130}$$

which is always negative, irrespective of the sign of w (i.e. whether $\epsilon_{12} > \tfrac{1}{2}(\epsilon_{11} + \epsilon_{22})$ or *vice versa*). This is a serious restriction on the applicability of the method, since many systems neither have $s^{\mathrm{E}} = 0$ arising from random mixing and required for regular solutions, nor show a negative value, but rather have positive excess entropies.

The excess Gibbs energy is now obtained from (4.128) and (4.130)

$$g^{\mathrm{E}} = x(1-x)(\tfrac{1}{2}\mathbf{N}w)[1 - x(1-x)(\tfrac{1}{2}\mathbf{N}w/Z\mathbf{R}T) + \ldots] \tag{4.131}$$

It is seen that to the extent that (4.129) is obeyed, $g^{\mathrm{E}} \simeq h^{\mathrm{E}}$, the difference appears only in the correction term in the square bracket. The excess chemical potentials and activity coefficients follow from (4.131). The interchange energy per mole of 1–2 contacts formed, $\tfrac{1}{2}\mathbf{N}w$, and the coordination number Z are the two parameters of this method, the latter appearing in a correction term which ordinarily is no more than 10% in homogeneous liquid phases. Larger values of w, making the correction involving Z more important, lead to demixing if $w \gg 0$, and to chemical bonding if $w \ll 0$.

References

1. E. A. Guggenheim, *Mixtures*, Clarendon Press, Oxford, 1952, p. 7.
2. Following, e.g., ref. 1, pp. 15–18.
3. Following, e.g., J. S. Rowlinson, *Liquids and Liquid Mixtures*, Butterworth, London, 2nd edn., 1969, pp. 111–113.
4. Following, e.g., A. G. Williamson, *An Introduction to Non-Electrolyte Solutions*, Oliver and Boyd, Edinburgh, 1967, pp. 66–71.
5. See ref. 4, pp. 74–81; J. M. Prausnitz, *Molecular Thermodynamics of Fluid-Phase Equilibria*, Prentice-Hall, Englewood Cliffs, N.J., 1969, pp. 197–199.
6. O. A. Redlich and A. T. Kister, *Ind. Eng. Chem.*, **40**, 345 (1948); see also ref. 5, p. 195.
7. Following, e.g., R. Haase, *Thermodynamik der Mischphasen*, Springer, 1956 p. 569.
8. J. H. Hildebrand, *J. Am. Chem. Soc.*, **51**, 66 (1929).
9. E. A. Guggenheim, *Proc. Roy. Soc., Ser. A.*, **148**, 304 (1935); see also ref. 1, p. 29.
10. Ref. 1, p. 183 *ff.*
11. K. Wohl, *Trans. Am. Inst. Chem. Eng.*, **42**, 215 (1946).
12. J. J. van Laar, *Z. Phy. Chem. Stoechiom. Verwandschaftslehre*, **72**, 723 (1910).
13. Following, e.g., H. Kehiaian, *Bull. Acad. Polon. Sci., Ser. Sci. Chim.*, **10**, 569 (1962).
14. Following, e.g., J. H. Hildebrand and R. L. Scott, *The Solubility of Nonelectrolytes*, Reinhold, New York, 3rd edn., 1950, pp. 124–126.
15. Following, I. Prigogine, *The Molecular Theory of Solutions*, North-Holland, Amsterdan, 1957.
16. H. C. Longuet-Higgins, *Proc. Roy. Soc., Ser. A*, **205**, 247 (1951).
17. W. B. Brown, *Proc. Roy. Soc., Ser. A*, **240**, 561 (1957).
18. Following A. Bellemans, V. Mathot and M. Simon, *Adv. Chem. Phys.*, **11**, 117 (1967).
19. E. A. Guggenheim, *Proc. Roy. Soc., Ser. A*, **148**, 304 (1935); G. S. Rushbrooke, *Proc. Roy. Soc., Ser. A*, **166**, 296 (1938); E. A. Guggenheim, *Proc. Roy. Soc., Ser. A*, **169**, 134 (1938); J. G. Kirkwood, *J. Phys. Chem.*, **43**, 97 (1939); E. A. Guggenheim, ref. 1.
20. W. L. Bragg and E. J. Williams, *Proc. Roy. Soc., Ser. A*, **145**, 699 (1934); **151**, 540 (1935); E. J. Williams, *Proc. Roy. Soc., Ser. A*, **152**, 231 (1935); H. Bethe, *Proc. Roy. Soc., Ser. A*, **150**, 552 (1935); R. H. Fowler and E. A. Guggenheim, *Proc. Roy. Soc., Ser. A*, **174**, 189 (1940).

Chapter 5

Mixtures of Molecular Liquids

A. Simple Mixtures

Simple mixtures will be defined as those in which the nominal components are identical with the actual species in the pure liquids as well as in the mixture. This excludes all the cases where the molecules of one or more of the components associate either among themselves (self association), or mutually with those of another component (solvation, adduct formation), or dissociate – the latter case being rare in mixtures of molecular liquids. This leaves under consideration all those cases where the interactions between the molecules are governed by dispersion forces and by weak dipole interactions, and where the assumption of *random mixing* is valid. Even without specific chemical interactions, however, the physical interactions dealt with in this section are sufficient to cause severe deviations from ideality, leading in extreme cases to phase separation (p. 149).

a. Ideal mixtures

The decision on whether a given system is ideal, and obeys the laws of ideal mixing (p. 140 ff), depends on the precision with which the experimental data can be obtained. It is often valid to state that within the experimental errors a system is ideal, but closer examination may reveal slight deviations. Thus, the system 1,2-dibromoethane + 1,2-dibromopropane[1] at 85 °C was often quoted as an example for an ideal system[2], but it does show deviations[3]. A system may be (nearly) ideal at one temperature and total pressure, and show deviations under different conditions. A necessary (but not sufficient) condition for ideal mixing (if the vapours mix ideally) is

$$\Delta p/p = [p_{\mathrm{obs}} - (x_1 p_1^* + x_2 p_2^*)]/p_{\mathrm{obs}} = 0 \qquad (5.1)$$

This condition is satisfied within very narrow limits for mixtures of isotopically substituted liquids with moderate mass differences, but is valid also within ±0.005 for other mixtures, some examples of which are shown in Table 5.1. Even if (5.1) is obeyed, it is possible that $g^E \neq 0$, since the vapour is not a perfect gas mixture, as has been shown for the benzene + 1,2-dichloroethane system ($\hat{g}^E$ = 33 J mol^{-1} at 20 °C).[4]

b. Athermal mixtures

Most real systems, which deviate from the ideal mixing laws, have non-zero enthalpies of mixing, but there is a class, called *athermal mixtures*[5], which obey (within the experimental errors) the law

$$h^E(x_i) = 0 \qquad (5.2)$$

at least over a certain temperature range. For such systems the excess Gibbs energy is given entirely by the entropy contribution

$$g^E = -Ts^E \qquad (5.3)$$

Even if there are no real systems obeying (5.2) and (5.3), these expressions may be used as points of reference, to which systems may be conveniently referred which

Table 5.1. Some systems which nearly obey Raoult's law, eq. (5.1).

Component 1	Component 2	Temp. (°C)	Ref.
Benzene	1,2-Dichloroethane	50.0	a
Benzene	Toluene	79.6	b
n-Heptane	n-Butane		c
n-Heptane	Cyclohexane	20.0	d
n-Heptane	Methylcyclohexane	97.2	b
n-Heptane	2,2,4-Trimethylpentane	97.2	b
1,2-Dibromoethane	1,2-Dibromopropane	85.1	a
Methanol	Ethanol	25.0	e

[a]J. v. Zawidzki, *Z. Phy. Chem. Stoechiom. Verwandschaftstehre*, **35**, 129 (1900).
[b]H. A. Beatty and G. Calingaert, *Ind. Eng. Chem.*, **26**, 504 (1934).
[c]G. Calingaert and L. B. Hitchcock, *J. Am. Chem. Soc.*, **49**, 750 (1927).
[d]J. L. Crutzen, R. Haase and L. Sieg, *Z. Naturforsch., A*, **5**, 600 (1950).
[e]G. C. Schmidt, *Z. Phys. Chem. Stoechiom. Verwandschaftstehre*, **99**, 71 (1921).

have small h^{E} but relatively large s^{E}. Such a situation may arise when the interactions between the molecules of each of the components are very similar, so that they are similar also in the mixture, but there is a large discrepancy in sizes between the molecules, so that their mixing is not completely random, the maximum entropy of mixing is not achieved, and s^{E} differs from zero.

First, it is convenient to demonstrate that it follows from (5.2), for the random mixing model, to a good approximation[6], that $v^{E} = 0$. The volume change of mixing is given by

$$V^{M} = (\partial G^{M}/\partial P) \simeq (\partial U^{M}/\partial P) - T(\partial S^{M}/\partial P) \quad (\text{at } T, P = P_0) \tag{5.4}$$

where the approximate equality is due to G^{M} and A^{M} differing only to the second order in V^{E}. The last term in (5.4) equals zero, since for random mixing the entropy depends only on the composition. The first term can be written as

$$\begin{aligned}(\partial U^{M}/\partial P) &= (\partial V_0/\partial P)(\partial U^{M}/\partial V_0) = -\kappa_0 V_0 (U^{M}/\partial V_0) \\ &= -\kappa_0 V_0 [(\partial U_m/\partial V_0) - (\Sigma \partial U_i^*/\partial V_0)] \\ &\simeq -\kappa_0 V_0 [-U_m/V_0 + \Sigma U_i^*/V_0] = \kappa_0 U^{M}\end{aligned} \tag{5.5}$$

at constant T, and at low P ($= P_0$), where the compressibility is κ_0 and the volume is V_0. The approximate substitution $-U/V$ for $(\partial U/\partial V)_T$ has been justified on empirical grounds[6]. The volume change on mixing is thus proportional to the internal-energy change, and since the latter is zero, by (5.2), the former is also zero. The following discussion will treat v^{E} as equal to zero, even though the derivation is not rigorous. Thus g^{E} and $s^{E}_{|P}$ will be obtained from statistical calculations yielding in fact a^{E} and s^{E}_{V}.

The simplest way to obtain s^{E} is from considerations of free volumes[7] (section 2C(a)). The entropy change for any expansion from a free volume V^{f}_{initial} to a free volume V^{f}_{final}, where the free volume is the volume accessible to the molecules, is

$$\Delta S = n\mathbf{R} \ln(V^{f}_{\text{final}}/V^{f}_{\text{initial}}) \tag{5.6}$$

Therefore, in a mixing process, where molecules of component 1, which were initially confined to the free volume $n_1 v_1^{f}$, have access finally to the volume $n_1 v_1^{f} + n_2 v_2^{f}$, and similarly for component 2, the entropy of mixing is

$$S^{M} = -\mathbf{R}\left[n_1 \ln \frac{n_1 v_1^{f}}{n_1 v_1^{f} + n_2 v_2^{f}} + n_2 \ln \frac{n_2 v_2^{f}}{n_1 v_1^{f} + n_2 v_2^{f}}\right] \tag{5.7}$$

or, per mole of mixture

$$s^{M} = -\mathbf{R}[x_1 \ln x_1/(x_1 + rx_2) + x_2 \ln rx_2/(x_1 + rx_2)] \tag{5.8}$$

where $r = v_2^{f}/v_1^{f}$ is the ratio of the free volumes. Now, the free volume is the total volume minus the excluded, or occupied volume. If the free volume is assumed to be proportional to the total volume, and the same proportionality factor is assumed to apply to the two components, it is possible to set $r = v_2^*/v_1^*$, and then (5.8) reduces to (4.30)

$$s^M = -\mathbf{R}[x_1 \ln \varphi_1 + x_2 \ln \varphi_2] \tag{4.30}$$

Otherwise, r may be taken as an arbitrary parameter, and (5.8) can be stated in terms of z-fractions (4.74)

$$s^M = -\mathbf{R}[x_1 \ln z_1 + x_2 \ln z_2] \tag{5.9}$$

The expressions (4.30) and (5.9) lead to $s^E > 0 > g^E$. The partial molal entropy for component 1 is

$$s_1 = -\mathbf{R} \ln \varphi_1 \text{ (or } z_1) + \mathbf{R}\varphi_2 \text{ (or } z_2)(1 - 1/r) \tag{5.10}$$

and this leads directly to the configurational activity coefficient

$$\begin{aligned} \ln f_{1,\text{conf}} &= -\ln \varphi_1 \text{ (or } z_1)/x_1 + \varphi_2 \text{ (or } z_2)(1 - 1/r) \\ &= \ln(1 - \varphi_2 \text{ (or } z_2)(1 - 1/r)) + \varphi_2 \text{ (or } z_2)(1 - 1/r) \end{aligned} \tag{5.11}$$

For component 2, (5.10) and (5.11) apply by changing the subscripts, and substituting r for $1/r$. If component 2 is the solute, then at infinite dilution

$$\ln f_{2,\text{conf}}^{\infty 0} = \lim_{x_2 \to 0} \ln f_{2,\text{conf}} = \ln r + (1 - r) \tag{5.12}$$

As required, for $r = 1$ these formulas reduce to $s_i = -\mathbf{R} \ln x_i$ and $\ln f_{i,\text{conf}} = 0$. These equations may be used as corrections for unequal sizes of the molecules in other expressions, where the restriction $h^E = 0$ has been removed.

The next step is to show how the configurational partition function (4.23) may be calculated, yielding the configurational (excess) entropy, for chain-like molecules mixed with small (solvent) molecules. It is assumed that the internal partition function may be factorized out, and its contribution cancels out when the entropy of mixing is calculated. However, the internal partition function includes contributions from rotational levels, but with flexible chains the configurational and rotational contributions cannot be well separated. The calculations therefore should be applied rigorously only to rigid molecules. The theory is based on the lattice model for the liquid (cf. section 2C(a)), and it is assumed that the molecules occupy sites of the same lattice in the pure components and in the mixture.

Consider first the case of a mixture of N_1 monomers and N_2 dimers[8] (Fig. 5.1(a)). The dimers each occupy two neighbouring sites in the lattice, so that there is a total of $N = N_1 + 2N_2$ sites. The *coordination number*, i.e. number of nearest-neighbours is Z for monomers, but only $2Z - 2$ for dimers. Since the energy of mixing is zero, random mixing is assumed. A chosen site has a probability of being occupied by monomers of $N_1/N = \varphi_1 = (1 - \varphi_2)$ and by dimers of $2N_2/N = \varphi_2$, assuming a double molar volume for dimers with respect to monomers. If a given site is occupied by a monomer, the ratio of the probabilities of a given neighbouring site to be occupied by monomers and dimers is as ZN_1 to $(2Z - 2)N_2$. Thus its probability of being occupied by a monomer is $ZN_1/(ZN_1 + (2Z - 2)N_2)$, and the combined probability of both sites being

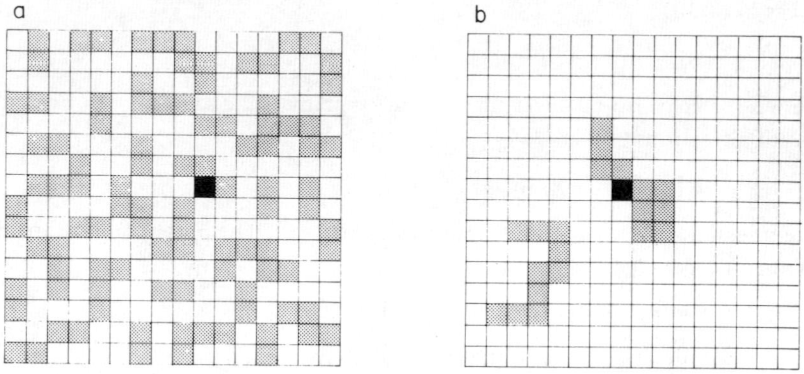

Figure 5.1. Schematic representations of lattice model for the mixtures of molecules of unequal sizes. (a) A solution of a dimer (component 2, grey) in a monomer (component 1, white squares), $x_2 = 0.243$, $\varphi_2 = 0.391$. The coordination number (two-dimensional) is $Z = 8$, and in the immediate surroundings of the blackened segment of the dimer there are 4 monomers, 1 segment of the same dimer, and 3 segments of other dimers. (b) A dilute solution of an oligomer, $r = 10$ (component 2) in a monomer (component 1), $x_2 = 0.0084$, $\varphi_2 = 0.078$. The coordination number is again $z = 8$, and $q = 7.75$. The surroundings of the blackened segment comprise 5 monomers and 3 segments of the *same* oligomer.

occupied by monomers is

$$P_{\mathrm{M}} = \frac{N_1}{N} \frac{ZN_1}{(ZN_1 + (2Z - 2)N_2)} \tag{5.13}$$

If a site is occupied by an element of a dimer, the ratio of the probabilities that a given neighbouring site is occupied by the other element of this dimer and by another molecule (monomer or another dimer) is as 1 to $Z - 1$. The combined probability that the two neighbouring sites are occupied by the two elements of the same dimer is then

$$P_{\mathrm{D}} = \frac{2N_2}{N} \frac{1}{((Z - 1) + 1)} = \frac{2N_2}{NZ} \tag{5.14}$$

The ratio of the probabilities P_{D} and P_{M} is then

$$\alpha = P_{\mathrm{D}}/P_{\mathrm{M}} = \frac{2N_2}{N_1} \frac{ZN_1 + (2Z - 2)N_2}{Z \qquad ZN_1}$$

$$= \frac{1}{Z} \frac{\varphi_2}{1 - \varphi_2} \frac{1 - \varphi_2/Z}{1 - \varphi_2} \tag{5.15}$$

A similar ratio α may be obtained for any r-mer, having r segments each occupying a lattice site, mixed with the monomer also occupying one lattice site. The increment of Gibbs energy in such a system is given by

$$\mathrm{d}G = \mu_1 \mathrm{d}(N_1/\mathbf{N}) + \mu_r \mathrm{d}(N_r/\mathbf{N}) \tag{5.16}$$

which for constant N can be written as

$$dG = (N/\mathbf{N}r)(\mu_r - r\mu_1)d\varphi_r \tag{5.17}$$

since $N_1 = (1 - \varphi_r)N$ and $N_r = (\varphi_r/r)N$. The chemical potentials are related to the probability ratio α and to the partition functions of the monomer and r-mer, Q_1 and Q_r, as[8]

$$\ln \alpha = (1/\mathbf{R}T)(\mu_r - r\mu_1) + (\ln Q_r - r \ln Q_1) \tag{5.18}$$

Therefore, by integration and calculation of the entropy of mixing the result

$$-s^M/\mathbf{R} = (r - (r-1)\varphi_r)^{-1}\left[\int_0^{\varphi_r} \ln \alpha \, d\varphi_r - \varphi_r \int_0^1 \ln \alpha \, d\varphi_r\right] \tag{5.19}$$

is obtained. For $r = 2$, the dimer, the value of α given by (5.15) may be inserted, to give an explicit function $s^M(Z,\varphi_2)$. For higher polymers (Fig. 5.1(b)), the parameter $q = (r(Z - 2) + 2)/Z$ is convenient, and an appropriate value of α may be calculated by arguments similar to those that led to (5.13), (5.14) and (5.15). The general solution for the entropy of mixing is then

$$-s^M/\mathbf{R} = x_1 \ln \varphi_1 + x_r \ln \varphi_r + \tfrac{1}{2}Z(x_1 + qx_r)\ln[(x_1 + rx_r)/(x_1 + qx_r)] - \tfrac{1}{2}Zqx_r \ln(r/q) \tag{5.20}$$

However, it is best to introduce at this stage Flory's approximation[9] and allow $Z \to \infty$, whence $q \to r$, so that the last two terms in (5.19) cancel out leaving the simple relationship (4.30)

$$-s^M/\mathbf{R} = x_1 \ln \varphi_1 + x_r \ln \varphi_r \tag{5.21}$$

In fact, for any realistic Z (between 6 and 12, the close-packing coordination number) and large enough r, Flory's approximation is a sufficiently good one, compared at least with the much more gross approximation of using a lattice model in the first place.

As mentioned above, there are hardly any athermal mixtures known so that it is impossible to test the equations by comparison with experiment. Even such systems as benzene + biphenyl[10] or heptane + hexadecane[11], which might be taken to correspond to the athermal model for dimers, have finite heats of mixing, for the former system at 70 °C $\hat{h}^E = 142$ J mol^{-1}, for the latter at 20 °C $\hat{h}^E = 109$ J mol^{-1}. On the other hand, certain polymer solutions have very low h^E, and may be described well by (5.20), while for others, (5.20) may be used with an h^E term to calculate realistic g^E values (see below).

c. *Regular mixtures*

Another simplified case is that in which random mixing is assumed but a finite energy of mixing is allowed. There must be some upper limit to this energy change since for strong interactions which are different in the pure components and in the mixture the assumption of random mixing is unrealistic. For the following, the stipulation

$$s_V^E(x_i) = 0 \tag{5.22}$$

is made, and a finite, but small v^E is permitted. The theory of *regular mixtures* has very little to say about the magnitude of v^E, but only those cases where second-order terms in the excess volume are negligible will be considered so that

$$g_P^E = a_V^E - (2\kappa_m v_m)^{-1} v^{E^2} + \ldots \simeq a_V^E = u_V^E \tag{5.23}$$

but

$$h_P^E = u_V^E + T(\alpha_m/\kappa_m)v^E + \ldots \tag{5.24}$$

$$s_P^E = s_V^E + (\alpha_m/\kappa_m)v^E + \ldots \simeq (\alpha_m/\kappa_m)v^E \sim 0 \tag{5.25}$$

the last approximate equality holding only for negligible v^E. In that case, $g^E = h^E$ and $s^E = 0$. If the approximations (5.4) and (5.5) are accepted, and v^E set equal to $\kappa_m u^E$, then for not too large u^E

$$h^E = u^E(1 + \alpha_m T) \quad \text{and} \quad s^E = \alpha_m u^E \tag{5.26}$$

where, at room temperature, α_m is of the order of 10^{-3} K^{-1} for many liquid mixtures. Calorimetric enthalpies of mixing must therefore be corrected by (5.26),[6] when they are to be compared with calculated values of u^E.

In the special case, where u^E is of the form

$$u^E = b(T)x_1x_2 = g^E \tag{5.27}$$

i.e. a two-suffix Margules equation (eqs. (4.65) to (4.68) with $b_j = 0$ for $j \geqslant 2$) holds for the activity coefficients, the mixture is called a *regular mixture.* Hildebrand and Wood[12] and Scatchard[13] have shown that (5.27) results in the random-mixing case if $v^E = 0$ and $v_1^* = v_2^* = v^*$ according to eqs. (4.98) to (4.100), or, relaxing the last requirement, if b is given the value of the expression in square brackets in (4.102), from arguments involving the pair correlation function and mutual pair potential energies. A similar result is obtained from the first-order conformal solution theory[14], eq. (4.111). A recapitulation of the requirements for these equations to be valid is in order here.

(1) Random mixing, which restricts the treatment to spherical, non-polar molecules with a central (as opposed to peripheral) force field.

(2) The potential energy of the system is additive in the mutual pair potentials, which depend only on the distance between the centres of the molecules: $u_m = \frac{1}{2}\Sigma_i\Sigma_j\epsilon_{ij}(r)$ where the summation extends over all pairs in one mole of mixture.

(3) Complete separation of the internal and configurational partition functions: $Q = Q_{\text{internal}} \cdot Q_{\text{translational}} \cdot Q_{\text{configurational}} = q^N \cdot \Lambda^{-3N} \cdot Z$.

(4) Absence of volume changes on mixing, that is the partial volumes equal the molar volumes: $v_i = v_i^*$ and $v^E = 0$.

A comparison of (4.100), (4.98) and (4.94) shows that the parameters C_{ij} are equal to $u_i^* v_i^*$. Insertion of these into (4.102) leads to

$$u^E = (x_1 v_1^* + x_2 v_2^*)\varphi_1 \varphi_2 [2C_{12}(v_1^* v_2^*)^{-1} - u_1^* v_1^{*-1} - u_2^* v_2^{*-1}] \tag{5.28}$$

The last two terms in the square brackets are called the *cohesive energy densities* of the the two pure components. The Berthelot geometric mean assumption is now made for C_{12} (see also Fig. 4.8), i.e.

$$C_{12} = (C_{11}C_{22})^{1/2} = (u_1^* u_2^*)^{1/2}(v_1^* v_2^*)^{1/2} \tag{5.29}$$

and since the potential energy of a liquid may be equated to the negative value of its energy of evaporation, $-u^V$

$$u^E = (x_1 v_1^* + x_2 v_2^*)\varphi_1 \varphi_2 [(u_1^{V*}/v_1^*)^{1/2} - (u_2^{V*}/v_2^*)^{1/2}]^2 \tag{5.30}$$

The square root of the negative of the cohesive energy density has received the special appellation *solubility parameter*

$$\delta_i = (u_i^{V*}/v_i^*)^{1/2} \tag{5.31}$$

Inserted into (5.30), this yields the compact equation

$$u^E = (x_1 v_1^* + x_2 v_2^*)\varphi_1 \varphi_2 (\delta_1 - \delta_2)^2 \tag{5.32}$$

showing the excess energy to depend only on the properties (δ_i and v_i^*) of the pure components, and on the composition. The use of (5.32) and partial differentiation leads to the excess chemical potentials and the activity coefficients

$$\mu_i^E = \mathbf{R}T \ln f_i = v_i^*(1 - \varphi_i)^2 (\delta_1 - \delta_2)^2 \tag{5.33}$$

It should be noted that since v_i^* (hence φ_i) and δ_i are functions of the temperature, $\ln f_i$ does not have a simple reciprocal dependence on T, as is obtained from the simple definition of regular solutions with $s^E = 0$, and (5.27) with b independent of T. For many purposes, however, the variation of v_i^* and δ_i with the temperature may be neglected.

The solubility parameter of a liquid can be determined from (5.31), noting that $u_i^V = h_i^V - \mathbf{R}T^2(\mathrm{d}(\ln\ p_i)/\mathrm{d}T)Z$, where $Z = p_i(v_i^g - v_i^l)/\mathbf{R}T$ is the compressibility factor, correcting for imperfection of the vapour. Inherently, δ_i values should be obtained, and used in (5.33), only for substances which obey the criteria enumerated above. This precludes their estimation from, and use for, solubility determinations, since incomplete miscibility presumes differences in properties and non-random mixing incompatible with these criteria. Nevertheless, as a good first approximation, even this estimate will be valid for non-polar, or slightly polar, liquids. Equating the chemical potentials of each component in the two phases $'$ and $''$ yields from (5.33)

$$\mathbf{R}T \ln(x_1'/x_1'') = v_2^*(\varphi_2''^2 - \varphi_2'^2)(\delta_1 - \delta_2)^2 \tag{5.34}$$

and similarly for component 2. If the mutual solubility is very small, and $x_1' \ll 1$, $x_1'' \sim 1, \varphi_2'' \sim 0$ and $\varphi_2' \sim 1$

$$-\ln x_1' \sim v_2^*(\delta_1 - \delta_2)^2/\mathbf{R}T \tag{5.35}$$

If the solubility parameter of the solvent (or solute) is known, that of the solute (or solvent) may be calculated.

Table 5.2 contains an extensive list of solubility parameters for non-polar or slightly polar liquids. These are valid at room temperature, say 20–25 °C, but their temperature-dependence has not been determined systematically. The accuracy is about ± 0.1 $J^{1/2}$ $cm^{-3/2}$, and sometimes severe discrepancies occur between data reported by different authors[15–17]. Note that δ ranges from about 15 to about 20 $J^{1/2}$ $cm^{-3/2}$ for the common solvents, and that the square of the difference between two δ values is involved, so from error propagation $(\delta_1 - \delta_2)^2$ is hardly ever known to better than ± 10 %, and very often the precision is much worse. This explains, in part, the breakdown of quantitative correlations based on (5.33) or (5.34), even for cases that obey the criteria for their application, although qualitatively the correlation is sound (Fig. 5.2).

Table 5.2. Solubility parameters of non-polar (or slightly polar) liquids at 20–25 °C, $\delta(J^{1/2}\ cm^{-3/2})$ (refs 15–17).

Perfluoro-n-hexane	12.1	1-Bromobutane	17.8
Perfluorotri-n-butylamine	12.1	4-Chlorotoluene	18.0
Perfluoro-n-heptane	12.3	Silicon tetrabromide	18.0
Perfluoromethyl cyclohexane	12.4	Decahydronaphthalene (decalin)	18.0
Perfluorocyclohexane	12.5	Mesitylene,	18.1
2,2-Dimethylpropane (neopentane)	12.8	*m*-, or *p*-Xylene	18.1
Pentadecafluoro-n-heptane	12.9	Bromoethane	18.2
2-methylbutane (isopentane)	13.9	Uranium hexafluoride	18.2
2,2,4-Trimethylpentane (iso-octane)	14.1	Toluene	18.3
1,1,2-Trifluoro-1,2,2-trichloroethane	14.5	*o*-Xylene	18.4
n-Pentane	14.7	1,1-Dichloroethane	18.6
1-Hexene	14.9	Benzene	18.8
n-Hexane	15.0	Ethyl sulphhydride	18.8
n-Heptane	15.3	Trichloroethylene	18.8
n-Octane	15.4	Chloroform	18.9
1-Octene	15.6	Styrene	19.0
Bromopentane	15.6	Tetrachloroethylene	19.0
Silicon tetrachloride	15.6	Iodoethane	19.3
Methylcyclohexane	16.0	Dimethyl sulphide	19.3
n-Decane	16.1	Pentachloroethane	19.3
Methylcyclopentane	16.2	Chlorobenzene	19.5
n-Hexadecane	16.4	Tetrahydronaphthalene	
Ethylcyclohexane	16.4	(tetralin)	19.5
1-Chloro-2-methylpropane	16.6	1,1,2-Trichloroethane	19.7
Germanium tetrachloride	16.6	1,1,2,2-Tetrachloroethane	19.8
Cyclopentane	16.7	Dichloromethane	20.0
Cyclohexane	16.8	Carbon disulphide	20.4
Chloroethane	17.0	1,2-Dichlorobenzene	20.4
1-Chloropentane	17.0	Iodomethane	20.4
1-Iodopentane	17.2	1,2-Dichloroethane	20.4
2-Bromobutane	17.2	1,2-Dibromoethane	20.7
Bis-(cyclohexyl)	17.4	Bromoform	21.5
1,1,1-Trichloroethane	17.4	Bromonaphthalene	21.7
1-Iodobutane	17.6	Bromine	23.5
Carbon tetrachloride	17.6	Tin tetraiodide	24.0
n-Propylbenzene	17.6	Di-iodomethane	24.2
Tin tetrachloride	17.8	Osmium tetroxide	25.8

It has been shown that δ is an additive property for non-polar molecules as far as the contributions of atoms and their bonding towards u^V is concerned[18]

$$\delta = (u^V v^*)^{1/2} v^{*-1} = (\Sigma G_j)/v^* \tag{5.36}$$

Table 5.3 shows the individual contributions $G_j(J^{1/2}\ cm^{3/2})$ for common groups and atoms. The contributions of highly polar groups, such as =CO, −CN or $-NO_2$, or of hydrogen-bonded groups, such as −OH, −COOH or $-NH_2$, should be regarded with great reservation.

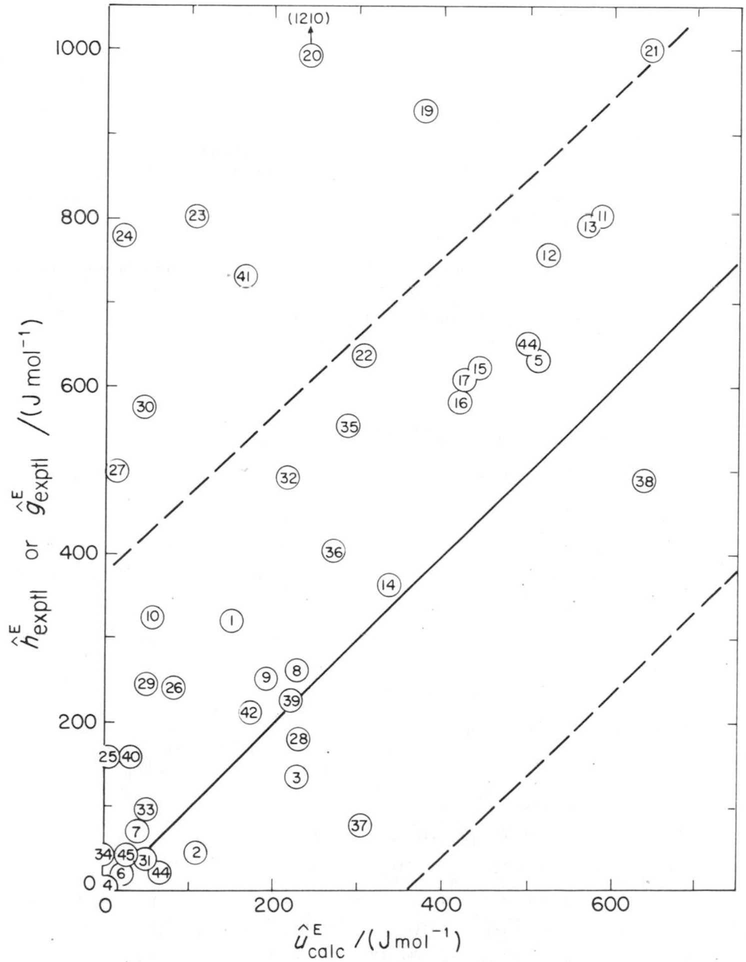

Figure 5.2. An evaluation of eq. (5.32), which compares $\hat{g}^E$ or $\hat{h}^E$ with $\hat{u}^E$ calculated from the regular solution treatment for 44 systems which should obey the equation, within ±0.15**R**T (the broken lines) as stated by R. L. Scott, *J. Phys. Chem.*, **62**, 136 (1958). The systems shown here were published subsequent to a similar comparison made by M. L. McGlashan, *Ann. Reports Chem. Soc., London*, **59**, 73 (1962), (*cf.* ref. 17, p. 98), and are all hydrocarbons and halogen-substituted hydrocarbons of low polarity from I. A. McClure, J. E. Bennet, A. E. P. Watson and G. C. Benson, *J. Phys. Chem.*, **69**, 2753, 2759 (1965) and G. W. Lundberg, *J. Chem. Eng. Data*, **9**, 193 (1964).

Table 5.3. Group contributions to solubility parameters, G_j ($J^{1/2}$ $cm^{3/2}$).

Group	Ref.[a]	Ref.[b]	Group	Ref.[a]	Ref.[b]
$-CH_3$	437	405	$-CF_3$	560	
$-CH_2-$	272	287	$-CF_2-$	307	
$-CH<$	57	121	$-Cl$	532	498
$>C<$	190		$-Br$	695	
$-CH{=}CH-$	617	610	$-I$	868	
$=CH_2$	388		$-SH$	644	
$-CH=$	227		$-S-$	460	
$>C=$	39		$-O-$	143	
$-CH(CH_3)_2$	930	930	$-OH$		910
$CH{\equiv}C-$	583		$-CHO$		710
$-C{\equiv}C-$	454		$>CO$	563	
$-C_6H_5$	1505	1495	$-C(O)O-$	634	700
$>C_6H_4$	1348		$-OCOCH_3$		1108
$-C_{10}H_7$	2380		$-COOH$		1355
5-ring	225		$-CN$	838	
6-ring	205		$-NO$	~900	
$-H$	165–205		$-NH_2$		477
$-C_6H_{11}$(cyclo)	1622	1610			

[a]P. S. Small, *J. Appl. Chem.*, **3**, 75 (1953).
[b]A. E. Reineck and K. F. Lin, *J. Paint Technol.*, **40**, 611 (1968).

d. General phenomenology of non-ideal mixtures

The approximations of athermal solutions ($h^E = 0$) and of regular solutions ($s^E = 0$) are idealized limiting cases, and real solutions usually have non-zero values of both s^E and h^E. Several empirical equations have been proposed, recognizing the effects of unequal sizes of the molecules and of their mutual interactions. Generally, the excess Gibbs energy is written as the sum $g^E = g^E_{conf} + g^E_{inter}$, which leads to corresponding values of the excess chemical potential and the activity coefficient

$$\ln f_i = \ln f_{i,\,conf} + \ln f_{i,\,inter} \tag{5.37}$$

One way to deal with the interaction energy is to start from the *two-liquid approximation* (section 4B(c)) and include it in a Boltzmann factor that weights the probability of finding a molecule of component 1 or of component 2 in the immediate vicinity of a given molecule. According to Wilson[19], the ratio of probabilities of finding the same kind of molecule, x_{ii}, to that of finding a different molecule, x_{ij}, equals the ratio of the bulk mole fractions weighted appropriately

$$\frac{x_{ii}}{x_{ij}} = \frac{x_i \exp(-\langle \epsilon_{ii} \rangle/kT)}{x_j \exp(-\langle \epsilon_{ij} \rangle/kT)} \tag{5.38}$$

where the average potential energies $\langle \epsilon \rangle$ are assumed to be independent of r and treated as free parameters. *Local* volume fractions Φ are then defined as

$$\Phi_i = x_{ii}v_i^*/(x_{ii}v_i^* + x_{ij}v_j^*) \tag{5.39}$$

and these replace the mole fractions in the expression for the entropy of mixing (5.9), yielding for the excess Gibbs energy

$$g^E = RT[x_1 \ln(\Phi_1/x_1) + x_2 \ln(\Phi_2/x_2)] \tag{5.40}$$

and upon substitution of the values of x_{ii} and x_{ij} from (5.38) in Φ_i of (5.39) and (5.40), the expression

$$g^E = -RT[x_1 \ln(x_1 + x_2\Lambda_{12}) + x_2 \ln(x_2 + x_1\Lambda_{21})] \tag{5.41}$$

is obtained, where

$$\Lambda_{12} = (v_2^*/v_1^*)\exp[(\langle\epsilon_{11}\rangle - \langle\epsilon_{12}\rangle)/kT] \tag{5.42a}$$

$$\Lambda_{21} = (v_1^*/v_2^*)\exp[(\langle\epsilon_{22}\rangle - \langle\epsilon_{12}\rangle)/kT] \tag{5.42b}$$

The parameters Λ are thus functions of r and of T with two free parameters $(\langle\epsilon_{11}\rangle - \langle\epsilon_{12}\rangle)/k$ and $(\langle\epsilon_{22}\rangle - \langle\epsilon_{12}\rangle)/k$. Wilson's equation is thus useful for fitting data, but not for predicting activity coefficients of new systems.

Wilson's treatment is in effect a non-random (NR) form of the two-liquid (TL) approximation, commonly abbreviated as the NRTL approximation. Another development of the NRTL idea, due to Renon and Prausmitz[20], has added a further parameter α_{12} to deal with the non-randomness of the distribution. This multiplies the average potential energies $\langle\epsilon\rangle$ in (5.38). Instead of calculating local volume fractions, these workers used the local mole fractions x_{ij}, and calculated residual Gibbs energies, defined as $g_i/N = x_{ii}\langle\epsilon_{ii}\rangle + x_{ij}\langle\epsilon_{ij}\rangle$, and hence for the excess Gibbs energy

$$\begin{aligned} g^E &= x_1(g_1 - N\langle\epsilon_{11}\rangle) + x_2(g_2 - N\langle\epsilon_{22}\rangle) \\ &= x_1 x_{21} N(\langle\epsilon_{21}\rangle - \langle\epsilon_{11}\rangle) + x_2 x_{12} N(\langle\epsilon_{12}\rangle - \langle\epsilon_{22}\rangle) \\ &= x_1 x_2 N\left[\frac{(\langle\epsilon_{21}\rangle - \langle\epsilon_{11}\rangle)G_{12}}{x_1 + x_2 G_{21}} + \frac{(\langle\epsilon_{12}\rangle - \langle\epsilon_{22}\rangle)G_{12}}{x_1 G_{12} + x_2}\right] \end{aligned} \tag{5.43}$$

where $G_{ij} = \exp(-\alpha_{ij}(\langle\epsilon_{ij}\rangle - \langle\epsilon_{ii}\rangle)/kT)$. Renon's equation is in fact a three-parameter equation, adding to those of Wilson the non-randomness parameter α_{ij}, which has values between 0.20 and 0.47 for many binary systems. For the simple case of $\alpha_{ij} = 0$, $G_{12} = G_{21} = 1$, and g^E is reduced to the simple form of $g^E = x_1 x_2 b$ (eq. 4.65 with all $j > 1$ equalling zero), with $b = (2\langle\epsilon_{12}\rangle - \langle\epsilon_{11}\rangle - \langle\epsilon_{22}\rangle)$ independent of the temperature. If the substitution $\tau_{ij} = N(\langle\epsilon_{ij}\rangle - \langle\epsilon_{ii}\rangle)/RT$ is made in (5.43), the activity coefficients become

$$\ln f_1 = x_2^2[\tau_{21}G_{21}^2(x_1 + x_2G_{21})^{-2} + \tau_{12}G_{12}(x_2 + x_1G_{12})^{-2}] \tag{5.44}$$

and for f_2 the suffixes 1 and 2 are interchanged in all the terms. In the same way as for Wilson's equation, Renon's equation is useful for correlating data but since the parameters cannot be calculated from the properties of the pure components it has no predictive value.

Returning now to (5.37) some other approaches have been proposed to make possible evaluation of the contributions to the excess Gibbs energy in terms of the properties of the pure components. For the first term on the right-hand side, the value of (5.11) is used, and for the second, the form of (5.33) is retained, but the expression $v_1^*(\delta_1 - \delta_2)^2/\mathbf{R}T$, specific for the solubility parameter approach, is exchanged for the non-committal χ, which is a non-dimensional parameter. Then

$$\ln f_1 = \ln(1 - z_2(1 - 1/r)) + z_2(1 - 1/r) + z_2^2\chi \qquad (5.45\text{a})$$

$$\ln f_2 = \ln(1 - z_1(1 - r)) + z_1(1 - r) + z_1^2\chi \qquad (5.45\text{b})$$

which is a two-parameter equation (r and χ) or, if $r = v_2^*/v_1^*$ is used, $z = \varphi$, and (5.45) is a one-parameter equation (χ). For those cases where the regular solution—solubility parameter treatment breaks down, as when non-central forces prevail between the molecules (for globular molecules, such as carbon tetrachloride, or chain-like molecules, such as hexane), or for polar molecules, other attempts have been made to relate χ to the properties of the pure components (see sections 5A(e) and 5A(f)).

Equation (5.45) has been applied specifically to *polymer solutions*, and for non-polar polymers in non-polar solvents χ has been given the regular solution value $v_{\text{solvent}}^*(\delta_{\text{solvent}} - \delta_{\text{polymer}})^2/\mathbf{R}T$. The solubility parameter of the (amorphous) polymer cannot, of course, be obtained directly from h^{V}, since it is non-volatile. It has been estimated from the swelling behaviour of cross-linked polymers and the solubility behaviour of linear polymers. The swelling and solubility are maximal for a solvent which matches δ_{polymer} most closely. A comparsion with (4.88) and (4.89) shows that demixing will occur if $\chi > 0.5$, provided that r is very large. For smaller polymers, the critical value of χ is increased by the factor $(1 + r^{-1/2})^2$.

Solutions of polar polymers, or polymers in general in polar solvents, show more complicated relationships, discussed in section 5A(f).

e. Molecules with non-central forces

The intermolecular forces assumed for the mixtures treated above were central forces; that is, the potential energy between two molecules was assumed to depend on the distance between their centres, and each molecule was represented by a sphere, of a diameter $\sigma = 2^{1/6}r/r_1^*$, where $2^{1/6}r/r_1^*$ is the distance where the potential energy is zero. This treatment is well suited to small molecules such as Ar or CH_4, but it is questionable whether it applies to molecules constituting substances which are liquid at room temperature, such as CCl_4, or n-C_6H_{14}. Of these examples, the former is more or less spherical, but the latter is a flexible chain. It has been argued that some of the shortcomings of the solubility parameter approach are due to the disregard of the non-central character of the actual intermolecular forces between molecules which are larger than, say, methane (section 3B(b)).

The same difficulty is, of course, encountered with pure liquids, and it is possible to use the approach taken with these fluids and define a suitable *acentric factor*[21], ω_{ij}, to take care of the deviations from pure central forces. For a pure fluid

$$\omega = -\log_{10}(p(T = 0.7T_c)/P_c) - 1.000 \qquad (5.46)$$

where p is the vapour pressure at the reduced temperature $T = 0.7T_c$. For a gas mixture the linear relationship

$$\omega_{ij} = x_i\omega_i + x_j\omega_j \tag{5.47}$$

has been found to hold well for the purpose of the calculation of the second virial coefficient. For condensed fluids, the parameter k_{ij}, based on critical temperatures

$$k_{ij} = 1 - T_{c(ij)}/(T_{c(i)}T_{c(j)})^{1/2} \tag{5.48}$$

has been proposed[22], with numerical values ranging from 0.01 to 0.20. It is related in an indirect and complicated way to ω_i and ω_j and the critical constants of the components. Since there is a direct relationship between the parameter C_i in (5.29), the scaling factor of the potential energy minimum ϵ_i^*, and the critical temperature $T_{c(i)}$, one may use k_{ij} as a measure of the deviation of C_{12} from the Berthelot rule of the geometric mean, and replace (5.32) by

$$u^E = (x_1v_1^* + x_2v_2^*)\varphi_1\varphi_2[(\delta_1 - \delta_2)^2 + 2k'_{12}\delta_1\delta_2] \tag{5.49}$$

where k'_{12} is related to, but not identical with, k_{12} above. There is no simple way to calculate k'_{12} from the properties of the two components – the application of the London formula for dispersion forces, which requires knowledge of ionization potentials, is not useful, and neither is reliance on critical data. If, on the other hand, k'_{12} is taken as an empirical correction, then the whole term in square brackets in (5.49) may be treated as such, devoid of predictive value leading back to the empirical (5.27), in terms of volume, rather than mole fractions.

More fruitful is another approach[23], which has been shown recently to give meaningful results for chain-molecules, and can also be extended to polar molecules (see next section), namely the contact point[24] and segment interaction[25] treatments. Research in this field has received new impetus, when it was shown that gas-liquid chromatography can give accurate values for the activity coefficient of a volatile solute at infinite dilution in a relatively non-volatile solvent

$$\ln f_2^{\infty 0} = \lim_{\varphi_2 \to 0} f_2 = \lim_{\varphi_2 \to 0} (\ln f_{2,\mathrm{conf}} + \varphi_1^2\chi) = \ln f_{2,\mathrm{conf}}^{\infty 0} + \chi \tag{5.50}$$

where the value given in (5.12), with $r = v_2^*/v_1^*$, is a valid measure of $\ln f_{\mathrm{conf}}$, and χ is the interactional contribution to the activity coefficient[26]. Both treatments imply the lattice theory for liquids (section 2C(a)), and consider the interactions of atoms, or small groups of atoms, only with those with which they are in immediate contact, since the dispersion forces are very short-range forces[27]. The general equation for χ is

$$\chi = C_2 \sum_{a,b} (\theta_{a1}\theta_{b2} + \theta_{a2}\theta_{b1} - \theta_{a1}\theta_{b1} - \theta_{a2}\theta_{b2})w_{ab}/kT \tag{5.51}$$

and the two above-named approaches give different interpretations to the various terms in this equation, as follows.

(i) The *contact point* approach[24] assigns to each atom, or small group of atoms, in the molecule an arbitrary number of contact points with neighbouring molecules.

Various assignments have been proposed but the concensus of views among the proponents of this approach is that for aliphatic and alicyclic hydrocarbons each hydrogen atom bonded to a skeletal carbon gives that carbon one contact point, that each CH group in an aromatic ring has two contact points, and that fluorine, chlorine and bromine atoms have each 2, 3 and 4 contact points, respectively. All the $-CH_3$ groups in different molecules are treated as equivalent, and so are the normal $-CH_2-$ groups, the cyclic $-CH_2-$ groups and the aromatic $-CH-$ (cf. section 3B(b)). These are then classified into types a,b . . . , so that θ_{a1} is the fraction of contact points of type a in a molecule of component 1, and so on. For example n-pentane has two $-CH_3$ groups and three $-CH_2-$ groups, and a total of 12 contact points, so that $\theta_{CH_3,\text{pentane}} = 6/12 = 0.5$. The energy parameter w_{ab} is the interchange energy, i.e. half the energy change that takes place in the process a—a + b—b → 2a—b

$$w_{ab} = \tfrac{1}{2}(2\epsilon_{ab} - \epsilon_{aa} - \epsilon_{bb}) \tag{5.52}$$

where the atoms (groups) are in contact. Since, however, the distances and orientations in this contact are vague, the quantity w_{ab} is best treated as a free parameter for a given pair a—b of atoms (or groups), independent of the molecules in which they occur. The parameter C_2 is the total number of contact points in a molecule of component 2, and the summation in (5.51) is taken over all different pairs of atoms (or groups) a, b, . . . in the two components. In a mixture of benzene and cyclohexane, for instance, there are only two kinds of groups, and only one interaction term w_{ab}. A similar situation exists in a mixture of pentane and decane, but in a mixture of benzene and decane there are three kinds of groups: a = aromatic $-CH-$, b = $-CH_3$, and c = $-CH_2-$; and three pairs of interactions: w_{ab}, w_{ac} and w_{bc}. The success of this approach may be assessed by the degree of agreement obtained by different workers for estimates of a given w_{ab}/k, as shown in Table 5.4.

Once w_{ab} values have been established for the groups a, b, . . . , it is possible to calculate χ for any molecules that contain only these groups, and that have not been studied experimentally, just by counting the contact points and applying (5.51). In another modification of this approach[29], groups such as $-CH_3$, $-CH_2-$, $-O-$, $-Cl$, etc. are assigned a unit surface area each, minus a fraction of the area through which the atom or group is bonded chemically to adjacent atoms or groups. This fraction was selected as 1/8, assuming a liquid lattice with a

Table 5.4. Interchange energy parameters w_{ab}/k for the contact point theory

a	$-CH_3$	$-CH_3$	$-CH_2-$	$-CH_3$ aromatic	$-CH_2-$ aromatic	$-CH_3$	$-CH_2-$
b	$-CH_2-$	$-CH{=}CH_2$	$-CH{=}CH_2$	$-CH{=}$	$-CH{=}$	$-Cl$	$-Cl$
w_{ab}/k (K)	62[a], 54[b]	77[a]	41[a]	50[a], 54[b]	16[a], 15[b]	313[c]	228[c]

[a] Y. B. Tewari, D. E. Martire and J. P. Sheridan, *J. Phys. Chem.*, 74, 2345 (1970).
[b] B. W. Gainey and R. L. Pecsole, *J. Phys. Chem.*, 74, 2548 (1970).
[c] Y. B. Tewari, J. P. Sheridan and D. E. Martire, *J. Phys. Chem.*, 74, 3263 (1970).

coordination number $Z = 8$. The net contact areas per $-CH_3$, $-CH_2-$, etc. groups are therefore 0.875, 0.750, etc. Some other arbitrary selections are made in order to achieve best agreement with experiment: benzene is assigned a contact area of 3.800 (therefore a phenyl group has a value of 3.675), and carbon tetrachloride an area of 3.600, while ethereal oxygen has systematically an area of 0.750. In order to remove some of the arbitrariness with which the contact points have been assigned, it has been proposed[28] to make only one arbitrary assignment, namely 14 contact points to n-hexane, and apply the relationship $C_2 = C_{C_6H_{14}}(v_2^*/v_{C_6H_{14}}^*)$, ensuring an equal volume per contact-point (see Fig. 5.3). This leads to the second approach mentioned above, in terms of segments.

(ii) The *segment interaction* treatment[25] considers that each molecule of component i occupies r_i sites in the liquid lattice, and hence may be considered to be made up of r_i segments of approximately equal volume. For hydrocarbons, these would be $-CH_3$ and $-CH_2CH_2-$ for aliphatic and alicyclic molecules, and 1/3 of benzene or $-CHCH-$, for aromatic molecules, each occupying roughly 30 cm^3 mol^{-1}. For normal aliphatic hydrocarbons there are therefore 'end groups' $-CH_3$ and 'middle groups' $-CH_2CH_2-$, and $\theta_n = \theta_{CH_3}$ in $C_nH_{2n+2} = 3/(n+1)$, and there are a total of $1 + n/2$ segments. For a mixture of normal hydrocarbons, therefore, (5.51) becomes

$$\chi = (1 + n_2/2)(\theta_{n1} - \theta_{n2})^2 w_{em}/kT \tag{5.53}$$

where w_{em} is the 'end group' – 'middle group' interchange energy, a free parameter independent of n, and $w_{em}/kT = 0.498 \pm 0.018$ for a large number of hydrocarbons (see Fig. 5.3). In branched hydrocarbons the branching point may be neglected and a segment $=CHCH_2-$ may be treated as $-CH_2CH_2-$, and similarly aromatic $=CCH-$ as $-CHCH-$. The main assumptions in this treatment are that the segments interact independently and that the number of segments in a molecule, r_i, can be derived simply from the molar volume, say $r_i = v_i^*$ (cm^3 mol)$^{-1}$/30.

Since the molar volumes vary with temperature, both r_1 and r_2, and $r = r_1/r_2$, for the two components 1 and 2, vary with the temperature. For a mixture of n-hexadecane and n-hexane (Fig. 5.3), if r is calculated from the number of segments it has the fixed value 9/4 = 2.25, but the molar volume ratio has a −5 % slope between 20 °C and 60 °C, whereas the value of r that best fits $\ln f_2$ increases from 2.05 ± 0.08 at 20 °C to 2.22 ± 0.00 at 60 °C (for the Flory approximation, $Z \to \infty$, which gives the simplest equation for $\ln f_{2,conf}$, which is also quite adequate). The variation of r with T contributes to the heat of mixing, and has been given as an adequate reason for h^E changing sign at elevated temperatures for hydrocarbon mixtures[30]. In order to circumvent this difficulty, it has been proposed that the properties of homologous liquids should be compared at corresponding temperatures, chosen so that Δv^* per $-CH_2CH_2-$ segment is a constant. Equal increments of the cohesive energy, because of dispersion forces, should occur for equal average distances of the interacting segments, ensured by a constant volume per segment[31]. It is difficult, however, to see how this principle can be applied to mixtures.

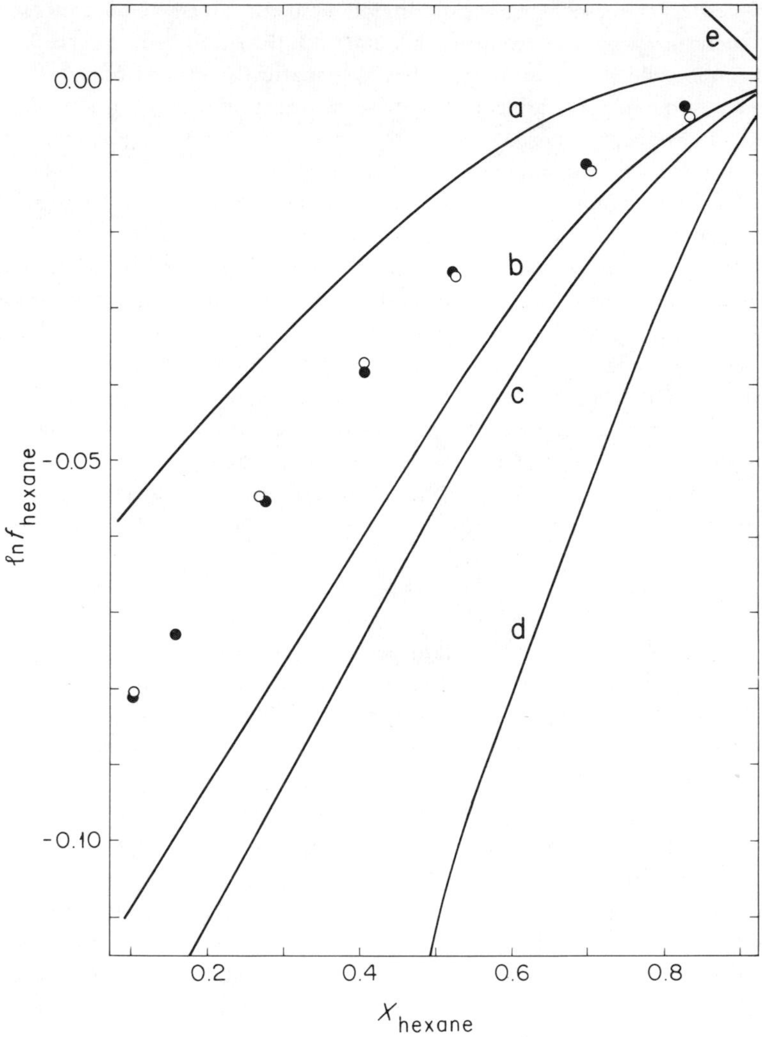

Figure 5.3. Activity coefficients of n-hexane in its mixtures with n-hexadecane at 20 °C. Experimental data: (●) from M. L. McGlashan and A. G. Williamson, *Trans. Faraday Soc.*, **57**, 588 (1961); (○) from C. P. Hicks and C. L. Young, *Trans. Faraday Soc.*, **64**, 2679 (1968). Calculated curves: (a) contact point theory, $w_{CH_3/CH_2}/\mathbf{k} = 62$ K; (b) segment interaction theory, $w_{CH_3/CH_2}/\mathbf{k}T = 0.498$; (c) regular solution theory, $\delta_{hexane} = 15.0$ and $\delta_{hexadecane} = 16.4\ J^{1/2}\text{-}cm^{-3/2}$, corrected for unequal volumes, $r = v_{hexadecane}/v^*_{hexane} = 2.23$; (d) athermal solution theory, $r = 2.23$; (e) regular solution theory, without correction for unequal volume, *positive* values for $\ln f_{hexane}$ are predicted.

f. Polar liquids

The molecular theories examined in the previous sections cannot inherently be extended to cover also polar liquids. The assumption of random mixing, leading to $s_V^E = 0$ on the thermodynamic level, and to $g_{11}(r) = g_{22}(r) = g_{12}(r)/\varphi_2$, (4.97), on the statistical level, breaks down for polar liquids interacting via permanent dipoles, whence orientational effects become important. As long as the polarity is low, so that $N\mu_i^2/3kTv_i^* \ll 1$, the orientation effects are swamped by the thermal randomization, so that the treatment in terms of the solubility parameters δ is valid. Thus, 'slighly' polar hydrocarbons, such as toluene, or halogen-substituted hydrocarbons, such as dichloromethane, can be accommodated by this model. 'Strongly' polar liquids, obviously, require a different treatment. A major difficulty is that the molecules cannot as a rule be considered to be point-dipoles, since the intermolecular distances are not large compared with the lengths of the dipoles.

A general assumption used in dealing with polar molecules is that it is possible to separate the orientation and induction forces on the one hand, and the dispersion forces on the other. It is then assumed that the dispersion interactions with the polar molecules are the same as with a *homomorph*, that is a non-polar hydrocarbon with the same structure and the same molar volume at the same reduced temperature. The dispersion energy contribution to the cohesive energy may be calculated for polar molecules as follows[32]. Corresponding temperatures are selected from thermal expansion data such that each member of a homologous series has, per incremental CH_2 segment, a constant volume of 19.08 $cm^3\ mol^{-1}$ (that calculated for the standard liquid, n-hexane at 0 °C). Then the average intersegment distances are constant, and the cohesive energies C_i/v_i^* at these temperatures have a constant contribution from each CH_2 group. A plot of $C_i/v_i^* = \tilde{u}_i^V$ against the number n of CH_2 groups should be a straight line. The corresponding plot for polar molecules will be displaced from that of the hydrocarbons because of induction and orientation effects. At high values of n, the contribution from orientation is very small, since the chances for two dipoles to be sufficiently near for effective interaction is very small in view of the inverse sixth-power distance-dependence. In fact, at $n > 7$, $\tilde{u}_{\text{polar}}^V$ is displaced in parallel from $\tilde{u}_{\text{hydrocarbons}}^V$. The parallel displacement is the contribution from induction forces, which should, indeed, be independent of n since the polar group sees around it only a limited number of CH_2 segments, and not the whole chain. At low n there is an important contribution of orientation effects for polar molecules. This can be estimated from the deviation of $\tilde{u}_{\text{polar}}^V$ from the straight line extrapolated down from the higher n region. In the same way as the solubility parameter for a non-polar molecule is defined as $\delta = (u^V/v^*)^{1/2}$, for polar molecules the definition is

$$\lambda^2 + \tau^2 = \tilde{u}^V/\tilde{v}^* \tag{5.54}$$

where λ^2 is the contribution from dispersion, and τ^2 is that from induction and orientation. The values obtained by Weimer and Prausnitz[16] in a similar manner, using the homomorph concept (Fig. 5.4) are shown in Table 5.5. It should be noted

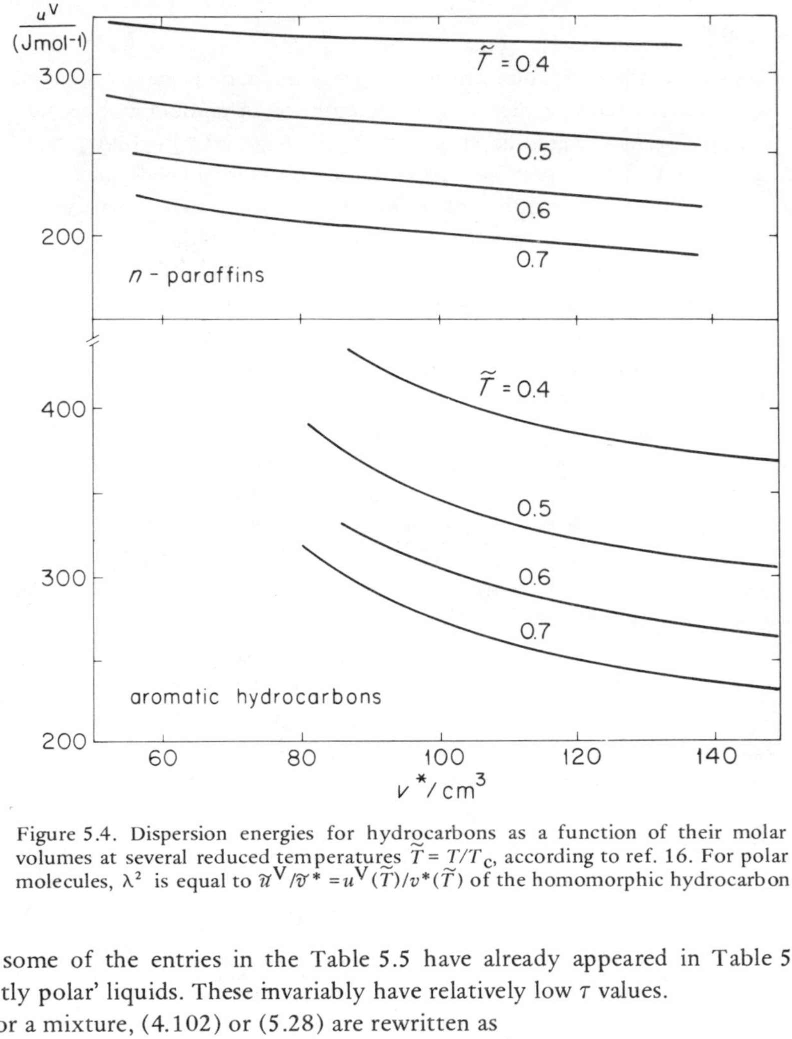

Figure 5.4. Dispersion energies for hydrocarbons as a function of their molar volumes at several reduced temperatures $\tilde{T} = T/T_c$, according to ref. 16. For polar molecules, λ^2 is equal to $\tilde{u}^V/\tilde{v}^* = u^V(\tilde{T})/v^*(\tilde{T})$ of the homomorphic hydrocarbon.

that some of the entries in the Table 5.5 have already appeared in Table 5.2 as 'slightly polar' liquids. These invariably have relatively low τ values.

For a mixture, (4.102) or (5.28) are rewritten as

$$u^E = (x_1v_1^* + x_2v_2^*)\varphi_1\varphi_2[2C_{12}(v_1^*v_2^*)^{-1} - C_{11}v_1^{*-2} - C_{22}v_2^{*-2}]$$
$$= (x_1v_1^* + x_2v_2^*)\varphi_1\varphi_2[(\lambda_1 - \lambda_2)^2 + (\tau_1 - \tau_2)^2 - 2\psi_{12}] \qquad (5.55)$$

where the binary interaction parameter ψ_{12} is defined as

$$\psi_{12} = \lambda_1\lambda_2 + \tau_1\tau_2 - C_{12} \qquad (5.56)$$

In the case of a mixture of a polar component 1 with a non-polar component 2, $\tau_2 = 0$ and $\lambda_2 = \delta_2$, the ordinary solubility parameter. There is no immediately obvious way for estimating ψ_{12}, since the geometric mean principle is clearly inapplicable. For series of solutions of hydrocarbons (component 2) in polar liquids, the approximate relationship $\psi_{12} = 0.4\tau_1^2$ has been found, but there is no indication of the generality of this finding. It is thus seen that this approach has not

Table 5.5. Polar (τ (J cm^{-3})$^{1/2}$) and non-polar (λ (J cm^{-3})$^{1/2}$) solubility parameters for polar molecules[16,17] at 25°C and for hydrogen-bonded molecules[35] at 60°C.

Liquid	τ (J cm^{-3})$^{1/2}$	λ (J cm^{-3})$^{1/2}$	Liquid	τ (J cm^{-3})$^{1/2}$	λ (J cm^{-3})$^{1/2}$
At 25°C			1-Nitropropane	13.1	16.6
Acetophenone	7.6	19.4	*N*-Methylpyrrolidone	13.4	18.8
Tetrahydrofurane	7.6	17.1	Acetic anhydride	14.6	16.1
Pyridine	7.6	20.3	Propionitrile	14.7	16.4
Cyclohexanone	8.3	18.2	Furfural	15.6	18.6
Chloroethane	8.9	15.2	Nitroethane	15.7	16.4
3-Pentanone	9.1	15.9	Dimethylacetamide	15.8	17.0
Diethyl carbonate	9.2	16.2	γ-Butyrolactone	16.4	19.5
Bromoethane	9.9	15.7	Dimethylformamide	16.6	17.0
Nitrobenzene	10.0	19.9	3-Chloropropionitrile	17.9	17.3
Bis-(2-chloroethyl) ether	10.7	17.1	Acetonitrile	18.4	16.5
Trimethyl phosphate	10.7	17.4	Ethylenediamine	19.3	16.6
Iodoethane	10.7	15.7	Nitromethane	19.4	16.6
2-Butanone	11.0	15.7	Dimethylsulphoxide	19.4	17.6
Cyclopentanone	11.0	17.9	Methanol	24.7	16.4
2,4-Pentadione	11.7	16.6			
2,5-Pentadione	12.0	17.3	*at 60°C*		
Diethyl oxalate	12.0	17.2	Di-n-propylamine	5.5	
2-Nitropropane	12.4	16.3	Diethylamine	7.0	
Methoxyacetone	12.5	16.2	n-Hexylamine	8.1	
Acetone	12.6	15.7	n-Butylamine	9.7	
Dimethyl carbonate	12.7	15.9	2-Butanol	14.9	
Butyronitrile	12.9	16.4	Ethanol	19.6	
2,3-Butadione	13.0	15.9	Methanol	23.0	
Aniline	13.1	20.2	Ethylene glycol	28.2	

provided yet a means for predicting the properties of the mixture from those of the components.

A completely different treatment considers the dipole–dipole and dipole–induced-dipole interactions directly[33], rather than via the cohesive energies. Onsager's theory[34] for mixtures gives the following relationship for the dielectric constant

$$(\epsilon - 1)(2\epsilon + 1)/9\epsilon = \varphi_1 P_1^* + \varphi_2 P_2^* \quad (5.57)$$

where the molar polarization P_i^* is given by (cf. eq. (1.34))

$$P_i^* = (4\pi/3)\mathrm{N}v_i^{*-1}[r_i\alpha_i + (r_i\mu_i)^2/3\mathrm{k}T] \quad (5.58)$$

where α_i is the polarizability, μ_i is the intrinsic dipole moment (in vacuum) of the molecules of i, and the parameter r_i corrects for the reaction field of the environment on the dipole. This parameter is given by

$$r_i = [(2\epsilon + 1)/(2\epsilon + n_i^2)][(n_i^2 + 2)/3] = [1 - 2(\epsilon - 1)\alpha_i/(2\epsilon + 1)a_i^3]^{-1} \quad (5.59)$$

where n_i is the refractive index, and a_i the radius of the (assumed) spherical cavity, which is formed by the dipolar molecule owing to its thermal rotation, and on which the reaction field acts. This radius is defined operationally by (5.59). If n_i^2 is small compared to ϵ, r_i may be adequately represented by $(n_i^2 + 2)/3$, and for sufficiently high ϵ, also the left-hand side of (5.57) may be represented by $2(\epsilon - 1)/9$. When the energy of the polar molecule, which is the resultant of the interaction energies of the dipole in the field, the polarization of the environment, and the polarization of the molecule, is used in the expressions for the partition function for polar effects, which, it is assumed, may be factorized out, the excess Helmholtz-energy due to polarization, a_P^E is given by

$$a_P^E = \frac{1}{9}\mathrm{N}\left[\frac{P_1^* - P_2^*}{3 + 2(\varphi_1 P_1^* + \varphi_2 P_2^*)}\right]\left[\frac{x_1\varphi_2 r_1\mu_1^2}{a_1^3(1 + 3P_1^*)} - \frac{x_2\varphi_1 r_2\mu_2^2}{a_2^3(1 + 3P_2^*)}\right] \quad (5.60)$$

The P_i^* are temperature-dependent, according to (5.58), hence the derivative of (5.60) with respect to the temperature will give $-s_P^E$ and the contribution to the energy of mixing due to polarization is $a_P^E - Ts_P^E$. If component 2, let us say, is non-polar, $\mu_2 = 0$ and unless $\alpha_2 \gg \alpha_1$, $P_1^* > P_2^*$ and a_P^E is positive. It can be shown that s^E will be s-shaped, and cross the abscissa at the point $\varphi_2 = (P_2^* + 1/3)^2/(P_1^* - P_2^*)^2$, being positive on the low x_2 side of this point. Although g_P^E ($\approx a_P^E$) may be nearly symmetrical, h_P^E ($\approx u_P^E$) will be skew. If both components are polar, u_P^E may very well be negative, and all functions asymmetrical with respect to composition. No direct comparison (5.60) with experiment has yet been made, although the curves calculated from the parameters of the ethanol + heptane system[33] have the correct shape. It remains to be seen whether the polar and dispersion effects are additive, i.e. whether $g^E = a_P^E + (x_1 v_1^* + x_2 v_2^*)\varphi_1\varphi_2(\lambda_1 - \lambda_2)^2$, i.e. g^E may be determined completely by the properties of the pure components: n_i, μ_i, α_i, λ_i and v_i^*, with no cross-term for the mixture.

B. Associated Mixtures

Associated mixtures are defined as those in which there is chemical evidence (preferably extra-thermodynamic, such as spectroscopic, evidence) for the association of the particles of at least one of the components either with each other (*self association*) or with those of another component (mutual association or *adduct formation*). It is possible to approach the description of such mixtures in terms of the properties of the components in a formal thermodynamic manner, or with respect to the molecular interactions, such as dipole interactions, hydrogen bonding, coordinative bond formation, etc. It is usually possible to treat the association in terms of a definite chemical reaction (or a set of reactions, cf. p. 137), to which an equilibrium constant (or constants) and standard changes of thermodynamic quantities, such as the enthalpy, per mole of reaction can be assigned.

It is usually possible to employ the same models that have been found useful for simple mixtures also for associated ones, and to impose on the particles further restrictions, corresponding to the specific interactions. It is however useful to refer to *ideal associated mixtures* (which should occur in practice as rarely as ideal simple mixtures) in order to study the consequence of the association reaction per se, without the complications of non-specific interactions which occur in real systems. *Non-ideal associated mixtures* will show also the effects of the non-specific interactions between the species.

In the following, mainly nominally binary mixtures of component 1 (having molecules of type A) and of component 2 (having molecules of type B) will be discussed. Extension to multicomponent systems is straightforward. In the general case of association to species A_i, B_j and A_mB_n the total number of moles will be given by

$$n_1 = \Sigma_i i n_{A_i} + \Sigma_m \Sigma_n m n_{A_m B_n} \tag{5.61a}$$

$$n_2 = \Sigma_j j n_{B_j} + \Sigma_m \Sigma_n n n_{A_m B_n} \tag{5.61b}$$

A fundamental relationship, which is valid in all cases of association, relates the chemical potentials μ_1 and μ_2 of the nominal components to those of the corresponding non-associated ('monomeric') molecular species A and B[36]. This is based on the fact that equilibrium occurs among the species so that for $A_i \rightleftarrows iA$, $\mu_{A_i} = i\mu_A$, etc. The total differential of the Gibbs energy at constant temperature and pressure is

$$\begin{aligned} dG &= \Sigma_i \mu_{A_i} dn_{A_i} + \Sigma_j \mu_{B_j} dn_{B_j} + \Sigma_m \Sigma_n \mu_{A_m B_n} dn_{A_m B_n} \\ &= \mu_A \Sigma_i i dn_{A_i} + \mu_B \Sigma_j j dn_{B_j} + \mu_A \Sigma_m \Sigma_n m dn_{A_m B_n} \\ &\quad + \mu_B \Sigma_m \Sigma_n n dn_{A_m B_n} = \mu_A dn_1 + \mu_B dn_2 \end{aligned} \tag{5.62}$$

On the other hand, for a binary mixture at constant T and P

$$dG = \mu_1 dn_1 + \mu_2 dn_2 \tag{5.63}$$

which leads to the identities

$$\mu_1 = \mu_A \quad \text{and} \quad \mu_2 = \mu_B \tag{5.64}$$

The chemical potentials of the components are thus equal to those of the monomers, a result which is independent of the ideality or otherwise of the mixture and of the nature of the association. A further result from expressing the chemical potential in terms of the nominal components and in terms of the monomeric species is

$$\mu_1 = \mu_1^0 + RT \ln x_1 f_1 = \mu_A = \mu_A^0 + RT \ln x_A f_A \tag{5.65}$$

and similarly for component 2 (B). Elimination of the standard chemical potentials is made by setting $x_1 = 1$ for the pure component 1, and denoting $x_A f_A$ as $x_A^* f_A^*$ in this state. Then follow

$$f_1 = x_A f_A / x_1 x_A^* f_A^* \tag{5.66a}$$

$$f_2 = x_B f_B / x_2 x_B^* f_B^* \tag{5.66b}$$

as the general expressions for the activity coefficients of the nominal components. For ideal associated mixtures $f_A = f_A^* = f_B = f_B^* = 1$, so that f_1 (ideal associated solution) $= x_A / x_1 x_A^*$ etc., while for non-ideal mixtures, the activity coefficients of the monomeric species must be assigned proper values.

a. Ideal associated mixtures

Dolezalek[37] has shown that deviations from Raoult's law for liquid mixtures can often be explained by association, assuming the association products and the other particles to mix ideally. Kehiaian and Sosukowska–Kehiaian[38] examined in detail the formal thermodynamic consequences of such an assumption. Here only two simple cases will be treated.

Case (1) Self association of A to dimers A_2, so that the mixture involves the species A, A_2 and B.

Case (2) Mutual association of A and B to form AB, so that the mixture involves the species A, AB and B.

The results may be generalized to other values of *i*, *j*, *m* and *n* in (5.61) without difficulty, even to infinite series of species, if simple laws governing the equilibrium constants are assumed.

Case (1). The chemical equilibrium $2A \leftrightarrows A_2$ is characterized by the equilibrium constant $K_2 = x_{A_2}/x_A^2$ which can be written in terms of the mole fractions of the nominal components, i.e. x_1, as

$$K_2 = \xi / [x_1 - \xi(2 - x_1)]^2 \tag{5.67}$$

where $\xi \equiv x_{A_2}$ for short. Hence is obtained the explicit relation for ξ in terms of K_2 and the nominal composition

$$\xi = \frac{x_1}{2 - x_1} - \frac{[4K_2 x_1(2 - x_1) + 1]^{1/2} - 1}{2K_2(2 - x_1)^2} \tag{5.68}$$

For pure component 1, $x_1 = 1$ and ξ^* of the dimer and x_A^* of the monomer are obtained as

$$\xi^* = 1 - [(4K_2 + 1)^{1/2} - 1]/2K_2 \tag{5.69a}$$

$$x_A^* = 1 - \xi^* = [(4K_2 + 1)^{1/2} - 1]/2K_2 \tag{5.69b}$$

For increasing values of K_2 the maximum possible value of ξ is obtained for any x_1 from (5.68) as $x_1/(2 - x_1)$ for $K \to \infty$, and the slope $(\partial\xi/\partial x_1)_{T,p}$ is always positive, except at $x_1 = 0$, where it is zero.

The molar Gibbs energy of the mixture is

$$\begin{aligned} g_m &= x_1(\mu_A^0 + \mathbf{R}T \ln x_A) + x_2(\mu_B^0 + \mathbf{R}T \ln x_B) \\ &= x_1(\mu_1^0 + \mathbf{R}T \ln x_1) + x_2(\mu_2^0 + \mathbf{R}T \ln x_2) + \mathbf{R}Tx_1 \ln(x_A/x_A^* x_1) \\ &\quad + \mathbf{R}Tx_2 \ln(x_B/x_2) \end{aligned} \tag{5.70}$$

since $\mu_2^0 = \mu_B^0$ (as $x_B^* = 1$) but $\mu_1^0 = \mu_A^0 + \mathbf{R}T \ln x_A^*$. The first two terms of (5.70) are the ideal Gibbs energy of mixing, and the last two are the excess Gibbs energy. The activity coefficients of the *nominal* components are

$$f_1 = x_A/x_A^* x_1 = [x_1 - \xi(2 - x_1)]/x_1(1 - \xi^*) > 1 \tag{5.71a}$$

$$f_2 = x_B/x_2 = 1 + \xi > 1 \tag{5.71b}$$

and since $x_A \geqslant x_A^* x_1$ and $x_B > x_2$, both activity coefficients are always larger than unity and $g^E > 0$ over the whole composition range. The value of ξ from (5.68) may be substituted in (5.71) to give f_1 and f_2 in terms of x_1 and K_2, as shown in Fig. 5.5. The limiting values of the activity coefficients at infinite dilution are

$$\lim f_1(x_1 \to 0) = 1/(1 - \xi^*) \tag{5.72a}$$

$$\lim f_2(x_1 \to 1) = 1 + \xi^* \tag{5.72b}$$

As K_2 becomes larger g^E also increases and reaches the limit

$$\lim_{K_2 \to \infty} (g^E/\mathbf{R}T) = \tfrac{1}{2}[2(1 - x_1) \ln 2 - x_1 \ln x_1 - (2 - x_1) \ln(2 - x_1)] \tag{5.73}$$

which is asymmetric with respect to the composition, with a maximum value of 0.224 at $x_1 = 0.4$.

If $\Delta h_{A_2}^0$ is the standard molar enthalpy change for the dimerization reaction, $\Delta h_{A_2}^0 = h_{A_2}^* - 2h_A^*$, and since $h_2^* = h_B^*$ but $h_1^* = h_A^* + \xi^*(1 + \xi^*)^{-1} \Delta h_{A_2}^0$ for the pure components, the molar enthalpy of the mixture is

$$\begin{aligned} h_A &= x_1 h_m^* + x_2 h_B^* + \xi(1 + \xi)^{-1} \Delta h_{A_2}^0 \\ &= x_1(h_1^* - \xi^*(1 + \xi^*)^{-1} \Delta h_{A_2}^0) + x_2 h_2^* + \xi(1 + \xi)^{-1} \Delta h_{A_2}^0 \end{aligned} \tag{5.74}$$

so that the excess enthalpy is

$$h^E = [\xi(1 + \xi)^{-1} - x_1 \xi^*(1 + \xi^*)^{-1}] \Delta h_{A_2}^0 \tag{5.75}$$

where the term in the brackets can be shown to be negative. Hence h^E and $\Delta h_{A_2}^0$

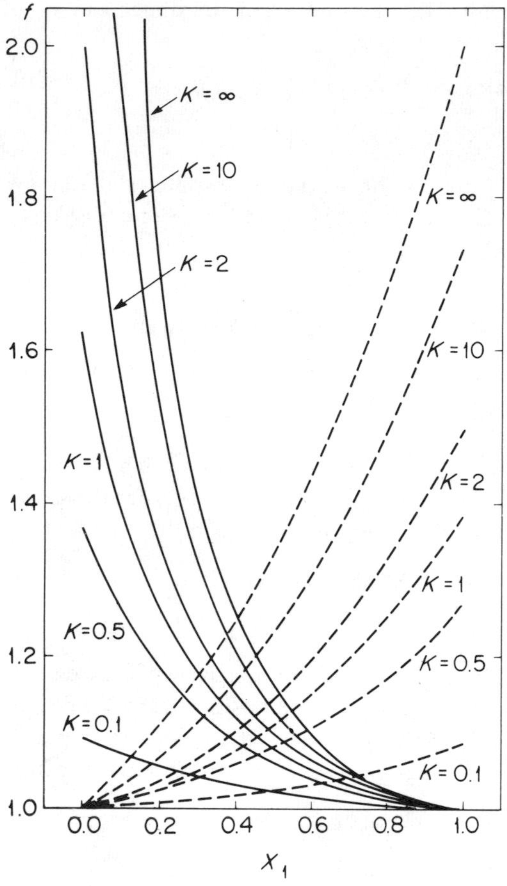

Figure 5.5. Activity coefficients of the nominal components in an *ideal* associated mixture of type $A + A_2 + B$ (self dimerization), plotted against the nominal composition x_1, with K_2 as the parameter, having the values 0.1, 0.5, 1.0, 2, 10 and ∞; (——) f_1, (- - - -) f_2. Note that the plot is asymmetric with respect to x_1, and that $f > 1$.

have always opposite signs, and since the latter can be expected to be negative, the former would be positive. The ratio $h^E/\Delta h^0_{A_2}$ depends on both x_1 and K_2, and its extreme value is obtained at $K_2 \sim 1.1$ and $x_1 \sim 0.4$, being -0.039. Obviously, for $K_2 = \infty$, i.e. stable dimers, there would be no excess enthalpy of mixing. The excess entropy of mixing depends on $\Delta h^0_{A_2}$, on K_2 and on x_1, and for any K_2, may be negative for a $\Delta h^0_{A_2}$ value less negative than some critical value, positive if it is more negative than another critical value, and changing from positive to negative as x_1 increases for in-between values of $\Delta h^0_{A_2}$.

Case (2). The chemical equilibrium $A + B \rightleftarrows AB$ is symmetrical in the components, hence it is expected that all the excess functions will also be symmetrical. The equilibrium constant K_{AB} can be written in terms of the nominal components, i.e. x_1

$$K_{AB} = \zeta/[x_1(1 - x_1)(1 + \zeta)^2 - \zeta] \tag{5.76}$$

where $\zeta \equiv x_{AB}$ for short. Hence

$$\zeta = \frac{K_{AB} + 1}{2K_{AB}x_1(1 - x_1)} - 1 - \left[\left(\frac{K_{AB} + 1}{2K_{AB}x_1(1 - x_1)} - 1\right)^2 - 1\right]^{1/2} \tag{5.77}$$

For the pure components, obviously, $\zeta = 0$ and $x_A^* = x_1^* = 1$ and similarly $x_B^* = x_2^* = 1$. For any K_{AB} the maximal ζ is obtained at $x_1 = 0.5$, while for any x_1, ζ increases with increasing K_{AB} to the maximal value $x_1/(1 - x_1)$ for $x_1 \leqslant 0.5$ and $(1 - x_1)/x_1$ for $x_1 \geqslant 0.5$ as $K_{AB} \rightarrow \infty$.

The molar Gibbs energy is given by

$$\begin{aligned} g_m &= x_1(\mu_A^0 + \mathbf{R}T \ln x_A) + x_2(\mu_B^0 + \mathbf{R}T \ln x_B) \\ &= x_1(\mu_1^0 + \mathbf{R}T \ln x_1) + x_2(\mu_2^0 + \mathbf{R}T \ln x_2) + \mathbf{R}Tx_1 \ln(x_A/x_1) \\ &\quad + \mathbf{R}Tx_2 \ln(x_B/x_2) \end{aligned} \tag{5.78}$$

since $\mu_1^0 = \mu_A^0$ and $\mu_2^0 = \mu_B^0$ in this case, so that

$$\begin{aligned} g^E/\mathbf{R}T &= x_1 \ln(x_A/x_1) + x_2 \ln(x_B/x_2) \\ &= x_1 \ln[1 - (1 - x_1)\zeta/x_1] + (1 - x_1) \ln[1 - x_1\zeta/(1 - x_1)] \end{aligned} \tag{5.79}$$

Since $x_A < x_1$ and $x_B < x_2$, the excess Gibbs energy is always negative, and so are the deviations from Raoult's law for the nominal components (shown in Fig. 5.6 for several K values).

$$f_1 = x_A/x_1 = 1 - \zeta(1 - x_1)/x_1 < 1 \tag{5.80a}$$

$$f_2 = x_B/x_2 = 1 - \zeta x_1/(1 - x_1) < 1 \tag{5.80b}$$

If Δh_{AB}^0 is the standard molar enthalpy change for the adduct formation reaction, $\Delta h_{AB}^0 = h_{AB}^* - h_A^* - h_B^*$, the molar enthalpy of the mixture is

$$h_m = x_1 h_A^* + x_2 h_B^* + \zeta(1 + \zeta)^{-1} \Delta h_{AB}^0 \tag{5.81}$$

and since $h_A^* = h_1^*$ and $h_B^* = h_2^*$, the excess enthalpy is

$$h^E = \zeta(1 + \zeta)^{-1} \Delta h_{AB}^0 \tag{5.82}$$

and has the same sign as Δh_{AB}^0, and is therefore usually negative. The excess entropy may have either sign, and may even change the sign twice over the x_1 range, depending on the magnitudes of K_{AB} and Δh_{AB}^0, but it has an extremum at $x_1 = 0.5$

$$\hat{s}^E/\mathbf{R} = (\Delta h_{AB}^0/\mathbf{R}T)\hat{\zeta}(1 - \hat{\zeta})^{-1} - \ln(1 - \hat{\zeta}) \tag{5.83}$$

where $\hat{\zeta}$, i.e. the value at $x_1 = 0.5$, is obtained from (5.79) as $1 - \exp(\hat{g}^E/\mathbf{R}T)$, and is related to K_{AB} by $K_{AB} - 4\hat{\zeta}(1 - \hat{\zeta})^{-2}$.

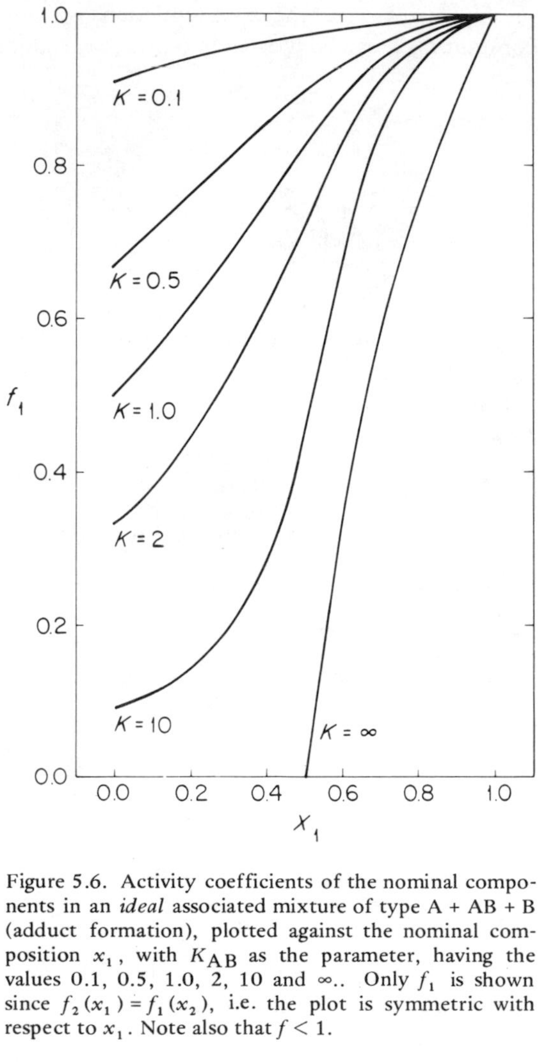

Figure 5.6. Activity coefficients of the nominal components in an *ideal* associated mixture of type A + AB + B (adduct formation), plotted against the nominal composition x_1, with K_{AB} as the parameter, having the values 0.1, 0.5, 1.0, 2, 10 and ∞.. Only f_1 is shown since $f_2(x_1) = f_1(x_2)$, i.e. the plot is symmetric with respect to x_1. Note also that $f < 1$.

The acetone–chloroform system approximates this type of ideal associated mixtures. The data[39] conform to the expectations from eqs. (5.79) and (5.82) with $K_{AB} = 0.77$ and $\Delta h^0_{AB} = -11.3$ kJ mol^{-1}.

It should be noted generally that although the mixing process is ideal, there exist in these mixtures finite excess enthalpies (and volumes, where in the appropriate equations everywhere v replaces h), which originate from the chemical reactions and their standard heat (and volume) changes. A change in the nominal composition causes a change in the concentrations of the actual species, so that a net chemical reaction occurs and with it the heat (and volume) effects.

b. Athermal associated mixtures

If the effects of the difference in size of the molecular species is recognized in addition to the effects of association, but the mixing process itself is assumed to proceed without a thermal effect beyond that of the chemical reactions, athermal associated mixtures result[40]. The general associated species is A_mB_n, and the equilibrium condition $\mu_{A_mB_n} = m\mu_A + n\mu_B$ holds, and further, $h^M(A_mB_n) = 0$ and Flory's approximation is used for the molar entropy (4.30, 5.21) $s^M(A_mB_n) = -R\Sigma x_{A_mB_n} \ln \varphi_{A_mB_n}$, with the additional assumption that the standard volume change of the reaction $\Delta v^0_{mn} = v^*_{A_mB_n} - mv^*_A - nv^*_B = 0$ (this implies that $v^E = 0$). From the latter relationship follows $\varphi_{A_mB_n}/x_{A_mB_n} = m(\varphi_A/x_A) + n(\varphi_B/x_B)$. The equilibrium constant in terms of volume fractions is

$$K_{mn(\varphi)} = \varphi_{A_mB_n}/\varphi^m_A\varphi^n_B = K_{mn}e^{m+n-1} \tag{5.84}$$

where K_{mn} is the thermodynamic constant. From (5.84), $K_{mn(\varphi)}$ is seen to be independent of the composition.

The activity coefficient of the actual general species is no longer unity as for the ideal mixtures, but is

$$\ln f_{A_mB_n} = \ln[m(\varphi_A/x_A) + n(\varphi_B/x_B)] + 1 - [m(\varphi_A/x_A) + n(\varphi_B/x_B)] \tag{5.85}$$

Expression (5.85) with the values $(m,n) = (1,0)$ for the monomer A, and (0,1) for the monomer B, may be used with (5.65) and the equilibrium condition for the chemical potentials, to give the activity coefficients of the nominal components as

$$f_1 = \varphi_A(x_1\varphi^*_A)^{-1} \exp[(\varphi^*_A/x^*_A) - (\varphi_A/x_A)] \tag{5.86}$$

and *mutatis mutandis* for component 2. At infinite dilution of component 1, $\lim (\varphi_A/x_A)(x_1 \to 0) = v^*_A\varphi^*_B/v^*_Bx^*_B$, which may be introduced into (5.86) to give the limiting value of the activity coefficient.

Of particular interest in this respect is the case of self association, i.e. where $n = 0$, and m may have suitable values. If only dimers are formed ($A + A_2 + B$ systems) $m = 2$ and $K_{(\varphi)} = K_2e$. For pure component 1, $\varphi^*_A = 2(1 + (1 + 4K_{(\varphi)})^{1/2})^{-1}$, and at any concentration

$$\varphi_A = 2\varphi_1(1 + (1 + 4K_{(\varphi)}\varphi_1)^{1/2})^{-1} \tag{5.87}$$

Since the chemical reaction is the only source for any heat effect, the expression obtained for ideal associated mixtures (5.75) may be used, substituting for ξ, with $r = v^*_B/v^*_A$

$$\xi = r\varphi_{A_2}/(2 + (r-2)\varphi_1 + r\varphi_A); \quad \xi^* = \varphi^*_{A_2}/(\varphi^*_{A_2} + 2\varphi^*_A) \tag{5.88}$$

to give

$$h^E = -\tfrac{1}{2}[(\varphi_A/\varphi_1) - \varphi^*_A]x_1\Delta h^0_2 \tag{5.89}$$

The activity coefficient for component 1 is obtained from (5.86) as

$$\ln f_1 = [\ln(\varphi_1/x_1) + 1 - (\varphi_1/x_1)] + \ln(\varphi_A/\varphi_1\varphi^*_A) - \tfrac{1}{2}(\varphi^*_{A_2} - \varphi_{A_2}) \tag{5.90}$$

The first term, that in the square brackets, is the contribution from the non-associated – athermal interactions, cf. (5.11).

Another case of particular interest is that where the self association of component A leads to a series of species A, A_2, A_3 . . . , while B remains monomeric. The assumption that for the chemical equilibria $A_{m-1} + A \rightleftarrows A_m$ the following relationships hold (the Mecke–Kempter model)[41]

$$\Delta h^0_{m-1,\,m} = \Delta h^0_2 \quad \text{(independent of } m\text{)} \tag{5.91a}$$

$$\Delta s^0_{m-1,\,m} = \Delta s^0_2 \quad \text{(independent of } m\text{)} \tag{5.91b}$$

hence

$$K_{m-1,\,m} = K_2 \quad \text{(independent of } m\text{)} \tag{5.91c}$$

leads to particularly simple expressions. The volume fraction equilibrium constant is

$$K_{m(\varphi)} = \varphi_{A_m}/\varphi_A^m = (eK_2)^{m-1} \tag{5.92}$$

From $\varphi_1 = \varphi_A + \sum_{m=2}^{\infty} \varphi_{A_m} = \varphi_A + \sum_{m=2}^{\infty} (eK_2)^{m-1}\varphi_A^m$ follows

$$\varphi_A = \varphi_1/(1 + (eK_2)\varphi_1); \quad \varphi_A^* = 1/(1 + (eK_2))$$

$$\varphi_B = \varphi_2 \tag{5.93}$$

and the mole fractions are given by

$$x_A = reK_2\varphi_A/[eK_2(1 - \varphi_1) + r\ln(1 + eK_2\varphi_1)] \tag{5.94a}$$

$$x_B = eK_2(1 - \varphi_1)/[eK_2(1 - \varphi_1) + r\ln(1 + eK_2\varphi_1)] \tag{5.94b}$$

From these expressions and (5.86) the activity coefficients of the nominal components may be calculated, with r and K_2 as parameters (Fig. 5.7).

Certain actual mixtures, such as n-alcohols–n-alkanes or n-amines–n-alkanes have been shown[42,43] to behave as expected for athermal associated mixtures to a good approximation.

c. *Regular associated mixtures*

An approximation to the behaviour of actual systems, better than that of athermal associated mixtures, is that of the regular associated mixture, which allows for finite enthalpies of mixing of the actual components of the associated mixture. A detailed examination has been made of the simplified case, where only the mutual mixing of monomeric A and B species but not that of any associated species among themselves or with monomeric A or B, has a finite heat effect[44]. The non-specific interaction is characterized by the energy α (cf. eq. (4.69)).

In the system A + A_2 + B, eq. (5.67) for the equilibrium constant in ideal mixtures must be modified for the regular mixtures to give

$$K_2 = \xi[x_1 - \xi(2 - x_1)]^{-2}\exp[-\alpha(1 - x_1)(2 - x_1)(1 + \xi)^2/RT] \tag{5.95}$$

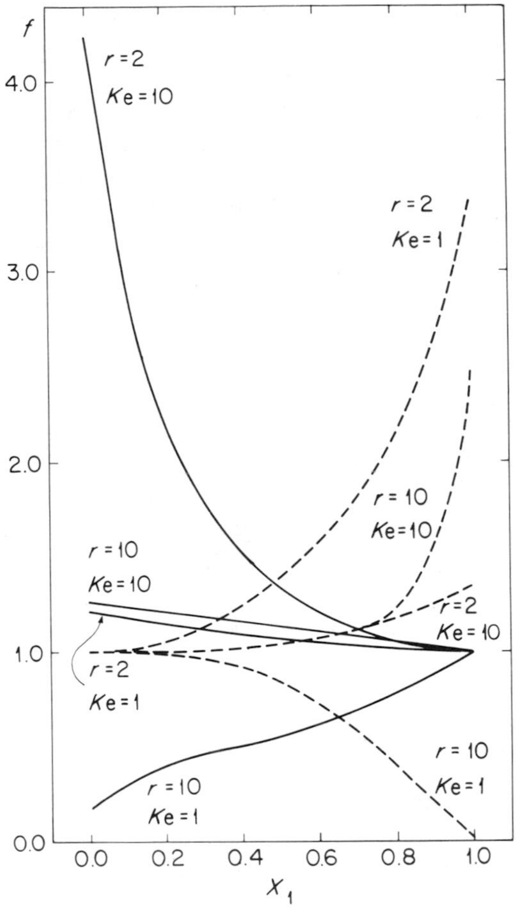

Figure 5.7. Activity coefficients of the nominal components in *athermal* associated mixtures of the type $A + A_2 + B$ (self dimerization), plotted against the nominal composition x_1, with both K_2 and $r = v_2^*/v_1^*$ as parameters, having the values $K_2 = 1/e$ and $10/e$ and $r = 2$ and 10; (———) f_1, (– – – – –) f_2. Note that $f < 1$ may be obtained, contrary to the ideal associated mixture case, Fig. 5.5, provided K_2 is small and r is large.

and in the system A + AB + B, eq. (5.76) must be correspondingly modified to

$$K_{AB} = \zeta[x_1(1 - x_1)(1 + \zeta)^2 - \zeta]^{-1} \exp[-\alpha(1 - x_1(1 - x_1)(1 + \zeta)^2)/RT] \tag{5.96}$$

The exponential terms thus correct for the non-ideality of the interactions, namely for the non-specific interaction of A and B. These non-specific interactions have interesting consequences. For instance, ideal mixtures, even if associated, cannot split into two immiscible phases[45]. Suppose that an ideal mixture contains species A, B and A_mB_n in two phases $'$ and $''$. Since $\mu'_A = \mu'_1 = \mu''_1 = \mu''_A$, and since in ideal solutions $\mu'_A = \mu^{0\prime}_A + \mathbf{R}T \ln x'_A$ and $\mu''_A = \mu^{0\prime\prime}_A + \mathbf{R}T \ln x''_A$ and $\mu^{0\prime}_A = \mu^{0\prime\prime}_A = \mu^0_A$, it follows that $x'_A = x''_A$. Similar considerations apply to component 2, and $x'_B = x^0_B$, so that necessarily also $x'_{A_mB_n} = x''_{A_mB_n}$ for all m and n, as $x_{A_mB_n} = K_{mn}x^m_A x^n_B$ and K_{mn} is independent of the composition in ideal mixtures. But if the mole fractions of all the species are identical in the two phases $'$ and $''$, these are identical in all respects, and constitute but one homogeneous phase. If non-ideality is permitted, on the other hand, demixing may occur, as discussed in section 4A(c).

Another interesting consequence of the non-ideality is the possibility for an associating substance to show *increasing* association when *diluted* with another, inert substance, contrary to the expectation from ideal mixtures, where dissociation always follows dilution. Consider the system $A + A_2 + B$, Fig. 5.8, where pure component 1 contains a mole fraction ξ^* of dimers. On dilution with component 2 (B), if the association equilibrium were 'frozen', the mole fraction of dimers would become $\xi' = \xi^*(1 - x_B) = \xi^* x_1/(1 + (1 - x_1)\xi^*)$. The real dimer mole fraction depends on x_1 in such a manner (derivable from 5.95) that

$$\lim (\partial\xi/\partial x_1)_{x_1=0} = 0$$

$$\lim (\partial\xi/\partial x_1)_{x_1=1} = \xi^* [2 - \alpha(1 - \xi^{*2})/\mathbf{R}T] \qquad (5.97)$$

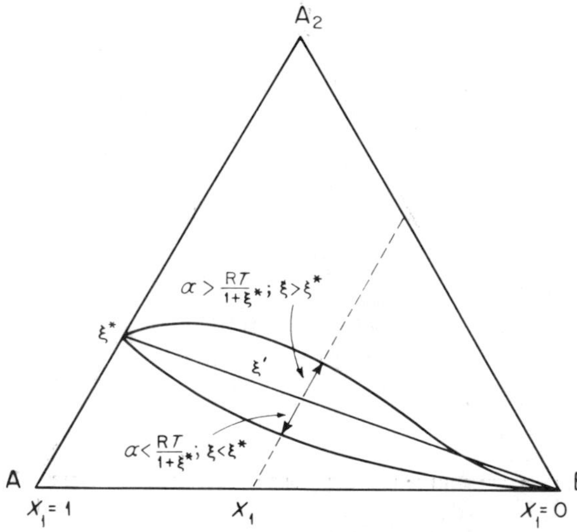

Figure 5.8. The dimer fraction ξ in a *regular* self associated mixture of type $A + A_2 + B$, as a function of the nominal composition x_1 and the energy parameter α. For pure component 1, the dimer fraction is ξ^*, and the line ξ' indicates this fraction if the equilibrium were frozen on dilution with component 2 = B. Depending on the value of α, more dimers are formed, $\xi > \xi^*$ or fewer are formed, in the real equilibrium mixture, at a given nominal composition x_1.

while for the 'frozen' dimers

$$\lim (\partial\xi'/\partial x_1)_{x_1=0} = \xi^*/(1 + \xi^*)$$

$$\lim (\partial\xi'/\partial x_1)_{x_1=1} = \xi^*(1 + \xi^*) \qquad (5.98)$$

In dilute solutions of component 1, therefore, the equilibrium line lies below the 'frozen' line, and the dimers must dissociate. In concentrated solutions, however, if $\alpha > \mathbf{R}T/(1 + \xi^*)$, the 'frozen' line lies below the equilibrium line, so that more dimers must be formed by association, and $\xi > \xi^*$ when some diluent B is added to pure component A.

A further consequence of the non-ideality of the mixtures is the non-constancy of the equilibrium quotients $K_{(x)}$ as the composition changes. For the system A + AB + B, it is possible to show that $K_{(x)}$ (i.e. K calculated for the ideal case from (5.76)) varies from a maximal value at $x_1 = 0$ to a minimal value at $x_1 = 0.5$, with the ratio between them of $\alpha(1 + \hat{\xi})^2/4\mathbf{R}T$. For the particular case of $K_{AB} = 1$ and $\alpha = \mathbf{R}T$ (Fig. 5.9), $K_{(x)}$ varies from 2.72 at $x_1 = 0$ to 1.83 at $x_1 = 0.5$, but within the range $0.2 < x_1 < 0.8$, the variation of $K_{(x)}$ is much smaller, with $K_{(x)} = 2.0 \pm 0.2$ expressing the data well. Therefore in the absence of data of very high precision covering also the dilute range, a wrong conclusion on the ideality of the system and the magnitude of the association constant may be reached. For the system $A + A_2 + B$, $K_{(x)}$ varies similarly from $K_2 \exp(2\alpha/\mathbf{R}T)$ at $x_1 = 0$ to K_2 at $x_1 = 1$, the variability clearly increasing with increasing non-ideality (the case $K_2 = 1$ and $\alpha = \mathbf{R}T$ is shown in Fig. 5.9). In a Mecke–Kempter system, $A + A_2 + \ldots + A_\infty + B$, with non-specific interactions between B and each A_i species according to a parameter $\alpha_i = \alpha(1 + m(i - 1))$, with m some constant, the association quotients are defined by $\xi_i = K_{(x)}^{-1}(K_{(x)}x_A)^i$, hence $K_{(x)} = K_2 \cdot \exp(\alpha(1 - m/x_2)x_B^2/\mathbf{R}T)$, again depending on the composition (the cases $K_2 = 1$, $\alpha = \mathbf{R}T$, $m = 0$ or 1 are shown in Fig. 5.9). Wrong conclusions about the type of the system and the interactions between the species may therefore again be arrived at by considering only $K_{(x)}$, or its variation over a too narrow concentration range.

It must be stressed, however, that the particular assumption followed by Kehiaian and co-workers in their work on regular associated mixtures[44], i.e. that only the monomers A and B interact non-specifically, but not the monomers A with the oligomers A_i, or with adducts A_mB_n, etc., is rather restrictive. It is more reasonable to take α as representing an average between the homomolecular and heteromolecular interactions (e.g. $\alpha_{A,A}$, $\alpha_{B,B}$, $\alpha_{A,B}$, $\alpha_{A,AB}$, $\alpha_{B,AB}$, $\alpha_{AB,AB}$ in the system A + AB + B, which may be considered as a ternary system). Following Prigogine and Defay[45], this average non-specific interaction energy is defined so that it will contribute a term $\alpha x_1 x_2$ to the excess Gibbs energy of the system, which may be written as

$$g^E = x_1 \mathbf{R}T \ln(x_A/x_1 x_A^*) + x_2 \mathbf{R}T \ln(x_B/x_2 x_B^*) + \alpha x_1 x_2 \qquad (5.99)$$

Therefore the expressions for the activity coefficients of the nominal components become (cf. (5.66a))

$$f_1 = (x_A/x_1 x_A^*) \exp(\alpha x_2^2) \qquad (5.100)$$

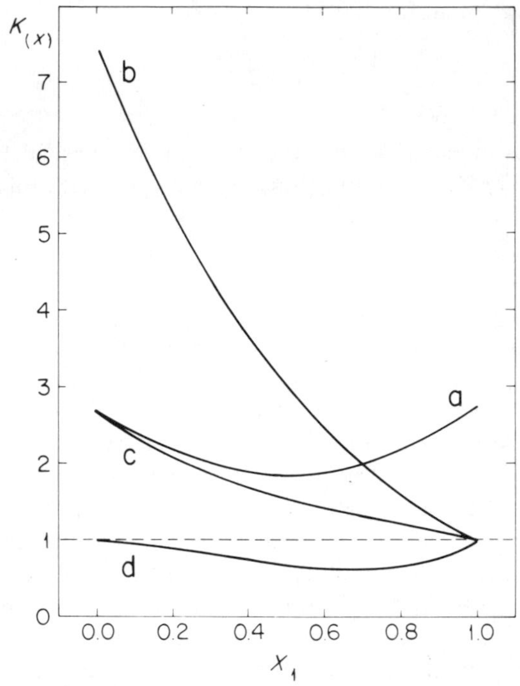

Figure 5.9. The dependence of the mole-fraction equilibrium quotient $K_{(x)}$ (so-called concentration constant) on the composition x_1 for several *regular* associated mixtures. In every case, the true equilibrium constant is $K = 1$, and the energy parameter is $\alpha = \mathbf{R}T$. Curve a, system A + AB + B; curve b, system A + A_2 + B; curve c, Mecke–Kempter system A + A_2 + A_3 + . . . + B, with $K_i = K_2 = K$ and $\alpha_i = \alpha$; curve d, Mecke–Dempter system as above but with $\alpha_i = i\alpha$.

and similarly for f_2, i.e. equating f_A with $f_A^* \exp(\alpha x_2^2)$. Empirically, these are two parameter equations, since x_A^* and x_B^*, as well as the variables x_A and x_B, can be obtained from K and α. The parameter α is related to the molar volumes of the components and to their solubility parameters, or other interaction parameters, according to the considerations discussed previously (sections 5A(c) to 5a(f)). An application of this approach has been made recently to the system HF–$(HF)_i$–HR, where HR represents the aliphatic hydrocarbons from propane to n-heptane[46]. A further correction, for the disparity in size of the components, according to the athermal mixture model, has also been incorporated[45,46]. Another illustration is the system ethanol–methylcyclohexane[47], which will be discussed in detail below (Fig. 5.10).

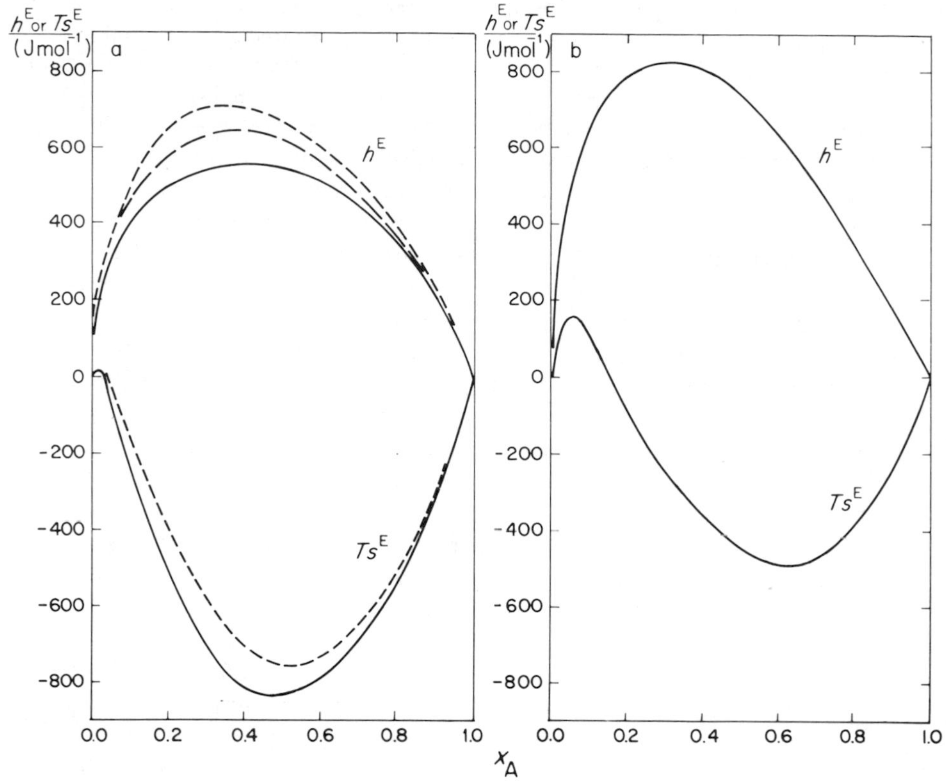

Figure 5.10. Excess thermodynamic functions h^E (J mol^{-1}) and Ts^E (J mol^{-1}) for ethanol (A)–hydrocarbon (B) mixtures as a function of the alcohol mole fraction x_A. (a) ———, B = methylcyclohexane, 25 °C(ref. 47); – – – – – B = methylcylohexane, 35 °C(ref. 63) —— —— B = n-hexane, 30 °C (ref. 58);(b) b = toluene, 25 °C(ref. 47).

d. Alcohols in inert solvents

It is impossible to examine in the present volume even a fraction of the cases of mixtures of associated liquids. One case that has received detailed consideration from many authors must suffice to illustrate the applications of the above formal thermodynamic approaches to real cases. This is the system of normal aliphatic alcohols in inert solvents: hydrocarbons or halogenated hydrocarbons. The alcohols undergo self association by virtue of their hydrogen-bonding properties (Fig. 3.2a)

$$\text{ROH} + (\text{ROH})_{i-1} \rightleftarrows (\text{ROH})_i \qquad (i = 2, 3, \ldots) \qquad (5.101)$$

and further interaction with the solvent may, or may not, occur.

There is no doubt that the pure, undiluted alcohols themselves are extensively associated, both in the vapour phase (P, V, T and heat capacity data[48]) and as liquids (correlation factors g^* from dielectric polarization data[49], section 3B(d)). At the other extreme, properties of monomeric methanol were studied by the matrix isolation technique[50], and species showing the same infrared absorption frequency were identified at high dilutions in solvents such as carbon tetrachloride, decane, heptane, benzene or toluene[51–57]. It is thus evident that considerable dissociation must occur when the alcohol is mixed with the diluent; that is, hydrogen bonds are broken, and this explains the postive heats of mixing (Fig. 5.10). For a detailed interpretation, however, some further facts must be considered.

Alcohol–hydrocarbon systems may show phase separation: methanol and n-hexane separate below 33.7 °C, and methanol and n-heptane at 51.2 °C (refs. 58, 59), but even ethanol is immiscible with a hydrocarbon, methylcyclohexane, at a sufficiently low temperature[47], about −83 °C. According to the discussion in the previous section, this means that there are in these systems non-specific interactions, beyond any association reactions. These are the major interactions at low temperatures[60], but contribute also at higher temperatures, contrary to the assumptions of some authors who have neglected these interactions.

Dielectric polarization studies[60] show that in dilute solutions of alcohols in carbon tetrachloride, aromatic and aliphatic hydrocarbons, the average dipole moment is smaller than that of the monomer, and the correlation factor $g < 1$. This means that association leads to cyclic species, where the dipoles are antiparallel (structure II) or have a very small resultant (structure V). At higher concentrations, though, higher dipole moments than the monomer and $g > g^*$ are observed, showing that open-chain structures (such as III or IV) are predominant. There is no escaping the conclusion that both types of species must be formed.

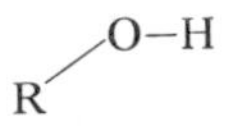

(I)

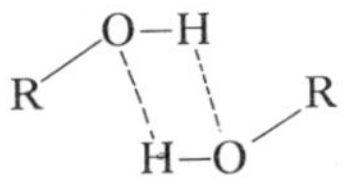

(II)

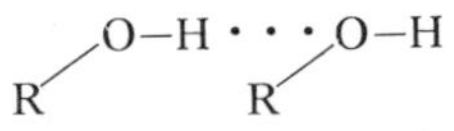

(III)

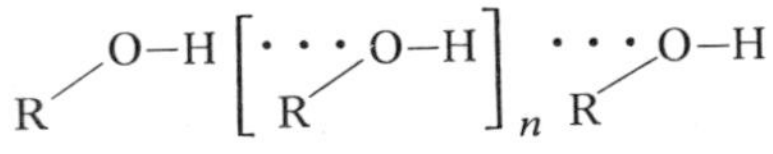

(IV)

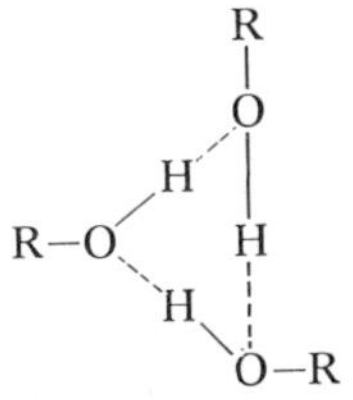

(V)

There is good evidence that carbon tetrachloride and aromatic hydrocarbons are not inert towards the alcohols (apart, of course, from ordinary dispersion forces), and that adducts or 'solvates' are formed. The former diluent forms an R–O–H . . . Cl–CCl_3 hydrogen bond[57,61] and the latter show an interaction of the R–O–H dipole with the π-electron system[47,62]. This interaction is usually not expressed in terms of definite stoichiometric adducts, but in terms of a negative contribution to the heat of mixing, or a smaller α parameter for non-specific interactions in these solvents, compared with the more inert aliphatic hydrocarbons. Even these latter show a slight variation among themselves[63], although the data for n-hexane at 30 °C are seen to fit in quite well between those for methylcyclohexane at 25 °C and 35 °C (Fig. 5.10(a)).

Finally, there is some evidence that the mixing of the alcohols and the diluents is accompanied by some volume changes[64]. These are quite small for the aromatic diluents: $\hat{v}^E = -0.070$ cm^3 for ethanol–toluene mixtures and $+0.024$ cm^3 for ethanol–*m*-xlyene mixtures at 25 °C, somewhat larger at 45 °C, but always $< 0.2\%$ of v_m. For aliphatic diluents the volume changes are not so small: $\hat{v}^E = +0.566$ cm^3 for ethanol–cyclohexane mixtures and $+1.22$ cm^3 for 1-propanol–heptane mixtures at 25 °C, always $< 1\%$ of v_m. This shows that setting $v^E = 0$ is an acceptable first approximation for derivations of g^E of h^E in terms of volume fractions.

The major controversy converges on the extent of association and the models used to relate the assumed species and interactions to the observed quantities. The methods that have yielded most of the data on this problem are infrared spectrophotometry mainly in the overtone region of the O–H stretching vibration, vapour–liquid equilibrium measurements, and heat of mixing measurements. Some other methods, such as nuclear magnetic resonance[65], ultrasonics[66], and dielectric constant measurements[60], have provided data which are not sufficiently detailed to permit the decision in favour of any particluar one of the models.

The infrared absorption shows distinct peaks at three frequencies, which are assigned to 'free' hydroxyl groups (such as in monomers or at the end of chain structures III and IV), and to two types of hydrogen-bonded hydroxyl groups, which may be single-bonded R–O–H . . . O and double-bonded H . . . O(R)–H . . . O, or unstrained (open-chain) and strained (cyclic) groups, or any other combination. No final uncontested assignment has been made. Even the correction of the 'monomer' peak for chain-end group absorption is not quite clear. At best, these measurements may give the fraction of the monomer species as a function of the concentration and of the temperature. Various computation methods are then used to obtain equilibrium constants for association reactions (5.101), and from their temperature coefficients the standard heats of the reactions. There is, however, no unique set of oligomers that produces an impressively 'best' least-squares fit[54], and the disregard of non-specific interactions could improve the fit of a set which is not indeed the true one. There is no inherent reason why certain species should be disfavoured (except, perhaps, the cyclic dimer, structure II, with highly strained hydrogen bonds), and the ability to fit the data with few parameters is no valid argument against the presence of several more species. Thus, in addition to the monomer, a dimer alone[62], perhaps cyclic[51], a tetramer alone, both linear and cyclic[57], a set of two species, dimer and trimer[54] or dimer and tetramer[51,54], a set of three species, dimer trimer and cyclic tetramer[53] or dimer, trimer and a higher oligomer[55], a dimer and a Kempter–Mecke series of oligomers[52], have been claimed to represent the data adequately, but they cannot do this uniquely[54]. Perhaps the best interpretation is to forego any attempt to assign a definite stoichiometry[56].

The situation is similar in the thermodynamic studies. Here the data range from excess heat capacities c_p^E as a function of the temperature and the concentration[47], to heats of mixing $h^M = h^E$ at one or several temperatures,[24,47,58,63–71] to activity coefficients and excess Gibbs energies, obtained from vapour–liquid equilibria.[24,47,48,59,61,67,68,72–74] Heats of solution at infinite dilution and limiting activity coefficients have also been obtained from these data. The main problem with these thermodynamic data is that they may be fitted with several plausible expressions for g^E, h^E and c_p^E, based on somewhat different models, and there is no good way to decide in favour of one or another, although some other models may be discounted. Thus, the quasi-lattice contact–point approach, which does not take into account the association of the alcohol beyond a single kind of OH–OH interaction[24,60] is incompatible with other evidence, such as the above mentioned dipole moment and spectroscopic data. The alcohol vapour probably is associated to only small oligomers, such as dimers or tetramers[48], but only very few studies apply the ideal association model with two or three oligomers[61] to g^E or h^E data of methanol mixtures with carbon tetrachloride, hexane or hexadecane. As argued above, non-specific interactions should not be neglected, so that the ideal association model should not be used. The concensus among the other studies is that an infinite series of associated species (of the Kempter–Mecke type) is formed, the differences reducing to the problems of whether the dimer has a different association constant[47], whether mole fractions, volume fractions or molar concentrations should be used as the concentration scale[68,71], and how the non-specific interactions should be handled[59,71]. As regards the latter, a regular solution term $\chi v\ \varphi_1\varphi_2$ where 1 designates the alcohol and 2 the diluent (cf. eqs. 5.30 and 5.45), has been found adequate. The parameter χ is highest for methanol–heptane (~7.3 $\mathbf{R}T\,\mathrm{dm}^{-3}$), where phase separation occurs, and becomes smaller as the chain-length of the alcohol increases, or when aromatic hydrocarbons or carbon tetrachloride are substituted for the aliphatic hydrocarbon.[47,68,71,73] The use of the mole fraction concentration scale[67,71] leads to difficulties, and the association constant $K_{(\varphi)i} = c_i/c_{i-1}c_A v_A^* = \varphi_i/\varphi_{i-1}\varphi_A$, where A is the monomeric alcohol, seems to express the association equilibrium best. Its value depends on the alcohol, the diluent and the temperature but not on the composition. The association reaction (5.101) is characterized by a standard enthalpy change Δh_i^0 of −18 to −25 kJ (mole i-mer)$^{-1}$, which depends on the temperature in one approach[47], but not in the other[59]. Since the temperature-dependence may also be ascribed to the non-specific interactions, there is no basis to decide the point.

The problem of the consistency between the spectroscopic and the thermodynamic results has of course received attention. Neither the formation of a discrete set of oligomers, nor the formation of an infinite association series can be discounted by the results of either approach if sufficient parameters are used (at least three seem to be required if composition- and temperature-dependencies are to be accounted for), although the former interpretation has found favour with the 'spectroscopists' and the latter with the 'thermodynamicists'. An awareness of this situation may encourage an alternative approach which will forego any assignment of definite stoichiometries to the association reactions, and deal with the system in terms of 'structure'. Such an approach seems to be the most appropriate for aqueous mixtures, to be discussed below.

e. *Aqueous mixtures*

Mixtures of molecular liquids where one of the components is water have, naturally, received a great deal of attention. It is impossible to deal in the scope of this book with all the important features of such mixtures, and only a few of them will be discussed.

One interesting problem is the mutual solubility of water and the so called 'inert' solvents, such as hydrocarbons or halogen-substituted hydrocarbons. A recent comprehensive study is that of McAuliffe[75], who measured the aqueous solubilities of hydrocarbons by means of gas chromatography. Appendix 5.1 tabulates the solubilities at room temperatures in terms of mole fractions. The temperature-dependence of the solubility is also of interest; as regards the solubility of hydrocarbons in water, data for the liquid hydrocarbons are scarce[76], but those for the lower hydrocarbons should be representative. The solubilities decrease with increasing temperatures (Δh° for RH(l) → RH(saturated aqueous solution) is negative, about −5 to −10 kJ mol^{-1}). These data have a bearing on the structure of water (section 3B(e)) through the concept of *hydrophobic interaction*[76–78]. Data for gas solubilities have been mainly used to demonstrate the hydrophobic bonding effect, but recent data for solubilities of aromatic hydrocarbons[79] may serve as well. Since there is very little direct interaction between the hydrocarbon molecules and the water molecules (through dispersion forces, and dipole–induced-dipole interactions), the low solubility is governed by the increased amount of structure induced by the solute in the water. The solubility trends show that the increased structuredness is higher at higher temperatures, where intrinsically there is less structure than at lower temperatures. A convenient measure of the increased structuredness is the Gibbs hydrophobic interaction energy[79], given by

$$\delta G_{AA}^{HI}(\sigma) = \Delta G_{AA}(\sigma) - U_{AA}(\sigma) \tag{5.102}$$

where $\Delta G_{AA}(\sigma)$ is the change in Gibbs energy occurring when two particles of A move from infinite distance in water to the distance σ, while $U_{AA}(\sigma)$ is the work required for the corresponding process in vacuum. The distance σ is selected so that it corresponds to the distance between the centres of the two A particles in a 'fused' molecule A–A (e.g. ethane for A = methane, or biphenyl for A = benzene) (Fig. 5.11). This concept points the way for the experimental measurement of $\delta G_{AA}^{HI}(\sigma)$, since it should be equal to the difference between the standard Gibbs energy of transfer of A–A from the gas phase to water and twice this quantity for the transfer of A.

$$\delta G_{AA}^{HI}(\sigma) = \Delta\mu_{A-A}^0 - 2\Delta\mu_A^0 \tag{5.103}$$

At 20 °C, the values of $G_{AA}^{HI}(\sigma)$/(kJ per mole of A–A) are −8.85 for

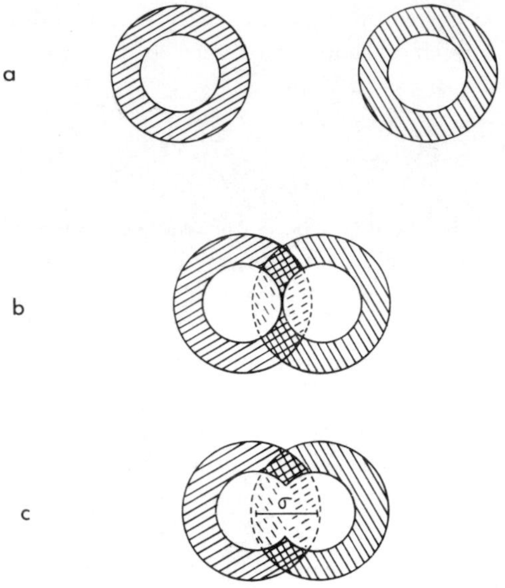

Figure 5.11. Loss of water structuredness near hydrophobic solutes when two solute particles A 'fuse', from state (a) to state A–A (c). Structuredness in the cross-hatched and the dash-hatched regions is lost.

A–A = ethane, −6.28 for A–A = butane and −4.76 for A–A = biphenyl, all becoming more negative with increasing temperatures.

The solution of water in the 'inert' solvents, on the other hand, is accompanied with monomerization of the water molecules[80]. In solvents of some polarity, where the solubility is somewhat higher (Appendix 5.1), there is a fraction of up to 15 % of associated water molecules. Disregarding the small asssociation, the solubility at room temperature has been accounted for by the expression (5.45), where χ is given by the appropriate regular-solution solubility-parameter expression $v_{H_2O}(\delta_{H_2O} - \delta_{HR})^2/\mathbf{R}T$, and $z = \varphi$. Empirically[81] it was found that $\delta_{H_2O} = (u^V_{H_2O}/v^*_{H_2O})^{1/2} = 48.7\ J^{1/2}\ cm^{-3/2}$, together with the usual δ_{HR} of the hydrocarbons (Table 5.2) fits the solubility data very well ($\ln x_{H_2O\ in\ HR} = \ln(a_{H_2O}/f_{H_2O\ in\ HR}) = -\ln f_{H_2O\ in\ HR}$, since in equilibrium with (essentially pure) water, $a_{H_2O} \simeq 1$). It is amazing, though, that $(u^V_{H_2O}/v^*_{H_2O})^{1/2}$ is a good measure for δ_{H_2O} in the above expression, in view of the polar nature and hydrogen-bonding properties of water. Indeed, the fortuitous nature of the fit of the solubility data with the above expression is shown by its complete failure to express the right trend of the temperature coefficient of the solubility, which is strongly positive.

In solvents with higher water solubilities water associates[80], first to small oligomers (possibly cyclic trimers, but the uncertainty here is the same as discussed for the alcohols, p. 197), and with increasing concentrations, say beyond 1 M or

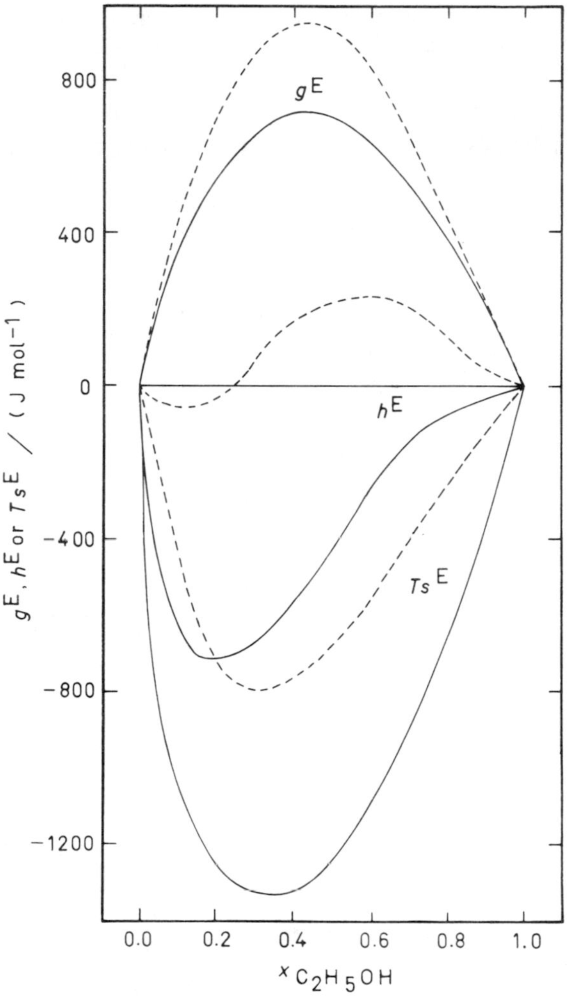

Figure 5.12. Excess thermodynamic functions of water–ethanol mixtures as a function of composition at 298 K ———, and 348 K – – – – . Data from F. Franks, '*Physico-Chemical Properties in Mixed Aqueous Solvents*', edited by F. Franks, Heinemann, London, 1967, p. 57.

$x_{H_2O} = 0.1$ to structures resembling those in liquid water. Although monomeric water or small oligomers are compatible with polar liquids with large hydrocarbon radicals, these larger aggregates are not, and limited solubilities result again, Appendix 5.2. In most cases, because of the smallness of the water molecules, $x_{W \text{ in } S}$ is one to two orders of magnitude larger than $x_{S \text{ in } W}$, where W represents water and S the other liquid. As the water structure gets destroyed with increasing temperatures, the mutual solubility often increases and an UCST (p. 150) is observed. Many such binary water–solvent mixtures have been found, involving a variety of functional groups (Appendix 5.3). The appearance of an LCST (or a closed solubility loop) has also been observed in aqueous systems, while it is a very rare occurrence in non-aqueous systems. Most of the aqueous systems showing an LCST have a secondary or tertiary (or aromatic) amine nitrogen (amines, piperidine or pyridine derivatives) or are polyfunctional ethers (glycol mono- or

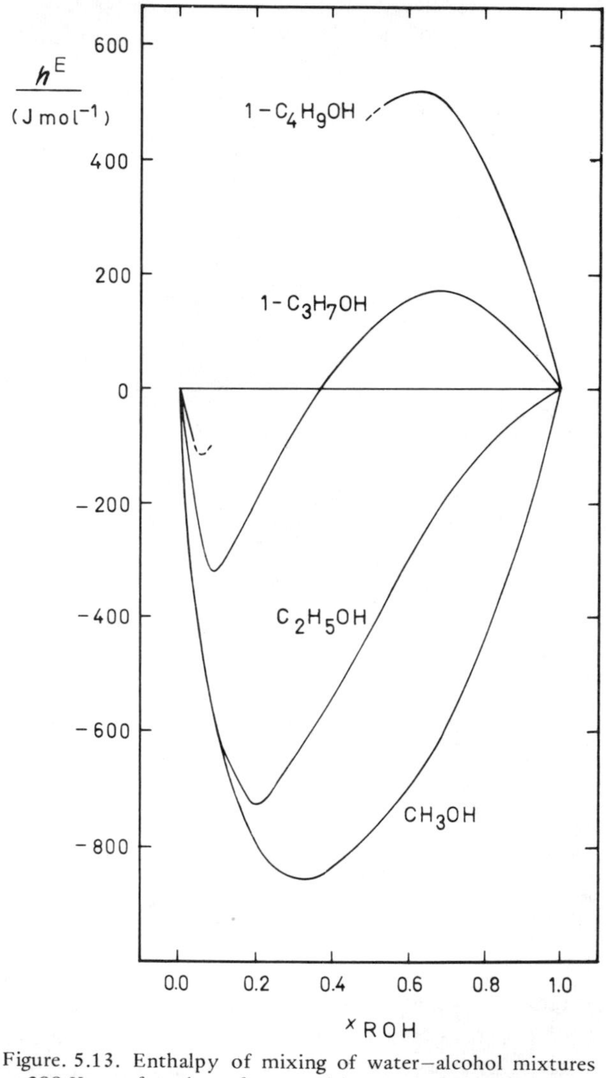

Figure. 5.13. Enthalpy of mixing of water–alcohol mixtures at 298 K as a function of composition. Data for methanol: L. Benjamin and G. C. Benson, *J. Phys. Chem.*, **67**, 858 (1963); ethanol: A. G. Mitchell and W. F. K. Wynne-Jones, *Discuss. Faraday Soc.*, **15**, 161 (1953); 1-propanol: R. F. Lama and C. Y. Lu, *J. Chem. Eng. Data*, **10**, 216 (1965); 1-butanol: S. R. Goodwin and D. M. T. Newsham, *J. Chem. Thermod.*, 3, 325 (1971).

di-ethers). The reasons for the appearance of an LCST in such systems are not clear. They are described formally by the requirements $T(\partial^2 s^E/\partial x^2) > (\partial^2 h^E/\partial x^2) > 0$, and if g^E is given by (4.88), then $\alpha < 0$, $\beta > 2\mathbf{R}$, and $T_c = \alpha/(2\mathbf{R} - \beta)$. Then h^E and s^E are both negative, and a strong mutual association of the amine or the polyether with water is inferred. The product of this association is apparently less soluble in the structured liquid (the water-rich phase) than are the unassociated components, but no complete theory for this phenomenon has been presented.

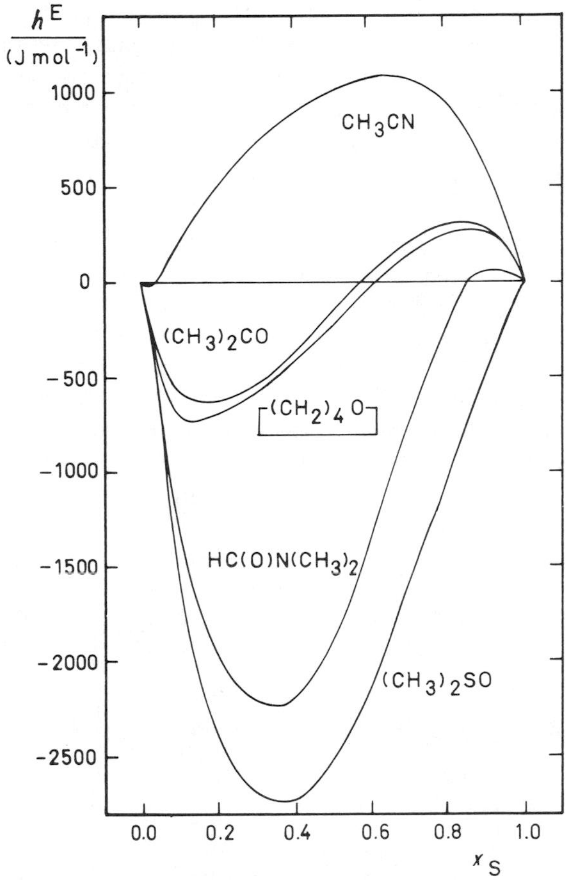

Figure 5.14. Enthalpy of mixing of water–aprotic solvent mixtures at 298 K as a function of the composition. Data for: acetonitrile: K. W. Morcom and R. W. Smith, *J. Chem. Thermod.*, **1**, 503 (1969); tetrahydrofuran: H. Nakayama and K. Shinoda, *J. Chem. Thermod.*, 3, 401 (1971); dimethylsulphoxide: H. L. Clever and S. P. Pigott, *J. Chem. Thermod.*, 3, 221 (1971); acetone: D. O. Hanson and M. Van Winkle, *J. Chem. Eng. Data*, **5**, 30 (1960); dimethylformamide (300 K): H. Peters and E. Tappe, *Monatsber. Deut. Akad. Wiss. Berlin*, **9**, 692 (1967).

The above presentation involves the concept of water structure (section 3B(e)), and this is discussed best on the basis of the thermodynamic data for completely water miscible liquids. These are the lower homologues of the monofunctional solvents, and many polyfunctional solvents, with roughly up to three carbon atoms per oxygen or nitrogen atom (Appendix 5.2). Excess enthalpy, entropy and Gibbs energy for mixtures of ethanol and water at two temperatures are shown in Fig. 5.12, and the excess enthalpies for water mixtures with methanol, ethanol and l-propanol are compared in Fig. 5.13. The behaviour of l-propanol is typical also of such liquids as tert-butanol, dioxane[82], acetone, tetrahydrofuran (Fig. 5.14) and other liquids in binary mixtures with water. It is thus not specific to liquids which are themselves hydrogen-bonded (as the alcohols), since also aprotic solvents show this behaviour. On the other hand, other aprotic solvents, such as dimethylsulphoxide, dimethylformamide or acetonitrile in mixtures with water, have either almost completely endothermic enthalpies of mixing (acetonitrile, which shows phase separation at lower temperatures, UCST at −0.9 °C) or completely exothermic ones (dimethylsulphoxide or dimethylformamide), with only a slight inversion at an extreme end of the composition range. The balance between the self association of water, the self association of the other liquid, and their mutual association is very delicate. The effect of small additions of 'inert' molecules or parts of molecules to enhance the structure of water in their dilute solutions is certainly operative at low x_S, and finds expression in numerous properties of such solutions. The effect increases in magnitude, and occurs at lower x_S, the larger the 'inert' portion of the solute with respect to its polar or hydrogen–bonding functional group.[82–84] The structure-enhancing effect is shown most clearly in the negative s_W^E in ethanol at low x_S, and its negative initial slope[84]. It is however necessary to find ad hoc reasons for the behaviour of each aqueous–solvent system, and there is no way as yet to predict the properties of the mixtures from those of the components to even a much cruder approximation than is today possible for mixtures of non-polar or even slightly polar molecular liquids.

C. Appendixes: Mutual solubilities of water and organic liquids

Appendix 5.1. Mutual solubilities of some hydrocarbon and halocarbon organic liquids (component 2) and water (component 1) at 298 K.

Liquid 2	$10^6 x_2$ (in 1)	$10^3 x_1$ (in 2)	Liquid 2	$10^6 x_2$ (in 1)	$10^3 x_1$ (in 2)
Pentane	9.5	0.480	Toluene	100.5	1.71
Hexane	1.98	0.351[a]	*o*-Xylene	29.6	
Heptane	0.57	0.461	*m*-Xylene	33.2	2.37[a]
Octane	0.096	0.650[a]	*p*-Xylene	32	
Decane		0.572	Ethylbenzene	25.8	2.54
Dodecane		0.615	Butylbenzene		3.06
Cyclopentane	40.8	0.553[a]	Mesitylene	2.9[a]	1.97[a]
Cyclohexane	11.8	0.470[a]	1-Chloropropane	620[a]	
Decalin	<26	0.485[a]	1-Chlorobutane	214[a]	
Benzene	409.5	2.74	1-Chloropentane	34[a]	

Appendix 5.1. (*continued*)

Liquid 2	10^6x_2 (in 1)	10^3x_1 (in 2)	Liquid 2	10^6x_2 (in 1)	10^3x_1 (in 2)
Chlorobenzene	78[b]	2.04	1,1,1,2-Tetrachloro-		
Dichloromethane	2800	9.3	ethane	117[a]	
1,1-Dichloroethane	920	<11	1,1,2,2-Tetrachloro-		
1,2-Dichloroethane	1490[a]	8.2[a]	ethane	308[a]	10.6
1,1-Dichloroethylene	39	1.9	Tetrachloroethylene	16	0.92
trans-1,2-Dichloro-			Bromoethane	1520[a]	
ethylene	1180	29	1-Bromopropane	337[b]	
cis-1,2-Dichloro-			1-Bromobutane	80[b]	
ethylene	650	19	Bromobenzene	51[b]	
o-Dichlorobenzene	49	24.7	1,2-Dibromoethane	412[b]	7.4
m-Dichlorobenzene	13.6[a]		Bromoform	227[b]	5.96
Chloroform	1240[a]	4.8[c]	1,1,2,2-Tetrabromo-		
1,1,1-Trichloroethane	178[a]	2.5	ethane	34[b]	
Trichloroethylene	150	2.3	Iodobenzene	30[b]	
Carbon tetrachloride	90	0.54[d]	Di-iodomethane	83[b]	

Appendix 5.2. Mutual solubilities of some oxygen- and nitrogen-containing organic liquids (component 2) and water (component 1) at 298 K.

Liquid 2	10^3x_2 (in 1)	x_1 (in 2)	Liquid 2	10^3x_2 (in 1)	x_1 (in 2)
Methanol	miscible		*p*-Cresol	3.82	
Ethanol	miscible		1,2-Ethanediol	miscible	
1-Propanol	miscible		1,2-Propanediol	miscible	
2-Propanol	miscible		1,3-Propanediol	miscible	
1-Butanol	19.2	0.515	1,3-Butanediol	miscible	
2-Butanol	33.6[a]	0.765	1,2,3-Propanetriol	miscible	
2-Methylpropan-1-ol	26.3	0.456	Diethyl ether	15.4	0.0577
2-Methylpropan-2-ol	miscible		Methylpropyl ether	7.6	
1-Pentanol	4.56	0.284	Dipropyl ether	0.86	0.0250
2-Pentanol	9.50	0.396	Di-isopropyl ether	2.14[a]	0.031[a]
3-Pentanol	11.1	0.308	Dibutyl ether	0.041[a]	0.0136[a]
2-Methylbutan-1-ol	5.85	0.310	Dihexyl ether	0.01	0.012
2-Methylbutan-2-ol	24.7	0.600	Tetrahydrofuran	miscible	
3-Methylbutan-1-ol	5.56	0.343	Tetrahydropyran	17.9	0.135
3-Methylbutan-2-ol	11.9	0.399	Dibenzyl ether	0.0036	
2,2-Dimethylpropanol	7.35	0.310	Dimethoxymethane	81.2	0.153
Cyclohexanol	7.0	0.426[a]	1,2-Dimethoxyethane	miscible	
1-Hexanol	1.23[a]	0.313[a]	1,1-Diethoxyethane	8	
2-Methylcyclohexanol	1.9	0.163	1,4-Dioxolane		
4-Methylpentan-2-ol	2.94	0.278	(dioxane)	miscible	
2-Ethylbutan-1-ol	0.94	0.214	Acetaldehyde	miscible	
1-Heptanol	0.28		propionaldehyde	120.5	0.326
2-Heptanol	0.6	0.259	Benzaldehyde	0.5	0.0174
1-Octanol	0.0745		Acetone	miscible	
2-Octanol	0.08	0.007	2-Butanone	73.2[a]	0.318[a]
2-Ethylhexan-1-ol	0.097[a]	0.162[a]	2-Pentanone	12.0	0.147
Benzyl alcohol	0.63[a]	0.354[a]	3-Pentanone	7.3[a]	0.113[a]
Phenol	17.8	0.677	2-Hexanone	6.5	0.175
o-Cresol	5.26		Cyclohexanone	4.3[a]	0.322[a]
m-Cresol	4.26	0.181	4-Methylpentan-2-one	3.1	0.097

Appendix 5.2. (*continued*)

Liquid 2	$10^3 x_2$ (in 1)	x_1 (in 2)
2-Heptanone	0.69	0.0835
2-Methylcyclohexanone	5	
2,6-Dimethylheptanone	0.08	0.033
2,4-Pentadione	31.2	0.226
Formic acid	miscible	
Acetic acid	miscible	
Propionic acid	miscible	
Butyric acid	miscible	
Isobutyric acid	65.0[a]	0.797
Valeric acid	4.3[a]	0.460
Hexanoic acid	1.50[a]	0.273
Octanoic acid	0.99	
Methyl formate	82	
Ethyl formate	31.5	0.456[a]
Methyl acetate	71.5[a]	0.264[a]
Ethyl acetate	17.7	0.129
Butyl acetate	0.67[a]	0.109[a]
Methyl propionate	10.1[c]	
Ethyl propionate	3.4	0.066[a]
Propyl propionate	0.94	
Methyl butyrate	3.0[c]	
Ethyl butyrate	0.76[a]	0.047[a]
Methyl benzoate	0.28[a]	0.053[a]
Ethyl benzoate	0.060[a]	0.04[a]
1,2-Ethanediol-diacetate	32.3[a]	0.380[a]
Ethylene carbonate	miscible[e]	
1,2-Propylene carbonate		0.312
Dibutyl phthalate	<0.007	0.067
Tributyl phosphate	0.026	0.497
Nitromethane	35.3	0.0675
Nitroethane	11.7	0.0423
1-Nitropropane	3.07	0.030
Nitrobenzene	0.30[b]	0.0162[a]
Acetonitrile	miscible	
Propionitrile	36.3	
Butyronitrile	8.9	
Benzonitrile	0.4	0.054
Succinonitrile	miscible[f]	
Adiponitrile	miscible	
Ethylamine	miscible	
1-Propylamine	miscible	
2-Propylamine	miscible	
1-Butylamine	miscible	
2-Methylpropyl-1-amine	miscible	
2-Methylpropyl-2-amine	miscible	

Liquid 2	$10^3 x_2$ (in 1)	x_1 (in 2)
Cyclohexylamine	miscible	
2-Ethylhexyl-1-amine	0.035[a]	0.708[a]
Diethylamine	miscible	
Dipropylamine	0.46[a]	0.694[a]
Di-isopropylamine	miscible	
Dibutylamine	0.066[a]	0.320[a]
Triethylamine	10.2[a]	0.214
Tributylamine		0.0124
Ethylenediamine	miscible	
1,2-Propylenediamine	miscible	
1,3-Propylenediamine	miscible	
Piperidine	miscible	
Diethylenetriamine	miscible	
Triethylenetetramine	miscible	
Aniline	6.7	0.205
o-Toluidine	2.83[a]	0.096[a]
Pyrrole	12.5	
Pyridine	miscible	
Quinoline	0.85[a]	
Carbon disulphide	0.280	0.00060
Thiophene	miscible	
Dimethylsulphoxide		
Sulpholane	miscible[b]	
2-Chloroethanol	miscible	
2-Bromoethanol	miscible	
2-Cyanoethanol	miscible	
Bis-2-chloroethyl ether	1.28[a]	0.008[a]
2-Methoxyethanol	miscible	
2-Ethoxyethanol	miscible	
2-Butoxyethanol	miscible	
3,6-Dioxaheptan-1-ol	miscible	
γ-Butyrolactone	miscible	
Ethanolamine	miscible	
Diethanolamine	miscible	
Triethanolamine	miscible	
Morpholine	miscible	
Formamide	miscible	
N-Methylformamide	miscible	
Dimethylformamide	miscible	
N-Methylacetamide	miscible	
Dimethylacetamide	miscible	
Diethylpropionamide	miscible	
Tetramethylurea	miscible	
2-Pyrrolidone	miscible	
1-Methyl-2-pyrrolidone	miscible	
N-Methylcaprolactam	miscible	
Hexamethyl phosphoramide	miscible	

Appendix 5.3. Some binary aqueous-organic mixtures with critical solution temperatures (CST) between 250 K and 500 K.

Organic component	LCST (K)	UCST (K)	Organic component	LCST (K)	UCST (K)
1-Butanol		400	Adiponitrile		374
2-Butanol		385	1-Aza-2-methylcyclo-hexane	352	500
2-Methylpropan-1-ol		402	1-Aza-4-methylcyclo-hexane	358	463
Phenol		339	1-Aza-1-ethylcyclo-hexane	281	
o-Cresol		439	1-Azacycloheptane	341	
m-Cresol		421	1-Azacyclo-octane	258	
p-Cresol		416	3-Methylpyridine	322	426
1,2-Dihydroxybenzene		308	2,3-Dimethylpyridine	290	
2-Butanone		416	2,4-Dimethylpyridine	296	
Cyclopentanone		378	2,5-Dimethylpyridine	286	
2,4-Pentadione		362	2,6-Dimethylpyridine	307	
Butyric acid		270	3,4-Dimethylpyridine	269	
1-Methylpropionic acid		295	3,5-Dimethylpyridine	260	
2-Methylbutyric acid		368	Nitromethane		376
3-Oxaheptan-1-ol	322	401	Bromal hydrate	324	380
3-Oxa-5-Methylhexan-1-ol	298	423	Methyl acetate		381
3-Oxa-1-Methylhexan-1-ol	316		Furfural		395
5,8-Dioxadodecane	322	403	4-Chlorophenol		402
3,6,9-Trioxaheptadecan-1-ol	281		Phenylhydrazine		328
Diethylamine	417		4,6-Dimethyl-1,2-pyrone	333	359
Dipropylamine	268		1-Methyl-2-acetonyl-pyridine	309	
Methyldiethylamine	322		Nicotine	335	506
Triethylamine	292				
Acetonitrile		272			
Propionitrile		385			
Succinonitrile		332			

Remarks and references for appendixes

[a]293 K; [b]303 K; [c]295 K; [d]313 K; [e]289 K; [f]333 K.

J. A. Riddick and W. B. Bunger, *Organic Solvents*, Wiley–Interscience, New York, 1970.

Y. Marcus and A. S. Kertes, *Ion Exchange and Solvent Extraction of Metal Complexes*, Wiley–Interscience, London, 1969.

J. R. Christian, *Quart. Rev.*, **24**, 20 (1970).

A. W. Francis, *Critical Solution Temperatures*, American Chemical Society, Washington, D.C., 1961.

A. W. Francis, *Liquid–Liquid Equilibriums*, Wiley–Interscience, New York, 1963.

References

1. J. Y. Zawidzki, *Z. Phys. Chem. Stoechiom Verwandochaftslehre*, **35**, 129 (1900).
2. E. A. Guggenheim, *Mixtures*, Clarendon Press, Oxford, 1952, p. 27.
3. A. G. Williamson, *An Introduction to Non-Electrolyte Solutions*, Oliver and Boyd, Edinburgh, 1967, pp. 31–34.
4. J. S. Rowlinson, *Liquids and Liquid Mixtures*, Butterworth, London, 2nd edn., 1969, p. 137.
5. E. A. Guggenheim, *Mixtures*, Clarendon Press, Oxford, 1952, p. 183 ff.
6. J. H. Hildebrand and R. L. Scott, *The Solubility of Nonelectrolytes*, Reinhold, New York, 1950, pp. 137–140.
7. J. H. Hildebrand, *J. Chem. Phys.*, **15**, 225 (1947).
8. Ref. 2, pp. 184–193; T. S. Chang, *Proc. Roy. Soc., Ser. A*, **169**, 512 (1939); *Proc. Camb. Phil. Soc.*, **35**, 265 (1939).
9. P. J. Flory, *J. Chem. Phys.*, **10**, 51 (1942).

10. D. S. Adcoetz and M. L. McGlashan, *Proc. Roy. Soc., Ser. A,* **226**, 206 (1954).
11. J. H. Van der Waals and J. J. Hermans, *Rec. Trav. Chim. Pays-Bas,* **68**, 181 (1949).
12. J. H. Hildebrand and S. E. Wood, *J. Chem. Phys.*, **1**, 817 (1933).
13. G. Scatchard, *Chem. Rev.,* **8**, 321 (1931).
14. H. C. Longuet-Higgins, *Proc. Roy. Soc., Ser. A,* **205**, 247 (1951).
15. H. Burrell and B. Immergut, in *Polymer Handbook*, edited by J. Brandrup and E. H. Immergut, Wiley–Interscience, New York, 1966, p. IV–341–358.
16. R. F. Weimer and J. M. Prausnitz, *Hydrocarbon Processing,* **44(9)**, 237 (1965).
17. J. H. Hildebrand, J. M. Prausnitz and R. L. Scott, *Regular and Related Solutions*, Van Nostrand Reinhold, New York, 1970, pp. 213–215.
18. P. A. Small, *J. Appl. Chem.*, **3**, 71 (1953); A. E. Reineck and K. F. Lin, *J. Paint Technol.*, **40**, 611 (1968).
19. G. M. Wilson, *J. Am. Chem. Soc.*, **86**, 127 (1964).
20. H. Renon and J. M. Prausnitz, *A.I.Ch.E. Journal,* **14**, 135 (1968).
21. K. S. Pitzer, D. Z. Lippman, R. F. Curl, Jr., C. M. Huggins and D. E. Petersen, *J. Am. Chem. Soc.*, **77**, 3427, 3433 (1955); **79**, 2369 (1957); *Ind. Eng. Chem.*, **50**, 265 (1958).
22. P. L. Chueh and J. M. Prausnitz, *Ind. Eng. Chem., Fundam.*, **6**, 492 (1967).
23. I. Langmuir, *Third Colloid Symposium Monograph*, 48 (1925).
24. J. A. Barker, *J. Chem. Phys.*, **20**, 1526 (1952); **21**, 1391 (1953).
25. H. Tompa, *Polymer Solutions*, Butterworth, London, 1956, Ch. 4; *Trans. Faraday Soc.*, **45**, 101 (1949).
26. See, e.g., C. L. Young et al., *Trans. Faraday Soc.*, **64**, 337, 349, 1537, 2675 (1968).
27. O. Redlich, E. L. Derr, and G. J. Pierotti, *J. Am. Chem. Soc.*, **81**, 2283 (1959); M. N. Papadopoulos and E. L. Derr, *J. Am. Chem. Soc.*, **81**, 2285 (1959).
28. Y. B. Tewari, J. P. Sheridan and D. E. Martire, *J. Phys. Chem.*, **74**, 3263 (1970).
29. H. V. Kehiaian et al., *J. Chim. Phys.*, **68**, 922, 935 (1971).
30. D. H. Everett and R. J. Munn, *Trans. Faraday Soc.*, **60**, 1951 (1964).
31. M. L. McGlashan, K. W. Morcom and A. G. Williamson, *Trans. Faraday Soc.*, **57**, 601 (1961).
32. A. Bondi and D. J. Simkin, *J. Chem. Phys.*, **25**, 1073 (1956); *A.I.Ch.E. Journal,* **3**, 473 (1957); R. Anderson and J. M. Prausnitz, *A.I.Ch.E. Journal,* **7**, 96 (1961); E. F. Meyer and R. E. Wagner, *J. Phys. Chem.*, **70**, 3162 (1966); **75**, 642 (1971).
33. R. W. Haskell, *J. Phys. Chem.*, **73**, 2916 (1969).
34. L. Onsager, *J. Am. Chem. Soc.*, **58**, 1486 (1936); cf. also J. R. Weaver and R. W. Parry, *Inorg. Chem.*, **5**, 703 (1966).
35. J. L. Humphrey and M. van Winkle, *Ind. Eng. Chem., Process Des. Develop.*, **7**, 581 (1968).
36. I. Prigogine and R. Defay, *Chemical Thermodynamics*, Longman, London, 1954, p. 410.
37. F. Dolezalek, *Z. Phys. Chem. Stoechiom. Verwandschaftslehre,* **64**, 727 (1908).
38. H. Kehiaian and K. Sosukowska–Kehiaian, *Bull. Acad. Polon. Sci., Ser. Sci. Chim.*, **11**, 549, 583, 591 (1963); H. Kehiaian, *Bull. Acad. Polon. Sci., Ser. Sci. Chim.*, **12**, 497, 567 (1964).
39. J. v. Zawidski, *Z. Phys. Chem. Stoechiom. Verwandschaftslehre,* **35**, 129 (1900); H. Hirobi, *J. Fac. Sci. Tokyo,* **1**, 155 (1925) (cf. ref. 36, p. 428); L. Sarolea-Mathot, *Trans. Faraday Soc.*, **49**, 8 (1953).
40. H. Kehiaian and A. Treszczanowicz, *Bull. Acad. Polon. Sci., Ser. Sci. Chim.*, **14**, 891 (1966); H. Kehiaian, *Bull. Acad. Polon. Sci., Ser. Sci. Chim.*, **15**, 367 (1967); A. Treszczanowicz and H. Kehiaian, *Bull. Acad. Polon. Sci., Ser. Sci. Chim.*, **15** 495 (1967); **16**, 171 (1968); A. Treszczanowicz, *Bull. Acad. Polon. Sci., Ser. Sci. Chim.*, **16**, 439 (1968); A. Kehiaian and A. Treszczanowicz, *Bull. Acad. Polon. Sci., Ser. Sci. Chim.*, **16**, 445 (1968); R. Lacmann, *Z. Phys. Chem. (Frankfurt am Main),* **23**, 313, 324 (1960).
41. H. Kempter and R. Mecke, *Naturwissenschaften,* **34**, 583 (1939); *Z. Phys. Chem., Abt. B,* **46**, 229 (1940); cf. also E. N. Lassettre, *J. Am. Chem. Soc.*, **59**, 1383 (1937); O. Redlich and A. T. Kister, *J. Chem. Phys.*, **15**, 849 (1947).
42. H. Kehiaian, *Bull. Acad. Polon. Sci., Ser. Sci. Chim.*, **14**, 703 (1966).
43. I. A. Wiehe and E. B. Bagley, *Ind. Eng. Chem. Fundam.*, **6**, 209 (1967).
44. H. Kehiaian, *Bull. Acad. Polon. Sci., Ser. Sci. Chim.*, **12** 77 (1964); H. Kehiaian and A. Fajans, *Bull. Acad. Polon. Sci., Ser. Sci. Chim.*, **12**, 255 (1964); A. Treszczanowicz and H. Kehiaian, *Bull. Acad. Polon. Sci., Ser. Sci. Chim.*, **14**, 413 (1966).
45. I. Prigogine and R. Defay, *Chemical Thermodynamics*, Longman, London, 1954, pp. 432–433, 518–519; cf. also R. Haase, *Discuss. Faraday Soc.*, **15**, 271 (1953).

46. Y. Marcus, J. Shamir and J. Soriano, *J. Phys. Chem.*, **74**, 133 (1970).
47. S. C. P. Hwa and W. T. Ziegler, *J. Phys. Chem.*, **70**, 2572 (1966).
48. E. Steurer and K. L. Wolf, *Z. Phys. Chem., Abt. B*, **39**, 101 (1938); N. S. Berman and J. J. McKetta, *J. Phys. Chem.*, **66**, 1444 (1962).
49. G. Oster and J. G. Kirkwood, *J. Chem. Phys.*, **11**, 175 (1943).
50. M. Van Thiel, E. D. Becker and G. C. Pimentel, *J. Chem. Phys.*, **27**, 95 (1957).
51. J. J. Fox and A. E. Martin, *Trans. Faraday Soc.*, **36**, 897 (1940); U. Liddel and E. D. Becker, *Spectrochim. Acta*, **10**, 70 (1957).
52. N. D. Coggeshall and E. L. Saier, *J. Am. Chem. Soc.*, **73**, 5414 (1951).
53. W. C. Coburn, Jr. and E. Grunwald, *J. Am. Chem. Soc.*, **80**, 1318 (1958).
54. H. Dunken and H. Frizsche, *Spectrochim. Acta*, **20**, 785 (1964).
55. L. J. Bellamy and R. J. Pace, *Spectrochim. Acta*, **22**, 525 (1966).
56. H. C. Van Ness, J. Van Winkle, H. H. Richtol and H. B. Hollinger, *J. Phys. Chem.*, **71**, 1483 (1967).
57. A. N. Fletcher and C. A. Heller, *J. Phys. Chem.*, **71**, 3742 (1967); A. N. Fletcher, *J. Phys. Chem.*, **73**, 2217 (1969); **74**, 216 (1970).
58. C. G. Savini, D. R. Winterhalter and H. C. Van Ness, *J. Chem. Eng. Data*, **10**, 168 (1965); R. W. Kiser, G. D. Johnson and M. D. Shetlar, *J. Chem. Eng. Data*, **6**, 339 (1961).
59. I. A. Wiehe and E. B. Bagley, *Ind. Eng. Chem., Fundam.*, **6**, 209 (1967).
60. G. Oster, *J. Am. Chem. Soc.*, **68**, 2036 (1946); P. Huyskens, R. Henry and G. Gillerot, *Bull. Soc. Chim. Fr.*, **1962**, 720; J. Malecki, *J. Chem. Phys.*, **43**, 1351 (1965); P. E. Gold and R. L. Perrine, *J. Chem. Eng. Data*, **12**, 4 (1967); D. A. Ibbitson and L. F. Moore, *J. Chem. Soc., B*, **1967**, 76, 81.
61. H. Wolf and H. E. Höppel, *Ber. Bunsenges., Phys. Chem.*, **72**, 710, 1173 (1968); E. E. Tucker, S. B. Farnham and S. D. Christian, *J. Phys. Chem.*, **73**, 3820 (1969).
62. S. Singh and C. N. R. Rao, *J. Phys. Chem.*, **71**, 1074 (1967).
63. I. Brown, W. Fock and F. Smith, *Aust. J. Chem.*, **8**, 361 (1955); **9**, 364 (1956); **10**, 417 (1957); **14**, 387 (1961); **17**, 1106 (1964).
64. F. Pardo and H. C. Van Ness, *J. Chem. Eng. Data*, **10**, 163 (1965); H. C. Van Ness, C. A. Soczek, G. L. Peloquin and R. L. Machedo, *J. Chem. Eng. Data*, **12**, 217 (1967); H. C. Van Ness, C. A. Soczek and N. K. Kochar, *J. Chem. Eng. Data*, **12**, 346 (1967).
65. E. D. Becker, U. Liddel and J. N. Shoolery, *J. Mol. Spectrosc.*, **2**, 1 (1958); M. Saunders and J. B. Hyne, *J. Chem. Phys.*, **29**, 1319 (1958); J. C. Davis, K. S. Pitzer and C. N. R. Rao, *J. Phys. Chem.*, **64**, 1744 (1960); A. B. Littlewood and F. W. Willmont, *Trans. Faraday Soc.*, **62**, 3287 (1966); W. L. Chandler and R. H. Dinius, *J. Phys. Chem.*, **75**, 1597 (1969).
66. R. S. Musa and M. Eisner, *J. Chem. Phys.*, **30**, 227 (1959).
67. O. Redlich and A. T. Kister, *J. Chem. Phys.*, **15**, 849 (1947).
68. C. B. Kretschmer and R. Wiebe, *J. Chem. Phys.*, **22**, 1697 (1954).
69. J. R. Goates, R. L. Snow and M. R. James, *J. Phys. Chem.*, **65**, 335 (1961).
70. J. E. Otterstedt and R. W. Missen, *Trans. Faraday Soc.*, **36**, 897 (1940).
71. R. W. Haskell, H. B. Hollinger and H. C. Van Ness, *J. Phys. Chem.*, **72**, 4534 (1968).
72. H. Renon and J. M. Prausnitz, *Chem. Eng. Sci.*, **22**, 299 (1967).
73. D. Papoušek, Z. Papoušková and L. Pago, *Z. Phys. Chem. (Leipzig)*, **211**, 231 (1959).
74. C. B. Kretschner and R. Wiebe, *J. Am. Chem. Soc.*, **71**, 1793, 3176 (1949); J. E. Sinor and J. H. Weber, *J. Chem. Eng. Data*, **5**, 243 (1960); G. C. Paraskevopoulos and R. W. Missen, *Trans. Faraday Soc.*, **58**, 869 (1962).
75. C. McAuliffe, *J. Phys. Chem.*, **70**, 1267 (1966).
76. G. Nemethy and H. A. Scheraga, *J. Chem. Phys.*, **36**, 3401 (1962).
77. W. Kauzmann, *Adv. Protein Chem.*, **14**, 1 (1959).
78. A. Ben-Naim, *J. Chem. Phys.*, **54**, 1387, 3696 (1971).
79. A. Ben-Naim, J. Wilf and M. Yacobi, *J. Phys. Chem.*, **77**, 95 (1973).
80. S. D. Christian, A. A. Taha and B. W. Marsh, *Quart. Rev.*, **24**, 20 (1970).
81. C. Black, G. G. Joris and H. S. Taylor, *J. Chem. Phys.*, **16**, 537 (1948); J. H. Hildebrand and R. L. Scott, *The Solubility of Nonelectrolytes*, Reinhold, 3rd edn., 1950, pp. 266–268.
82. F. Franks and D. J. G. Ives, *Quart. Rev.*, **20**, 1 (1966).
83. F. Franks, *Physicochemical Processes in Mixed Aqueous Solvents*, Heineman, London, 1967; A. K. Covington and P. Jones *Hydrogen-Bonded Solvent Systems*, Taylor and Francis, London, 1968; Y. Marcus and J. Naveh, *J. Phys. Chem.*, **73**, 593 (1969).
84. G. Bertrand, F. Millero, C. Wu and L. Hepler, *J. Phys. Chem.*, **70**, 699 (1966).

Chapter 6

Solutions of Electrolytes in Dielectric Media

A. Introduction

a. Unsymmetrical mixtures

Electrolyte solutions are essentially unsymmetrical mixtures, since the whole composition range can very seldom be realized at given temperatures and pressures. Examples to the contrary, such as the nitric acid–water[1] system at 25 °C and the tetra-n-pentylammonium thiocyanate–nitrobenzene[2] system at 52 °C, are rare, and for a system such as ammonium nitrate-water[3], the temperature must be varied step-wise in order that the whole composition range may be studied. In general, then, it is recognized in electrolyte–dielectric medium systems that the former is the solute and the latter the solvent, and attention is focused on dilute solutions. It is therefore useful to use concentration scales appropriate for dilute solutions, such as the molal m (4.4) or molar c (4.5) systems. The latter suffers from the dependence of the solution volume on the temperature and the pressure, so that derivative functions at constant composition are not easy to obtain. For the molal scale, unit mass of the total solvent is definable rigorously, without recourse to often ambiguous assignments of molar masses (ambiguous in the case of self association, for example), and irrespective of solvent composition, i.e. whether it is a pure solvent or a mixture. For the latter case, the weight (mass) fractions w define the composition of the solvent unambiguously.

If the molal scale is used, *specific* (per kg)[4] thermodynamic quantities are the intensive quantities for the solvent, whereas those pertaining to the solute are *molal* ones (per mole solute). It is recognized however that the electrolyte dissociates, and for strong electrolytes, $C_{\nu^+}A_{\nu^-}$, dissociation is complete according to

$$C_{\nu^+}A_{\nu^-} \to \nu_+ C^{z^+} + \nu_- A^{z^-} \tag{6.1}$$

with the stoichiometric parameters

$$\nu_+ + \nu_- = \nu \tag{6.2}$$

If the molality of the electrolyte is m_s, the molality of the ions is

$$m_C + m_A = \nu m_s \tag{6.3}$$

The charges on the ions are z_+e and z_-e, and the quantities z_+ and z_- are to be taken in the algebraic sense. The electroneutrality condition is then

$$\nu_+ z_+ + \nu_- z_- = 0 \tag{6.4}$$

The chemical potential of the electrolyte is

$$\mu_s = \nu_+ \mu_+ + \nu_- \mu_- = \nu \mu_\pm \tag{6.5}$$

where $\mu_\pm$ is the *mean ionic* chemical potential.

The main difficulty arises from the concept of the standard state of the solute in an unsymmetrical solution. Clearly, the standard state should have the property that it eliminates the composition variable from the expression for the chemical potential. Since the molality scale has been selected, the expression

$$\mu_s = \mu_s^0 + \nu RT \ln m_s + \mu_s^E \tag{6.6}$$

requires an arbitrary division of $\mu_s - \nu RT \ln m_s$ between the standard chemical potential μ_s^0 and the excess chemical potential μ_s^E. This has been done by assigning to μ_s^E the value zero at the limit of infinite dilution of the electrolyte in the solvent, so that μ_s^0 is defined by

$$\mu_s^0 = \lim_{m_s \to 0} (\mu_s - \nu RT \ln m_s) \tag{6.7}$$

Since μ_s^0 is defined with reference to the state of infinite dilution of the electrolyte, this state is often called the *reference state* of the electrolyte solution. The particular state where $\nu RT \ln m_s + \mu^E = 0$, so that $\mu_s = \mu_s^0$ is *not* the standard state, implying that although the *standard chemical potential* is operationally defined for electrolyte solutions (eq. 6.7), the standard *state* is not realizable.

The excess chemical potential of the solute is related simply to the mean ionic molal activity coefficient $\gamma_\pm$

$$\mu_s^E = \nu\mu_\pm^E = \nu \mathbf{R}T \ln \gamma_\pm \tag{6.8}$$

and the individual ionic activity coefficients, by

$$\gamma_\pm = \gamma_+^{\nu_+/\nu}\gamma_-^{\nu_-/\nu} \tag{6.9}$$

Once the molal scale has been selected, the concentration variable for the solvent is eliminated, and all the expressions refer to a constant amount of solvent. The chemical potential of the solvent is its partial specific Gibbs energy, and the standard state of the solvent is the pure solvent, which is identical with the reference state of the electrolyte.

$$G_w = (\partial G/\partial W_w)_{T,P,n_s} = G_w^0 - \mathbf{R}T\nu m_s\phi \tag{6.10}$$

The last equation of (6.10) defines the *osmotic coefficient* ϕ of the solvent, which is related to its excess specific Gibbs energy by

$$G_w^E = \mathbf{R}T\nu m_s(1 - \phi) \tag{6.11}$$

fixing the value of ϕ at the limit of infinite dilution at 1. The total excess Gibbs energy of the solution containing unit mass (1 kg) of solvent and m_s moles of electrolyte is

$$G^E = G_w^E + m_s\mu_s^E = \mathbf{R}T\nu m_s(1 - \phi + \ln \gamma_\pm) \tag{6.12}$$

which is a particularly simple expression. Other excess quantities, total or partial, are readily obtained from (6.12) by differentiation: Since the molality is independent of the temperature or the pressure, simple equations result, e.g.

$$H^E = [\partial(G^E/T)/\partial(1/T)]_{P,m_s} = H_w^E + m_s h_s^E \tag{6.13}$$

or

$$\begin{aligned} h_s^E &= \mathbf{R}\nu[\partial(\ln \gamma_\pm)/\partial(1/T]_{P,m_s} \\ &= -\mathbf{R}\nu T^{-2}(\partial(\ln \gamma_\pm)/\partial T)_{P,m_s} \end{aligned} \tag{6.14}$$

A common alternative[5] symbol used for h_s^E is $\bar{L}_s$, and similarly, for $v_s^E = \mathbf{R}T\nu(\partial(\ln \gamma_\pm)/\partial P)_{T,m_s}$, the symbol $(\bar{V}_s - \bar{V}_s^0)$ is often used in the literature.

From (6.6) it follows that quite analogous expressions can be obtained for the thermodynamic quantities in the reference state by setting $\mu_s^E = 0$ and differentiating μ_s^0 with respect to temperature and pressure, to yield the standard partial molal entropy

$$s_s^0 = -(\partial\mu_s^0/\partial T)_P = \lim_{m_s \to 0} (s_s - \nu\mathbf{R} \ln m_s) \tag{6.15}$$

the standard partial molal enthalpy

$$h_s^0 = \mu_s^0 - Ts_s^0 = [\partial(\mu_s^0/T)/\partial(1/T)]_P = \lim_{m_s \to 0} h_s \tag{6.16}$$

and the standard partial molal volume

$$v_s^0 = (\partial \mu_s^0 / \partial P)_T = \lim_{m_s \to 0} v_s \tag{6.17}$$

and by further differentiation the heat capacity, compressibility and thermal expansivity. The physical significance of these quantities for the reference state is of interest since they relate directly to the interactions of the electrolyte with the solvent, with no complications from ion-ion interactions (which are described by the excess functions). Writing y_s^0 for any of the above standard partial molal thermodynamic quantities, one can define standard partial *ionic* quantities

$$y_s^0 = \nu_+ y_+^0 + \nu_- y_-^0 = \Sigma \nu_i y_i^0 \tag{6.18}$$

and each of these y_i^0 is **N** (Avogadro's number) times that quantity for a single ion i, interacting with the solvent. These interactions are of a short range relative to the coulombic ion–ion interactions, and are discussed in terms of solvation and effects of the ion on the solvent structure.

B. Ion–Solvent Interactions

a. Structural aspects of ion–solvent interactions

There are two main structural aspects to ion–solvent interactions which are considered in those treatments that do not consider merely the interactions of charged spheres in a dielectric continuum. One of these is the orientation effects of the electric field on the discrete dipoles of the molecules of which the polar solvent is composed. This leads to a loss of mobility and rotational freedom of these molecules in the vicinity of the ion. The other aspect applies only to those solvents possessing an inherent structure which may be broken or enhanced near the ions. Water is, of course, the foremost example of such solvents, and most of the following discussion will be in terms of hydration and effects on the structure of water. *Mutatis mutandis*, the discussion applies to any structured solvent.

Consider first the orientation of dipolar solvent molecules in the field of an ion. This has a strong bearing on the problem of the equivalence or otherwise of the solvation of a cation and an anion of a given size or, in other words, on the effect on the parameters of solvation of a flip of the charge of a solvated ion to the opposite sign. If the ion and the polar solvent molecule are isolated, configurations such as in Fig. 6.1 are conceivable. A spherically symmetrical solvent molecule with a central point-dipole, solvating a cation, is depicted in Fig. 6.1(a). Free rotation around the axis between the centres of the two molecules is expected. This could be the case with solvents such as acetonitrile, pyridine, or tertiary amines, where the nitrogen has a single 'lone-pair' of electrons. With solvents containing oxygen as the donor atom, including water, interaction of the two 'lone-pairs' would cause immobilization, and when the solvent has an appreciable quadrupole moment, a configuration such as in Fig. 6.1(b) may have the lowest energy. Rotation around the ion–molecule axis is then strongly hindered. Solvation of the anion may differ

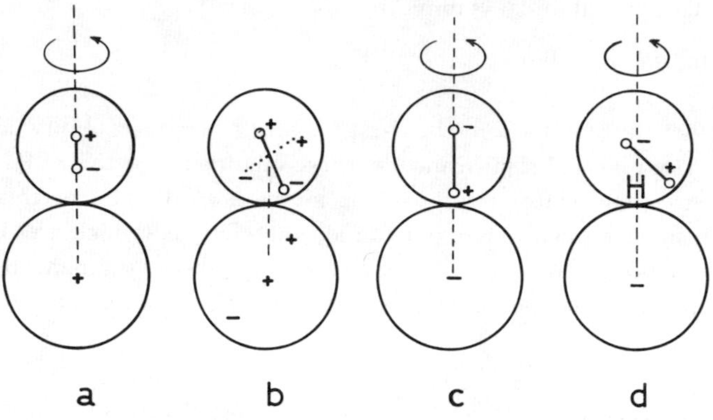

Figure 6.1. Some possible mutual configurations of ions and solvent molecules. (a) A point dipole in the solvent oriented towards a cation; rotation of the solvent around the dipole axis is free. (b) The solvent has a large quardrupole moment, so that the induced charges fix the orientation at an angle to the dipole axis. (c) An anion solvated by an aprotic dipolar solvent. (d) Hydrogen bonding, as it affects the solvation of an anion by a relatively weakly polar protic solvent.

from that of the cation, at a given size, for several reasons. The dipole may be located so that the positive end is nearer the surface than the negative end (Fig. 6.1(c)). This may be the case with water, but aprotic solvents are apt to have the opposite arrangement, and therefore solvate anions only very slightly, in comparison with their solvating-ability for cations. Dimethylsulphoxide is a good example for such a solvent[6]. Protic solvents, which are capable of hydrogen-bonding to anions, will seek a configuration with minimal free energy, in balance between the hydrogen bonding and the ion–dipole interaction, such as in Fig. 6.1(d). Water or alcohols would probably orient themselves in this fashion, rather than as in Fig. 6.1(c), with the hydroxylic hydrogen atom (or one of them in the case of water) along the ion–molecule axis. Rotation around this axis may occur, provided the ion–dipole and ion–quadrupole interactions are not too strong.

The ion–solvent pairs depicted in Fig. 6.1 are however not isolated but rather surrounded with other solvent molecules. Those adjacent to the ion will all be oriented by the central spherical field, but they will also interact with each other, not least through repulsion forces of their hard cores – steric hindrance in other words. Attractive dispersion forces, induction forces and mutual orientation forces will also be operative. Molecules further away from the ion will contribute their share, and hydrogen bonding – especially with water as the solvent – is of great importance. The effect of the ion on the structure of a structured solvent therefore needs consideration.

Operationally, the effects on the structure of water are otained from two kinds of measurements: from the partial specific entropy of the water,

$$S_{\mathrm{w}}^{\mathrm{E}} = \mathbf{R}\nu m_{\mathrm{s}}(\phi - 1 + (\partial\phi/\partial(\ln T))_{P,m_s})$$

and from the B coefficient in the Jones–Dole expression for the viscosity of the solution

$$\eta = \eta_0 + Ac^{1/2} + \Sigma B_i c_i \tag{6.19}$$

A *structure-breaking ion* has $(\partial S_w^E/\partial m_i) > 0$ and $B_i < 0$, while an ion is said to be *structure-making* if $B_i > 0$ and $(\partial S_w^E/\partial m_i)$ is negative. There exists a good correlation between these two measurable properties of the solution[7]: an ion increases the fluidity, and is more mobile if the water around it is less structured, i.e. is more disordered, has a larger entropy than in pure water, and vice versa for a structure-making ion. By these operational criteria, large single-charged monatomic cations and anions (K^+,Rb^+,Cs^+,Cl^-,Br^-,I^-,ClO_4^-, etc.) are net structure-breakers; small single-charged, and multi-charged cations are net structure-makers (Li^+,Na^+, Mg^{2+},Ca^{2+},La^{3+}, etc.), large hydrophobic cations or anions are structure-makers ($(C_3H_7)_4N^+$, $(C_4H_9)_4N^+$, $(C_6H_5)_4B^-$, etc.), and some ions are borderline cases, neither strong makers nor breakers of the structure of water (NH_4^+,$(C_2H_5)_4N^+$, $(HOC_2H_4)_4N^+$, etc.). There are, however, no quantitative rules which relate the operational criteria for water-structure modifications to the properties of the ions, such as charge, size, polarizability, geometry, etc. In order to gain more insight into this problem, it is useful to consider the concentric-shell model of Frank and co-workers[8].

The concentric-shell model has been developed specifically for aqueous solutions, the argument being that the central field of the ion causes the water molecules to be oriented in a manner incompatible with the tetrahedral structure of bulk water, which is 'ice-like'. Therefore, there exists a region of 'thawed' water where the structure is broken, the water molecules being partly oriented but not immobilized. This is depicted schematically in Fig. 6.2. An important feature is the change of the dielectric permittivity of the medium with the distance from the centre of the ion. This arises from the dependence of the dielectric constant ϵ on the field strength E (cf. eq. (1,32)) through[9,10]

$$\epsilon = n^{*2} + (\epsilon^* - n^{*2})/(1 + bE^2) \tag{6.20}$$

where n^* is the refractive index of water, and b, a coefficient that depends on the pressure and the temperature but not on the field strength, $b \simeq 1.1 \times 10^{-16}\ m^2\ V^{-2}$ at room conditions. The dielectric constant of (6.20) is inserted in the relationship between the electric displacement, $D(r) = zer^{-2}$ at a distance r from a point charge ze, and the field strength at this point, $E(r)$: $dD(r) = \epsilon dE(r)$. The integrated equation can be solved numerically[10] to yield the data $E(r)$ and $\epsilon(r)$, the latter shown in the upper part of Fig. 6.2 for $z = 1$. At small r, i.e. $r(nm) \leqslant r_p = 0.18z^{1/2}$, dielectric saturation occurs and $\epsilon \approx n^{*2}$. At large r, i.e. $r(nm) \geqslant r_s = 0.80z^{1/2}$, $\epsilon = \epsilon^*$, the value in bulk water. In the region $r_p < r < r_s$ the dielectric constant varies strongly with the distance.

These concentric shells are shown in Fig. 6.2(a) for a univalent ion. Figure 6.2(b) shows schematically the situation for a small ion, which has an effective radius (see below) $r_e < r_p$. Between r_e and r_p there is *primary hydration*, by immobilized oriented water molecules. These are held strongly according to the scheme in

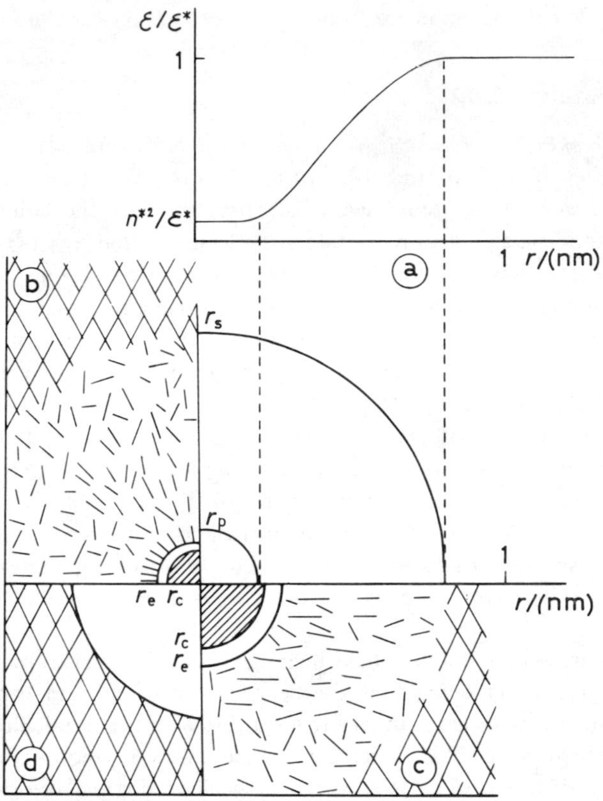

Figure 6.2. Upper part: the dielectric constant of the medium around an ion, as a function of the distance from the centre of the ion. The regions of dielectric saturation, of broken solvent structure, and of bulk solvent correspond to those in segment (a) of the lower part, $r < r_p$, $r_p < r < r_s$ and $r > r_s$ for primary solvation, secondary solvation and bulk solvent, respectively. (b) Solvation of a small ion. (c) Solvation of a large ion, where r_c is the intrinsic (crystal) radius, r_e the effective radius, the dashes represent oriented water dipoles, and the diamond-shaped region the structure of bulk water. (d) Structure-enhancement around a hydrophobic ion.

Fig. 6.1(b). From r_p to r_s there occurs *secondary hydration* of partly oriented water molecules, which have translational freedom, and $S_W < S_W^*$ (the orientation-produced order is stronger than the lost structural order). Beyond r_s the structure of bulk water prevails. It should be noted that the strongly oriented primary hydration water helps in this case to orient the water molecules in the secondary hydration shell, and these have postive activation energies for exchange with the bulk water, hence the alternative name[11] *positive hydration* for this case. There is a *net structure-making* effect, although the structure-making, defined operationally above, does not produce the structure of bulk water. A different situation occurs in the case of a large ion with $r_e > r_p$, depicted schematically in Fig. 6.2(c).

Here there is no primary hydration and in the secondary hydration shell the weak orientation of the water molecules by the central field only, with no help from primary hydration water, cannot overcome the incompatible orientation by the structured bulk water, hence $S_W > S_W^*$ in this region, and the water molecules have negative activation energies for exchange with bulk water. *Negative hydration* and *net structure-breaking* in the operationally defined sense occur.

For very large ions, with a weak central field, where the exterior surface is only weakly polarized, *hydrophobic hydration* occurs, as shown in Fig. 6.2(d). Excluded are large anions such as perchlorate or hexacyanoferrate, where the exterior surface is strongly hydrogen-bonding, or the polar part of organic ions such as butyrate or butylammonium, and included are ions with hydrocarbon exterior surfaces (tetraalkyl- or aryl- ammonium, phosphonium, borate, etc.) and the hydrocarbon parts of ions such as butyrate or butyl-ammonium. As discussed for non-electrolyte solutes in water (p. 199), these enhance the water structure around them, with little orientational effects of the central field, and are therefore structure-makers in both the operational sense, and as regards the tetrahedral structure itself, which is more regular in the vicinity of the hydrophobic ion than in bulk water.

b. Thermodynamics of ion solvation

The main problem with any discussion of ion solvation is that whereas the thermodynamic functions, such as μ_s^0, s_s^0, v_s^0, etc. apply to the electrolyte as a whole (Fig. 6.3), theoretical calculations based on molecular models deal with the ions individually. A helpful feature of the standard partial thermodynamic functions is that they are additive with respect to the ions. For instance, for two 1:1 electrolytes (i.e. $\nu_+ = \nu_- = 1$) designated by 1 and 2 with a common anion

$$y^0_{+(1)} - y^0_{+(2)} = y^0_{\pm(1)} - y^0_{\pm(2)} \tag{6.21}$$

If y_i^0 is known for any one ion, those for other ions can then be readily obtained from eq. (6.21). Lacking more information, *conventional* ionic thermodynamic functions $y^0_{i\,con}$ have been defined, based on the following conventions, regarding partial molar entropies and partial molar volumes

$$s^0_{i\,con}(H^+\ aq) = 65.296\ J\ K^{-1}\ mol^{-1}\ \text{at 298.15 K} \tag{6.22}$$

$$v^0_{i\,con}(H^+\ aq) = 0\ cm^3\ mol^{-1}\quad \text{at all } T \text{ and } P \tag{6.23}$$

Convention (6.22) is tantamount to assigning aqueous hydrogen ions half the entropy of hydrogen gas at the same temperature and at standard pressure (1.01325×10^5 Pa = 1 atm.) (ref. 12). Values of $s^0_{i\,con}$ and $v^0_{i\,con}$ based on conventions (6.22) and (6.23) are listed in Appendix 6.1, p. 250.

Of further interest are the thermodynamic functions $y^0_{i\,hyd}$ for the solvation of ions, based on the reaction

$$Z^z(g) \rightarrow Z^z \quad \text{(infinitely dilute solution)} \tag{6.24}$$

where z is to be taken algebraically. These functions apply to the transfer of an ion

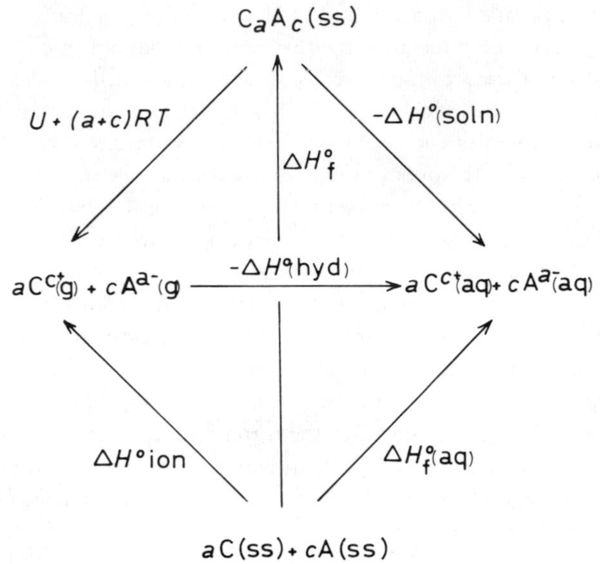

Figure 6.3. Processes leading to the formation of hydrated ions in aqueous solution, and the standard enthalpy changes involved: ΔH^0_{ion} is the sum of the sublimation, dissociation and ionization enthalpies, U is the lattice energy, ΔH^0_{soln} is the heat of solution, ΔH^0_{faq} is the heat of formation of the aqueous ions, ΔH^0_{f} is the heat of formation of the pure electrolyte, and ΔH^0_{hyd} is the heat of hydration (of the gaseous ions).

from the gas phase to the solution, disregarding the surface potential effects. Reaction (6.24) cannot be studied directly, but the reactions

$$Z(ss) \rightarrow Z^z(g) + ze^-(g) \tag{6.25}$$

the production of the gaseous ion from the standard state atom, with generalized thermodynamic functions $y^0_{\text{i ion}}$, and

$$Z(ss) + zH^+(aq) \rightarrow Z^z(aq) + (z/2)H_2(g) \tag{6.26}$$

the exchange of Z and hydrogen between the elemental and aqueous solution standard states, characterized by $y^0_{\text{i aq}}$, can be studied directly, at least for monatomic ions. The difference of (6.26) and (6.25) gives the left-hand side of

$$y^0_{\text{i aq}} - y^0_{\text{i ion}} = y^0_{\text{i hyd}} + zy^0_{\text{H}} \tag{6.27}$$

and is therefore also known, and leads to $y^0_{\text{i hyd}}$ if values for y^0_{H}, for the reaction

$$H^+(aq) + e^-(g) \rightarrow \tfrac{1}{2}H_2(g) \tag{6.28}$$

are known. These have been assigned conventionally, as follows[12].

(a) The molar Gibbs energy and enthalpy of formation of all elements in their standard states is zero at all temperatures.

(b) The standard molar Gibbs energy and enthalpy of formation of aqueous hydrogen ions are zero at all temperatures.

(c) The standard partial molar entropy of hydrogen ions is half the molar entropy of hydrogen gas at the same temperature and at standard pressure (convention (6.22)).

(d) The molar enthalpy of gaseous electrons is given by its kinetic energy: $h^0(e^-(g)) = (5/2)\mathbf{R}T$, and at 298.15 K $h^0 = 6196.5$ J mol^{-1}. (The molar entropy is obtained from the Sackur–Tetrode equation and is $s^0(e^-(g)) = 20.895$ J K^{-1} mol^{-1} at 298.15 K.)

From these self-consistent conventions[12] (i.e. upholding the requirement that $s^0 = (h^0 - g^0)/T$ for any process), the values for $y^0_{\mathrm{H\ con}}$ to be used in (6.27) to give $y^0_{\mathrm{i\ hyd\ con}}$ are obtained as

$$g^0_{\mathrm{H\ con}} = 33.3 \text{ J mol}^{-1} \text{ at } 298.15 \text{ K} \tag{6.29}$$

$$h^0_{\mathrm{H\ con}} = -6196.5 \text{ J mol}^{-1} \text{ at } 298.15 \text{ K} \tag{6.30}$$

$$s^0_{\mathrm{H\ con}} = -20.895 \text{ J K}^{-1} \text{ mol}^{-1} \text{ at } 298.15 \text{ K} \tag{6.31}$$

The values of $g^0_{\mathrm{i\ hyd\ con}}$, $h^0_{\mathrm{i\ hyd\ con}}$ and $s^0_{\mathrm{i\ hyd\ con}}$ calculated from (6.27) with these conventions for many ions i are listed in Appendix 6.1, p. 250.

The problem of replacing the conventional values by absolute values has been attacked by numerous authors. In some cases direct experiments to obtain the non-thermodynamic data have been made, and in other cases theoretical expressions have been derived for suitable models to yield this information. The reviews of Rosseinsky[13], Desnoyers and Jolicoeur[14], and Millero[15] discuss the direct experimental methods (such as thermocells for $s^0_{\mathrm{H^+}}$, the Volta potential between aqueous potassium chloride and an isolated mercury jet for $h^0_{\mathrm{H^+\ hyd}}$, and ionic vibration potentials for $v^0_{\mathrm{H^+}}$), and it is clear from these and other discussions that the values obtained are not necessarily nearer the truth than values based on theoretical models, because of inherent experimental inaccuracies.

The theoretical models are based ultimately on electrostatics, and consider the Gibbs free energy change of a fluid (the dielectric medium) in an electric field (of the ion being charged)[10]

$$\Delta G = (1/4\pi)\int\int \vec{E}\, d\vec{D}\, dV \tag{6.32}$$

where $\vec{E}$ and $\vec{D}$ are the vectors describing the electric field and the electric displacement, repectively, in the volume element dV under the influence of the field. For spherical ions $4\pi r^2 dr$ is substituted for dV, and $\frac{1}{2}\epsilon d(E^2(r))$ for $\vec{E}d\vec{D}$ (cf. the text after (6.20)). For the process (6.24), the expression

$$\Delta g_{\mathrm{i\ aq}} = \tfrac{1}{2}\mathbf{N}\int\int \epsilon d(E^2(r))r^2 dr \tag{6.33}$$

for the aqueous phase, must be compared with

$$\Delta g_{\mathrm{i\ gas}} = \tfrac{1}{2}\mathbf{N}z^2\mathbf{e}^2\int r^{-2} dr \tag{6.34}$$

for the gas phase, since there $\epsilon \equiv 1$ and $E(r) = ze/r^2$. There are now two problems that must be considered: the lower integration limit for r (the upper being infinity, for infinitely dilute solutions), and the form of the dependence of ϵ on $E(r)$ (cf. (6.20)). If r_{aq} is the lower limit for the aqueous phase, and r_{gas} for the gaseous phase, and if a step function is selected for ϵ, $\epsilon = 0$ for $r < r_{aq}$ and $\epsilon = \epsilon_{aq}$ for $r \geqslant r_{aq}$, so that $E(r)_{aq} = ze/\epsilon_{aq} r^2$, the result is[16]

$$\Delta g_{i\,(Born)} = \Delta g_{i\,aq} - \Delta g_{i\,gas} = \tfrac{1}{2} N z^2 e^2 [(r_{aq}\epsilon_{aq})^{-1} - r_{gas}^{-1}] \quad (6.35)$$

which is the Born-charging expression for the Gibbs energy of hydration. Various estimates have been made of the three parameters in the square bracket of (6.35). Use of the crystal-radius r_c for r_{aq} and r_{gas}, and ϵ_W^*, the bulk dielectric constant of water, for ϵ_{aq} lead to grave disagreement with experiment (for the sum for oppositely charged ions). An effective aqueous radius $r_{aq} = r_c + \Delta$ or $r_{aq} = kr_c$ (with Δ given by some authors as 55 or 80 pm, and k given as 1.25) can be justified on the grounds that dielectric saturation occurs near the ion and that void space occurs near the ion because of the particulate nature of the solvent molecules. An effective r_{gas} value, namely the van der Waals' radius, has been proposed for the ions in the gas phase. Effective dielectric constant values, independent of $E(r)$ and r, but dependent on z and r_c, have also been suggested[17]. For comparison with experiment, it has been argued that a term must be added which describes the work required for the insertion of the uncharged sphere, corresponding to the ion, from the gas phase into the solvent. This term includes the energy of the cavity which is formed in the solvent (this depends on r_{aq}, possibly in the form $Ar_{aq}^{-1} + B + Cr_{aq}^2$) and the compression work from the molar volume the ion may occupy in the gas ($0.02447\ m^3$ at 298.15 K and atmospheric pressure) to that in the solution ($1/d^* = 1.003 \times 10^{-3}\ m^3$ at the same conditions for the solvent water)[16–18]. The molar entropy of hydration is obtained from differentiation of (6.35) — ϵ_{aq} is temperature dependent — together with suitable cavity and compression terms, and the molar enthalpy from the Gibbs energy and the entropy.

The above Born-charging process assumes as a model a continuous dielectric medium, and a step function of ϵ. If a continuous function such as (6.20) is used instead, and the pressure-derivative of (6.33) (in effect replacement in (6.33) of $\epsilon(E,r)$ with $(\partial\epsilon/\partial P)_{T,E}$) is used for $\Delta v_{i\,aq}$, a fit of partial molar ionic volumes gives a parameter r_e, the lower integration limit, which can be used with (6.33) to give consistent Gibbs energy and entropy of hydration values[10].

A quite different approach, which does not assume the unrealistic continuous dielectric medium model, strives to divide the thermodynamic function for hydration between cation and anion by searching for an electrolyte with one unhydrated ion, so that the total effect is carried by the other ion. This can be done by extrapolating the conventional ionic thermodynamic functions for a series of electrolytes with a common ion and various counter-ions as a function of ionic size to the limit of infinite size, where ions should be non-solvated. The procedure of Halliwell and Nyburg[19] is best suited for this purpose: they extrapolate

$$h^0_{+hyd\,con}(r) - h^0_{-hyd\,con}(r) = \Delta h^0_{hyd\,abs}(r) - 2h^0_{H\,hyd\,abs} \quad (6.36)$$

(where the conventional values for a given r are obtained by interpolation in $h^0_{\text{ihyd con}}(r)$ curves) against a suitable function of r to infinite r, where the difference between the absolute hydration enthalpies for cations and anions should vanish, $\Delta h^0_{\text{hyd abs}}(\infty) = 0$. This then leaves as an intercept minus twice the absolute value of the enthalpy of hydration of the hydrogen ion, $h^0_{\text{H hyd abs}}$. The function of the radius for extrapolation is $[(r_c(\text{pm})) + 138]^{-3}$, where the exponent has been selected on the basis of ion–dipole interactions (ion–quadrupole interactions come in as a correction to deviations from linearity at smaller sizes) and 138 pm is the diameter of a water molecule, but the selection of values for r_c is controversial. Crystalline radii obtained by the minimum electron density criterion[20] lead to values somewhat different from those resulting from the usual crystalline radii[14].

Critical examination of all these approaches[13–20] leads to the following set of 'absolute' values for $y^0_{\text{i abs}}(H^+ \text{ aq})$

$$s^0_{\text{i abs}}(H^+ \text{ aq}) = -22.2 \pm 1.3 \text{ J K}^{-1} \text{ mol}^{-1} \tag{6.37}$$

$$v^0_{\text{i abs}}(H^+ \text{ aq}) = -4.7 \pm 1.1 \text{ cm}^3 \text{ mol}^{-1} \tag{6.38}$$

and for $y^0_{\text{H abs}}$ for eq. (6.28)

$$h^0_{\text{H hyd abs}} = -1100 \pm 6 \text{ kJ mol}^{-1} \tag{6.39}$$

$$g^0_{\text{H abs}} = 475 \pm 6 \text{ kJ mol}^{-1} \tag{6.40}$$

$$h^0_{\text{H abs}} = 436 \pm 6 \text{ kJ mol}^{-1} \tag{6.41}$$

$$s^0_{\text{H abs}} = -131 \pm \text{ J K mol}^{-1} \tag{6.42}$$

all for 298.15 K. The values in (6.37) and (6.38) can be used with (6.27), and those of (6.39) to (6.42) with

$$y^0_{\text{i abs}} = y^0_{\text{i con}} + z(y^0_{\text{H abs}} - y^0_{\text{H con}}) \tag{6.43}$$

to obtain 'absolute' thermodynamic quantities for all the ions for which conventional values are available. Corresponding data for non-aqueous solutions of electrolytes are generally lacking (but see p. 229).

c. Stoichiometry of ion solvation

Structural considerations discussed above show that a certain number of solvent molecules are immobilized in the field of an ion, and some other solvent molecules are oriented by it. Numerous attempts have been made to define these numbers operationally or conceptually in terms of solvation numbers. It is clear now that different approaches yield different numbers, and there is no unique stoichiometry for the solvation process, except in the cases where solvent molecules form coordinate bonds, between lone-pair electrons of suitable donor atoms, and empty low-energy orbitals of suitable metal cations. For water as a solvent, definite stoichiometries have been shown for certain kinetically inert cations, mainly by nuclear magnetic resonance methods, e.g. four for beryllium, six for aluminium and for chromium. Ions with a lower electrostatic field, or those which are not inert, cannot be assigned definite hydration numbers at infinite dilution by such methods.

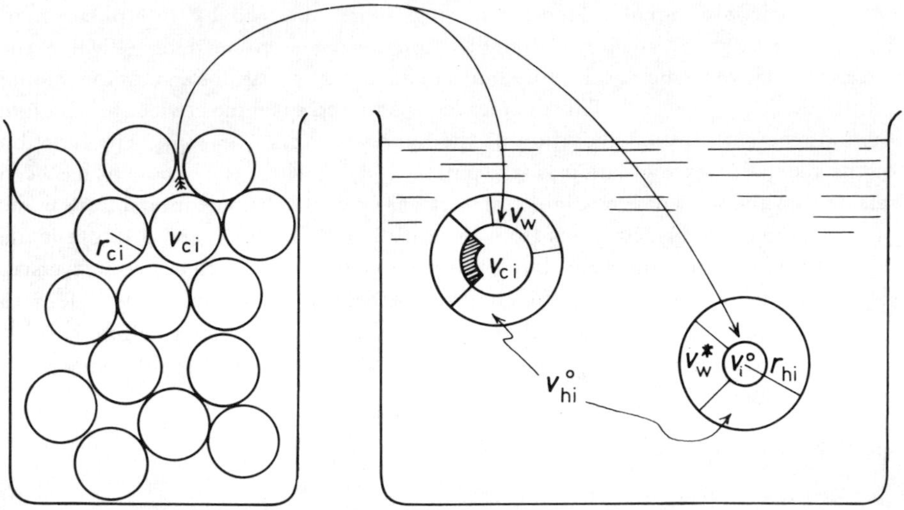

Figure 6.4. Schematic representation of the solvation of ions (having intrinsic volume v_{ci} and intrinsic radius r_{ci}) in the left-hand vessel, to give solvated ions of volume v^0_{hi} and radius r_{hi} in the right-hand vessel. In the solvated ion on the left, the ion retains its intrinsic volume, and the solvent is electrostricted down to volume v_W, its molar electrostriction being **N** times the hatched area. In the solvated ion on the right, the partial molar volume of the ion v^0_i is presented, the solvent retaining its bulk volume v^*_W.

Of the many possible approaches for such ions, only one will be discussed here: that based on considerations of the volume and size of the ions in solution and their dependence on the pressure[15,21]. (Another approach, useful for situations where ion–ion interactions are important, will be discussed in section 6C.) The following concepts now need definition. The standard molar *solvated volume,* i.e. the volume of the solvated ion, v^0_{hi}, is related to the radius of a solvated ion r^0_{hi} (Fig. 6.4) by

$$v^0_{hi} = \mathbf{N}k_p(4\pi/3)(r^0_{hi})^3 \tag{6.44}$$

where k_p is a packing coefficient, arising from the assumed spherical nature of the ions and the consequent void space between the spheres. Geometrical considerations, or an empirical determination with steel balls, lead to $k_p = 1.725$. For v^0_{hi} in $cm^3\ mol^{-1}$ and r^0_{hi} in pm, the numerical factor $\mathbf{N}k_p(4\pi/3)$ is 4.35×10^{-6}. The radius of the solvated ion is defined operationally, for example as the 'Stokes radius'

$$r^0_{hi} = |z_i|\mathbf{F}^2/6\pi\mathbf{N}\eta^*\lambda_i^\infty = 8.20|z_i|/\eta^*\lambda_i^\infty \tag{6.45}$$

the numerical factor applying for r^0_{hi} in pm, η^*, the dynamic viscosity of the solvent, in Pa·s, and λ_i^∞ the (infinite dilution) equivalent conductivity of the ion in $\Omega^{-1}\ cm^2\ equiv^{-1}$. This definition, however, holds only for large ions (e.g. $(C_5H_{11})_4N^+$), while for smaller ions (e.g. $(C_2H_5)_4N^+$) conductivities are so that $r^0_{hi\,(Stokes)} < r_{ci}$ and corrections must be applied, e.g. reading r^0_{hi} from the extrapolated portion of the $r^0_{hi\,(Stokes)}(r_{ci})$ curve for the large ions[22] (Fig. 6.5).

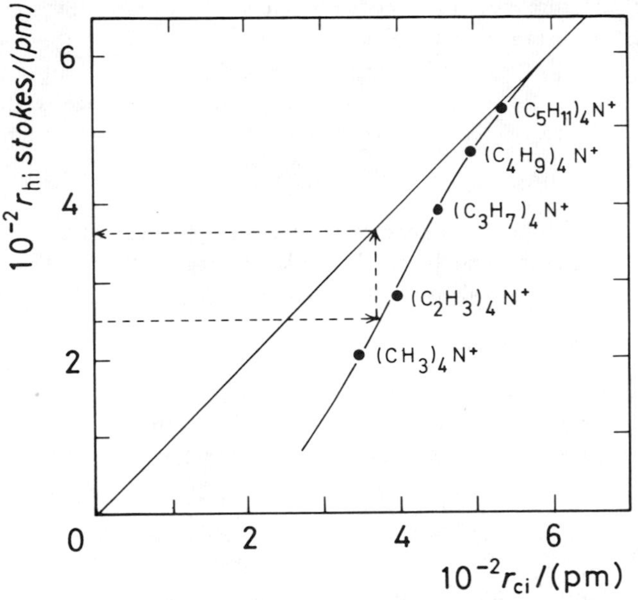

Figure 6.5. The relationship between Stokes radii of hydrated ions and their crystal radii. To find the *corrected Stokes radius* of an ion, the experimental value is located on the ordinate, the value read on the curve for the tetraalkylammonium ions, and the value of the corresponding crystal radius is equal to the corrected Stokes radius.

The molar volume of the solvated electrolyte v^0_{hs} may be obtained from compressibility data

$$v^0_{hs}(\mathrm{cm}^3) = \lim_{c_s \to 0} 1000\, c_s^{-1}\, (1 - \kappa/\kappa^*_W) \tag{6.46}$$

where κ is the compressibility $-V^{-1}(\partial V/\partial P)_T$ of a c_s molar solution and κ^*_w is the compressibility of the solvent. In this operational definition it is assumed that the solvent, already compressed by the field of the ion, is not compressible by external pressure, and neither is the ion itself. Obviously, there need be no direct relationship between volumes derived from the kinetic (Stokes law) and compressibility data. The latter must also be apportioned between cation and anion, most easily on the basis of an electrolyte with a non-solvated ion, of known volume v_{ci}.

This quantity, v_{ci}, is called the *intrinsic volume* of the ion (Fig. 6.4). It can be related by an equation analogous (6.44) to the intrinsic radius of the ion, which may be set equal to the crystal radius r_{ci} (and then the packing coefficient k_p is given values ranging from the geometrical figure, 1.725, to $k^3 = 1.955$, where $k = r_{aq}/r_{ci}$, cf. p. 220), or some effective radius r_{ei} (e.g. $r_{ci} + \Delta$) is used with $k_p = 1$, arbitrarily. Conceptually, the intrinsic volume is the molar volume of a hypothetical liquid state of the pure electrolyte at the temperature of interest, where the ions are closely packed, at distances apart as in the crystal.

The difference $v_{hi}^0 - v_{ci}$ is the measure of the volume of solvent in the solvation shell, which is under the influence of the field of the ion, hence is compressed, or has undergone *electrostriction.* The *solvation number* is then

$$h_i^0 = (v_{hi}^0 - v_{ci})/v_{W\ el} \tag{6.47}$$

where $v_{W\ el}$ is the molar volume of the electrostricted solvent, which equals $v_W^* - \Delta v_{W\ el}$, the subtrahend being the molar electrostriction of the solvent. This can be calculated from the pressure-derivative of (6.33) (cf. p. 219), or from

$$\Delta v_{W\ el} = v_W^* \kappa_W^* S_V / S_K \tag{6.48}$$

where $S_V/S_K = (\phi_V - \phi_V^\infty)/(\phi_K - \phi_K^\infty)$ is the ratio of the slopes of plots of the apparent molar volume ϕ_V and the apparent molar compressibility ϕ_K against the square root of the molar concentration $c_s^{1/2}$. Another way to obtain the solvation number is from

$$h_i^0 = (v_{hi}^0 - v_{i\ abs}^0)/v_W^* \tag{6.49}$$

where the standard absolute partial ionic molar volume $v_{i\ abs}^0$, obtained from the entries in Appendix 6.1, p. 250, for water as the solvent, and eqs. (6.38) and (6.43), where $v_{i\ con}^0$ equals the apparent ionic molar volume at infinite dilution ϕ_{Vi}^∞, i.e. the volume of the solution less the volume of the solvent, reckoned at the molar volume of the bulk solvent. The molar volume of the solvated ions is therefore larger by $h_i^0 v_W^*$ than the partial molar ionic volume. For water as the solvent at 298.15 K, eq. (6.48) gives $\Delta v_{W\ el} = 2.1\ cm^3$ (hence $v_{W\ el} = 15.9\ cm^3$). A better expression, allowing for the variation of $\Delta v_{W\ el}$ with the field strength, is

$$\Delta v_{W\ el}(cm^3) = 559|z|^{1/2}[138 + r_c(pm)]^{-1} + 6.68 \times 10^4|z|[138 + r_c(pm)]^{-2} + 3.57 \times 10^8 z^2 [138 + r_c(pm)]^{-4} \tag{6.50}$$

Equations (4.46) to (4.49) are mutually consistent[23], within the uncertainty of the molar volume data and of (6.38).

The quantities appearing in the above discussion, r_{ci}, v_{ci}, r_{hi}^0, h_i^0 and v_{hi}^0 are listed in Appendix 6.2, p. 251, for many ions. The numerical values selected are based on a critical consideration of published data.[15,16,21–23]

The above considerations apply to a single ion in an infinite amount of solvent, that is to the limit of infinite dilution of the electrolyte. Hence they are described by the standard chemical potential μ_s^0 or by the standard molar volume v_s^0 (or its individual ionic component v_i^0). The considerations are however valid also for finite concentrations, and can be thought of as coming on top of any other interactions occurring at higher concentrations. As long as the solvation spheres of ions do not overlap, the picture of discrete molecular interactions up to the limit of secondary solvation, with a continuous dielectric medium beyond that, is valid. This should be true up to about $\nu c_s = 2\ mol\ dm^{-3}$, as seen from the v_{hi}^0 values in Appendix 6.2, p. 251. At higher concentrations the solvent becomes scarce, and full solvation spheres may no longer be attained. An empricial dependence of the solvation number on the concentration is derived from the density data

$$h_s = h_s^0 - h_s^0 S_V (v_{cs} - v_s^0)^{-1} c_s^{1/2} \tag{6.51}$$

where S_V is, as before, the slope of the dependence of ϕ_V on $c_s^{1/2}$.

The binding of a part of the solvent to the ions introduces a contribution to the configurational entropy of the solution, since there are $m_s' = m_s(1 - h_s m_s M_W)^{-1} > m_s$ moles of electrolyte per unit mass of *free* solvent[24]. A further effect that needs consideration is the discrepancy in size between the solvated ions and the solvent molecules[25], which contributes also to the entropy of mixing, and thus to the chemical potentials of the electrolyte and of the solvent, that is to $\gamma_\pm$ and to ϕ. Assuming additivity of these effects, and confining the applicability to the range of low concentrations, where $[M_W \text{ (kg mol}^{-1})] [m_s \text{ (mol kg}^{-1})] \ll 1$, so that $\log(1 + M_W m_s) \approx 0.4343 M_W m_s$, the contribution of the configurational entropy to $\log \gamma_\pm$ is

$$\log \gamma_{conf} = 0.4343 M_W [h_s + (r_V - \nu) r_V / \nu] m_s \tag{6.52}$$

where $r_V = v_{hs}^0 / v_W^0 = \Sigma \nu_i v_{hi}^0 / v_W^0$, and where for water as a solvent $0.4343 M_W = 0.0078$. The corresponding contribution to ϕ is

$$\phi_{conf} = 1 + M_W [r_V (h_s + r_V - \nu)/\nu] m_s \tag{6.53}$$

where, for water as a solvent, $M_W = 0.018$ kg mol^{-1}. Both configurational terms are linear in m_s, within the limits of applicability of these expressions[26]. The total excess Gibbs energy, from eq. (6.12) is

$$G_{conf}^E = -RTm_s^2 M_W [h_s (r_V - \nu)] \tag{6.54}$$

for an amount of solution containing 1 kg of solvent. Since normally, at least for water as a solvent, $r_V = \Sigma \nu_i (h_i + v_i^0 / v_W^0) > \Sigma \nu_i = \nu$, the square bracket in (6.54) is positive, hence G_{conf}^E is negative. In the case where $r_V = \nu$, G_{conf}^E vanishes, since if there is no size effect, the arbitrary apportionment of the solvent between the bulk and solvation shells does not affect the overall Gibbs energy of the system.

d. *Hydrophobic ions*

As already noted, water has unique properties as a solvent, shared only with a limited group of other solvents: water-rich mixtures with other solvents, deuterium oxide, and to a lesser degree, hydrogen peroxide and sulphuric acid. Some of the effects concerning non-electrolytes, described by the term 'hydrophobic interactions', have already been discussed (section 5B(e)), as have also structural effects on the solvent arising from 'hydrophobic ions'. These are ions with a weak field (low charge and large size), and an 'inert' outer surface which interacts only through weak dispersion forces with the polar solvent. On top of electrostatic interactions with other ions, these ions suffer 'water-structure-enforced' interactions with them, discussed further below. Here only effects at infinite dilution, arising from the interactions of a single ion, are discussed. The main ideas are due to

Frank and co-workers[8], who pointed out that although electrostriction is small, the standard partial molar volumes of tetraalkyl-ammonium salts are low, since the alkyl chains are situated in the voids of cages. These cages, of transient nature, are made up of water molecules in a pentagonal dodecahedron structure like that of the solid clathrates $(C_nH_{2n+1})_4N^+X^- \cdot mH_2O$ (where for $X^- = F^-$, $n = 4$, $m = 32$, and $n = 5$ (isopentyl), $m = 38$, and for $X^- = Br^-$, $n = 4$, $m = 30$, are examples). The entropies of fusion of these clathrates are low, signifying that a great deal of the structure of these cages in the solid persists in the melt (which is 1.5–1.8 m), and presumably also in dilute solutions, since the partial molar heat capacities of these electrolytes are abnormally high. Similar conclusions are reached from an examination of other properties, such as the expansability, or the *B* coefficients of the Jones–Dole viscosity expression (6.19): they are positive and large for the structure-making tetraalklyammonium ions, but have negative temperature coefficients. This is consistent also with conductivity results, where $\lambda_i^\infty \eta^*$ increases with temperature, and on going from deuterium oxide to light water[27], or other conditions where water has a smaller structuredness in any case (section 3B(e)). In the above sense of structure, the tetraalkyl cations with alkyl chains from propyl onwards are strong structure-makers, the tetramethylammonium cation is a weak structure-breaker (the orientational and electrostrictive forces of the charge prevail over the hydrophobic effects of the methyl groups), while the tetraethyl and tetraethanol ($N(-CH_2CH_2OH)_4^+$) cations have only slight net structural effects.

Further effects, arising from the hydrophobic interactions that cause enhanced structure of the solvent around the ion, are observed when the spheres of structured solvent around two neighbouring ions overlap. These effects will be discussed in section C on ion–ion interactions, since they are not observed at the limit of infinite dilution. It must be remembered, however, that these particular ion–ion interactions require the intermediary action of ion-solvent interactions.

e. Solvent-sorting by ions

The effect of ions on two or more different solvents in a mixture is closely related to the salting out of non-electrolytes from salt solutions. The salting-out (or -in) of non-electrolytes by ions is a phenomenon[28] observed at finite electrolyte concentrations, but the parameters describing it are best obtained on extrapolation to infinite dilution of the reactants in the solvent. The salting (Setchenov) coefficient is defined as

$$k_s = \lim_{c_s \to 0,\, c_n \to 0} d(\log f_n^\infty)/dc_s \qquad (6\text{–}55)$$

where n is the non-electrolyte, and f_n^∞ is its activity coefficient related to the infinite dilution reference state, obtained from

$$f_n^\infty = y_{n(s)} = (c_{n(s)}/c_{n(w)})_{a_n = \text{const.}} \qquad (6.56)$$

The ratio of concentrations of the non-electrolyte in the presence of electrolyte (s) and in its absence (w) at a constant activity of the non-electrolyte is often obtained from solubility data, but can also be obtained from distribution measurements, which facilitate extrapolation to $c_n = 0$ for appreciably soluble non-electrolytes. The integral forms of (6.55), $\log f_n = k_s c_s$ and $f_n = 1/(1 - 2.303 k_s c_s)$ are also valid at the limit of low concentrations. Other relationships that follow from (6.55) are

$$2.303\mathrm{R}Tk_s = \lim \mathrm{d}\mu_n^E/\mathrm{d}c_s = \lim \mathrm{d}\mu_n/\mathrm{d}c_s = \lim \mathrm{d}^2 G/\mathrm{d}c_n \mathrm{d}c_s$$
$$\text{(for all limits, } c_s \to 0 \text{ and } c_n \to 0) \qquad (6.57)$$

These expressions may be used to derive theoretical expressions for the salting phenomenon.

If the non-electrolyte is considered to be excluded from the solvated volume of the ions, i.e. excluded from ($c_s \times v_{hs}$ (cm^3)) cm^3 in a total volume of 1000 cm^3 of the solution, then $f_n^\infty = 1000/(1000 - c_s v_{hs}\,(\mathrm{cm}^3))$. Setting approximately the partial molar volumes $v_s^0 = \nu v_w^*$, the salting coefficient becomes, from (6.49),

$$k_s = (v_w^*/2303)(h_s^0 - \nu) = 0.0078(h_s^0 - \nu) \qquad (6.58)$$

No dependence of k_s on the properties of n is provided by this expression. A more theoretically based expression is obtained when the non-electrolyte is considered to be excluded from the electrostricted region around the ion[29]. The derivative $\mathrm{d}\mu_n/\mathrm{d}c_s = v_n^* \cdot (\mathrm{d}^2 G/\mathrm{d}V_n \mathrm{d}c_s)$ leads to

$$k_s = v_n^*(v_{cs} - V_s^0)/2.303\kappa_w^* \mathrm{R}T$$
$$\simeq 8 \times 10^{-4}\, v_n^*\,(\mathrm{cm}^3)\, h_s^0 \qquad (6.59)$$

where the last equality is based on (6.48) and (6.49). In this expression k_s is seen to depend both on the electrolyte (through h_s^0) and on the non-electrolyte (through v_n^*). With electrostrictive ions, (6.59) thus accounts for salting-out, but not for salting-in. The magnitude of k_s calculated from it is 3 to 10 times larger than experimental values.

A quite different approach estimates the difference in electrostatic work performed when ions are brought from a hypothetical discharged state to their charged state in the pure solvent and the non-electrolyte solution[30]. The dielectric constant of the solution is related linearly to the concentrations (at least at the limit of low concentrations) $\epsilon = \epsilon_w - \delta_n c_n - \delta_s c_s$, and this expression together with the Born equation

$$\Delta G_{el} = \tfrac{1}{2}\mathrm{N}e^2 \epsilon_0^{-1} \left(\frac{1}{\epsilon} - \frac{1}{\epsilon_w}\right) \sum_i c_i z_i^2 / r_i \qquad (6.60)$$

leads to

$$k_s = \mathrm{N}e^2 \epsilon_0^{-1} \delta_n (2.303\mathrm{R}T\epsilon_w^2)^{-1} \sum_i \nu_i z_i^2 / r_i \qquad (6.61)$$

The salting coefficient depends on the non-electrolyte through δ_n, the molar dielectric decrement, which is usually positive, leading to $k_s > 0$, i.e. salting-out, but for non-electrolytes more polar than water, with $\delta_n < 0$, salting-in is predicted (the prediction is correct for HCN). The salting coefficient depends on the electrolyte through ν_i, z_i and r_i, predicting that small, highly charged ions should be efficient for salting-out, which they indeed are.

A more sophisticated salting theory, proposed by Debye, and later modified, considers the sorting of non-electrolyte and solvent molecules in the field of the ions according to the dielectric constant of the volume elements the former occupy[31]. The mole fraction of non-electrolyte in the field of ion j, x_{n_j}, relates to that in the bulk at a distance $r \to \infty$ away from the ion, x_r^0, as

$$\ln \frac{x_{nj}}{x_n^0} = \frac{-z_j^2 e^2}{8\pi k T \epsilon^2 \epsilon_0} \left[\frac{v_n}{v_w} \frac{\partial \epsilon}{\partial \rho_w} - \frac{\partial \epsilon}{\partial \rho_n} \right]_{T;\, x_n^0 \to 0} r^{-4} \tag{6.62}$$

where ρ is the number density of the molecules, i.e. their number per unit volume. Summation over all the ions and integration over the total available volume then gives the activity coefficient of the non-electrolyte. A recent modification of the original treatment introduces the derivatives of the dielectric constant with respect to the number densities in the form $(v_n/v_w)(\partial\epsilon/\partial\rho_w)_{\rho_n,T} = 1000\mathbf{N}^{-1}\delta_s v_n^*(v_s^0 + h^0 v_w^*)^{-1}$ at the limit of infinite dilution, and $(\partial\epsilon/\partial\rho_n)_{\rho_w,T} = -1000\mathbf{N}^{-1}\delta_n = -\mathbf{N}^{-1} v_n^*(\epsilon_w - \epsilon_n)$. At 25 °C, the numerical value for the salting coefficient becomes (for v_n^* and v_s^0 in $cm^3\ mol^{-1}$)

$$k_s \approx 0.025 v_n^{*\,3/4} (\delta_s (v_s^0 + 18h^0)^{-1} + (78.54 - \epsilon_n)1000^{-1})^{3/4} \Sigma \nu_i z_i{}^{3/2} \tag{6.63}$$

This depends on the non-electrolyte through its volume ($v_n^{*\,3/4}$) and its dielectric constant ($-\epsilon_n$), and on the electrolyte through its molar dielectric decrement divided by its hydrated volume $[\delta_s(v_s^0 + 18h^0)^{-1}]$ and the charge on the ions ($\Sigma\nu_j z_j^{3/2}$). Salting-out is predicted to be the rule, but salting-in is permitted in principle under unusual circumstances of very large ions and high-dielectric-constant non-electrolytes. This corresponds to the circumstances under which salting-in has indeed been observed, but not numerically for the actual cases observed. For salting-out, eq. (6.63) predicts coefficients of the correct magnitude, but not in detailed agreement with the observed values.

The case of salting-out, for which (6.62) was derived with the understanding that as $x_n \to 0$ then $x_w \to x_w^0 \to 1$, is a sub-case of the more general problem of the sorting of the molecules of two solvents in the field of an ion at infinite dilution of the electrolyte only. If the two solvents are at comparable concentrations and need no longer be designated as n for non-electrolyte and w for water, but may be called a and b, the equation analogous to (6.62) that is obtained, provided the dielectric constant and solvent volumes are independent of the field strength and the solvents themselves mix ideally[32], is

$$\ln\left(\frac{x_a}{x_b}\right)_j - \ln\frac{x_a^0}{x_b^0} = \frac{-(x_a v_a^* + x_b v_b^*) z_j^2 e^2}{8\pi \mathbf{R} T \epsilon^2 \epsilon_0} \left(\frac{\partial\epsilon}{\partial x_a} - \frac{\partial\epsilon}{\partial x_b}\right)_T r^{-4} \tag{6.64a}$$

The left-hand-side may be rewritten as

$$\ln(x_{a_j}/x_a^0) - \ln(x_{b_j}/x_b^0) = 2.303\alpha_j \tag{6.64b}$$

where α is the (decadic) solvent-sorting coefficient. For more realistic conditions, that is in the vicinity of high-field ions where ϵ is field-dependent, but still for ideal mixing of the solvents, it was shown[32] that $\alpha_j = -(\Delta g_{ja} - \Delta g_{jb})/2.303RT$, where Δg_{ja} is the partial molar Gibbs energy of solvation of the ion j in the component a of the mixture. For an electrolyte s, $\alpha_s = \nu^{-1}\Sigma\nu_i\alpha_i$, and a further approximation is made by using Δg^0_{sa}, the standard molar Gibbs energy of solvation of the electrolyte s by the pure solvent a for the partial quantity, to yield finally

$$\alpha_{s(a/b)} \approx -(\Delta g^0_{sa} - \Delta g^0_{sb})/2.303\nu RT \tag{6.65}$$

for the solvent-sorting coefficient. Solvents that solvate the electrolyte better than others will have positive α values relative to them, that is, will accumulate preferentially near the ions. Thus water will be relatively more abundant than alcohols, acetone or dioxane near most electrolytes, but ammonia, hydrazine and formic acid will be enriched relatively to water near heavy-metal ions.[33a,33b] Equations (6.63) and (6.65) recognize the immobilization of solvents near high-field ions through solvation, which expresses itself also by the field-dependence of ϵ, eq. (6.20).

The problem of solvent-sorting by the ions of an electrolyte is closely related to the problem of the *transfer* of an electrolyte from one solvent to another. The standard Gibbs energy of transfer of s from a to b is given by

$$\Delta G^{0tr}_{s(a\to b)} = \Delta g^0_{sb} - \Delta g^0_{sa} = \mu^0_{sb} - \mu^0_{sa} \tag{6.66a}$$

The *transfer activity coefficient* $\gamma^{tr}_{s(a\to b)}$ is defined as $\exp(\Delta G^{0tr}_{s(a\to b)}/RT)$, and is related by (6.6), (6.7) and (6.8) to the activity coefficients of s in the solvents a and b, provided they are in equilibrium ($\mu_{s(a)} = \mu_{s(b)}$)

$$\gamma_s(b) = \gamma_s(a)m_s(a)/m_s(b)\gamma^{tr}_{s(a\to b)} \tag{6.66b}$$

Such a situation could occur, for example, at saturation (i.e. $m_s(a)$ and $m_s(b)$ are the solubilities), or with immiscible solvents (i.e. $m_s(a)/m_s(b) = D_s$, the distribution ratio). Equation (6.66b) permits the calculation of activity coefficients for the solvent b provided the data for solvent a and the transfer activity coefficient are known in addition to the equilibrium concentration data. Conversely, if the 'salt effect' activity coefficients $\gamma_s(a)$ and $\gamma_b(b)$ can be obtained independently, or from theory, the concentration ratio *or* the transfer activity coefficient can be calculated if the other quantity is known.

Derivative quantities, such as heats of transfer or entropies of transfer are defined in an obviously analogous manner. As with the corresponding quantities for water as solvent, theories and correlations apply only to single ions rather than to electrolytes. Extra thermodynamic assumptions or measurements must be made on every solvent (or solvent composition in the case of mixtures) separately, in order to obtain the absolute ionic thermodynamic quantities (cf. the discussion on pp. 217 to 221). The most promising approach divides the quantity equally between two large monovalent ions (cation and anion), where the charge lies buried within a non-polar envelope,[33b] such as in tetraphenylarsonium tetraphenylboride, $(C_6H_5)_4As^+\ (C_6H_5)_4B^-$, or triisopentyl-n-butylammonium tetraphenylboride, $(i\text{-}C_5H_{11})_3(C_4H_9)N^+\ (C_6H_5)_4B^-$.

Although equating $\gamma^{tr}_{+(a\rightarrow b)}$ with $\gamma^{tr}_{-(a\rightarrow b)}$ for these large ions, which should be solvated to an approximately equal extent by all solvents, is reasonable, this has not yet resulted in a universally accepted convention.

In general, the same theoretical considerations apply to non-aqueous solvents as those discussed for water in section (6B(b)). For solvent mixtures the situation is more complicated than for pure solvents, since solvent-sorting in the vicinity of the ions will produce conditions different from those in the bulk solvent. Therefore, theoretical equations, such as the Born equation (6.35), or its modifications, require realistic estimates of the local dielectric constant ϵ (solvent mixture), that replaces ϵ_{aq} in this equation. However the splitting of $\gamma^{tr}_{s(a\rightarrow b)}$ into equal parts for the large reference ions should cause no problems even for mixed solvents.

For the partial ionic entropies, a linear relationship has been found[34] between the quantity in a given non-aqueous solvent x and that in water, corrected for the differences in the molecular weight

$$s^0_{i(x)} - \mathbf{R}\ln(1000/M_x) = a + bs^0_{4\,(w)} - b\mathbf{R}\ln(1000/M_w) \tag{6.67}$$

For the alkali halides, the parameter b ranged from ·0.64 to 1.04 and a from -93.7 J K^{-1} mol^{-1} to $+3.3$ J K^{-1} mol^{-1} (on the basis of eq. (6.37), $s^0_{H^+(w)}$ = -22.2 J K^{-1} mol^{-1}) for the series of solvents: ammonia, dimethylformamide, ethanol, methanol, *N*-methylformamide, formamide, water, deuterium oxide. This relationship is believed[3b] to be general.

C. Ion–Ion Interactions

a. The Debye–Hückel limiting law

There is general agreement that the Debye–Hückel theory[35] gives the correct description of the ion–ion interactions at the limit of extremely dilute solutions. It yields a square-root concentration-dependence of the electrical work contribution to the Gibbs energy, which is the only ion–ion interaction of sufficiently long range to be of any importance at these dilutions. Hence it also yields square-root concentration-dependencies for derivative functions, such as the partial molar volume or heat content. The main features of the theory are a consideration of the solvent as a dielectric medium having a given dielectric constant ϵ, and of the ions as point charges z_ie, capable of approaching each other to a distance of closest approach a (which may be interpreted as the mean diameter of the ions). The main approximation in the theory is[36] that the electrostatic potential energy $|z_i|\,\mathrm{e}\psi_j(r)$ of an ion i at a distance r from an ion j is equal to the potential of mean force between the ions. Further approximations depend on the requirement that the concentration be very low, as will be described below.

Consider Fig. 6.6, where the ion j is arbitrarily selected as the centre of coordinates, and the volume element dV is at a distance r from this ion. Overall electrical neutrality requires that at zero potential, that is at infinite distance from

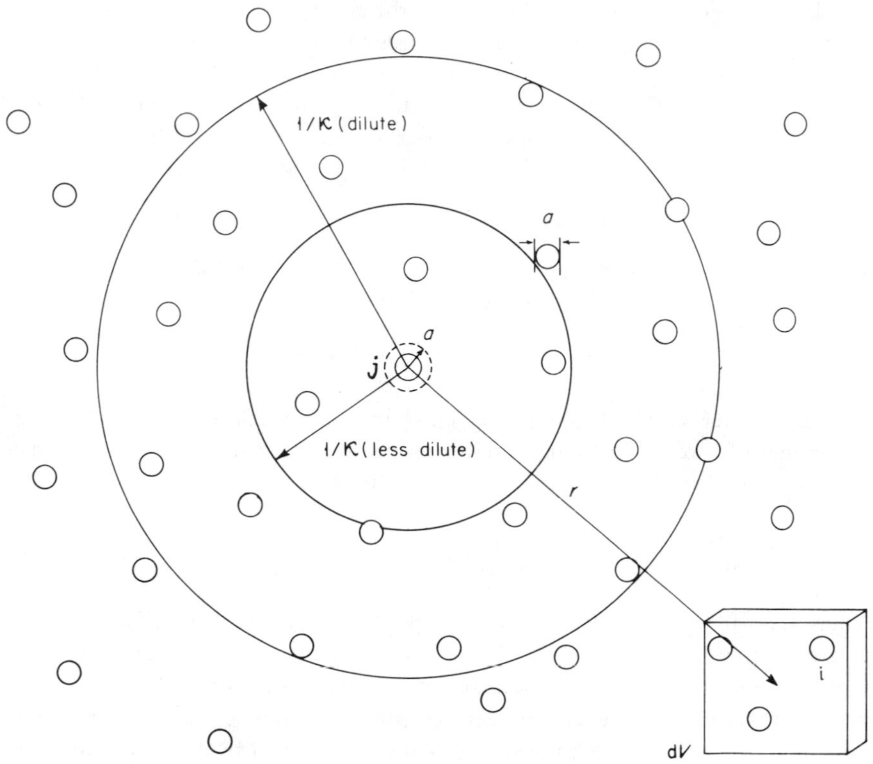

Figure 6.6. Schematic representation of ionic distribution around a central ion j in an electrolyte solution. Attention is directed to a volume element dV at distance r from ion j, containing ions of kinds i . . . , to the distance of closest approach a, and to the average radius of the ionic atmosphere, $1/\kappa$, in a dilute and a less dilute solution. In the latter case the ionic atmosphere is not very symmetrical. None of the sizes and distances are to scale.

the ion j, the mean charge density is zero, $\rho(\infty) = \Sigma_i \rho_i^0 z_i \mathrm{e} = 0$, where $\rho_i^0 = \mathbf{N} c_i$ is the bulk average number density of ions of kind i in the solution. The local mean charge density $\rho(r) = \Sigma_i \rho_i(r) z_i \mathrm{e}$ may however differ from zero at a point where the potential $\psi(r)$ does not vanish, and according to the Poisson law, the charge density $\rho(r)$ in the volume element dV is

$$4\pi \boldsymbol{\epsilon}_0^{-1} \epsilon^{-1} \rho(r) = -r^{-2} \mathrm{d}(r^2 \mathrm{d}\psi(r)/\mathrm{d}r)/\mathrm{d}r \qquad (6.68)$$

assuming spherical symmetry around the ion j. The assumption is reasonable for low concentrations (much lower indeed than depicted schematically in Fig. 6.6). It is necessary to take into account, however, also the thermal movement of the ions. The Boltzmann distribution $\rho_i(r) = \rho_i^0 \exp(-z_i \mathrm{e}\psi_j(r)/\mathbf{k}T)$ cannot be used directly, since the exponential relationship of ψ and ρ that it demands is incompatible with the Poisson equation and the theorem of the superposition of fields, which require a linear relationship between them. Expansion of the exponential expression helps to overcome this problem. Only in very dilute solutions would $z_i e\psi_j(r) \ll kT$ permit the truncation of the expanded form of the exponent after the second term. The summation over all kinds of ions i causes the first term to vanish, so that only the second term

$$\rho(r) = -\Sigma_i \rho_i^0 z_i^2 e^2 \psi_j(r)/kT \tag{6.69}$$

remains, which has the correct linear relationship between ρ and ψ. Thus a differential equation for $\psi_j(r)$ is obtained from (6.68) and (6.69)

$$r^{-2} d(r^2 d\psi_j(r)/dr)/dr = [4\pi e^2/\epsilon_0 \epsilon kT] \sum_i \rho_i^0 z_i^2 \psi_j = \kappa^2 \psi_j \tag{6.70}$$

where the coefficient of ψ_j on the right-hand side is designated by κ^2; κ has the dimension of a reciprocal length and is proportional to the square-root of the ionic concentration (or of ρ_i^0). The solution of (6.70) is

$$\psi_j(r) = \frac{z_j e}{\epsilon_0 \epsilon} \frac{e^{\kappa a}}{1 + \kappa a} \frac{e^{-\kappa r}}{r} \tag{6.71}$$

where the r-independent factors are obtained from the boundary condition that the electric induction just inside the limit $r = a$, where no other ions can enter, which is $z_j e/a^2$, equals that just outside it, which is given by $-\epsilon_0 \epsilon d\psi_j/dr = \epsilon_0 \epsilon a^{-1}(1 + \kappa a)\psi(a)$. The charge density $\rho(r)$ is then $-(\epsilon_0 \epsilon \kappa^2/4\pi)\psi(r)$, and the charge in a shell of thickness dr at a distance r is $4\pi r^2 dr$ times this, $-\kappa^2 z_j e$ $e^{\kappa a}(1 + \kappa a)^{-1} e^{-\kappa r} r dr$. This function has a maximum at $r = 1/\kappa$, and therefore κ is called the reciprocal of the thickness of the ionic atmosphere (Figs. 6.6 and 6.7). The total charge surrounding the ion j is the integral of this function from a to ∞, which comes out to be $-z_j e$, i.e. exactly cancelling the charge on the ion j itself, as expected.

In order to obtain the electrical energy due to ion–ion interactions, it is necessary first to realize that (assuming that $G^{el} = A^{el}$, i.e. that $V^{el} = 0$ and there are no contributions from these interactions to the volume of the solution)

$$dG^{el}/e dz_j = \psi_{aj} \tag{6.72}$$

Here ψ_{aj} is the electrostatic potential due to the remaining ions (the ionic atmosphere) at the position of the ion j, that is at $r \leqslant a$, which is the difference between the total potential given by (6.71) and that due to the ion j itself at $r = a$

$$\psi_{aj} = \psi_j(a) - z_j e/\epsilon_0 \epsilon a = -z_j e \epsilon_0^{-1} \epsilon^{-1} \kappa/(1 + \kappa a) \tag{6.73}$$

Summation over all the ions j and integration over dz_j will then give the electrical energy G^{el} of the solution which contains $V\rho_j^0$ ions of each type j. It is convenient for the integration to use the *Debye charging process*, that is increase simultaneously the fraction λ of the final charge of every ion from zero to infinity. Therefore each z_j in (6.72) and (6.73) is replaced by $z_j\lambda$ (and therefore κ becomes

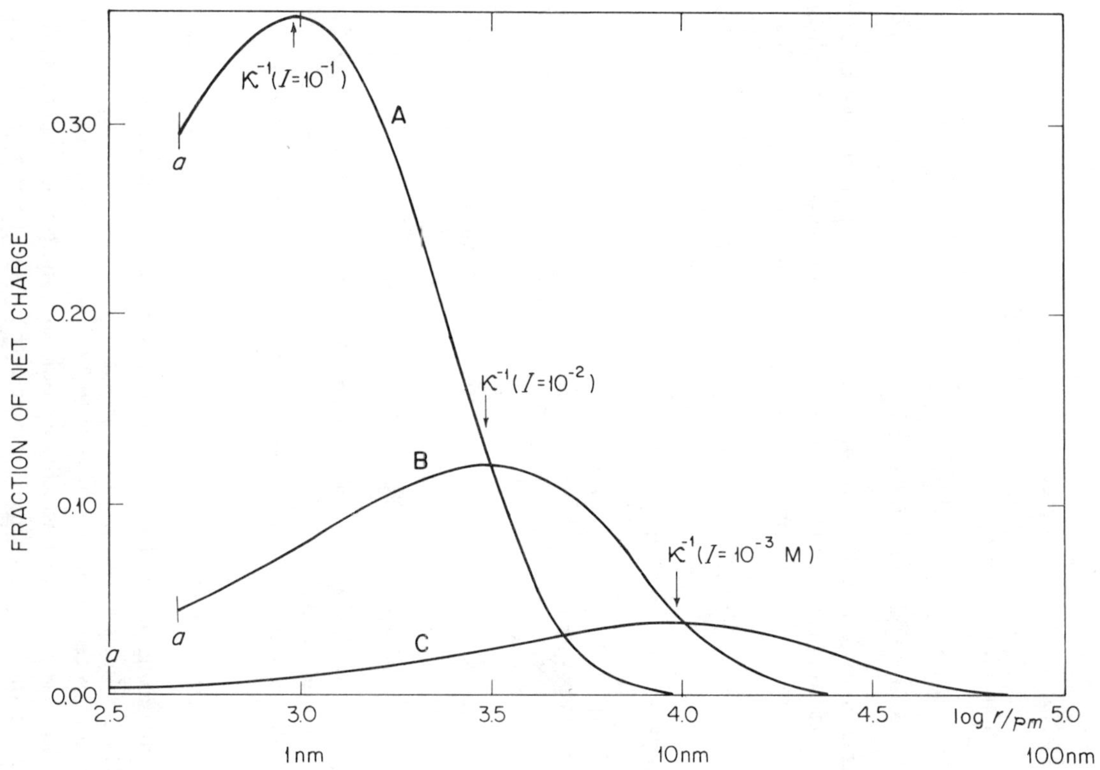

Figure 6.7. The fraction of the charge $-z_j e$ residing in the ionic atmosphere as a function of the distance r (*logarithmic scale*) from the ion j. Curve A: $a = 481.0$ pm, $I = 0.1$ M, $\kappa a = 0.50$; curve B: $a = 304.2$ pm, $I = 0.01$ M, $\kappa a = 0.100$; curve C: $a = 481.0$ pm, $I = 0.001$ M, $\kappa a = 0.050$. The positions of the maxima, corresponding to the average radius of the ionic atmosphere $1/\kappa$, are marked.

$\kappa\lambda$) to give

$$G^{el} = \int_0^1 V\sum_j \rho_j^0 e\psi_{aj}(\lambda) z_j d\lambda$$

$$= -V\sum_j \rho_j^0 z_j^2 e^2 \epsilon_0^{-1} \epsilon^{-1} \kappa \int_0^1 \lambda^2 (1 + \lambda\kappa a)^{-1} d\lambda$$

$$= -V\sum_j \rho_j^0 z_j^2 e^2 (3\epsilon_0 \epsilon)^{-1} \kappa\tau(\kappa a) \qquad (6.74)$$

where the function $\tau(\kappa a)$ is given by

$$\tau(t) = 3t^{-3}(\ln(1 + t) - t + 1/2t^2)$$

$$= 1 - 3t/4 + 3t^2/5 - 3t^3/6 + \ldots \qquad (6.75)$$

and approaches unity for sufficiently small κa, that is at high dilutions. The distance of closest approach a enters G^{el} only in $\tau(\kappa a)$, so that at sufficiently high dilutions, G^{el} is independent of a, and the limiting law expression

$$G^{el} = -Ve^2 \kappa(3\epsilon_0\epsilon)^{-1} \sum_j \rho_j^0 z_j^2 \qquad (6.76)$$

is obtained, or, from (6.70) replacing κ

$$\frac{G^{el}}{RT} = \frac{V}{12\pi N}\left(\frac{4\pi e^2}{k\epsilon_0 \epsilon T}\right)^{3/2} (\sum_j \rho_j^0 z_j^2)^{3/2}$$

$$= -2.4958 \times 10^{11}\, V(\epsilon T)^{-3/2} I^{3/2} \qquad (6.77)$$

where the numerical factor applies for V expressed in m^3, and where I is the ionic strength

$$I = \tfrac{1}{2}\sum c_j z_j^2 \simeq \tfrac{1}{2} d^* m_j z_j^2 \qquad (6.78)$$

and $c_j = 10^{-3} n_j/V$ is the molar concentration. (The last equality is valid for limiting-law dilutions only and d^* is the density of the solvent, differing appreciably from unity for non-aqueous solvents only.) For water at 298.15 K, the relationship

$$\kappa = 3.287 \times 10^9 I^{1/2} m^{-1} \qquad (6.79)$$

is useful for converting practical ionic strengths I, expressed in M = mol dm^{-3}, to the theoretically useful quantity κ, and *vice versa*.

An integration process as an alternative to (6.74) has been proposed through the *Güntelberg charging process*, in which the ionic atmosphere is assumed to be unchanged, while the ion j is charged from zero to its full charge z_j e. Thus the terms in κ are taken outside the integral, which remains as $\int_0^1 \lambda^2 d\lambda = 1/3$, and the final expression is $(1 + \kappa a)^{-1}$ times the expression on the right-hand side of (6.76), which at the limit of high dilutions is practically the same. The Debye process, leading to (6.74) is preferable for calculating the composition derivatives, i.e. the electrical contributions to the chemical potentials.

The electrical contribution to the chemical potential of the solvent is obtained by differentiation of (6.74) with respect to the mass of the solvent W_W, as

$$G_W^{el}/RT = (24\pi N)^{-1} V_W \kappa^3 \sigma(\kappa a) \tag{6.80}$$

where V_W is the specific volume, and the function $\sigma(\kappa a)$ is given by

$$\begin{aligned}\sigma(t) &= 3t^{-3}[1 + t - (1 + t)^{-1} - 2\ln(1 + t)]\\ &= 1 - 6t/4 + 9t^2/5 - 12t^3/6 + \ldots\end{aligned} \tag{6.81}$$

At very low concentrations, again $\sigma(\kappa a)$ approaches unity, though less rapidly than does $\tau(\kappa a)$: $\tau(0.02) = 0.985$, while $\sigma(0.02) = 0.971$, i.e. both functions are still significantly different from unity even at $\kappa a = 0.02$, corresponding for $a = 350$ pm to $I = 3 \times 10^{-4}$ M. For water as a solvent at 298.15 K, the relationship

$$\phi = 1 - 0.2764 m_s^{1/2} (\sum_i \nu_i z_i^2)^{3/2} \nu^{-1} \sigma(\kappa a) \tag{6.82}$$

describes the dependence of the osmotic coefficient on the molality of the solution in *dilute solutions*.

The electrical contribution to the chemical potential of the electrolyte is the sum of the corresponding contributions of the individual ionic species obtained by partial differentiation of (6.74) with respect to n_i

$$\mu_i^{el}/RT = -\frac{z_i^2 e^2}{2k\epsilon_0 \epsilon T}\frac{\kappa}{1 + \kappa a} + \frac{v_i \kappa^3 \sigma(\kappa a)}{24\pi N} \tag{6.83}$$

The sum $\mu_s^{el} = \Sigma \nu_i \mu_i^{el}$ is obtained by noting that $\Sigma \nu_i z_i^2$ for a single electrolyte is $\nu \mid z_+ z_- \mid$, and that $\Sigma \nu_i v_i = v_s$. The mean molal activity coefficient is given by $\ln \gamma_\pm = \mu_s^{el}/\nu RT$ so that for a solution of a sing electrolyte

$$\ln \gamma_\pm = -\frac{\mid z_+ z_- \mid e^2}{2k\epsilon_0 \epsilon T}\frac{\kappa}{1 + \kappa a} + \frac{v_s \kappa^3 \sigma(\kappa a)}{24\pi \nu N} \tag{6.84}$$

where the second term is negligible at all concentrations where the Debye–Hückel law is applicable. For water as a solvent at 298.15 K, eq. (6.84) becomes

$$\log \gamma_\pm = -0.509 \mid z_+ z_- \mid I^{1/2}/(1 + 3.29 \times 10^{-3} a\,(\text{pm})\, I^{1/2}) \tag{6.85}$$

It should be pointed out that in the limiting law region, i.e. at $I < 0.001$ M, the denominator of (6.85) is so close to unity that it may be neglected, and $\log \gamma_{\pm(DHLL)}$ (the Debye–Hükel limiting law activity coefficient) is directly proportional to the square-root of the ionic strength. The same limiting law is obtained from differentiation of (6.77), and the general expression is

$$\log \gamma_\pm = -S_\gamma \mid z_+ z_- \mid I^{1/2} \tag{6.86}$$

where the *limiting slope* $S_\gamma = 1.824 \times 10^6 (\epsilon T)^{-3/2}$. Limiting laws which have the same general form

$$t_s - t_s^0 = -S_t \mid z_+ z_- \mid I^{1/2} \tag{6.87}$$

Table 6.1. Some properties of water, and limiting slopes for electrolyte solutions (ref. 5, and for d^* and κ^* from *Water, a Comprehensive Treatise*, edited by F. Franks, Plenum Press, New York, 1972, p. 384).

Quantity	0 °C	20 °C	25 °C	40 °C
d^* (kg m^{-3})	999.840	998.204	997.045	992.215
ϵ^*	88.15	80.36	78.54	73.35
$\epsilon^* T$ (K)	20479	23559	23417	22970
$-(10^6\ d(\ln \epsilon^*)/dT)$ (K)	4657	4594	4579	4541
$10^6 \alpha^*$ (K^{-1})	−67.9	206.6	257.C	385.4
$10^5 \kappa^*$ (Pa^{-1})	50.886	45.894	45.249	44.243
c_p^* (J kg^{-1})	4217.4	4181.6	4179.3	4178.3
S_γ	0.4883	0.5046	0.5091	0.5241
S_h (J mol^{-1})	2130	2773	2962	3598
$10^6 S_v$ (m^3 mol^{-1})	3.7	3.7	3.8	4.0
S_{cp} (J K^{-1} mol^{-1})	27.2	36.4	38.5	45.6
$10^9 S_e$ (m^3 K^{-1} mol^{-1})	−5.0	7.1	9.6	17
$10^{11} S_\kappa$ (m^3 Pa^{-1} mol^{-1})	1.7	1.5	1.5	1.5

are obtained also for derivative functions of the Gibbs energy (eqs. (6.13) to (6.17)), where t_s is any partial molar quantity for the electrolyte at finitely low concentrations where ion–ion interactions are governed solely by the coulombic forces, and t_s^0 is the corresponding reference state (infinite dilution) quantity.

For $t = h_s$ (heat content), $S_h = -2.303\nu RT^2 S_\gamma {}^3\!/\!_2 (T^{-1} + \partial(\ln \epsilon)/\partial T + \alpha/3)$, where α is the coefficient of thermal expansion, arising from $-\partial(\ln I)/\partial T$. for $t = v_s$ (volume),

$$S_v = 2.303\nu RTS_\gamma(1/2)(3(\partial(\ln \epsilon)/\partial P) - \kappa),$$

where $\kappa = -(\partial(\ln v)/\partial P) = \partial(\ln I)/\partial P$ is the coefficient of compressibility. For

$$t = c_{ps} - c_{ps}^0 = [\partial(h_s - h_s^0)/\partial T]_{P,m} \quad \text{(heat capacity)},$$

$$t = e_s - e_s^0 = [\partial(v_s - v_s^0)/\partial T]_{P,m} \quad \text{(expansibility)},$$

and for

$$t = \kappa_s - \kappa_s^0 = [\partial(v_s - v_s^0)/\partial P]_{T,m} \quad \text{(compressibility)},$$

more complicated expressions are obtained, for which the literature should be consulted (see ref. 5, pp. 76–80).

Most of these derivative functions cannot be evaluated for most solvents since the derivatives of the dielectric constant with respect to the temperature and the pressure are unknown. For water, fortunately, the required quantities are known, and S_h, S_v, S_{op}, S_e and S_κ have been evaluated (see ref. 5, pp. 159–163, 171–175), as shown in Table 6.1.

b. Dilute electrolyte solutions

The dilute electrolyte solution range will be regarded as comprising the range from *ca.* 10^{-3} M to *ca.* 1 M in aqueous solutions. Below this range, the DHLL will be obeyed reasonably well for 1:1, 1:2 and 2:1 electrolytes (less well by 2:2 or higher-valency-type electrolytes), and above this range there will be an overlap of hydration shells of individual ions, and phenomena pertaining to concentrated solutions will set in. In general it may be stated that in the dilute range (in aqueous solutions), the following semiempirical expression for the activity coefficient will be valid

$$\log \gamma_\pm = -S_\gamma \mid z_+z_- \mid I^{1/2}(1 + BaI^{1/2})^{-1} + bI \tag{6.88}$$

The coefficients S_γ and B are given exactly by the Debye–Hückel theory (eq. 6.85), and a and b are disposable parameters. The latter are not completely independent, since expansion of the fraction in (6.88) will yield as the first term the limiting law (6.86), as the second term a linear expression $(S_\gamma \mid z_+z_- \mid Ba + b)I$, and then higher terms. It has been shown empirically that in the range considered a good choice for Ba is 1.5 $M^{-1/2}$, and that b is then able to carry the burden of fitting the activity coefficients of many electrolytes.

It has been shown theoretically that the linear coefficient b could arise from several factors. These include dispersion forces and a size effect, contributing non-electrolyte-type terms (cf. (5.11) and (5.33)), the effects of ions on the dielectric constant of the solvent (differentiation of (6.60) with respect to n_s at $c_n = 0$) and solvation effects. The fact that the ions bind some of the solvent, producing solvated ions, which in turn interact further, must modify the consideration of all the other factors, since the size and the surface of the solvated ions will differ from those of the unsolvated ions. The size effect depends on the relative volumes of the solvated ions and the solvent, and the distance of closest approach, a, also depends on the radii of the solvated ions. Furthermore, since the dielectric saturation of the solvent occurs within the solvation shell, and outside it the bulk dielectric constant prevails, there is no more need to consider the effects of the ions on the dielectric constant. The surface of the solvated ions is similar to the solvent, so that excess dispersion interactions should be minimal. Consequently, the Debye–Hückel term (6.85) may be taken to apply even in dilute solutions outside the DHLL range, representing the electrical ion–ion interactions, and a combined solvation-size term may be added to represent the configurational entropy contribution to the excess chemical potential.

The mere fact that h_i solvent molecules are bound to each ion i reduces the amount of *free* solvent[37] per unit mass by $(\Sigma \nu_i h_i) m_s M_w$, producing a term $-\log[1 - (h - \nu) m_s M_w]$, which may be expanded in dilute aqueous solutions to the linear contribution to $\log \gamma_\pm$ of $0.0078(h - \nu)m$. This term is of the correct form, sign and magnitude to represent the empirical term bI, but since h is operationally defined, the tabulated values (Appendix 6.2, p. 251) do not necessarily produce the best fit.

If the size effect is also taken into account, the expression

$$\log \gamma_\pm = 0.4343(\mu_s^{el}/RT) + \log[(1 - hm_sM_w)(1 + m_sM_w v_{hs} v_w^{-1})^{-1}]$$
$$+ 0.4343(h + v_{hs}v_w^{-1})\nu^{-1} m_sM_w(h - \nu + v_{hs}v_w^{-1})(1 + m_sM_w v_{hs} v_w^{-1})^{-1} \tag{6.89}$$

Table 6.2. Values of ionic hydration numbers h_i, and linear coefficients, 100*b* (mol kg^{-1}) from fitting activity coefficients at 25 °C of some uni-univalent electrolytes.

Ion	h_+	o.e.*	OH^-	F^-	NO_3^-	ClO_4^-	Cl^-	Br^-	I^-
h_-			4.0	3.0	0	0.3	0.9	0.9	0.9
d.e.*			— —	—	+	+	+	++	++
H^+	3.9	++			16.6	12.8	10.3	14.7	19.5
Li^+	3.4	++	−10.0		7.2	16.2	9.4	11.0	17.0
Na^+	2.0	+	2.3	−5.7	−6.5	0.1	2.4	3.8	7.0
NH_4^+	0.2				−12.3	−16.7	−2.3	−3.9	
K^+	0.6	—	4.5	1.0	−16.9		−1.4	−1.1	1.0

*o.e. = ordering effect, d.e. = disordering effect on water structure.

is obtained[38]. For the first term eq. (6.83) is used, in which $a = \Sigma r_{ci}(1 + 0.58 h_i v_w v_{hi}^{-1})^{1/3}$ applies the solvation correction to the crystalline radii, and the term involving $\sigma(\kappa a)$ may usually be neglected. If the volumes v_{hi} are taken as independent experimental information, eq. (6.89) becomes a one-parameter equation, the parameter $h = \Sigma \nu_i h_i$ being considered strictly additive. A fit to experimental data thus produces h_i values[38], shown in Table 6.2, that may be compared with those obtained mainly from compressibility and partial molar volumes, shown in Appendix 6.2, p. 251.

Also shown in Table 6.2 are *b* values obtained by fitting (6.88) to experimental data in the range 0.01 to 1.00 m, with the constant parameter $Ba = 1.5$. The fit is within 2 %, but it is seen that the *b* values are not additive with respect to the cation and anion and bear little relationship to the h_i values. (The choice $Ba = 1.5$, however, is consistent with the conventional *pH*.)

Fits obtained for other choices of *Ba* (e.g. 1.0[39a] or 1.6[39b]) also produce *b* values, which are specific for each electrolyte, non-additive for cations and anions, and for $Ba = 1$ have been shown to vary with the concentration[40]. The sign of *b* is of interest, and it can be correlated with the effects the ions have on the structure of water[41] (cf. section 6B(a)). Suppose that a positive quality may be assigned to the structure-ordering effect of cations and to the structure-disordering effect of anions, and vice versa for the negative quality. The sign and relative magnitude of *b* is then determined by the product of the qualities for the cation and the anion (Table 6.2). In other words, if cation and anion are either both structure-making or both structure-breaking, the activity coefficient is lowered, $b < 0$, while if one is structure-making and the other is structure-breaking, $b > 0$ and the activity coefficient is increased. In the latter case, the incompatibility of the hydration environments increases the energy of the system, and the chemical potential of the electrolyte.

In the case of hydrophobic ions, such as the tetraalkylammonium cations, water structure effects may lead to cation–cation interactions. These are manifested, for example, in the fact that $S_v = (\partial \phi_V / \partial m^{1/2})_T > 0$, contrary to ordinary electrolytes where $S_v > 0$. (Only in the DHLL range is S_v for the tetraalkylammonium ions positive and of the expected magnitude[42].) If the water structure enhancement is assumed to extend to, say, three layers of water molecules around each ion, this corresponds to a diameter of 1.5–2.0 nm, to be compared with the average distance apart in a 0.2 M solution of 1.6 nm. Overlap of these low-density structured water layers occurs with an increase of entropy and a decrease of partial molar volume. Since the water structure produced by these hydrophobic ions is incompatible with the oriented water around order-producing anions, such as fluoride, these electrolytes have highly positive *b* values, and conversely with iodides, where *b* values are highly negative. On the qualitative scale of Table 6.2, these hydrophobic cations rate a −−− designation.

A recent consideration[43] challenges (6.88), i.e. the notion that the difference ($\log \gamma_\pm - \log \gamma_\pm^{\mathrm{el}}$) should depend linearly on the concentration. That is, the short-range interactions, interpreted as virial coefficients expressing pair-wise ion interactions (also triple and higher interactions for additional terms at higher concentrations), have a form which makes *b* of (6.88) depend on *I*. The ion distribution function is written in terms of the radial distribution function $g_{ij}(r) = \exp(-q_{ij}(r))$ or in expanded form

$$g_{ij}(r) = 1 - q_{ij}(r) + \tfrac{1}{2}q_{ij}^2(r) - \ldots \tag{6.90}$$

where

$$q_{ij}(r) = (z_i \mathrm{e}/\mathrm{k}T)\psi_j(r) \tag{6.91}$$

and the potential $\psi_j(r)$ is given in (6.71). Rather than proceeding as before, p. 232, with a charging process, the pressure function (2.13) for a fluid is used, which in this case is the osmotic pressure, and gives the osmotic coefficient by division by $c_\mathrm{S}\mathrm{R}T$

$$\phi = 1 - (2/3)\sum_i\sum_j c_i c_j (\Sigma c_i)^{-1} \int_0^\infty r(\partial u_{ij}(r)/\partial r)(\mathrm{k}T)^{-1} g_{ij}(r) 4\pi r^2\, \mathrm{d}r \tag{6.92}$$

If a hard-core potential energy for $r < a$ and the coulombic potential energy for $r \geqslant a$ are used

$$u_{ij}(r < a) = \infty, \qquad \text{and} \qquad u_{ij}(r \geqslant a) = z_i z_j \mathrm{e}^2/\epsilon_0 \epsilon r \tag{6.93}$$

and the fact that $g_{ij}(r < a) = 0$, then for a symmetrical electrolyte, truncating the expansion (6.90) after the third term

$$\phi = 1 - z^2\mathrm{e}^2(6\epsilon_0\epsilon\mathrm{k}T)^{-1}(1 + \kappa a)^{-1} + (2\pi/3)a^3\mathrm{N}c_\mathrm{S} + \\ + (\pi/3)(z^2\mathrm{e}^2/\epsilon_0\epsilon\mathrm{k}T)^2 a(1 + \kappa a)^{-2}\mathrm{N}c_\mathrm{S} \tag{6.94}$$

In (6.94), ϕ is the sum of a first term, which corresponds to the long-range electrostatic effects, and two terms which arise from hard-core, i.e. repulsion, effects. The long-range term reduces to the familiar DHLL as κa becomes small compared with unity, and the other two terms vanish altogether. Equations were derived for ϕ and $\log \gamma_\pm$, which could be fitted to the experimental results of over two hundred electrolytes up to $I = 1$ M, with two universal and two disposable parameters.

A much more detailed statistical mechanical calculation has been developed by Friedman, Rasaiah and co-workers[44]. They employ the McMillan–Mayer theory, and essentially the HNC approximation, section 2B(b), to obtain from an assumed potential function, $u_{ij}(r)$, between pairs of ions (ij being ++, +– and –– for cation–cation, cation–anion and anion–anion pairs) the pair correlation function $g_{ij}(r)$. From this, through the compressibility equation (2.14), the Helmholz free energy of the system is obtained

$$(\partial P/\partial \rho)_{V,T} = kT[1 + 4\pi\rho \int (g(r) - 1)r^2 \, dr]^{-1} \tag{6.95a}$$

$$P(\rho) = \rho kT + \int_0^{\rho} [(\partial P/\partial \rho)_{\rho'} - kT] \, d\rho' \tag{6.95b}$$

$$A^{E}(N,V,T) = N \int_0^{\rho} [P(\rho) - \rho kT] \rho^{-2} \, d\rho \tag{6.95c}$$

However, it is necessary to convert from the McMillan–Mayer system to the practical (Lewis–Randall) system of variables. The former has as independent variables T, the molarity of solute $c_s = \rho/\mathbf{N}$, and the pressure on the solution $P = P_o + P_{osm}$, which is in osmotic equilibrium with the pure solvent, the pressure on which is P_0 (= 1 atmosphere). In the practical system, the independent variables are (T, m_s, P_0) and the appropriate equations for the conversion have been derived[45].

Various models have been tested, with different ion–ion pair potentials $u_{ij}(r)$. The *primitive model* employs a coulombic attraction term $z_i z_j e^2/r$ and a hard-sphere repulsion term COR_{ij} ($COR_{ij} = \infty$ for $r < a_{ij}$ and $COR_{ij} = 0$ for $r \geqslant a_{ij}$, a_{ij} being the sum of the crystal radii), which is the potential that is in fact employed by the Debye–Hückel theory. A simple modification recognizes the effect of solvation, by adding to the core-repulsion term COR_{ij} a term GUR_{ij}, representing the effect of the overlap of solvation spheres first discussed by Gurney[41]

$$\begin{aligned} GUR_{ij} &= A_{ij} \quad \text{for} \quad r < (a_{ij} + a_w) \\ &= 0 \quad \text{for} \quad r \geqslant (a_{ij} + a_w) \end{aligned} \tag{6.96}$$

This represents a square well (or a 'square mound') potential superimposed on the coulombic potential. For a given system, a_{ij} and a_w, the diameter of the solvent molecules, is given, so that only A_{ij} are free parameters, and since usually $A_{++} = A_{--} = 0$ is also specified, there remains a single disposable parameter A_{+-}. The simple three-term potential function

$$u_{ij}(r) = z_i z_j e^2/r + COR_{ij} + GUR_{ij} \tag{6.97}$$

was found to be quite successful for fitting osmotic coefficients, obtained by HNC calculations of P via (6.95) as $(P - P_0)/c_s\mathbf{R}T$ (but it must be converted to practical ϕ) and of activity coefficients.

More sophisticated potential functions have been proposed which correspond even better to physical reality and permit the calculation and prediction of a wide variety of functions for diverse electrolyte–solvent or mixed electrolyte systems. The general form is written symbolically as

$$u_{ij}(r) = z_i z_j e^2/r + COR_{ij} + GUR_{ij} + CAV_{ij} \tag{6.98}$$

$$COR_{ij} = B_{ij}(a_{ij}/r)^9 \tag{6.99}$$

$$GUR_{ij} = A_{ij} V_{ij}(r)/v_w \tag{6.100}$$

$$CAV_{ij} = (z_i r_j^3 + z_j r_i^3)e^2(\epsilon - \epsilon_c)/2\epsilon_0 \epsilon r^4 (2\epsilon + \epsilon_c) \tag{6.101}$$

The core potential COR_{ij} is softened compared with the hard-sphere potential by using a reciprocal power of the distance; the power 9 used is that which is useful for crystal lattice energies. The parameter B_{ij} is inversely proportional to a_{ij}, with a known proportionality constant. The parameter A_{ij} of the Gurney potential continues to be disposable – it is the free-energy change per mole of solvent going from the solvation sphere to the bulk solvent – and V_{ij} is the mutual volume, per mole of solute, of the overlapping hydration spheres, as a function of the distance r, and of the extents of these spheres (cf. Fig. 5.11(b)) obtained geometrically

$$V_{ij}(r) = \pi[-(4r)^{-1}(r_i'^2 - r_j'^2) + (2/3)(r_i'^3 + r_j'^3) - (1/2)r(r_i'^2 + r_j'^2) + (1/12)r^3] \tag{6.102}$$

where r_i' is the sum of r_{ci} and a multiple of r_w, selected most simply as $(r_{ci} + r_w)$, and similarly for r_j'.

There is a third effect, the cavity of low dielectric constant in which the ion is situated, given by CAV_{ij}. In eq. (6.101) ϵ_c is the dielectric constant of this cavity, which may be taken as n_D^2, and r_i and r_j are the crystal radii. The softer core repulsion is particularly evident in that range of r, where the population of pairs of ions of opposite signs $\rho_+\rho_- g_{+-}(r)$ is greatest. The CAV_{ij} term takes over some of the effect of the simpler GUR_{ij} term in (6.97).

c. *Ion association*

Some inadequacies of treatments of dilute electrolyte solutions may be removed by considering ion association, which is a reality if $g_{+-}(r)$ is larger than some critical value. There is good evidence from conductivity and spectrophotometric measurements that ion-pairing occurs for higher-valency-type electrolytes, such as 2:2, at room temperatures, and for lower-valency-type electrolytes at elevated temperatures in aqueous solutions, and for all electrolytes in lower-dielectric-constant solvents. If the electrolyte is dissociated to a degree α, then the molality of the ions i is $m_i' = \nu_i \alpha m_s$, that of the undissociated electrolyte is $m_u = (1 - \alpha)m_s$, and the ionic strength is $\frac{1}{2}\Sigma(\alpha c_i)z_i^2$. The activity of the ions equals the activity of the electrolyte; for a binary (symmetrical) electrolyte, the expression is

$$m_i'^2 \gamma_{i\pm}^2 = a_i = a_s = m_s^2 \gamma_\pm^2 \tag{6.103}$$

hence

$$\gamma_{i\pm} = \gamma_{\pm}/\alpha \tag{6.104}$$

so that the mean activity coefficient of the ionized part, γ_i, is larger than the stoichiometric activity coefficient $\gamma_{\pm}$. The *association constant* can be defined as $K_a = a_u/a_i$, where $a_u = m_u\gamma_u$ is the activity of the unionized part of the electrolyte. This may be rewritten as follows

$$K_a = m_u\gamma_u/m_i'^2\,\gamma_{i\pm}^2 \tag{6.105a}$$

$$= (1-\alpha)\gamma_u/m_s\gamma_{\pm}^2 \tag{6.105b}$$

$$= (1-\alpha)\gamma_u/m_s\alpha^2\,\gamma_{i\pm}^2 \tag{6.105c}$$

Independent experimental determination (e.g. by Raman spectrophotometry) of α, and estimates of γ_u and γ_i lead to a value of K_a, and conversely, a theoretical estimate of K_a leads to values of $\alpha(m_s)$. The activity coefficient γ_i may be estimated from those of completely dissociated electrolytes, and γ_u from the salting properties of non-electrolytes.

The simplest theoretical evaluation of K_a, which is also an operational definition of the undissociated electrolyte or *ion-pair*, considers as paired only ions in actual contact, that is those whose centres are exactly a distance a apart[46]. The electrostatic work $-\,|z_+z_-|\,\mathbf{e}^2/\boldsymbol{\epsilon}_0\epsilon a$ required to bring them from infinity to this distance in the dielectric medium is $\mathbf{N}^{-1}$ times the change in Gibbs energy on association, hence

$$K_a = \exp(-\Delta G_a^0/\mathbf{R}T) = \exp(|\,z_+z_-\,|\,\mathbf{e}^2/\boldsymbol{\epsilon}_0\epsilon a\mathbf{k}T) \tag{6.106}$$

although (6.106) does not take into account the entropy changes accompanying the association. Note that K_a in (6.106) is dimensionless, contrary to that defined in (6.105) and ordinarily employed.

Another approach broadens the concept of ion-pairs to include not only ions in contact, but also solvent-shared ion-pairs (that is oppositely charged ions separated by one solvent molecule), and solvent-separated ion-pairs (where more than one solvent molecules separate the ions), in fact all the oppositely charged ions the mutual interaction energy of which is larger than $2\mathbf{k}T$. Consider Fig. 6.8, where the radial distribution ρ_+r^2 of like-charged ions i and ρ_-r^2 of unlike-charged ions i around a positive ion j of charge $z_j\mathbf{e}$ is plotted as a function of the reduced distance $r\,|\,z_iz_j\,|^{-1}$ (for water as the solvent at 25 °C). The mutual electrostatic effects of the ions i is ignored for this purpose, and only the interaction with the ion j is considered

$$\rho_{\pm}r^2\,\mathrm{d}r = \rho_i^{\infty}\,\exp(-z_jz_{\pm}\mathbf{e}^2/\boldsymbol{\epsilon}_0\epsilon r\mathbf{k}T)r^2\,\mathrm{d}r \tag{6.107}$$

Like-charged ions are seen to be very scarce at short distances, and unlike-charged ions are seen to show a minimum at $r\,|\,z_iz_j\,|^{-1} = \mathbf{e}^2/2\boldsymbol{\epsilon}_0\epsilon\mathbf{k}T$. At this distance the interaction energy equals $2\mathbf{k}T$, hence at shorter distances apart, this operational approach, due to Bjerrum[47], considers the ions to be paired. The fraction of the i

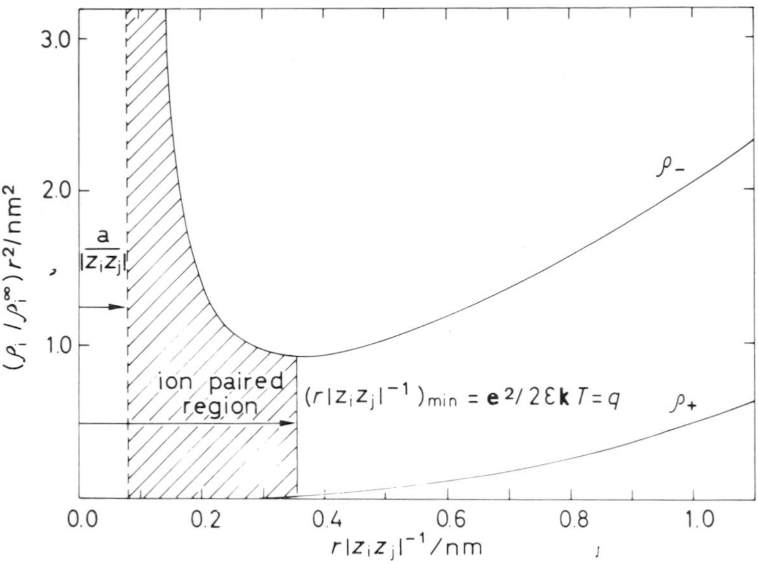

Figure 6.8. The distribution of ions i around a positive ion j, expressed as $(\rho_i(r)/\rho_i(\infty))r^2$ (nm^2) as a function of the charge-normalized distance $r|z_iz_j|^{-1}$ (nm). Note for the unlike-charged (negative) ions the ion-paired region between $a/|z_iz_j|$ and q, the latter representing the abscissa at the minimum of ρ_-.

ions paired with the ion j is their probability of being within the volume extending from contact, at $r = a$, to the critical distance $q = |z_iz_j|\,\mathbf{e}^2/2\epsilon_0\epsilon\mathbf{k}T$, and since the concentration of ions j is $\rho_j = 1000\mathbf{N}c_s\ \mathrm{m}^{-3}$, the fraction of associated ions is

$$1 - \alpha = 4000\pi\mathbf{N}c_s \int_a^q \exp(-|z_+z_-|\,\mathbf{e}^2/\epsilon_0\epsilon r\mathbf{k}T)r^2\,\mathrm{d}r \tag{6.108}$$

For water at 25 °C $q = 0.357\,|z_+z_-|$ *nm;* for other conditions it is $8360\,|z_+z_-|/\epsilon T$ nm. It is now convenient to change variables: $u = |z_+z_-|\,\mathbf{e}^2/\epsilon_0\epsilon\mathbf{k}T$, $t = u/r$ and $b = u/a$ (and $2 = u/q$), so that (6.108) becomes

$$1 - \alpha = 4000\pi\mathbf{N}c_s u^3 \int_2^b t^{-4}\exp(t)\mathrm{d}t$$

$$= 4000\pi\mathbf{N}c_s u^3 Q(b) \tag{6.109}$$

The fraction $1 - \alpha$ may be converted into K_a by the use of (6.105), since at infinite dilution γ_u, α and γ_i are all expected to approach unity, and m_s to approach c_s/d^*, so that $K_a = (1 - \alpha)d^*/c_s = 4000\pi\mathbf{N}d^*u^3Q(b)$. At 25 °C

$$\log b = \log|z_+z_-| + 1.746 - \log\epsilon - \log a\ (a \text{ in nm}) \tag{6.110a}$$

$$\log K_a\ (K_a \text{ in kg mol}^{-1}) = 3\log|z_+z_-| + 6.120 + \log d^* - 3\log\epsilon + \log Q(b) \tag{6.110b}$$

Values of $\log Q(b)$ have been tabulated, and are shown graphically in Fig. 6.9. Thus,

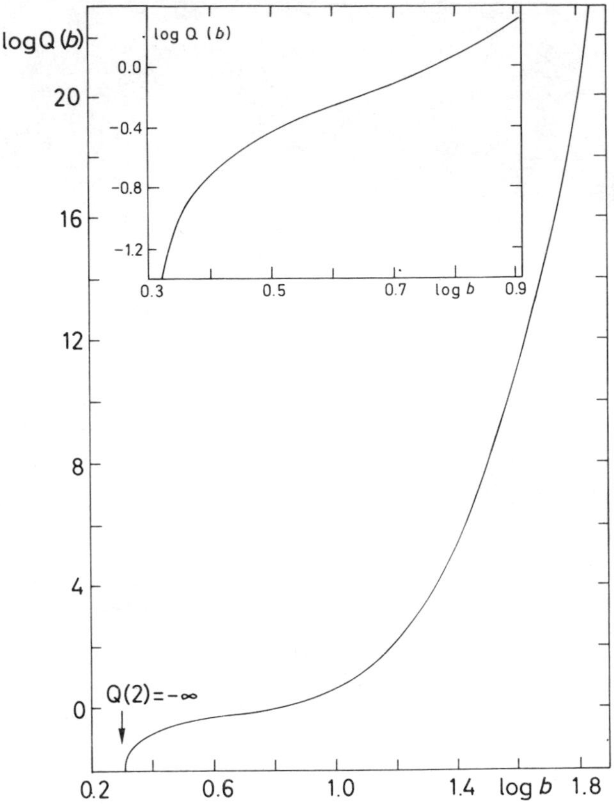

Figure 6.9. The function log Q (log b) for the integral of eq. (6.109). The insert is a magnification for low values of b.

for a given solvent (fixing ϵ) and temperature, the association constant of an electrolyte is determined by the charges of the ions and the distance parameter a. It should be noted that for water at 25°C q for 1:1 electrolytes is smaller than all reasonable a values, hence these would not be associated. Since the product ϵT falls with increasing temperatures, association of even 1:1 electrolytes becomes possible in water at elevated temperatures. Association ceases when b falls to 2 or below, with increase in either a or ϵ, and for room temperature and reasonable $a \geqslant 0.5$ nm the limit of ϵ, above which there is no association, is about 56 for a 1:1 electrolyte.

Contrary to the region $r < q$, where ion-pairing occurs, in the region $r > q$ the ordinary Debye–Hückel ionic atmosphere prevails, that is for the free ions

$$\log \gamma_{i\pm} = -S_\gamma |z_+z_-|(\alpha I)^{1/2}/(1 + Bq(\alpha I)^{1/2}) + bI \tag{6.111}$$

replaces (6.88). The whole of I, not only αI, is expected to contribute to the configurational entropy term. The activity of the non-ionized part is expected to be given by a Setchenov-type equation (cf. (6.55)): $\log \gamma_u = b'I$. Several iterations may be necessary to obtain α from K_a, (6.105), (6.111) and the last expression, with reasonable estimates of b and b', for any given m_S.

Since the operational definition of ion-pairs depends on the arbitrary choice of q, the value of r at the minimum of the unlike-charged ion distribution curve (Fig. 6.8), where the electrostatic energy is twice kT, the calculated values of K_a and α suffer from the same abitrariness. They may be upper limits for the actual ion-pairing in the solution, in the same sense as contact pairs, eq. (6.100), may be considered the lower limit. Experimental techniques provide various intermediate answers depending on their sensitivity to the closeness of interaction of the ions assumed to be paired and on the properties attributed to the free ions. Stronger interaction than electrostatic ion-pairing, i.e. complex formation, cannot be treated here (see Chapter 3 of reference 16).

d. Concentrated aqueous electrolyte solutions

Concentrated electrolyte solutions, above say 1 M, have not been studied extensively in any solvent besides water, and even in this solvent much information is still lacking. In water, a 1 M solution contains a large excess of solvent molecules; even for quite voluminous electrolytes, with ϕ_S^∞ (cm^3) = 400, there would be 600 cm^3 or ca. 33 moles of water per mole of electrolyte. This is, however, not so for solvents with larger molar volumes: a 1 M solution of tetrapentylammonium thiocyanate in *p*-cymene, for example, contains fewer than 4 moles of solvent per mole of electrolyte. The question of whether volume concentrations (volume fractions) or mole fraction statistics should be applied to the configurational entropy in such solutions is thus acute, but has unfortunately received no adequate answer yet.

Returning to aqueous solutions, the hydration approach has been shown to give reasonable fits of activity coefficients up to *ca.* 4 M solutions with one concentration-independent parameter h, and independent (infinite-dilution) hydrated molar volume data. However, in a 4 M solution ions are on average 0.6 nm apart, and diameters of hydrated ions are 0.4–0.8 nm, so that at these and at higher concentrations hydration shells overlap considerably and h can no longer be considered a constant (cf. (6.51)). Two other approaches have shown promise, but have not yet been applied extensively.

The cube-root approach, based on a disordered-lattice model, actually predated the Debye–Hückel theory, and is found to express the *electrostatic* ion–ion interactions better than the Debye–Hückel equation in the range 0.1–4 M, while the other interactions must again be taken into account with a linear term[48]. It should be realized that at lower concentrations the cube-root law must change into a square-root law of the DHLL form. It has been shown recently[49] that a mathematical approximation permits replacement of (6.74) for the electrostatic contribution to the Gibbs energy by

$$G^{el} = -V\Sigma\rho_j^0 z_j^2 e^2 (12\epsilon_0\epsilon a)^{-1}(\kappa a)(4 + \kappa a)(1 + \kappa a)^{-1} \tag{6.112}$$

This expression becomes identical with (6.74) at low (κa), but at $\kappa a > 2$, that is at

25 °C for $I > 0.37\ a^2$ (a in nm)

$$G^{el} = -(200/3)\pi^{1/3}\mathbf{e}^{8/3}\mathbf{N}^{8/3}\mathbf{R}^{-1/3}V\nu(\Sigma\nu_i z_i^2)^{4/3}(\epsilon_0\epsilon)^{-4/3}T^{-1/3}a^{-1/3}c_s^{4/3} \quad (6.113)$$

Differentiation of (6.112) or of (6.113) with respect to n_s and to W_w gives the corresponding values of log $\gamma_\pm^{el}$ and of ϕ^{el}, and (6.113) clearly shows the cube-root dependence of log $\gamma_\pm^{el}$ on the concentration. The expressions for $I > 0.37\ a^2$ (a in nm) are, at 25 °C

$$\log \gamma_\pm^{el} = -0.1916|z_+z_-|(Ia^{-1})^{1/3}(0.75 + 0.25c_s m_s^{-1}) \quad (a \text{ in nm}) \quad (6.114)$$

where the last factor is near unity, and

$$\phi^{el} = -(1/9)|z_+z_-|(Ia^{-1})^{1/3}c_s m_s^{-1} \quad (a \text{ in nm}) \quad (6.115)$$

These expressions were shown[49] to be applicable also for higher-valency-type electrolytes (such as $MgSO_4$, $K_3Fe(CN)_6$ and $Al_2(SO_4)_3$) and for elevated temperatures (up to 275 °C for NaCl) which are troublesome with Debye–Hückel expressions. The test of the applicability however, is that they yield *linear* terms for the dependence of (log $\gamma_\pm$ – log $\gamma_\pm^{el}$) on the concentration, for which there exist good reasons for the above-mentioned range, as discussed above for dilute solutions.

The second approach starts from the multilayer adsorption theory (BET)[50], and considers hydration layers as if they were adsorbed on the ions[51], recognizing the decrease in hydration as the concentration increases (cf. eq. (6.51)). The ratio of the pressure to the saturation pressure in the BET theory is replaced by a_w, and the monomolecularily sorbed volume is replaced by $h_1 m_s$, where h_1 is the (fixed) hydration number of the first layer. The constants $C = \exp[(E_1 - E_w^L)/\mathbf{R}T]$ and $K = \exp[-(E_h - E_w^L)/\mathbf{R}T]$, where E_1 is the binding energy of water in the first layer, E_h that in any subsequent hydration layer, and $E_w^L = u_w^V$ is the liquefaction energy of water, adopted from the BET theory, are used as arbitrary parameters to fit the data to the resulting equation

$$0.018m_s a_w/(1 - a_w) = (CKh_1)^{-1} - (C - 1)(Ch_1)^{-1}a_w \quad (6.116)$$

Here the hydration number h_1 is fixed at an arbitrarily selected value. This is 4 for univalent cations and 8 for *bona fide* (i.e. non-associating) divalent cations, while anions are not considered hydrated. In the fits attempted[51] good linearity of the left-hand side of (6.116) with a_w was obtained for values $a_w < 0.4$, corresponding to molarities in the range 8–20 for 1:1 and 4–10 for 2:1 electrolytes. Fits over a temperature range for HCl, LiBr, $LiNO_3$ and $Ca(NO_3)_2$ show the validity of the approach, and yield reasonable values of $E_1 - E_w^L$ of 8–12 kJ (mol water)$^{-1}$. The constant K is near unity and corresponds to a bonding of water beyond the first layer weaker by only *ca.* 0.5 kJ (mol water)$^{-1}$. This approach is thus seen to be empirically satisfactory, bridging the gap between moderately concentrated solutions, where hydration treatments with constant h work, and molten hydrated salts, where the quasi-lattice treatments of molten salts (section 7D(a)) works best.

e. Mixed electrolytes

The best way to deal with mixtures of electrolytes is to consider the excess thermodynamic functions in the mixture over and above the excess functions in the binary (single electrolyte–solvent) solutions. The mixing can take place under several conditions: mixing at constant total molality $m = \Sigma m_i$, osmolality $m' = \Sigma(\nu_i/\nu)m_i$ or ionic strength $I = \frac{1}{2}\Sigma m_j z_j^2$ (i refers to electrolyte and j to individual ions), or at constant solvent activity. The latter condition is observed in the common isopiestic experimental method, where the molalities of solutions of single electrolytes (reference solutions) are compared with compositions and molalities of mixtures at a common equilibrium solvent vapour pressure. In the following, only ternary (mixed) solutions of 1:1 electrolytes with a common ion will be considered: mixtures of MX and NX. Broadening of the considerations to include higher-charge types, unsymmetrical electrolytes, or mixtures with no common ions is no trivial matter, since some interactions and terms which cancel for symmetrical common-ion ternary systems must be included, and these complicate the equations considerably.[4,5,22,43,52,53]

At constant ionic strength I, and for a fraction $y = m_{MX}/(m_{MX} + m_{NX}) = I_{MX}/I$ of electrolyte MX, the excess Gibbs energy of mixing is

$$\Delta g^E = g^E(y, I) - y g_{MX}^{E\circ}(I) - (1 - y) g_{NX}^{E\circ}(I) \tag{6.117}$$

This can be expressed as a power series in y, similarly to the Redlich–Kister equation for non-electrolytes (4.65)

$$\Delta g^E = \mathbf{R}Ty(1 - y)I^2 \Sigma g_p (1 - 2y)^p \quad (p = 0, 1, \ldots) \tag{6.118}$$

The accuracies of experimental data rarely warrant terms with $p \geqslant 2$. The excess Gibbs energy of the solvent is given[4,45,52] by the difference in osmotic coefficients

$$\begin{aligned}\Delta\phi &= \phi - y\phi_{MX}^{\circ} - (1 - y)\phi_{NX}^{\circ} \\ &= \tfrac{1}{2}y(1 - y)I[g_0 + (\partial g_0/\partial(\ln I)) + \{g_1 + (\partial g_1/\partial(\ln I))\}(1 - 2y)]\end{aligned} \tag{6.119}$$

The excess Gibbs energies of the two electrolytes are given by their mean activity coefficients, e.g. for NX.

$$\begin{aligned}\ln \gamma_{\pm NX} = \ln \gamma_{\pm NX}^{\circ} &+ y(\phi_{NX}^{\circ} - \phi_{MX}^{\circ}) + \{\tfrac{1}{2}[(g_0 + g_1) + \partial(g_0 + g_1)/\partial(\ln I)] \\ &- \tfrac{1}{2}y[\partial(g_0 + 3g_1)/\partial(\ln I)] - y^2[g_1 - \partial g_1/\partial(\ln I)]\}yI\end{aligned} \tag{6.120}$$

and similarly for $\gamma_{\pm MX}$ only with a negative second term on the right and $(1 - y)$ substituted everywhere for y. In the special case where all $g_p = 0$ for $p \geqslant 1$, the expressions become considerably simplified

$$\Delta\phi = \tfrac{1}{2}y(1 - y)\{g_0 + (\partial g_0/\partial(\ln I))\}I \tag{6.121}$$

$$\log \gamma_{\pm MX} = \log \gamma_{\pm MX}^{\circ} + 0.4343(1 - y)(\phi_{MX}^{\circ} - \phi_{NX}^{\circ}) + \frac{0.4343}{2} g_0(1 - y)I \tag{6.122a}$$

$$\log \gamma_{\pm NX} = \log \gamma^{\circ}_{\pm NX} + 0.4343y(\phi^{\circ}_{NX} - \phi^{\circ}_{MX}) + \frac{0.4343}{2} g_0 yI \tag{6.122b}$$

In this case it can be shown that if α_M and α_N are defined so that $(\alpha_M + \alpha_N) = -0.4343g_0$ and $(\alpha_M - \alpha_N) = 2 \times 0.4343(\phi^{\circ}_{NX} - \phi^{\circ}_{MX})/I$, then α_M and α_N serve as the Harned coefficients[51]

$$\log \gamma_{\pm MX} = \log \gamma^{\circ}_{\pm MX} - \alpha_M(1-y)I \tag{6.123a}$$

$$\log \gamma_{\pm NX} = \log \gamma^{\circ}_{\pm NX} - \alpha_N yI \tag{6.123b}$$

The coefficients g_p are obtained from experimental data $\Delta\phi$ or $\ln \gamma_{\pm i}$ as power series in I, and if either $\Delta\phi$ or $\ln \gamma_{\pm i}$ for one electrolyte has been measured as a function of y and I, the other two quantities can be obtained via the g_p coefficients, or by a method involving an integration from $I = 0$ to I (the McKay–Perring method or its modifications[5,22,53]).

Other thermodynamic functions, such as the heat of mixing or the volume change of mixing of two electrolyte solutions, can be cast into a similar form as (6.118) to give[54].

$$\Delta\psi^E = \psi^E - y\psi^E_{MX} - (1-y)\psi^E_{NX} = y(1-y)I\Sigma\psi_p(I)(1-2y)^p \tag{6.124}$$

with $\psi/\mathbf{R}T$ representing g, $h = [\partial(g/T)/\partial(1/T)]_P$, $v = (\partial g/\partial P)_T$, etc.

The molecular significance of g_p or its derivatives is of interest since g_p describes the $\geqslant (p+2)$th order interactions of ions in the electrolyte mixture. That is, if, as commonly observed, $g_p = 0$ for $p > 1$, only second- and third-order interactions need be taken into account. Second-order interactions involve the ion combinations MM, MN and NN, while third-order ones include MMX, MNX, NNX, MXX and NXX*. For the case $g_p = 0$ for $p \geqslant 1$, θ of the Pitzer treatment[43] becomes $\frac{1}{2}g_0$ of the Friedman treatment[4] (eq. (6.119)) or $\frac{1}{2}B^{\circ}/I$ of the Scatchard treatment[52] for $\Delta\phi$. If also g_1 is finite, ternary terms as listed above must be considered (the last two cancel out in the excess Gibbs energy of mixing or $\Delta\phi$ expressions). In rare cases where g_p with $p>1$ must be included in order to describe the $\Delta g^E(y, I)$ behaviour, higher-order interactions must occur, as is indeed found at high concentrations for solutions involving tetraalkylammonium ions. The cluster expansion theory of McMillan and Mayer, as developed by Friedman[4,44] gives explicit expressions for g_p, and for the hard-core potential (6.93) (the 'primitive model') yields a limiting slope lim $(I \to 0)$ $d(\ln g_0)/d(I^{1/2}) = 2.356$ for water at 25 °C, i.e. a definitely positive quantity. Measurements do not reach to sufficiently low concentrations to test this, indeed many mixtures show negative slopes in the accessible range $I > 0.1$ M.

*Fourth- and higher-order interactions need rarely be considered at the accessible I values. Even for symmetrical electrolytes, however, terms in third-order interactions enter g_0, although they are unimportant at low I. Interactions of three ions of the same sign (e.g. MMM or NNN) seem to be unimportant in all cases. For non-symmetrical electrolyte mixtures (e.g. a 2:1 and a 1:1 salt), the terms in the triplet interactions become relatively more important in g_0.

On a more empirical level, a quite different approach is based on the hydration of dilute eletrolyte solutions, p. 237[37,55]. For a binary (single) electrolyte of 1:1 type in aqueous solutions the equation

$$\log \gamma_\pm = \log f_{DH} - \tfrac{1}{2}h \log a_w - \log(1 + 0.018(2 - h))m \tag{6.125}$$

is written, where the first term on the right represents the Debye–Hückel term (6.85) for the electrostatic interactions, the second represents the second term on the right-hand side of (6.84), assuming $v_{hs} = hv_w^*$ (cf. 6.80), and the third arises from the binding of a part of the water. For a ternary electrolyte solution, the left-hand side of (6.125) is replaced by $y \log \gamma_{MX} + (1 - y) \log \gamma_{NX}$. The individual ionic activity coefficient (note that $\tfrac{1}{2}\log a_w = -\tfrac{1}{2}\, 0.4343 \times 0.018 vm\phi = -0.00782 m\phi$)

$$\log \gamma_M = y \log \gamma_{MX} + (1 - y)\log \gamma_{NX} + 0.00782\,[(2 - y)h_M - (1 - y)h_N - h_X]\, m\phi$$

$$\log \gamma_N = y \log \gamma_{MX} + (1 - y)\log \gamma_{NX} + 0.00782 - yh_M + (1 + y)h_N - h_X]\, m\phi$$

$$\log \gamma_X = y \log \gamma_{MX} + (1 - y)\log \gamma_{NX} + 0.00782[-yh_M - (1 - y)h_N + h_X]m\phi \tag{6.126}$$

are obtained, which can be combined to give

$$\log \gamma_{MX} = \log \gamma_{NX} + 0.00782(h_M - h_N)m\phi \tag{6.127}$$

Harned's coefficient α_M and α_N (6.123) can be obtained by going to the limits $y = 0$ and $y = 1$, whereby ϕ is eliminated

$$\alpha_M = m^{-1} \log(\mathring{\gamma}_{MX}/\mathring{\gamma}_{NX}) - 0.00782 h_X \mathring{\phi}_{NX} \tag{6.128}$$

and α_N is obtained by replacing MX everywhere by NX and *vice versa*. Since ϕ of the mixture can be calculated from the α values

$$\phi = \mathring{\phi}_{NX} + (2.303/2)ym[y(\alpha_M + \alpha_N) - 2\alpha_M] \tag{6.129}$$

it is seen that the activity coefficients for the electrolytes in the mixture are expressed solely in terms of data for the binary solutions: $\mathring{\gamma}_{MX}$, $\mathring{\gamma}_{NX}$, $\mathring{\phi}_{MX}$ and $\mathring{\phi}_{NX}$ at concentration m, and the ionic hydration numbers h_M, h_N and h_X. In certain cases, it was assumed[55] that one of the hydration numbers h_M (or h_N) or h_X vanishes, and simpler equations result. Similar equations to (6.126) hold for common cation mixtures, MX–MY (with replacements of M by X, N by Y and X by M) with a change of sign before the last term.

D. Appendixes: Some properties of aqueous ions

Appendix 6.1. Conventional partial molar entropy s_i^0, volume v_i^0, Gibbs energy of hydration, $g_{i\,hyd\,con}^0$, enthalpy of hydration $h_{i\,hyd\,con}^0$ and entropy of hydration $s_{i\,hyd\,con}^0$ for selected ions at 298.15 K (25 °C).

Ion (aq)	$s_{i\,con}^0$ (J K^{-1} mol^{-1})	$v_{i\,con}^0$ (cm^3 mol^{-1})	$g_{i\,hyd\,con}^0$ (kJ mol^{-1})	$h_{i\,hyd\,con}^0$ (kJ mol^{-1})	$s_{i\,hyd\,con}^0$ (J K^{-1} mol^{-1})
H^+	65.3	0	−1522.91	−1536.20	−43.55
Li^+	79.5	−0.88	−944.0	−959.9	−53.39
Na^+	125.7	−1.21	−844.5	−850.7	−20.42
K^+	167.8	9.02	−770.1	−776.2	13.35
Rb^+	185.3	14.07	−748.9	−741.1	25.1
Cs^+	198.4	21.34	−716.7	−708.4	28.0
Cu^+	105.9		−1003.3	−1039.2	−120.5
Ag^+	138.0	−0.7	−912.0	−920.3	−27.9
Au^+			−1040.1	−1020.	
Tl^+	192.5	10.6	−776.2	−771.2	17.70
NH_4^+	178.3	17.86		−761.	
$(CH_3)_4N^+$		89.57	−615.6	−625.0	
$(C_2H_5)_4N^+$		149.12			
$(C_3H_7)_4N^+$		214.44			
$(C_4H_9)_4N^+$		275.66			
Be^{2+}		−12.0	−3307.9	−3377.5	−230.
Mg^{2+}	12.6	−21.17	−2771.9	−2812.6	−136.0
Ca^{2+}	75.4	−17.85	−2459.2	−2482.9	−79.29
Sr^{2+}	91.2	−18.16	−2292.8	−2335.2	−142.3
Ba^{2+}	143.1	−12.47	−2185.3	−2193.0	−29.3
Mn^{2+}	56.9	−17.7	−2697.4	−2735.4	−127.6
Fe^{2+}	−7.4	−29.7	−2761.9	−2810.4	−162.8
Co^{2+}	17.6	−24.0	−2883.	−2945.	−207.5
Ni^{2+}	1.6	−24.0	−2933.4	−2996.2	−210.5
Cu^{2+}	30.8	−27.76	−2947.5	−2990.5	−144.1
Zn^{2+}	23.7	−21.6	−2893.6	−2934.8	−136.82
Cd^{2+}	69.5	−20.0	−2667.0	−2696.3	−98.24
Hg^{2+}	98.4	−19.3	−2691.3	−2711.1	−66.53
Sn^{2+}	113.		−2425.5	−2442.7	−57.7
Pb^{2+}	141.1	−15.5	−2363.2	−2370.2	−23.4
Al^{3+}	−117.6	−42.2	−5914.	−5995.	−269.4
Sc^{3+}			−5237.	−5296.	−197.
Y^{3+}			−4904.	−4955.	−172.
La^{3+}	33.	−39.10	−4569.	−4618.	−166.9
Ce^{3+}				−4888.	
Cr^{3+}	−112.	−39.5	−5648.	−5735.	−293.
Fe^{3+}	−120.	−43.7	−5632.	−5712.	−268.
Ga^{3+}	−135.		−5927.	−6021.	−314.2
In^{3+}	45.		−5410.	−5479.	−231.4
Tl^{3+}	3.		−5393.	−5519.	−423.8
Ce^{4+}				−8269.	
Th^{4+}		−53.5		−8201.	
UO_2^{2+}	59.			−2233.	
F^-	−74.9	−1.16	69.5	3.4	−221.8
Cl^-	−9.9	17.83	125.5	76.6	−164.0
Br^-	15.5	24.71	154.4	110.5	−147.3
I^-	43.9	36.22	191.6	153.9	−126.4
OH^-	−75.8	−4.04	−5.	−74.	−232.
SH^-	−4.1			108.	
BF_4^-	102.	44.18		219.	

Appendix 6.1. (*continued*)

Ion (aq)	s_i^0 con ($J\ K^{-1}\ mol^{-1}$)	v_i^0 con ($cm^3\ mol^{-1}$)	g_i^0 hyd con ($kJ\ mol^{-1}$)	h_i^0 hyd con ($kJ\ mol^{-1}$)	s_i^0 hyd con ($J\ K^{-1}\ mol^{-1}$)
NO_3^-	81.1	29.00		112.7	
NO_2^-	59.8	26.2		76.	
SCN^-	144.3	35.7		132.	
HCO_2^-	26.3	26.27		19.	
$CH_3CO_2^-$	21.7	40.46		19.	
HCO_3^-	29.7	23.4		49.	
ClO_3^-	97.9	36.66		127.4	
BrO_3^-		35.3		66.	
IO_3^-		25.3		−33.0	
ClO_4^-	115.4	44.12		199.	
HSO_4^-	62.4	35.67		153.	
$H_2PO_4^-$	23.8	29.1			
S^{2-}	−108.4	−8.2	−403.8	−482.1	−262.7
SO_4^{2-}	−126.5	13.98			
CO_3^{2-}	−183.7	−4.3		−109.	
HPO_4^{2-}	−166.6	7.7			
$C_2O_4^{2-}$	−86.2	16.0			
SO_3^{2-}		8.9			
$S_2O_3^{2-}$		34.0			

Appendix 6.2. Crystal radii r_{ci}, molar intrinsic ionic volumes v_{ci}, absolute partial molar ionic volumes $v_{i\,abs}^0$, radii of hydrated ions r_{hi}^0, molar hydrated ionic volumes v_{hi}^0, and hydration numbers h_i^0 of selected ions at infinite dilution at 298.15 K.

Ion (aq)	r_{ci} (pm)	r_{hi}^0 (pm)	v_{ci} (cm^3)	$v_{i\,abs}^0$ (cm^3)	v_{hi}^0 (cm^3)	h_i^0
H^+			(0.0)	−4.7		
Li^+	60	200	2.0	−5.6	34.6	2.2
Na^+	95	228	6.4	−5.9	52.1	3.2
K^+	133	217	15.3	4.3	44.2	2.2
Rb^+	148	224	19.4	9.4	49.4	2.2
Cs^+	169	207	25.5	16.6	38.7	1.2
Cu^+	96					
Ag^+	126	197	10.1	−5.4	34.2	2.2
Tl^+	144	218	21.5	5.9	46.3	2.2
NH_4^+	148	159	17.4	13.2		2.0
$(CH_3)_4N^+$	347	367		84.9		
$(C_2H_5)_4N^+$	400	400		144.4		
$(C_3H_7)_4N^+$	452	452		209.3		
$(C_4H_9)_4N^+$	494	494		271.0		
Be^{2+}	31			−21.4		
Mg^{2+}	65	309	2.1	−30.6	126.0	8.7
Ca^{2+}	99	310	7.3	−27.3	128.6	8.7
Sr^{2+}	113	310	11.2	−27.6		8.7
Ba^{2+}	135	313	18.8	−21.9	134.4	8.7
Mn^{2+}	80		4.0	−27.1		5.2
Fe^{2+}	75		3.3	−34.1		6.1
Co^{2+}	72		2.9	−33.4		5.9
Ni^{2+}	70		2.8	−33.4		5.8

Appendix 6.2. (*continued*)

Ion (aq)	r_{ci} (pm)	r^0_{hi} (pm)	v_{ci} (cm^3)	$v^0_{i\,abs}$ (cm^3)	v^0_{hi} (cm^3)	h^0_i
Cu^{2+}	70			−37.2		
Zn^{2+}	74		3.1	−31.0		5.5
Cd^{2+}	97	264	6.9	−29.4	80.0	6.1
Hg^{2+}	110			−28.7		
Sn^{2+}	93					
Pb^{2+}	132		13.5	−24.9		8.2
Al^{3+}	50		1.0	−56.3		7.4
Sc^{3+}	81					
Y^{3+}	93					
La^{3+}	115	332	11.6	−53.2	158.8	11.8
Ce^{3+}	107					
Cr^{3+}	62		2.1	−53.6		7.4
Fe^{3+}	60	457	1.7	−57.8		7.8
Ga^{3+}	62					
In^{3+}	81					
Tl^{3+}	95					
Ce^{4+}	94					
Th^{4+}	99			−72.3		
UO_2^{2+}						
F^-	136	306	7.7	3.5	125.5	6.8
Cl^-	181	255	20.7	22.5	72.9	2.8
Br^-	195	242	27.2	29.4	61.9	1.8
I^-	216		37.1	40.9		
OH^-	140	318	6.4	0.7	139.5	7.7
SH^-	195					
BF_4^-	228			48.9		
NO_3^-	189	223	32.0	33.7	48.1	0.8
NO_2^-	155			30.9		
SCN^-	195	212	41.3	40.4	41.3	~0
HCO_2^-	158			31.0		
$CH_3CO_2^-$	159	296	35.9	45.2	113.3	3.8
HCO_3^-		241	24.5	28.1	60.5	1.8
ClO_3^-	200	256	38.7	41.4	73.2	1.8
BrO_3^-	191	272	40.8	40.0	88.2	2.7
IO_3^-	182	327	30.4	30.0	152.8	6.8
ClO_4^-	236	243		48.8		
HSO_4^-				40.4		
$H_2PO_4^-$				33.8		
SO_4^{2-}	230	364	38.4	23.4	214.0	7.2
CO_3^{2-}	185	394	34.	5.1	267.2	13.8
HPO_4^{2-}				17.1		
$C_2O_4^{2-}$				25.4		
SO_3^{2-}				18.3		
$S_2O_3^{2-}$				43.4		

References

1. W. Davis, Jr. and H. J. DeBruin, *J. Inorg. Nucl. Chem.*, **26**, 1069 (1964).
2. F. R. Longo, P. H. Daum, R. Chapman and W. G. Thomas, *J. Phys. Chem.*, **71**, 2755 (1969).
3. J. Braunstein, in *Ionic Interactions, Vol. 1*, edited by S. Petrucci, Academic Press, New York, 1971, p. 180.

4. H. L. Friedman, *J. Chem. Phys.*, **32**, 1351 (1960).
5. H. Harned and B. B. Owen, *The Physical Chemistry of Electrolyte Solutions*, Reinhold, New York, 3rd ed., 1958.
6. A. J. Parker, *Quart. Rev.*, **16**, 163 (1962).
7. R. H. Gurney, *Ionic Processes in Solution*, McGraw-Hill, New York, 1953.
8. H. S. Frank and M. W. Evans, *J. Chem. Phys.*, **3**, 507 (1945); H. S. Frank and W.-Y. Wen, *Discuss. Faraday Soc.*, **24**, 133 (1957); H. S. Frank, *Z. Phys. Chem. (Leipzig)*, **228**, 367 (1965).
9. D. C. Grahame, *J. Chem. Phys.*, **21**, 1054 (1953); F. Booth, *J. Chem. Phys.*, **19**, 391, 1327, 1615 (1951); J. Malsh, *Phys. Z.*, **29**, 770 (1928); **30**, 837 (1929).
10. J. Padova, *J. Chem. Phys.*, **39**, 1552 (1963); *Electrochim. Acta*, **12**, 1227 (1967); cf. also H. S. Frank, *J. Chem. Phys.*, **23**, 2023 (1955); K. J. Laidler and C. Pegis, *Proc. Roy. Soc., Ser. A*, **241**, 80 (1957); B. E. Conway, R. E. Verrall and J. E. Desnoyers, *Z. Phys. Chem. (Leipzig)*, **230**, 157 (1965); K. J. Laidler and J. S. Muirhead-Gould, *Trans. Faraday Soc.*, **63**, 953 (1967).
11. O. Ya. Samoilov, *Structure of Electrolyte Solutions and the Hydration of Ions*, (English translation), Consultants Bureau, New York, 1965.
12. R. M. Noyes, *J. Chem. Educ.*, **40**, 2 (1963).
13. D. R. Rosseinsky, *Chem. Rev.*, **65**, 467 (1965).
14. J. E. Desnoyers and C. Jolicoeur, in *Modern Aspects of Electrochemistry, Vol. 5*, edited by J. O'M. Bockris and B. E. Conway, Butterworths, London, 1969, p. 1.
15. F. J. Millero, *Chem. Rev.*, **71**, 147 (1971).
16. Y. Marcus and A. S. Kertes, *Ion Exchange and Solvent Extraction of Metal Complexes*, Wiley–Interscience, London, 1969, pp. 10–33.
17. R. M. Noyes, *J. Am. Chem. Soc.*, **84**, 573 (1962); **86**, 971 (1964); R. H. Stokes, *J. Am. Chem. Soc.*, **86**, 979 (1964).
18. E. Glueckauf, *Trans. Faraday Soc.*, **60**, 572, 1637 (1964); **61**, 914 (1965); **64**, 2423 (1968).
19. H. F. Halliwell and S. C. Nyburg, *Trans. Faraday Soc.*, **59**, 1126 (1963).
20. B. S. Gourary and F. J. Adrian, *Solid State Phys.*, **10**, 127 (1970); M. J. Blandamer and M. C. R. Symons, *J. Phys. Chem.*, **67**, 1304 (1963); D. F. C. Morris, *Structure and Bonding*, **4**, 63 (1968).
21. K. H. Stern and E. S. Amis, *Chem. Rev.*, **59**, 1 (1959).
22. R. A. Robinson and R. H. Stokes, *Electrolyte Solutions*, Butterworth, London, 2nd ed., 1959; cf. also R. E. Nightingale, *J. Phys. Chem.*, **63**, 1381 (1959).
23. J. Padova, *J. Chem. Phys.*, **40**, 691 (1964).
24. N. Bjerrum, *Z. Anorg. All. Chem.*, **109**, 275 (1920); G. Scatchard, *J. Am. Chem. Soc.*, **43**, 2406 (1921); **47**, 2098 (1925).
25. E. Glueckauf, *Trans. Faraday Soc.*, **51**, 1235 (1955); D. G. Miller, *J. Phys. Chem.*, **60**, 1296 (1956); B. E. Conway and R. E. Verrall, *J. Phys. Chem.*, **70**, 1473 (1966).
26. Ref. 16, pp. 49–54.
27. E. R. Nightingale, Jr., *J. Phys. Chem.*, **66**, 894 (1962); R. L. Kay and D. F. Evans, *J. Phys. Chem.*, **69**, 4216 (1965); **70**, 2325 (1966); R. L. Kay, T. Vituccio, C. Zawoyski and D. F. Evans, *J. Phys. Chem.*, **70**, 366 (1966); F. J. Millero and W. Drost-Hansen, *J. Phys. Chem.*, **72**, 1758 (1968); C. V. Krishnan and H. L. Friedman, *J. Phys. Chem.*, **74**, 2356 (1970).
28. Ref. 16, pp. 74–85.
29. W. F. McDevit and F. A. Long, *J. Am. Chem. Soc.*, **74**, 1773 (1952).
30. P. Debye and J. McAulay, *Z. Phys.*, **26**, 22 (1925).
31. P. Debye, *Z. Phys. Chem. Stoechiom. Verwandschaftsleture*, **130**, 56 (1927); M. Givon, Y. Marcus and M. Shiloh, *J. Phys. Chem.*, **67**, 2495 (1963).
32. H. S. Frank, *J. Chem. Phys.*, **23**, 2023 (1955); J. Padova, *J. Phys. Chem.*, **72**, 796 (1968).
33a. J. Padova, in *Water and Aqueous Solutions*, edited by R. A. Horne, Wiley, New York, 1972, p. 109.
33b. O. Popovych, *Crit. Rev. Anal. Chem.*, **1**, 73 (1970).
34. C. M. Criss, R. P. Held and E. Luksha, *J. Phys. Chem.*, **72**, 2970 (1968).
35. P. Debye and E. Hückel, *Phys. Z.*, **24**, 185 (1923).
36. R. H. Fowler, E. A. Guggenheim, *Statistical Thermodynamics*, Cambridge University Press, London, 1939, pp. 385–403.
37. N. Bjerrum, *Z. Anorg. All. Chem.*, **109**, 275 (1920).
38. R. H. Stokes and R. A. Robinson, *Trans. Faraday Soc.*, **53**, 301, (1957); E. Glueckauf, *Trans. Faraday Soc.*, **51**, 1235 (1955).

39. (a) E. Guggenheim, *Phil. Mag.*, **19**, 588 (1935);(b) V. P. Vasilev, *Zh. Neorg. Khim.*, **7**, 1788 (1962).
40. K. S. Pitzer and L. Brewer, *Thermodynamics*, (Revision of Lewis and Randall), McGraw-Hill, New York, 2nd Ed., 1961.
41. R. W. Gurney, *Ionic Processes in Solution*, McGraw-Hill, New York, 1953.
42. L. H. Laliberté and B. E. Conway, *J. Phys. Chem.*, **74**, 4116 (1970); W. Y. Wen and S. Saito, *J. Chem. Phys.*, **68**, 2639 (1964); B. E. Conway and R. Verrall, *Trans. Faraday Soc.*, **62**, 2738 (1966); H. E. Wirth, *J. Phys. Chem.*, **71**, 2922 (1967).
43. K. S. Pitzer, *J. Phys. Chem.*, **77**, 268 (1973); K. S. Pitzer and G. Mayorga, *J. Phys. Chem.*, **77**, 2300 (1973).
44. J. C. Rasaiah and H. L. Friedman, *J. Phys. Chem.*, **72**, 3352 (1968); *J. Chem. Phys.*, **48**, 2742 (1968); J. C. Rasaiah, *J. Chem. Phys.*, **52**, 704 (1970); H. L. Friedman, *Modern Aspects of Electrochemsitry, Vol. 6*, edited by J. O'M. Bockris and B. E. Conway, Butterworth, London, 1971, pp. 1–90; P. S. Ramanathan and H. L. Friedman, *J. Soln. Chem.*, **1**, 237 (1972).
45. H. L. Friedman, *J. Soln. Chem.*, **1**, 387, 413, 419 (1972).
46. J. T. Denison and J. B. Ramsay, *J. Am. Chem. Soc.*, **77**, 2615 (1955); W. R. Gilkerson, *J. Chem. Phys.*, **25**, 1199 (1965).
47. N. Bjerrum, *Kgl. Danske Videnskab. Selskab.*, **7**, 9 (1926).
48. J. C. Ghosh, *J. Chem. Soc., London*, **113**, 449, 707 (1918); N. Bjerrum, *Z. Elektrochem.*, **24**, 321 (1918); *Z. Anorg. Allg. Chem.*, **109**, 275 (1920); J. G. Kirkwood, *Chem. Rev.*, **19**, 275 (1936); H. S. Frank and P. T. Thompson, *J. Chem. Phys.*, **31**, 1086 (1959); in *Structure of Electrolyte Solutions*, edited by W. J. Hamer, Wiley, New York, 1959, Ch. 8; R. A. Robinson and R. H. Stokes, *Electrolyte Solutions*, Butterworth, London, 2nd edn., 1959; J. E. Desnoyers and B. E. Conway, *J. Phys. Chem.*, **68**, 2305 (1964).
49. E. Glueckauf, *Proc. Roy. Soc., Ser. A*, **310**, 449 (1969).
50. S. Brunauer, P. H. Emmett and E. Teller, *J. Am. Chem. Soc.*, **60**, 309 (1938); R. B. Anderson, *J. Am. Chem. Soc.*, **68**, 686 (1946).
51. R. H. Stokes and R. A. Robinson, *J. Am. Chem. Soc.*, **70**, 1870 (1948); A. N. Campbell and G. Oliver, *Can. J. Chem.*, **47**, 2671 (1969); H. Braunstein and J. Braunstein, *J. Chem. Thermod.*, **3**, 419 (1971).
52. G. Scatchard, *J. Am. Chem. Soc.*, **83**, 2636 (1961); **90**, 3124 (1968); **91**, 2410 (1969); Y. C. Wu, R. M. Rush and G. Scarchard, *J. Phys. Chem.*, **72**, 4048 (1968); **73**, 2047, 4433 (1969); G. Scatchard, R. M. Rush and J. S. Johnson, *J. Phys. Chem.*, **74**, 3786 (1970).
53. Ref. 16, pp. 63–71; cf. also J. N. Brønsted, *J. Am. Chem. Soc.*, **44**, 877 (1922); E. A. Guggenheim, *Phil. Mag.*, **19**, 588 (1935); H. S. Harned, *J. Am. Chem. Soc.*, **51**, 1865 (1935).
54. T. F. Young, *Rec. Chem. Progr.*, **12**, 81 (1951); T. F. Young and M. B. Smith, *J. Phys. Chem.*, **58**, 716 (1954); T. F. Young, Y. C. Wu and A. A. Krawetz, *Discuss. Faraday Soc.*, **24**, 37 (1957); Y. C. Wu, *J. Phys. Chem.*, **74**, 3781 (1970).
55. R. A. Robinson and R. G. Bates, *Anal. Chem.*, **45**, 1666, 1684 (1973); cf. also R. G. Bates, B. R. Staples and R. A. Robinson, *Anal. Chem.*, **42**, 867 (1970); R. A. Robinson, W. C. Duer and R. G. Bates, *Anal. Chem.*, **43**, 1862 (1971).

Chapter 7

Molten Salt Mixtures

A. Introduction

Molten salt mixtures have in common with electrolyte solutions in dielectrical media dealt with in Chapter 6 the long-range coulomb forces, which affect strongly the properties of the liquid mixture. However in contrast, and in common with molecular mixtures dealt with in Chapter 5, they are symmetrical in terms of the components, and in all but dilute solutions there is no distinction between solute and solvent. The rational (mole fraction) concentration scale is therefore most useful for describing compositions, but it has to be modified in order to recognize the ionic nature of the mixtures.

Molten salts and their mixtures are normally studied sufficiently near the melting point of the relevant crystalline salts to make the concept of the quasi-lattice useful (section 2C). The strong coulomb interactions will cause a strong tendency for cations to be surrounded essentially only by anions, and *vice versa*, anions by cations. Nearest-neighbour interactions will essentially be those of oppositely charged ions, while next-nearest-neighbour interactions will essentially be those of like-charged ions. However, this ordering effect is not expected to prevail over more than two ionic layers. It provides in essence two interlocking lattices, one for each charge type. This quasi-lattice model will often be invoked, but is by no means required for the description and understanding of the properties of molten salt mixtures.

The main distinguishing feature of molten salt mixtures is that they consist of ions – the effect of uncharged species and covalent bonding will be discussed later in this Chapter – so that an equilibrium mixture of cations A^{z_A+}, B^{z_B+}, C^{z_C+}, . . . and of anions X^{z_X-}, Y^{z_Y-}, Z^{z_Z-}, . . . will be considered. The condition of electroneutrality of the system as a whole imposes the restriction

$$\Sigma n_i z_i = n_A z_A + n_B z_B + \ldots = \Sigma n_j z_j = n_X z_X + n_Y z_Y + \ldots \tag{7.1}$$

(here and in the following i will represent cations and j anions). Further restrictions may be imposed by the choice of the components. A *reciprocal* salt *mixture*, containing A^{z_A+}, B^{z_B+}, X^{z_X-} and Y^{z_Y-}, may be made up from the components $A_{z_X}X_{z_A}$ and $B_{z_Y}Y_{z_B}$, so that $n_A = (z_X/z_A)n_X$ and $n_B = (z_Y/z_B)n_Y$. The same mixture can, however, be also made up in a different way by mixing a different number of moles of $A_{z_X}X_{z_A}$, $B_{z_X}X_{z_B}$ and $A_{z_Y}Y_{z_A}$, with no $B_{z_Y}Y_{z_B}$ at all, the mixture consisting of three independent components only. Mole fractions of the components are defined in the usual way, but *ion fractions* will be defined separately for cations and for anions. For instance

$$x_A = n_A/\Sigma n_i \quad (i = A, B, \ldots) \tag{7.2a}$$

$$x_X = n_X/\Sigma n_j \quad (j = X, Y, \ldots) \tag{7.2b}$$

For a binary *common-ion mixture*, e.g. AX + BX, $x_A = 1 - x_B$ and $x_X = 1$, while, for example, for an additive ternary common cation system AX + AY + AZ, $x_X + x_Y + x_Z = x_A = 1$.

As for mixtures of other kinds of liquids, it is useful also for molten salts to define an ideal mixture. Cognizance is given to the strong coulombic forces that make cations and anions not readily interchangeable in their positions, by using the Temkin model, rather than the Haase one[1]. The latter assumes complete dissociation of the salts to ions, which are randomly distributed, regardless of sign, while the Temkin model allows the cations A, B, . . . to mix among themselves on the cation positions i of a *quasi-lattice*, and the anions X, Y, . . . to mix randomly on the anion positions j. Thus, the sign, not the magnitude of the charges on the ions, is taken into account, as is expressed by the ion fractions (7.2). From this follows (cf. (4.25) and (4.29))

$$-S^M_{id}/R = \Sigma n_i \ln x_i + \Sigma n_j \ln x_j \tag{7.3}$$

The requirements of vanishing heat and volume change of mixing $H^M = 0$ and $V^M = 0$ are inherent in this model. The chemical potential of a component

$ij = A_{z_X}X_{z_A}$ in a mixture becomes therefore

$$\mu_{ij} = \mu^{\circ}_{ij} + z_X RT \ln x_A + z_A RT \ln x_X \tag{7.4}$$

and its activity is $a_{ij} = x_A^{z_X} x_X^{z_A}$. For a common-ion mixture, such as AX + BX, the activity becomes $a_{AX} = x_{AX} = x_A$, since $x_X = 1$, for the ideal mixture.

However, charge-unsymmetrical systems present a special problem, since even for the case of random mixing, Temkin's model is not necessarily obeyed[2]. If, for instance, AX is mixed with BX_{z_B}, cation vacancies from the latter salt (p. 126) may be introduced into the mixture, to an extent from zero (when Temkin's model is obeyed) to $x_B(z_B - 1)N$ vacancies per mole of mixture. In the latter case, the vacancies are considered associated with the polyvalent ions, forming 'z_B-mers', occupying the corresponding number of sites. An ideal mixture may then be specified if instead of ion fractions, (7.2), equivalent fractions are employed[2]

$$x'_A = n_A/(n_A + z_B n_B) \tag{7.5a}$$

$$x'_B = z_B n_B/(n_A + z_B n_B) \tag{7.5b}$$

This is, in fact, equivalent to using volume fractions or z-fractions (cf. (4.74)) with the volume ratio r specified as equalling z_B. Ideal mixing can now be written as before, in terms of x'_A and x'_B in (7.3) and (7.4) instead of the ion fractions. The unsatisfactory nature of this procedure arises from the possibility of vacancies coalescing when present at a high concentration, so that their number is $< x_B(z_B - 1)N$ except at high dilutions of B^{z_B}, and the expected statistics are not followed. Since this definition of ideality, in terms of equivalent fractions, is even more arbitrary than Temkin's model, the latter will normally be used in the following.

B. Common-Ion Mixtures

In common-ion mixtures, nearest-neighbour interactions involve the same pairs of ions as in the pure components, Fig. 7.1. Thus in an AX + BX mixture, A has always X ions around it, as in pure AX, and so has B. The magnitude of these interactions, however, may be influenced by the presence of the foreign ions further off. On the other hand, there will occur in the mixtures new next-nearest-neighbour interactions, A—X—B, which are absent in the pure components. A first approximation treatment will recognize only these next-nearest-neighbour interactions as contributors to the excess thermodynamic functions. Refinements will then deal with the effects of ions farther apart.

Molten salts mix in general more nearly ideally than other types of liquids, and among common-ion binary mixtures, cases of phase separation are rare indeed. Phenomenologically, regular behaviour, $h^E = g^E = b x_A x_B$, with b temperature-independent (p. 148, and eq. 5.27) is the rule, but exceptions are common. Still, excess volumes and entropies are small[2a], that is, mixing is fairly random, as shown in Table 7.1. Deviations from regular behaviour are shown mainly by the asymmetry of h^E or g^E with respect to composition. The interaction parameter

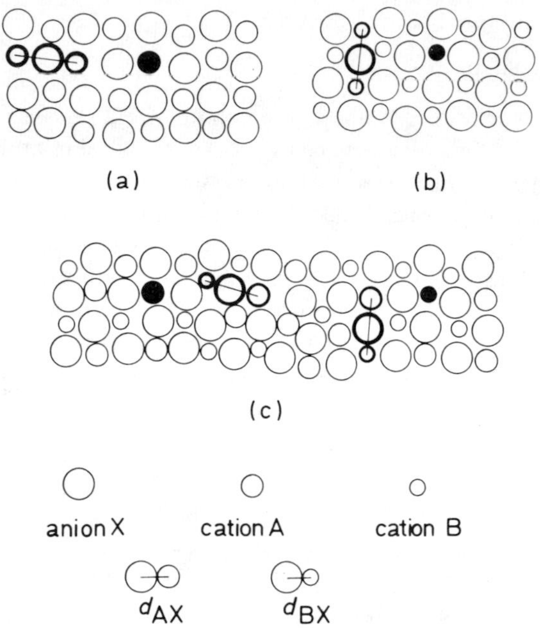

Figure 7.1. Quasi-lattice (two-dimensional) representation of charge-symmetrical molten salts: (a) AX; (b) BX; (c) mixed AX + BX. The ion marked in black in (a) has the same nearest environment as that in (c), as has the ion marked in black in (b). Next-nearest-neighbour interactions are shown by the heavy-rimmed circles. The interchange energy $\Delta\epsilon$ is the change in potential energy when the two pairs AXB in (c) are made from the pairs AXA in (a) and BXB in (b).

$h^E/x_A x_B$ conforms often to the equation

$$h^E(\text{kJ mol}^{-1})/x_A x_B = a + bx_A + cx_A x_B \tag{7.6}$$

where b introduces the asymmetry, as shown in Appendix 7.1, p. 290. For systems with multivalent ions, h^E refers to one equivalent of the mixture, but the data were recalculated, where necessary, in terms of x, mole fractions, rather than equivalent fractions. A quantity called $\lambda = 4\hat{h}^E$, for the equimolar mixture is a sufficiently good approximation of the (composition-dependent) interaction parameter. This equals $a + 0.5b + 0.25c$ according to (7.6), and refers to the data in Appendix 7.1, p. 290.

The partial molal heats of mixing at infinite dilution are $h_A^{E\infty} = a$ and $h_B^{E\infty} = a + b$ if (7.6) is obeyed. The same quantities may be used to describe the excess Gibbs energies if the excess entropies are known or are assumed to be negligible.

Table 7.1. Excess molar entropies and volume changes of mixing of equimolar common-ion molten salt binary mixtures.

(A, B)X	$\hat{s}^E$ ($J\ K^{-1}\ mol^{-1}$)	$\hat{v}^E$ ($cm^3\ mol^{-1}$)	(A, B)X	$\hat{s}^E$ ($J\ K^{-1}\ mol^{-1}$)	$\hat{v}^E$ ($cm^3\ mol^{-1}$)
(Li, Na)F		+0.03[j]	(Li,Cs)Br		+0.45[j]
(Li, K)F	−1.4[a]	+0.20[j]	(Na,K)Br		+0.16[j]
(Li, Rb)F		+0.30[j]	(Na,Rb)Br		+0.21[j]
(Na, K)F		+0.15[j]	(Na,Ag)Br	0[g]	+0.17[g]
(Na, Rb)F		+0.25[j]	(K,Rb)Br		+0.08[j]
(K, Rb)F		+0.05[j]	(K,Ag)Br	0[g]	+0.27[g]
(Li, Na)Cl	−1.5[b]	+0.05[j]	(Rb,Ag)Br	0[g]	+0.42[g]
(Li, K)Cl	−1.5[a]	+0.30[j]	(Na,Zn)I	+3.0[f]	
(Li, Rb)Cl	−1.6[a]	+0.35[j]	(K,Ag)I	+1.9[c]	
(Li, Cs)Cl	−1.6[a]	+0.50[j]	(K,Cd)I	+2.8[f]	
(Li, Ag)Cl	0[c]		(Li,Na)NO_3		+0.06[h,k]
(Li, Cd)Cl		−0.05[d]	(Li,K)NO_3		+0.11[k]
(Li, Pb)Cl		+0.40[d]	(Li,Rb)NO_3		+0.13[k]
(Na, K)Cl	−0.3[b]	+0.20[j]	(Li,Ag)NO_3		+0.02[k]
(Na, Rb)Cl	+0.3[a], −0.3[b]	+0.25[j]	(Li,Tl)NO_3		+0.18[k]
(Na,Cs)Cl	+0.3[a], −0.6[b]	+0.45[j]	(Na,K)NO_3		+0.01[k], +0.07[h]
(Na,Ag)Cl	+0.5[c]		(Na,Rb)NO_3		+0.03[k], +0.20[h]
(Na,Mg)Cl	−1.2[c]		(Na,Cs)NO_3		+0.17[k], +0.39[h]
(Na,Ca)Cl	−1.0[c]		(Na,Ag)NO_3	−0.2[a]	+0.15[k]
(Na,Sr)Cl	−0.2[c]		(Na,Tl)NO_3		+0.19[k]
(Na,Ba)Cl	+0.2[c]		(K,Rb)NO_3		0.00[k]
(Na,Zn)Cl		−1.38[d]	(K,Cs)NO_3		+0.05[k]
(Na,Cd)Cl		+0.20[d]	(K,Ag)NO_3		+0.26[k]
(Na,Pb)Cl		+1.19[d], +0.40[c]	(K,Tl)NO_3		+0.18[k]
(K,Rb)Cl		+0.15[j]	(Rb,Cs)NO_3		+0.02[k]
(K,Zn)Cl		+0.35[d]	(Rb,Ag)NO_3		+0.23[k]
(K,Cd)Cl	+2.4[f]	+0.97[d]	(Rb,Tl)NO_3		+0.16[k]
(K,Pb)Cl		+1.66	(Ag,Tl)NO_3		−0.09[k]
(Rb,Cd)Cl		+2.50[e]	(F,Cl)Li		+0.31[j]
(Rb,Pb)Cl		+1.88[e]	(F,Cl)Na		+0.48[j]
(Cs,Cd)Cl	−3.8[f]	+2.50[e]	(F,Cl)K		+0.58[j]
(Cs,Pb)Cl		+2.50[e]	(F,Cl)Rb		+0.30[j]
(Ag,Tl)Cl	−1.0[a]		(Cl,Br)Li	−0.1[i]	+0.10[j]
(Li,Na)Br		+0.05[j]	(Cl,Br)Na	−0.2[i], −0.1[c]	+0.06[j]
(Li,K)Br	−0.7[a]	+0.12[j]	(Cl,Br)K	−0.5[i], −0.9[c]	+0.11[j]
(Li,Ag)Br	−0.2[c]	−0.13[g]	(Cl,I)Ag	+1.0[c]	
(Li,Rb)Br		+0.30[j]	(Br,I)Ag	+0.4[c]	

[a] J. Lumsden, *Thermodynamics of Molten Salt Mixtures*, Academic Press, London, 1966.
[b] L. U. Thulin, *Theoretical Treatment and Practical Applications of Galvanic Cells with Membranes,* NTH, Trondheim University, 1970.
[c] G. J. Janz and C. G. M. Dijkhuis, *Molten Salts, Vol. 2,* NSRDS-NBS 28, Washington, D.C., 1969.
[d] M. F. Lavrantov and A. F. Alabyshev, *Zh. Prik. Khim. (Leningrad),* **26**, 235, 321 (1953); **27**, 685 (1954).
[e] H. Bloom, P. W. D. Boyd, J. L. Laver and J. Wong, *Aust. J. Chem.,* **19**, 1591 (1966).
[f] B. Vos, Thesis, University of Amsterdam, 1970.
[g] J. H. Hildebrand and E. J. Salstrom, *J. Am. Chem. Soc.,* **54**, 4257 (1932).
[h] B. F. Powers, J. L. Katz and O. J. Kleppa, *J. Phys. Chem.,* **66**, 103 (1962).
[i] T. Førland, in *Fused Salts*, edited by B. R. Sundheim, McGraw-Hill, New York, 1964.
[j] J. L. Holm, *Acta Chem. Scand.,* **25**, 3609 (1971).
[k] B. Cleaver and B. C. J. Neil, *Trans. Faraday Soc.,* **65**, 2860 (1969).

a. Mixtures of hard ions only

Highly ionic melts will be dealt with first. In these the ions will be represented by hard spheres of a defined radius. The interactions are described by the potential functions, cf. p. 127.

$$\epsilon_{ij}(r \leqslant d_{ij}) = \infty \tag{7.7a}$$

$$\epsilon_{ij}(r > d_{ij}) = -Kz_i z_j e^2 / r \tag{7.7b}$$

$$\epsilon_{ii}(r > 0) = Kz_i^2 e^2 / r \tag{7.7c}$$

where j represents the common ion, and K is a proportionality constant (which represents the reciprocal of a mean dielectric constant). The distances $d_{AX} = r_A + r_X$ and $d_{BX} = r_B + r_X$ are taken as the sums of radii which, together with the charges z_A, z_B and z_X, are the sole characteristics of the ions, and are independent of pressure and temperature. Since like-charged ions will not approach each other closer than d_{ii} anyway, their mutual interaction energy covers the whole range of r values, from $r > 0$ on. unlike-charged ions will approach each other nearer than the distance d_{ij}, and the repulsion energy plays a major role in characterizing the system. If d_{AX} equals d_{BX} and $z_A = z_B$, unlike ions may mix randomly on their quasi-lattice positions, and the mixture follows the Temkin model.

Actual systems do not obey the Temkin model, and size disparities between the unlike ions cause deviations from randomness. A second-order perturbation treatment, involving conformal mixture theory (section 4B(b)) has been found[3] to be useful. It actually dispenses altogether with a model, and does not specify necessarily a coulomb-type potential (and does not require therefore the constant K). Rather, the treatment allows for the case $z_i = z_j = 1$, any potential depending on r only for the like-charge ions, and instead of a hard-sphere potential for the repulsive interaction, it specifies a more general potential (cf. (4.93))

$$\epsilon_{ij}(r) = d^{-1}\psi(r/d) \tag{7.8}$$

The length parameter d of the reference molten salt is selected so that the perturbation parameters

$$\rho_A = (d/d_{AX}) - 1, \quad \rho_B = (d/d_{BX}) - 1 \tag{7.9}$$

add up to zero: $\rho_A + \rho_B = 0$. The configurational part of the Helmholz energy (cf. (4.92) and the Appendix to Chapter 1) is

$$-kT \ln Z(\rho) = -kT \ln \int \ldots \int \exp(-U(\rho)/kT)\, d(\vec{r}^{2N}) \tag{7.10}$$

The integration is carried out over the volume V of the systems for the $2N$ (N cations and N anions for symmetrical salts) volume elements $d\vec{r}$. The potential energy U is given by

$$U(\rho) = \Sigma\Sigma(\rho + 1)\epsilon_{ij}((\rho + 1)r) + \Sigma\Sigma\epsilon_{ii}(r) + \Sigma\Sigma\epsilon_{jj}(r) \tag{7.11}$$

there being N^2 terms in the first double sum, and $\frac{1}{2}N(N - 1)$ terms in each of the other sums, since the same like-charged ions should not be counted twice. As noted above, like-charged interactions depend on r only, but the unlike-charged interactions depend on ρ: substitution of d_{AX} for the conformal salt A in (7.8) and noting that $(\rho_A + 1)/d = 1/d_{AX}$ yields $\epsilon_{AX} = (\rho_A + 1)\epsilon_{ij}((\rho_A + 1)r)$ and, similarly, for any other conformal salt or mixture. Expansion of Z as a power series in ρ to the second order is now carried out for the two pure components and the mixture. For the mixture, the sums in (7.11) run over $x_A N^2$ terms $(\rho_A + 1)\epsilon_{AX}$, $x_B N^2$ terms $(\rho_B + 1)\epsilon_{BX}$, $x_A^2 N(N-1)$ terms ϵ_{AA}, $x_B^2 N(N-1)$ terms ϵ_{BB}, and $2x_A x_B N (N-1)$ terms ϵ_{AB}. The excess Helmholz energy is calculated according to (4.39), yielding

$$A^E = x_A x_B A^\circ(Z(\rho = 0), N, T)(\rho_A - \rho_B)^2 \tag{7.12}$$

where all the first-order terms have been cancelled out in the excess function difference. The function A^0 for the reference salt does not depend on x or ρ, and involves sums of integrals which cannot be calculated explicitly. What is important is the functional dependence of A^E on x and ρ. Substitution of (7.9) into (7.12), together with the restriction $\rho_A + \rho_B = 0$ yields

$$A^E = x_A x_B A' \delta^2 \tag{7.13}$$

where $A' = 4A^\circ(Z(\rho = 0), N, T)$, and δ is the size parameter*

$$\delta = (d_{AX} - d_{BX})/(d_{AX} + d_{BX}) \tag{7.14}$$

Inclusion of higher-order terms in the expansion of $Z(\rho)$ introduces asymmetrical terms, so that

$$A'' = 1 + \alpha\delta^2 + (x_A - x_B)(\beta\delta + \gamma\delta^2) + \ldots \tag{7.15}$$

should replace A' in (7.13), the parameters α, β and γ being again complicated integrals involving $Z(\rho = 0)$, N and T, but being independent of x and ρ.

Only the first-order dependence of $A(T, V)$ on ρ equals that of $G(T, P)$ (cf. discussion above (4.111)), so that there is no direct way of changing from (7.13) for the Helmholz energy, which involves second- or higher-order terms, to a Gibbs energy relationship. For systems with $v^E = 0$, however, the transformation may still be made, and from g^E, the functional form for the molar heat of mixing

$$h^E = x_A x_B h'(T,P)\delta^2 \tag{7.16}$$

is obtained to the second order[3], while higher-order terms will involve $h''(T, P, \delta, x)$ of a form similar to (7.15).

For charge-unsymmetrical systems, namely of the $AX + BX_{z_B}$ type, it has been shown that a modification of the perturbation method[3] leads to an expression where first-order terms in $(\rho_A - \rho_B)$ do not cancel out in a^E. At a given composition, it is possible to designate the mixture as the reference liquid, and set

*Sometimes the size parameter is written $\delta' = (d_{AX} - d_{BX})/d_{AX}d_{BX}$, which has the dimension $(\text{length})^{-1}$ and makes $A' = d^2 A^0$ in (7.13). The present restriction of $\rho_A + \rho_B = 0$ produces instead the relationship $A' = 4A^0$, and a dimensionless δ, which seems preferable. Plots of h^E against either δ or δ' usually produce curves of similar quality (linearity, if so expected).

$\rho_B = 0$, so that $\rho_A = (d_{BX}/d_{AX}) - 1$. The result for the excess Helmholz energy is

$$a^E = -a_{AB}(T, d_{BX}, x) + x_A\delta\{[1 + x_B(z_B - 1)]\alpha_{AB}(d_{BX}, x) - \alpha_A(d_{BX})\}$$
$$+ x_A\delta^2\Xi(T, V, d_{BX}, z_B, x) \qquad (7.17)$$

The parameters a_{AB}, α_{AB} and Ξ are functions of x, hence cannot be used to give an explicit dependence of a^E (or h^E) on the composition. The limiting value of a^E, as $x_B \to 0$ can, however, be evaluated, yielding

$$\lim_{x_B \to 0} (h^E/x_B) = h^0 + h'\delta + h''\delta^2 \qquad (7.18)$$

where h^0, h' and h'' are functions of (T, P, d_{BX}, z_B) and of certain derivatives $(\partial/\partial x_B)_{x_B=0}$, but not of δ. Expression (7.18) may be used to compare the experimental dependence of the limiting (infinite dilution of the multivalent ion) heat of mixing on the size parameter δ with the expectation from theory.

The expressions obtained by perturbation theory[3] require the specification of interaction potentials, (7.7) and (7.8), where the ions are specified by one size parameter, δ (or, in addition, for charge-unsymmetrical systems also the distance d_{BX} and the stoichiometric parameter z_B), but do not depend on any model. An alternative approach starts from the quasi-lattice model, and applies the *quasi-chemical interaction theory*[4] (p. 158) to the next-nearest-neighbour ion interactions. A coordination number Z' is assigned to the next-nearest-neighbour sublattice; for molten salts $Z' \sim 10$. For one mole of mixture there are $x_A\mathbf{N}$ ions A and $x_B\mathbf{N}$ ions B, and between them there are $\frac{1}{2}Z'\mathbf{N}$ next-nearest-neighbour interactions. If the number of A–X–B interactions is denoted by $xZ'\mathbf{N}$, the following situation occurs

Pair	*Number of pairs*	*Energy per pair*	*Total energy*
AA	$\frac{1}{2}(x_A - x)Z'\mathbf{N}$	ϵ_{AA}	$\frac{1}{2}(x_A - x)Z'\mathbf{N}\epsilon_{AA}$
AB	$xZ'\mathbf{N}$	ϵ_{AB}	$xZ'\mathbf{N}\epsilon_{AB}$
BB	$\frac{1}{2}(x_B - x)Z'\mathbf{N}$	ϵ_{BB}	$\frac{1}{2}(x_B - x)Z'\mathbf{N}\epsilon_{BB}$
All	$\frac{1}{2}Z'\mathbf{N}$		u

The total energy due to pair-wise ion-ion interaction is

$$u = \tfrac{1}{2}\mathbf{N}Z'[(x_A - x)\epsilon_{AA} + 2x\epsilon_{AB} + (x_B - x)\epsilon_{BB}]$$
$$= \tfrac{1}{2}\mathbf{N}Z'x_A\epsilon_{AA} + \tfrac{1}{2}\mathbf{N}Z'x_B\epsilon_{BB} + x\lambda \qquad (7.19)$$

where an energy parameter λ is defined as

$$\lambda = \tfrac{1}{2}\mathbf{N}Z'(2\epsilon_{AB} - \epsilon_{AA} - \epsilon_{BB}) = \tfrac{1}{2}\mathbf{N}Z'\Delta\epsilon_{coul} \qquad (7.20)$$

The quasi-chemical treatment then permits the elimination of x to give

$$h^E = x_A x_B\lambda[1 - x_A x_B 2\lambda/Z'\mathbf{R}T + \tfrac{1}{2}x_A x_B(x_A - x_B)^2(2\lambda/Z'\mathbf{R}T)^2 - \ldots] \qquad (7.21)$$

for the heat of mixing. If no interference from further neighbours is assumed, the coulombic interaction energy $\Delta\epsilon_{\text{coul}}$ for the situation[5] in Fig. 7.1 for $z_A = z_B = z_X$ is (cf. (7.7))

$$\Delta\epsilon_{\text{coul}} = -Kz_A^2e^2(d_{AA}^{-1} + d_{BB}^{-1})\delta^2 \tag{7.22}$$

Since $d_{AA}^{-1} + d_{BB}^{-1}$ is a slowly changing function of δ, introduction of (7.22) into (7.20) and the resulting expression into (7.21) yields a leading term identical in its dependence on x and δ to (7.16). For charge-unsymmetrical systems the appropriate form of (7.22) would be $Ke^2[2z_Az_B(d_{AX} + d_{BX})^{-1} - \tfrac{1}{2}z_X(z_Ad_{AX}^{-1} + z_Bd_{BX}^{-1})]$, which does not give a simple dependence on any power of δ. Furthermore, the next-nearest-neighbour coordination number Z' will become a function of the composition[5], which makes also λ composition-dependent. The first two terms in the Gibbs energy expression corresponding to (7.20) are

$$g^E = x_Ax_B\lambda[1 - x_Ax_B\lambda/Z'RT + \ldots] \tag{7.23}$$

Although the quasi-chemical treatment presumes λ to be independent of composition, the function (7.6), pertaining to the data in Appendix 7.1, p. 290, is actually found. The asymmetry of λ with respect to composition is in qualitative agreement with (7.15) and (7.21), but not in quantitative detail. It has been shown[5] that a treatment involving interaction energies dependent on the kind and number of ions surrounding a given central ion in both the first and second coordination spheres may lead to the observed asymmetric behaviour. This is the case for a second-power dependence of the energy on the number of ions present in the second coordination sphere, $i\text{A} + (Z' - i)\text{B} : \epsilon_{AA}/\epsilon_{AA}^0 = \epsilon_{BB}/\epsilon_{BB}^0 = i(2Z' - i)$.

b. Mixtures containing polarizable ions

For mixtures containing polarizable ions, the hard-sphere coulomb interaction potential (7.7) will no longer be adequate. A complication occurs relative to the previous discussion, where hard-sphere plus coulomb interactions (or an equivalent potential involving one parameter d) were adequate. The complication is that polarizable ions introduce scaling factors C_{AA}, C_{BB} and C_{AB}, which do not cancel out in the excess Helmholz energy. The potential functions are modified from (7.8) to give

$$\epsilon_{ij}(r \leqslant d_{ij}) = \infty \tag{7.24a}$$

$$\epsilon_{ij}(r > d_{ij}) = -d^{-1}\psi(r/d) + C_{ij}\Phi(r) \tag{7.24b}$$

$$\epsilon_{ii}(r > 0) = d^{-1}\psi(r/d) + C_{ii}\Phi(r) \tag{7.24c}$$

The perturbation theory for this case[6] performs a Taylor expansion in both d and C, the scaling parameters for the hard-sphere and the polarization interactions, leaving the coulombic interaction unscaled*. The end result is an expression similar

*A difficulty arises from the various possible contributions to $C\Phi(r)$, e.g. from ion–induced dipole and polarization interactions, where $\Phi(r) = -r^{-4}$, and from dispersion interactions, where $\Phi(r) = -r^{-6}$, which do not have the same or scaled C coupling parameters. For the sake of simplicity, all these interactions (non-coulombic and non-hard-sphere) have been included in a single term $C\Phi(r)$ (ref. 6).

to (7.16), with the addition of terms in C, making a^E (hence b^E) asymmetrical with respect to x, and at a given composition dependent on δ to the zeroth, first and second powers (to the second order). This is in line with the observed behaviour, but the expressions are so complicated with unknown integrals (depending on T, P, $Z(\rho = 0)$, C, d and partial derivatives with respect to x) that even the relative important of these dependencies on x and δ cannot be evaluated.

The quasi-lattice model provides more help here, and the contribution of the various interactions to the heat of mixing can be understood. A negative contribution arises from ion–induced-dipole interactions[2,7] when a polarizable ion is the common ion (the anion in AX + BX mixtures). This has a form similar to that of the ion–ion coulombic interaction (7.22), namely

$$\Delta\epsilon_{\mathrm{pol}} = -K\alpha e^2 (d_{\mathrm{AA}}^{-1} + d_{\mathrm{BB}}^{-1})^4 \delta^2 \qquad (7.25)$$

for univalent ions, where α is the polarizability of the common ion. As in (7.22), a one-dimensional interaction only is considered, and freedom from interference from further neighbours is assumed. If this is the only effect, $\lambda = \frac{1}{2} NZ'(\Delta\epsilon_{\mathrm{coul}} + \Delta\epsilon_{\mathrm{pol}})$ should replace (7.20), both contributions being negative.

On the other hand, dipole–dipole, i.e. dispersion, interactions may also be important, in particular when one of the non-common ions is polarizable, introducing a positive contribution to λ. The same factors which are operative for molecular liquids (Chapter 5) should be operative for ionic liquids too (cf. eqs. (4.100), (5.30)). If effective values of 75 % of the ionization potentials I and effective distances $r_{\mathrm{ii}} = \sqrt{2} d_{\mathrm{ij}}$ are used in the London dispersion energy expression, and if a Berthelot rule of mixing is adopted (cf. discussion above (4.112) and eq. (5.29))[8], the positive contribution to λ from dispersion forces becomes $\frac{1}{2}NZ'$ times

$$\Delta\epsilon_{\mathrm{disp}} = (3/8\sqrt{2})[\alpha_{\mathrm{A}} I_{\mathrm{A}}^{1/2} d_{\mathrm{AX}}^{-3} - \alpha_{\mathrm{B}} I_{\mathrm{B}}^{1/2} d_{\mathrm{BX}}^{-3}]^2 \qquad (7.26)$$

It must be recognized that $\Delta\epsilon_{\mathrm{disp}}$, through its dependence on d_{AX} and d_{BX}, is an implicit function of δ, but it is customary to separate the positive contribution of $\Delta\epsilon_{\mathrm{disp}}$ from the δ-dependent negative contributions appearing in (7.18).

It has been found that polarization and dispersion forces alone cannot account for the extracoulombic interactions for systems involving transition or post-transition metal ions. Some degree of covalency must be assumed, since the positive contribution of $\Delta\epsilon_{\mathrm{disp}}$ cannot compensate for the negative contribution $\Delta\epsilon_{\mathrm{coul}}$ for these systems, which show positive λ values. An extra term, obtained empirically[9] from heat of mixing data, of a form similar to the solubility-parameter term in the energy of mixing of molecular liquids, eq. (5.32), has been found useful.

All these treatments of the problem[6–10] place a heavy reliance on the interionic distances d_{AX} and d_{BX} or rather on their difference, that is, the parameter δ. However, these distances in the melt are not at all well known, and although temperature-dependent[11], are usually calculated as the sum of ionic crystal radii valid, with reservations, for room-temperature solids. Great uncertainty prevails for highly polarizable ions, such as silver, for which radii from 95 pm to 126 pm have been used, and for non-spherical ions, such as nitrate, for which radii from 182 pm to 216 pm have been used (Fig. 7.2). It is just these ions that have played a major role in developing the treatments of common-ion binary (or ternary) mixtures, by comparing the theory with experimental results for heats of mixing and electrochemical cell potentials and their temperature-dependencies.

An attempt to circumvent this difficulty has been made[12] by using, for characterizing the ions, instead of radii and interionic distances, rather their softness parameters, in addition to their charges. The softness parameters, σ_M for cations and σ_X for anions, have been obtained from data completely independent of the properties of molten salts. For cations, M^{z_m+}, the softness parameter is

$$\sigma_M = -1 + z_m^{-1} \left[\sum_{i=1}^{z_m} I_i + b^0_{hydM} \right] \Big/ [I_H + b^0_{hydH}] \qquad (7.27)$$

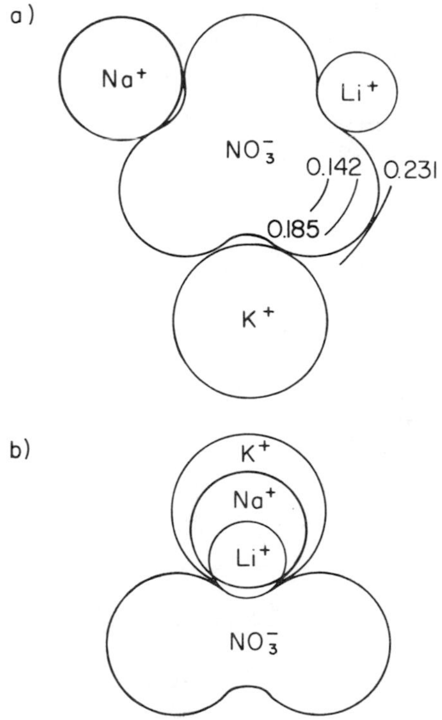

Figure 7.2. A representation of the nitrate ion in alkali nitrate melts. If the cations can take positions as in the upper part, (a), the radius of the nitrate ion is only $r_{NO_3} = d_{M-NO_3} - r_{M^+} = 0.142 \pm 0.002$ nm. The usually accepted radius, 0.185 nm, and the maximal extension, corresponding to the 'thermochemical' radius, 0.231 nm, are also indicated. Approach of the cations as in the lower part, (b), will reduce the 'radius' of the nitrate ion to the absurdly low value of 0.08 nm.

Table 7.2. Softness Parameters of Cations and Anions.

Softness Parameters of Cations[a]						Softness Parameters of Anions[a]			
M^+	σ_M	M^{2+}	σ_M	M^{m+}	σ_M	X^-	σ_X	X^{2-}	σ_X
H^+	0.00	Be^{2+}	−0.41	Al^{3+}	−0.25	F^-	−0.71	CO_3^{2-}	−0.37
Li^+	−0.95	Mg^{2+}	−0.37	Y^{3+}	−0.68	NO_3^-	−0.41	SO_4^{2-}	−0.31
Na^+	−0.75	Ca^{2+}	−0.65	La^{3+}	−0.65	NO_2^-	−0.24	S^{2-}	+1.02
K^+	−0.53	Sr^{2+}	−0.59	Pu^{3+}	−0.62	Cl^-	−0.16		
Rb^+	−0.49	Ba^{2+}	−0.60	Cr^{3+}	−0.06	OH^-	0.00		
Cs^+	−0.46	Mn^{2+}	−0.11	Fe^{3+}	+0.22	ClO_4^-	0.00		
Cu^+	+0.26	Fe^{2+}	−0.06	Ga^{2+}	+0.29	Br^-	+0.10		
Ag^+	+0.18	Co^{2+}	−0.18	In^{3+}	+0.44	I^-	+0.40		
Tl^+	+0.45	Ni^{2+}	−0.11	Tl^{3+}	+0.92	SCN^-	+0.84		
		Cu^{2+}	+0.39	Bi^{3+}	+0.61				
		Zn^{2+}	+0.37	Th^{4+}	−0.55				
		Cd^{2+}	+0.59	U^{4+}	−0.38				
		Hg^{2+}	+1.28	Pu^{4+}	−0.21				
		Sn^{2+}	+0.31						
		Pb^{2+}	+0.58						
		UO_2^{2+} [b]	−0.38						

[a] Y. Marcus, *Israel J. Chem.*, **10**, 659 (1972).
[b] Y. Marcus, *J. Inorg. Nucl. Chem.*, **37**, 493 (1975).

while for anions, X^{z_X-}, it is

$$\sigma_X = [-z_X^{-1} E_{aX} + E_{aOH} + z_X^{-1} h^0_{hydX} - h^0_{hydOH}]/[I_H + h^0_{hydH}] \tag{7.28}$$

where I_i is the ith ionization energy, E_a the electron affinity, and h^0_{hyd} the hydration enthalpy. Values of the softness parameters for many ions[12] appear in Table 7.2. It has been shown that the interaction parameter λ may be obtained from (7.20) if ϵ_{ij} is related to the softness parameters (rather than to length parameters as in (7.21), (7.25) and (7.26)) as follows

$$\epsilon_{AA} = p(\sigma_A, \sigma_X) + q\sigma_A^2(\sigma_A^2 - \sigma_A\sigma_X) \tag{7.29a}$$

and similarly for ϵ_{BB}, with B replacing A throughout, and

$$\epsilon_{AB} = p[(\sigma_A + \sigma_B)/2, \sigma_X] + q\sigma_A\sigma_B(\sigma_A\sigma_B - (\sigma_A + \sigma_B)\sigma_X/2) \tag{7.29b}$$

where p is *any* homogeneous first degree function of the softness parameters, and q is a constant. The terms in p cancel in the evaluation of λ from (7.20)

$$\lambda_{(A,B)X} = \tfrac{1}{2} NZ'q(\sigma_A^2 - \sigma_B^2)[(\sigma_A - \sigma_B)\sigma_X - (\sigma_A^2 - \sigma_B^2)] \tag{7.30}$$

The value $\tfrac{1}{2}NZ'q = 34$ kJ mol^{-1} was found to give a good fit for charge-symmetrical binary systems with either a common cation or a common anion, with $\lambda \sim 4\hat{h}^E$ (ref. 12). For charge-unsymmetrical systems (with a univalent common ion) $AX_{z_A} + BX_{z_B}$, the interaction parameter calculated on the basis of equivalent fractions (7.6)

$$\lambda = (4/z_A z_B) h^E (x' = 0.5) = \tfrac{1}{2} \mathbf{N} Z' (z_A + z_B)(2 z_A z_B)^{-1} \cdot$$
$$[(2/z_A z_B)\epsilon_{AB} - z_A^{-2}\epsilon_{AA} - z_B^{-2}\epsilon_{BB}] \quad (7.31)$$

$$\lambda(\text{kJ equiv}^{-1}) = 17(z_A + z_B)(z_A z_B)^{-1} [z_B^{-1}\sigma_B^2 - z_A^{-1}\sigma_A^2] [(z_B^{-1}\sigma_B - z_A^{-1}\sigma_A)\sigma_X$$
$$- (z_B^{-1}\sigma_B^2 - z_A^{-1}\sigma_A^2)] \quad (7.32)$$

Again, a good fit was found for mixtures of bi–univalent and bi–bivalent halides or nitrates. The expressions for ϵ_{ij}, containing fourth-degree terms in σ, have been designed empirically to allow both positive and negative values for λ, and are symmetrical with respect to the non-common ions, so that the different charges of A^{z_A+} and B^{z_B+} can also be accommodated as in (7.32). A positive value of λ results if one of the non-common ions is very soft, and the common ion is hard, as for instance in the (Pb,Ca)Cl_2, Na(Cl, Br) or Ca(F, I)$_2$ systems. Silver ions would fall into the pattern if $\sigma_M = 0.70$ were assigned to it, rather than the value obtained from the ionization potential and the heat of hydration (Table 7.2).

The softness parameters are a measure of the tendency of ions to bond covalently[12], so that it may be argued that the coulombic effect required by (7.19) and (7.21) is totally disregarded in the above approach. A combined approach, assigning each ion a size (radius), a charge and a softness parameter, all obtained independently, could combine (7.30) with (7.19) and (7.21) if two arbitrary parameters, K (a reciprocal dielectric constant, describing long-range many-body polarization effects) and q are chosen for fitting the data for all the ions considered. This would be an empirical correlation, but would have considerable predictive value.

c. *Associated common-ion mixtures*

In some binary common-ion mixtures interactions may become so strong that association results, and important contributions to the non-ideality of mixing arise from the effects that association produces. This is particularly the case with certain charge-unsymmetrical mixtures, where all of the non-ideality has been attributed to the formation of complex species (e.g. in the KCl + $MgCl_2$ or NaF + AlF_3 systems[13]). As for non-electrolyte mixtures (section 5B), this concept of ideal associated mixtures has been applied to many molten salt systems, but sometimes in an inappropriate way. Certainly the existing cases of phase separation[14] point to the occurrence of additional contributions to the non-ideality rather than to ideal association only (cf. p. 199).

In fact, some ideal association has been shown[15] to lead to no deviations from ideality at all. For instance, if the association is limited to the formation of pairs $AX^{z_A - 1}$ in a mixture AX_{z_A} + BX, the activity is given by $a_{BX} = x_B x_X = n_B/(n_A + n_B)$, as if no association occurred. The decrease of the ion fraction of X^- due to its association with A^{z_A+} is compensated by the increase of the ion fraction of the cations B^+ due to the removal of A^{z_A+}. Similar results are obtained for $AX + BX$ mixtures, if A_2X^+ is formed, mixing ideally with the other cations A^+ and B^+, whatever the value of the association constant.

Better approaches seem to be those which take into account non-ideal mixing, through energetic or entropic effects of the actual species on each other. Entropic effects take into account differences in size of associated and simple species. It is possible to take into account partial covalent binding, by ascribing to certain cation–anion interactions (such as the silver–chloride interactions in mixtures containing also alkali metal chlorides, AgCl + MCl) a dual character with two energy parameters describing the system[16]. One, $\Delta\epsilon^0$, relates to the process $AgCl_{ionic} \rightleftharpoons AgCl_{covalent}$ in pure AgCl melts, the other, $\Delta\epsilon^m$, to the same process in mixtures, where the alkali cation is regarded as affecting the interaction. Assuming a linear dependence of the energy on the composition, the fractions of $AgCl_{covalent}$ in pure and mixed melts can be calculated

$$\alpha^0 = \exp(-\Delta\epsilon^0/RT)/(1 + \exp(-\Delta\epsilon^0/RT)) \tag{7.33a}$$

$$\alpha = \exp[-(x_M\Delta\epsilon^m - x_{Ag}\Delta\epsilon^0)/RT]/[1 + \exp[-(x_M\Delta\epsilon^m - x_{Ag}\Delta\epsilon^0)/RT]] \tag{7.33b}$$

From these, the heat and entropy of mixing

$$h^E = x_{Ag}[(\alpha - \alpha^0)\Delta\epsilon^0 + \alpha x_M(\Delta\epsilon^m - \Delta\epsilon^0)] \tag{7.34}$$

$$s^E = -Rx_{Ag}[\alpha \ln \alpha + (1 - \alpha)\ln(1 - \alpha)] \tag{7.35}$$

can be obtained, with good fit to data from electrochemical and calorimetric measurements, and reasonable parameters $\Delta\epsilon^0$ and $\Delta\epsilon^m$. By ascribing all the deviations from ideal mixing to the ionic-covalent transition, the effects found for highly ionic melts (section 7A(a)) have been completely neglected, with no valid justification. As for molecular liquids, a combination of 'association' and 'physical non-ideality' (section 5B(c)) should work better than either approach alone.

In some other cases, association may proceed to such an extent that the systems switch from a common-anion to a common-cation configuration. For example, in the cryolite system $AlF_3 + NaF$, the actual (ideal) mixing is among the anions F^-, AlF_4^- and AlF_6^{3-} with the common cation Na^+. In other cases, with weaker association, a reciprocal mixture (as discussed in section 7B) is formed, as for example in the $KCl + MgCl_2$ system, where the two cations K^+ and Mg^{2+} and the two anions Cl^- and $MgCl_4^{2-}$ mix ideally on their respective sites, at least according to some of the evidence and its interpretation. It must be recognized that if association occurs in the mixture, some self-association may occur also in the parent components (cf. description for case 1 of molecular liquids, p. 184). This is the case, apparently, for the latter system, and the activity of magnesium chloride is

$$a_{MgCl_2} = x_{Mg}(x_{Cl}/x_{Cl}^0)^2 \tag{7.36}$$

where x_{Cl}^0 is the ionic fraction of chloride ions in pure molten magnesium chloride, which may be different from unity, because of the possible formation of $MgCl_4^{2-}$ ($x_{Mg}^0 = 1$, because Mg^{2+} is the only cation in this melt).

In the general case the ions and the associated species mix non-ideally so that both association constants and interaction parameters characterize the system. The Margules expression (4.66) may be applied to the stoichiometric activity coefficients. The leading term (that is b'_1 times the square of the ion fraction of the other salt) may be related to the association constants[15], yielding a negative value for b'_1 for any value of the latter for charge-symmetrical systems, and in general

$$-K_p\,[(p - z_A)K_p - 1]/(K_p + 1)^2 > b'_1 > -(p - z_A) \qquad (7.37)$$

for systems AX_{z_A} + BX where species $AX_p^{p-z_A}$ are formed. These values of b'_1 are numerically small, and in many cases smaller than the experimental values of b'_1 (obtained as $b_{j1} = (1/2)\lim(x_i \to 0)d^2(\ln f_j/dx_i^2)$, thus requiring the consideration of contributions from non-ideal mixing of the species, as specified in sections 7A(a) and 7A(b) above. The specification of the values b_{j1} depends on that of f_j, of μ_j^E or of g^E, and these in turn depend on the model selected for the ideal mixture. A system with large g^E based on Temkin ideality may have a negligible g^E based on equivalent fractions. The problem is compounded by the ambiguity of specifying p, the ligand number in the associated complex $AX_p^{p-z_A}$. The average nearest-neighbour coordination number in molten salts is $Z \sim 5 \pm 1$, which may accommodate any coordination from deformed tetrahedra to pyramids to deformed octahedra. It is very difficult to distinguish experimentally between directed coordination bonds and strong polarization or coulomb interactions in such situations.

These difficulties may be illustrated with the system MCl + $PbCl_2$, where M^+ is an alkali metal cation. A large amount of experimental information, obtained by a variety of methods (conductivity, transport numbers, surface tension, volume and enthalpy of mixing, chemical potentials from vapour pressures and e.m.f. measurements, Raman specta, etc.)[17], is available on these systems. There is general agreement that there is hardly any association in the cases $M^+ = Li^+$ or Na^+, and most of the evidence has been interpreted in terms of some association in the cases $M^+ = K^+$, Rb^+ or Cs^+. Even in these cases, though, it is agreed that the complex(es) formed is (are) not very strong. However, a controversy exists on the nature of the species present, and doubt on the presence of associated species at all, even in mixtures with the larger alkali metal cations, has been expressed[18]. The conductivity minimum and Raman spectrum point strongly to the presence of $PbCl_3^-$ or $PbCl_4^{2-}$ in KCl + $PbCl_2$ melts, and the strongly negative heats of mixing in this and the CsCl + $PbCl_2$ melts (Fig. 7.3) seem to confirm complex formation. The heats of mixing leads to λ with negative deviations from linearity which are maximal at 70–75 % CsCl, i.e. complex compositions $PbCl_4^{2-}$ or $PbCl_5^{3-}$. On the other hand, there is no definite information on the trend of λ with composition in charge-asymmetrical chloride systems where no complex formation is expected, e.g. the KCl + $CaCl_2$ system, but in other systems where no complex formation has been found, deviations from linearity (i.e., the cx_Ax_B term in λ′ of eq. (7.23)) have been recorded, e.g. for certain mixed alkali metal halides (Appendix 7.1, p. 290). The main difficulty, however, arises in the consideration of the Gibbs energies. In several studies excess Gibbs energies or chemical potentials have been

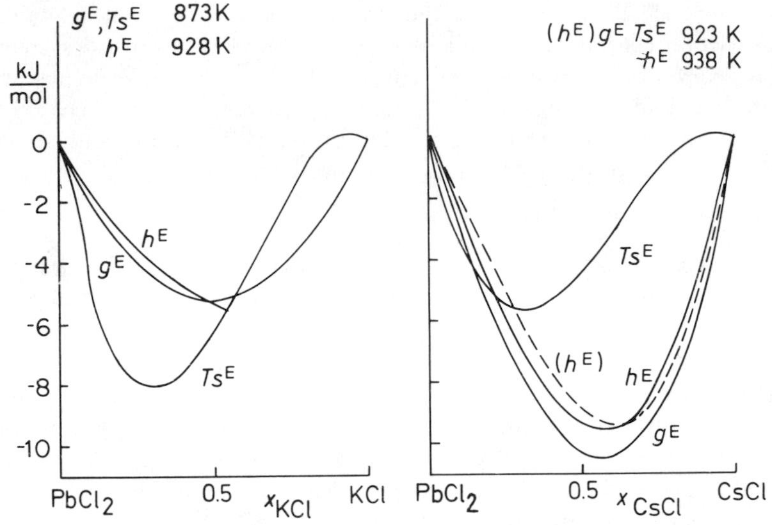

Figure 7.3. Excess functions for $PbCl_2$ + KCl (left hand) and $PbCl_2$ + CsCl (right hand) systems. The data for g^E and s^E are from Janz and Dijkhuis[17], those for h^E from McCarty and Kleppa[17]. Note that for the KCl system these data are not compatible with each other ($-Ts^E$ is negative), whereas for the CsCl system they are compatible (dashed curve for $(h^E) = g^E + Ts^E$).

calculated, based on Temkin ideality, eqs. (7.3) and (7.4), which are negative, and lead to positive partial excess entropies of lead chloride, contrary to the expectation on the basis of complex formation. However, if ideality on the basis of equivalent fractions is used as a basis for calculating excess functions (cf. the work of Bredig[18] and Vos[18]) the KCl + $PbCl_2$ system is nearly ideal: at 873 K e.m.f. data yield $g^E(\text{kJ mol}^{-1})/x'_D x'_{Pb} = -0.1 - 0.1x'_K$. A constant $K(PbCl_2 + CsCl \rightleftharpoons CsPbCl_3) = 30$ has been found to fit vapour pressure data, on the basis of Temkin activities and an ideal association system. On this basis a partial excess entropy of lead chloride $\bar{s}^E_{PbCl_2}$ which is positive at low and negative at high x_{Pb} is found. This is in agreement with data on $\bar{s}^E_{PbCl_2}$ derived from the temperature-dependence of the e.m.f. for the $PbCl_2$ + KCl system, but for the $PbCl_2$ + CsCl system a uniformly positive $\bar{s}^E_{PbCl_2}$ has been found. In conclusion, therefore, it is seen that some of the data are in conflict, that complex formation cannot account for data such as positive excess entropies, and that the species formed cannot be derived unambiguously from the data.

d. Ternary additive systems

The mixing properties of additive ternary systems with a common ion, such as AX + BX + CX, may be expressed simply as

$$h^E = x_A x_B \lambda_{AB} + x_B x_C \lambda_{BC} + x_A x_C \lambda_{AC} + x_A x_B x_C \lambda_{ABC} \tag{7.38}$$

and similarly for g^E and v^E, although data are scarce. The binary terms predominate, and the (composition-dependent) λ_{ik} values are assumed to be those for the corresponding binary mixture at the same n_i/n_k ratio as for the actual ternary mixture The last term, the ternary one, was found to be small in alkali nitrate + silver nitrate ternary mixtures, but not entirely negligible[19]. In the case of a dilute solution of one component, say AX, in mixtures of the other two, it is useful to transform the composition variables to

$$x = n_{BX}/(n_{BX} + n_{CX}) = x_{BX}/y \tag{7.39a}$$

$$y = (n_{BX} + n_{CX})/(n_{AX} + n_{BX} + n_{CX}) = 1 - x_{AX} \tag{7.39b}$$

As $y \rightarrow 1$ the system approaches the binary BX + CX. It has been shown[20] that

$$g^E = xg^E_{AB} + (1 - x)g^E_{AC} + y\left[g^E_{BC} + x(1 - x)\int_0^{1-y} g(x,y)\mathrm{d}y\right] \tag{7.40}$$

where g^E_{ik} are the excess molar Gibbs energies of the indicated binary systems, and g is a ternary deviation function defined by

$$\mu^E_A = x\mu^E_{A(AB)} + (1 - x)\mu^E_{A(AC)} + x(1 - x)y^2 g(x,y) \tag{7.41a}$$

with the important constraint that

$$\int_0^1 [x(1 - x)g - g^E_{BC}]\,\mathrm{d}y = 0 \tag{7.41b}$$

Independence of g from y is equivalent to a negligible ternary term b_{ABC} (corresponding to λ_{ABC} for excess enthalpies in (7.38)), and leads to $x(1 - x)g = - g^E_{BC}$, so that instead of (7.40)

$$g^E = xg^E_{AB} + (1 - x)g^E_{AC} - y^2 g^E_{BC} \tag{7.42}$$

The chemical potential of trace AX in the mixture can be written as

$$\lim_{y\rightarrow 1} \mu^E_A = x\mu^E_{A(B)} + (1 - x)\mu^E_{A(C)} - g^E_{BC} + [x(1 - x)g(x,1) + g^E_{BC}] \tag{7.43}$$

where, in view of the preceding, the term in square brackets may be negligible. If g^E_{BC} is very negative, $\mu^E_A(x, y \sim 1)$ will show a maximum, and this has been verified for several systems. The deviation of g from $-g^E_{BC}/x(1 - x)$ can be obtained from the quasi-chemical, quasi-lattice theory[21] as

$$g(x, 1) + g^E_{BC}/x(1 - x) \simeq -Z'\mathbf{R}T \ln \cosh [(g^E_{A(B)} - g^E_{A(C)})/2Z'\mathbf{R}T] \tag{7.44}$$

which is a negative quantity. If this is sufficiently large, it may counteract partly the effect of a negative g^E_{BC} in (7.43).

The ternary enthalpy term, λ_{ABC}, was found to be significant, though small, for silver nitrate+alkali nitrate ternary mixtures[19], and $g(x, 1) + g^E_{BC}/x(1 - x)$ was also found to be so for silver chloride+alkali chloride systems[20]. Larger deviations were found for systems with lead instead of silver. The detailed interactions responsible for these results have not yet been found.

C. Reciprocal Mixtures

The simplest reciprocal systems consists of two different cations and two different anions, all of the same charge. As mentioned in the Introduction, section 7A, one mole of such a mixture may be made up by mixing three components, making arbitrary choices of x_A and x_Y

$$x_A AX + x_Y BY + (x_B - x_Y)BX \rightarrow x_A A^+ + x_B B^+ + x_X X^- + x_Y Y^- \qquad (7.45)$$

where $x_X = x_A + x_B - x_Y$ and the restriction $x_A + x_B = x_X + x_Y = 1$ apply. When such a mixture is prepared from two components (Fig. 7.4), the nearest-neighbour interactions between A^+ and X^-, or between B^+ and Y^- are to some extent replaced by interactions $A^+ - Y^-$ and $B^+ - X^-$. Therefore, some contribution to the Gibbs energy of mixing (and the excess enthalpy) arises from the corresponding quantities (g^0 and h^0) for the metathetical reaction

$$AX + BY \rightleftharpoons AY + BX \qquad (7.46)$$

where AX, etc. are the pure liquid salts.

As a first approximation, all the contributions from the interactions of ions which are not nearest-neighbours may be ignored and completely random mixing of the ions of each kind of charge may be assumed[22]. Each ion then undergoes Z independent interactions with its neighbours. In one mole of mixture there would be $\mathbf{N}Zx_Ax_X$ interactions with the energy ϵ_{AX}, while in the x_A moles of pure AX which go into making the mixture according to (7.45) there are $\mathbf{N}Zx_A$ such interactions. The change in Gibbs energy or enthalpy on mixing according to (7.45) is on this account therefore

$$\begin{aligned} g^E = h^E &= \mathbf{N}Z[x_Ax_X\epsilon_{AX} + x_Ax_Y\epsilon_{AY} + x_Bx_X\epsilon_{BX} + x_Bx_Y\epsilon_{BY} \\ &\quad - x_A\epsilon_{AX} - x_Y\epsilon_{BY} - (x_B - x_Y)\epsilon_{BX}] \\ &= \mathbf{N}Zx_Ax_Y[\epsilon_{AY} + \epsilon_{BX} - \epsilon_{AX} - \epsilon_{BY}] = x_Ax_Yu^0 \end{aligned} \qquad (7.47)$$

where $u^0 = \mathbf{N}Z[\epsilon_{AY} + \epsilon_{BX} - \epsilon_{AX} - \epsilon_{BY}]$ is the molar energy change for the metathetical reaction (7.46). The mole fraction product x_Ax_Y in (7.47) refers to that component which was omitted in preparing the mixture according to (7.45). If the distribution of the ions on their quasi-lattices is completely random (i.e. for $u^0 \ll \mathbf{R}T$), (7.47) may suffice to describe the actual excess properties of the mixture, provided nearest-neighbour interactions predominate strongly. Regular-type behaviour (section 5A(c)) results. The partial molar quantities are obtained from (7.47) in a straightforward way, e.g.

$$\begin{aligned} \mu^E_{BY} &= \mathbf{R}T \ln f_{BY} = \partial ng^E/\partial n_{BY} = \partial ng^E/\partial n_{B^+} + \partial ng^E/\partial n_{Y^-} \\ &= (1 - x_B)(1 - x_Y)u^0 \end{aligned} \qquad (7.48)$$

where $n = n_{A^+} + n_{B^+} = n_{X^-} + n_{Y^-} = n_{AX} + n_{BX}$ is the total number of moles of salt present, the last two equalities referring to the mixing in (7.45).

A second approximation takes into account, in addition, contributions of next-nearest-neighbour interactions, assuming that in the ternary (reciprocal)

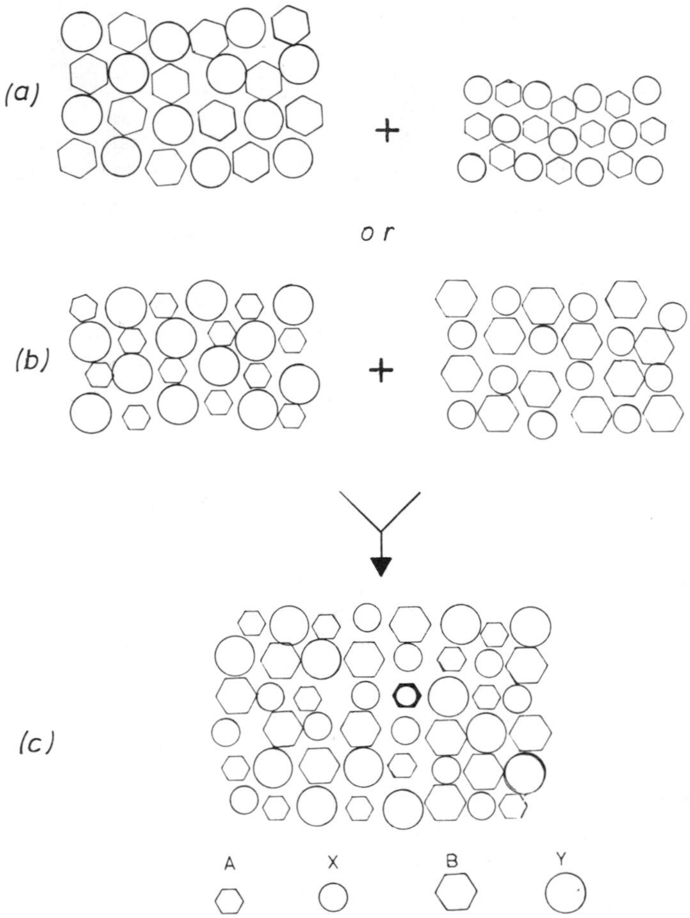

Figure 7.4. Quasi-lattice (two-dimensional) representation of a reciprocal charge-symmetrical molten salt system: (a) AX + BY; (b) BX + AY; (c) mixed BX, AY, BY, with n_X moles of BX, n_A moles of AY and $(n_Y\text{-}n_A)$ moles of BY. The marked cation A is surrounded at random with nearest-neighbour X and Y ions, and with next-nearest-neighbour A and B ions.

mixture, the contribution of each binary system, such as (A, B)X, is additive according to the mole fraction of its common ion: $x_X h^E_{X(A,B)}$ (and similarly for g^E). For binary interactions which are not excessively strong the first term in (7.21) or (7.23) is adequate, so that

$$g^E = h^E = x_A x_Y u^0 + x_A x_X x_Y \lambda_{A(X,Y)} + x_B x_X x_Y \lambda_{B(X,Y)} + x_X x_A x_B \lambda_{Y(A,B)} \tag{7.49}$$

results[2b]. However, even this expression does not pay attention to the ternary nature of the mixture, and, empirically, it is recognized that a ternary term, written $x_A x_B x_X x_Y \Lambda$ is necessary to describe the experimental results[23]. The equation becomes

$$h^E = x_A x_Y h^0 + \sum^4 x_i x_j x_k \lambda_{i(j,k)} + x_A x_B x_X x_Y \Lambda \tag{7.50}$$

for the enthalpy change of mixing, while for the excess Gibbs energy, g^0 is used in the first term, binary factors $b_{i(j,k)}$ replace the λ values in the second, and an appropriately modified Λ is substituted in the third[24]. The excess entropy of mixing is small, though not entirely negligible[2c], whereas a zero entropy change for (7.46) would have been expected, if configurational entropy only contributed.

Partial molar quantities are obtained from (7.50) in the same manner as for (7.48), resulting in rather complicated expressions. For the component BY, for example

$$\begin{aligned} \mathbf{R}T \ln f_{BY} = \mu^E_{BY} &= x_A x_X g^0 + x_A x_X (2x_X - 1) b_{A(X,Y)} \\ &+ x_X (x_A + x_X - 2x_A x_X) b_{B(X,Y)} + x_A (x_A + x_X - 2x_A x_X) b_{(A,B)Y} \\ &+ x_A x_X (2x_A - 1) b_{(A,B)X} + x_A x_X (2x_A + 2x_X - 1 - 3x_A x_X) \Lambda \end{aligned} \tag{7.51}$$

If pseudo-binary compositions only are considered, i.e. $x_A = x_X = x$ and $x_B = x_Y = 1 - x$, some simplification is possible, leading to

$$\begin{aligned} \mathbf{R}T \ln f_{BY} = \mu^E_{BY} &= x^2 g^\circ + x^2 [(b_{A(X,Y)} + b_{Y(A,B)} + 2(1 - x)\{b_{B(X,Y)} \\ &+ b_{X(A,B)} - b_{A(X,Y)} - b_{Y(A,B)}\}] + x^2 (1 - x)(3x - 1)\Lambda \end{aligned} \tag{7.52}$$

Similar expressions are obtained for the other components. Expressions similar in principle, i.e. involving a term in g^0, four terms in the b values and a term in Λ, but more complicated in detail because of the inclusion of stoichiometric factors, are obtained for charge-unsymmetrical reciprocal systems, e.g. $Ag_2SO_4 + NaNO_3$ (ref. 24b).

a. Theories of non-random reciprocal mixtures

Many of the considerations which were applied to binary mixtures with a common ion can be applied to reciprocal mixtures too. The conformal ionic mixture theory[3] can be applied as follows[25], using (7.10) for the configurational partition function, and recognizing four perturbation parameters $\rho_{ij} = (d/d_{ij}) - 1$ with i = A or B, and j = X or Y. The sums entering the potential energy $U(\rho)$ (cf. (7.11)) for reciprocal mixtures cover all the combinations of cations and anions. The excess Helmholz energy of mixing A^E now contains both first- and higher-order terms in ρ_{ij}. For one mole of mixture it can be written as

$$a^E = x_A x_Y a^0 + \sum^4 x_i a^E_{i(j,k)} + x_A x_B x_X x_Y P(\rho_{AY} + \rho_{BX} - \rho_{AX} - \rho_{BY})^2 + \ldots \tag{7.53}$$

where a^0 is the molar Helmholz energy change for reaction (7.46), $a^E_{i(j,k)}$ are binary terms given by (7.12) as $x_j x_k a^0_{i(j,k)}(\rho_j - \rho_k)^2$, and P is a difference between two undetermined complicated integrals, depending on the temperature. The similarity of (7.53) to the empirical (7.50) is obvious. However, the nature of the ternary term in the latter, Λ, is not given explicitly by the conformal mixture theory, since P cannot be determined.

More insight can be obtained from the quasi-chemical theory. The distribution of the ions on their quasi-lattices is not completely random and reciprocal mixtures have non-zero excess entropies of mixing. If these deviations from random mixing are not very large, the *symmetric approximation* of the quasi-chemical theory can be applied as follows[22,26]. Consider the symbols in (7.46) to represent not molar quantities but pairs of cations and anions. The molecular interchange energy is then* $w = u^0/\mathrm{N}Z$. The deviation from random ionic distribution, in which the probability of the occurrence of an AY pair would be $x_A x_Y$, will be designated by y, giving the probability $x_A x_Y + y$ for the non-random distribution. Application of the mass-action law to (7.46) yields then

$$(x_A x_Y + y)(x_B x_X + y)/(x_A x_X - y)(x_B x_Y - y) = K = \exp(-w/\mathrm{k}T) \qquad (7.54)$$

If both numerator and denominator on the left-hand side of (7.54) are divided by $x_A x_B x_X x_Y$, logarithms are taken and expanded, and the series truncated after the terms in the first power of y (for y much smaller than any $x_i x_j$ product), then

$$y = -x_A x_B x_X x_Y w/\mathrm{k}T \qquad (7.55)$$

results. That is, for small deviations from randomness, the parameter y is given by the interchange energy. For each AY + BX configuration produced over the random configuration, w is contributed to the excess energy of mixing, and there are $\mathbf{N}Zy$ such configurations, leading to the excess energy per mole of mixture

$$u^E \sim h^E = -x_A x_B x_X x_Y \mathrm{N}Zw^2/\mathrm{k}T = (x_A x_B x_X x_Y) - u^{o2}/Z\mathrm{R}T \qquad (7.56)$$

The excess entropy is $1/2T$ times h^E, and $g^E = \frac{1}{2}h^E$, as is obtained from the Gibbs–Helmholz relationship. This contribution should be added to those arising from the random mixing, embodied in (7.49), so that it is possible to identify the quantity Λ in (7.50), the ternary contribution, as

$$\Lambda = -u^{o2}/Z\mathrm{R}T \qquad (7.57)$$

Hereby the parameter Z of the quasi-lattice model, its coordination number, with values ranging typically from 4 to 6, is explicitly introduced into the excess functions for reciprocal mixtures.

Expansion of the logarithmic representation of (7.54) beyond first powers of y

*Actually, w is less than $u^0/\mathrm{N}Z$, since in u^0 are included not only nearest-neighbour interactions, which contribute exclusively to $w = \epsilon_{AY} + \epsilon_{BX} - \epsilon_{AX} - \epsilon_{BY}$. The existence of only small deviations from randomness in this treatment implies that nearest-neighbour interactions by far outweigh other effects.

leads to further terms in (7.56), or for that matter in (7.50), all of which retain symmetry with respect to the mole fractions.

The energy of the metathetical reaction (7.46) is seen in (7.47) to arise from changes in pair-wise cation–anion interactions. These, in turn, are dominated by coulombic interactions, the *reciprocal Coulomb effect* (rCe) which may be written as

$$u^0_{\mathrm{rCe}} = -\mathbf{N}Ke^2\,[d^{-1}_{\mathrm{AY}} + d^{-1}_{\mathrm{BX}} - d^{-1}_{\mathrm{AX}} - d^{-1}_{\mathrm{BY}}] \tag{7.58}$$

where K may be taken as a reciprocal effective dielectric constant (cf. (7.7b, c) and (7.22)) or to play the role of a Madelung constant. For mixtures containing highly polarizable ions, other contributions (cf. (7.25), (7.26) and (7.30)) will predominate over the Coulomb effect. For charge-unsymmetrical systems the terms in d^{-1}_{ij} should each be multiples by $z_i z_j$.

. The molar metathetical interchange energy u^0 may also be estimated from the softness parameters of the ions[12]. For the isolated ion-pairs participating in (7.46) and reacting in the gas phase, the molar energy change was found to be proportional to the product of the differences of the softness parameters of the two cations and of the two anions. For the condensed phase, the molten salt mixture, there are $Z/2$ as many cation–anion contacts as in the gas phase. The resulting expression is

$$u^0(\mathrm{kJ\ mol^{-1}}) = 230(\sigma_{\mathrm{A}} - \sigma_{\mathrm{B}})(\sigma_{\mathrm{X}} - \sigma_{\mathrm{Y}}) \tag{7.59}$$

where σ_{M} values from Table 7.2 are used for σ_{A} and σ_{B}, and σ_{X} values from this Table for σ_{X} and σ_{Y}. Good agreement was obtained for many systems with g^0 or h^0 values calculated for (7.46) from Gibbs energies (enthalpies) of formation and heats of fusion for the salts found in the literature.

Dilute solutions of one salt, e.g. AX, in another salt with no common ions, e.g. BY, which may be considered as the solvent, lend themselves to a different treatment. Experience shows that configurations XAX, especially for multivalent $\mathrm{A}^{z_{\mathrm{A}}}$ and monovalent X^-, may occur frequently, but configurations such as AXA or AYA would be very rare in such solutions, and may be discounted. This leads to the *asymmetric approximation*[27] for the treatment of deviations from randomness by means of the quasi-chemical theory. Two regions on the anion sublattice are distinguished: region a, which includes all those sites which are adjacent to one $\mathrm{A}^{z_{\mathrm{A}}}$ cation and $(Z-1)\mathrm{B}^+$ cations, and region b, which includes all other sites, having only B^+ as neighbours. The exchange of one anion X^- from the a region with an anion Y^- from the b region entails a change in energy of $\Delta\epsilon$. (If $\Delta\epsilon < 0$, X^- are stabilized in region a, that is, adjacent to an $\mathrm{A}^{z_{\mathrm{A}}}$ cation, relative to region b.) It is now expedient to calculate and maximize the number of ways of distributing the anions over the anion sites and the cations over the cation sites, rather than use the mass-action law expression employed in (7.54). Let x be the fraction of the a-type anionic sites occupied by X^- anions. There are altogether $\mathbf{N}Zx_{\mathrm{A}}$ a-type sites per mole of mixture, contributing $\mathbf{N}Zx_{\mathrm{A}}x\Delta\epsilon$ to the energy of the system, owing to occupancy by X^- anions, while X^- anions on b-type sites and Y^- anions do not contribute to the relative energy. The most probable distribution comes out to be given by

$$x/(1-x) = \exp(-\Delta\epsilon/kT)(x_X - Zxx_A)/(1 - Z(1-x)x_A - x_X) \qquad (7.60)$$

which can be solved for the limit $x_A \to 0$ and for $\Delta\epsilon \ll kT$ as $x = (1 - \Delta\epsilon/kT)x_X$. This expression clearly shows the excess stabilization of X^- on *a*-type sites for $\Delta\epsilon < 0$, relative to the random distribution. The excess chemical potential of the solute AX is given by

$$\mu^E_{AX} = RT[\ln(1 - Zxx_A/x_X) - \ln(1 - Z(1-x)x_A - x_X) + Z\ln(1-x) + (Z-1)\ln(1 + x/(1-x)\exp(-\Delta\epsilon/kT))] \qquad (7.61a)$$

which, at the limit $x_A \to 0$ reduces to

$$\mu^{E\infty}_{AX} = RT\ln f^{\infty}_{AX} = ZRT\ln[(1-x_X)x_X^{-1} - (1-x_X)\exp(-\Delta\epsilon/kT)] \qquad (7.61b)$$

The excess chemical potential of the other component containing A, AY, at the limit $x_A \to 0$ is

$$\mu^{E\infty}_{AY} = RT\ln f^{\infty}_{AY} = -ZRT\ln[(1-x_X) + x_X\exp(-\Delta\epsilon/kT)] \qquad (7.62)$$

while $\mu^{E\infty}_{BX} = \mu^{E\infty}_{BY} = 0$.

b. Associated reciprocal systems

It is difficult to distinguish between satibilization of X^- anions on sites adjacent to A^+ cations, discussed in the last paragraph, and bonding of A^+ and X^- ions. The former situation is discussed in terms of the asymmetric approximation to the quasi-chemical treatment of deviations from random distribution, the latter in terms of association or complex formation. This approach is feasible for dilute solutions of the associating ions in a solvent consisting of 'inert' ions, since then the associated aggregates are relatively improbable configurations, unless bonding occurs. A distinction may therefore be made on the basis of the occurrence of extra-coulombic forces leading to the bonding, but even these may be difficult to establish independently. Most of the quantitative work on complex formation in molten salts (dilute reciprocal systems), in fact, does not concern itself with the bonds formed, but infers them from the thermodynamic properties of the melts. The most successful and widely used method is the study of the electromotive force of suitable cells. It is convenient to use the *mole ratio* concentration scale, which at inifnite dilution coincides with the mole fraction scale. For the solute ij in the solvent BY

$$R_{ij} = n_{ij}/n_{BY} \qquad (7.63)$$

where ij may be AX, AY or BX. Data expressed on the molality scale are easily converted by $R_{ij} = m_{ij}M_{BY}^{-1}$, where M_{BY} is the molar mass (in kg) of the solvent, if necessary weighted according to its composition (e.g. in a (0.5 Na + 0.5 K)NO_3 solvent). The stoichiometric concentration R_{ij} is the sum of the mole ratios of the species, weighted according to their stoichiometries, e.g.

$$R_{AY} = R_{A^+} + R_{AX} + R_{AX_2^-} + 2R_{A_2X^+} + \ldots \tag{7.64}$$

It is assumed that the species behave ideally in dilute solutions, so that (charges on species omitted)

$$K_n = R_{AX_n}/R_{AX_{n-1}}R_X \tag{7.65}$$

The analysis of the data has been made mainly by one of two approaches[28]. In the first approach, the e.m.f. data give the stoichiometric activity coefficient of the component AY if the electrodes are reversible to A and the reference half-cell contains only AY, and no BX. The stoichiometric activity coefficients are also equal to the ratio of R of the 'free' ion to its stoichiometric R

$$f_{AY} = R_{A^+}/R_{AY} \tag{7.66}$$

Equations (7.64), (7.65) and (7.66) are used to relate f_{AY} to R_{BX} as follows (neglecting species such as A_2X^+)

$$-\ln f_{AY} = K_1 R_{BX} + (K_1 K_2 - \tfrac{1}{2}K_1^2)R_{BX}^2 + \ldots \tag{7.67}$$

so that

$$K_1 = \lim_{R_{AY}, R_{BX} \to 0} -(\partial(\ln f_{AY})/\partial R_{BX}) = \lim_{R_{AY}, R_{BX} \to 0} -(\partial(\ln f_{BX})/\partial R_{AY}) \tag{7.68}$$

and K_2 is obtainable from the second derivatives. The second equality in (7.68) applies when f_{BX} is obtainable from the e.m.f. data. In this approach the data are obtained as functions of both R_{AY} and R_{BX} and extrapolated to infinite dilution. Dependence of f_{AY} on R_{AY} signifies appreciable concentrations on the species A_2X^+ (or others containing more than one A), which are excluded from consideration in the asymmetrical approximation.

The second approach utilizes the conventional calculation of complex equilibrium constants familiar from aqueous solutions. If polynuclear species (such as A_2X^+) are excluded, this leads to

$$(1 - f_{AY})/f_{AY}R_{BX} = K_1 + K_1 K_2 R_{BX} + \ldots \tag{7.69}$$

where, again, K_1 is obtained by extrapolation to $R_{BX} = 0$ or numerically.

It is instructive to compare the association constants obtained from these treatments of the data, with the expectations from the quasi-lattice, quasi-chemical model[29]. A comparison (7.68) with (7.62) yields readily

$$K_1 = Z(\exp(-\Delta\epsilon/kT) - 1) \tag{7.70}$$

and it can be shown that for higher associations, the statistical ratios[29,30]

$$K_n = (Z - n + 1)K_1/Zn \tag{7.71}$$

apply, if it is assumed that $\Delta\epsilon$ is the same for all the steps (designated by index n) in the association $AX_{n-1} + X \rightleftharpoons AX_n$ (charges omitted). The quantity $A_1 = \mathbf{N}\Delta\epsilon$ is called the *specific molar Helmholz bond energy for the association* of A with X. It can be calculated for any association step from $A_n = -\mathbf{R}T \ln[nK_n(Z - n + 1)^{-1} + 1]$, and, according to the above assumption should be independent of n, i.e. $A_n = A_1$ for any n. This was confirmed for several solutes in nitrate media[31], at least as far as the second association step, and for silver bromide even for the association to Ag_2Br^+. Furthermore, for ions with no internal degrees of freedom, $\Delta\epsilon$ should be temperature-independent, and so should A_1 obtained from the temperature–dependence of K_1 according to (7.70) (provided Z does not vary with the temperature). This is, indeed, the case for many systems that have been studied. Appendix 7.2 p. 294, shows the data for nitrate solvents (BNO_3, where B is an alkali metal or alkaline-earth cation or their mixtures) for silver, cadmium and lead halide associations, and for several silver-polyatomic anion associations. The coordination number Z was arbitrarily fixed at 5, which is a reasonable mean value for nearest-neighbours in molten salts. (Table 3.10). The non-vanishing of dA_1/dT for the polyatomic anion cases has been ascribed to vibrational entropy contributions, and for the few cases involving monatomic anions to experimental error.

Very little reliable information on association constants and specific Helmholz bond energies in molten ionic solvents other than nitrates is available: certainly not sufficient for discussing effects of the composition of the solvent. For nitrate melts, and the association, say, of silver with chloride or of cadmium with bromide, data are available for several pure alkali nitrates, their mixtures, and some mixtures with alkaline-earth nitrates. To a first approximation a linear dependence of A_1 on the cation composition of the solvent is observed. However, for the single alkali nitrate solvents, the order of $|A_1(\mathrm{M})|$ with the alkali cations M depends on the associating ions, e.g. it is Na < Li < K < Cs for Ag–Cl, Na < K < Li for Cd–Br and Na > K for Ag–I. This has been explained in terms of the reciprocal coulomb effect[32], eq. (7.58), but for its application a value for $d_{\mathrm{MNO_3}}$, or the radius of the nitrate ion is required. This turns out to be a very elusive quantity, values for molten salts having been reported from 1.78 to 2.30, depending on the cation and perhaps even more on the situation a particular author wished to explain. A size $r_{\mathrm{NO_3}} \simeq r_{\mathrm{Br}}$ does explain in the present situation many of the data, including, for example, the inversion of Na relative to K for Ag–Cl and Ag–I, in terms of the reciprocal coulomb effect. The linear dependence of A_1 on the cation composition of the nitrate solvent has not remained unchallenged. Both the composition-dependence and the temperature-dependence observed for mixtures but not for pure components have been described in terms of the additivity of $(K(\mathrm{M}_i) + Z)/Z$ with x_{M_i} in the solvent (i.e. of $\exp(-A(\mathrm{M}_i)/\mathbf{R}T)$ rather than of $A(\mathrm{M}_i)$ itself), but the data are inconclusive and unable to determine which is the correct dependence. Both representations are empirical, so no decision on theoretical grounds is possible either. As the deviations are in all cases close to experimental error, this problem seems rather academic.

c. Miscibility gaps

In a manner similar to other liquid mixtures, miscibility gaps occur in mixed molten salts if the excess Gibbs energy of mixing is sufficiently large and positive (section 4A(c)). These gaps are rather rare for highly ionic common-ion binary systems (they do occur for systems such as $NaCl + AlCl_3$, where one component is a rather covalently bound salt), but are not uncommon in reciprocal systems[14]. The miscibility gap is often fairly symmetrical with respect to the composition (Fig. 7.5), and extends in the phase diagram along the *stable diagonal*, i.e. that pair of components which for the purpose of eq. (7.46) with a negative value of g^0 are designated by AY and BX.

It is possible to predict the occurrence of a miscibility gap from thermodynamic data: g^0 and $b_{(ij)k}$ (or $\lambda_{(ij)k}$) of the four binaries[24a], or from the softness parameters of the component ions[12]. This is done by calculating the upper critical consolute temperature T_c of the system, and comparing it with the liquidus temperatures. If the former is appreciably higher than the latter (in particular, higher than that of the highest melting salt), then a miscibility gap is assured. To that approximation, which permits the estimation of the excess Gibbs energy from (7.47), the upper critical consolute temperature of the miscibility gap is given by $T_c = |u^0|/4\mathbf{R}$ (ref. 22). Present theory does not permit the estimation of the extent of the gap, and a sensible further discussion can only be made if pseudo-binary compositions ($x_A = x_Y = x$ and $x_B = x_X = (1 - x)$) are considered and symmetry of the gap, i.e. occurrence of T_c at $x = 0.5$, is assumed.

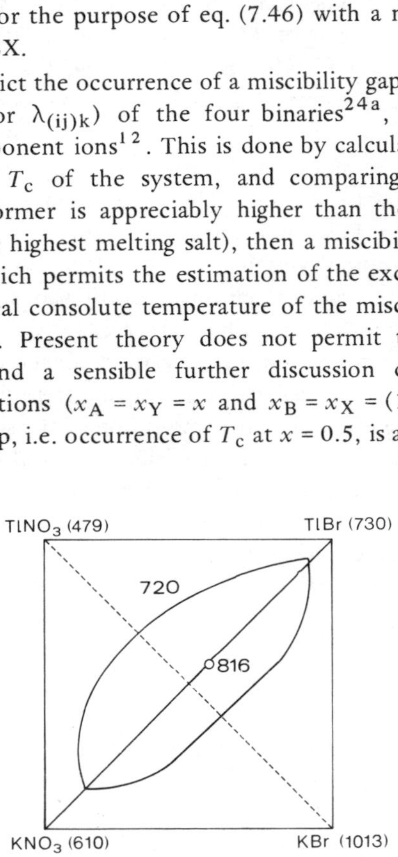

Figure 7.5. Phase diagram of the reciprocal system (K^+, Tl^+, Br^-, NO_3^-) The figures at the corners are the melting points (T_m(K)) of the pure components, the contour at 720 K shows the miscibility gap at that temperature, and the point indicates the composition at the upper critical consolute temperature of 816 K. The stable diagonal is continuous, the unstable one dashed (C. Sinistri, P. Franzosini, A. Timidei and M. Rolla, *Z. Naturforsch. A*, **20**, 561 (1965)).

According to the general requirements for the occurrence of a miscibility gap, eq. (4.83), $\partial a_{BY}/\partial(1-x) < 0$, may be selected as a criterion. At T_c this inequality turns into an equality, and with $a_{BY} = x(1-x)f_{BY}$ yields

$$(\partial(\ln f_{BY})/\partial(1-x))_{T_c} = -2/x \qquad (7.72)$$

This may now be applied to (7.51), which gives f_{BY} for the pseudo binary compositions along the stable diagonal

$$RT_c/x(1-x) = |g^0| + 1/2 \sum^4 b_{i(j,k)} + 3\,(½ - x)[b_{(A,B)Y} + b_{B(X,Y)} - b_{(A,B)X} - b_{A(X,Y)}] + (6x(1-x) - 1)\Lambda \qquad (7.73)$$

For symmetric gaps, as assumed above, $x(T_c) = ½$, so that the third term on the right-hand side of (7.73) vanishes and $T_c = g^0/4R + \Sigma^4 b_{i(j,k)}/8R + \Lambda/8R$ is obtained. This can be further simplified if a relationship between Λ and g^0 is assumed, e.g. (7.59), where g^0 replaces u^0, or for large g^0, simply $\Lambda \cong -3RT_c$ (ref. 24a). In the latter case (7.73) simplifies to the approximate expression

$$T_c = (2\,|g^0| + \sum^4 b_{i(j,k)})/11R \qquad (7.74)$$

This equation permits the use of tabulated Gibbs energies of formation of the salts (corrected for melting at the temperature considered if this refinement is wanted) and $b_{i(j,k)}$ (or the more readily available $\lambda_{i(j,k)}$ if the small difference is ignored) of the four binary mixtures to predict T_c, and hence the occurrence of a miscibility gap, as described above.

It has been shown[12] that to the same degree of approximation, it is possible to dispense with the thermodynamic data on the salts and binaries altogether, and make the prediction of T_c and of the occurrence of a miscibility gap from the properties of the ions themselves, namely from their softness parameters (Table 7.2). Equations (7.59) and (7.30) are used with (7.74) for this purpose for uni-univalent salts (replacing g^0 by u^0 and $b_{i(j,k)}$ by $\lambda_{i(j,k)}$)

$$T_c\,(K) = 5030\,|(\sigma_A - \sigma_B)(\sigma_X - \sigma_Y)| - 370\{2[(\sigma_A^2 - \sigma_B^2)^2 + (\sigma_X^2 - \sigma_Y^2)^2] - (\sigma_A + \sigma_B)(\sigma_X + \sigma_Y)[(\sigma_A - \sigma_B)^2 + (\sigma_X - \sigma_Y)^2]\} \qquad (7.75)$$

Good prediction of the occurrence of miscibility gaps in many systems was achieved by the use of (7.75), or the modifications of (7.74) and (7.75) for higher-valent salts[12]. The first term of (7.75) is from three to ten times larger than the second, and its use alone corresponds to an estimate 8/11 times that of the simplest approach, considering only nearest-neighbour interchange[22]. The relative magnitude of the terms in the braces (curly brackets) determine whether T_c would be lowered or raised by the next-nearest-neighbour interaction. The first term inside the braces is positive and lowers T_c; the second may be negative, if either the sums $(\sigma_A + \sigma_B)$ or $(\sigma_X + \sigma_Y)$ is negative, and may tend to raise T_c; both kinds of effects have been noted. For systems involving multivalent ions, the importance of the second term of (7.75) is relatively greater, and may be comparable to that of the first. In general, therefore, the magnitude of g^0 (corresponding to the first term of (7.75)) alone is insufficient for a valid prediction of the occurrence of a miscibility gap.

D. Mixtures of Ionic and Molecular Compounds

Several kinds of mixtures of molten ionic salts and covalent liquids are known. They include such different types as mixtures of the halides of mercury and alkali metals, aqueous melts or molten hydrated salts, molten organic salts mixed with organic liquids, mixtures of molten oxides involving silica or boron oxide, etc. No systematic discussion of all these systems can be included within the scope of this book but some of the more important ones will be dealt with. Mixtures of molten salts and metals are discussed in Chapter 8.

As an introduction, it is useful to consider the limit of low concentrations of the non-ionic component in the molten salt, and deal with the solubility of gases in molten salts. Several methods are available to measure these solubilities: equilibrium vapour pressure, pressure drop on saturation, transpiration methods, etc., and most of the data were obtained at low pressures 10^2-10^4 Pa, although some of the data were obtained in the 10^5-10^8 Pa range. Solubilities are usually sufficiently low for Henry's law to be followed, and it is also permissible to apply ideal gas laws to the equilibrium gas phase. The data are usually expressed in terms of the reciprocal of the Henry's law constant, K, expressed in moles gas (mole salt)$^{-1}$ Pa^{-1}. Conversion to the volume concentration solubility C, expressed in moles gas m^{-3} Pa^{-1}, is carried out by dividing K by v^* (m^3 mol^{-1}) of the salt (or salt mixture).

The data in the literature for some halide and nitrate salts and 'inert' gases (He, Ne, Ar, N_2, Cl_2) and some 'active' gases (HF, HCl, H_2O, NH_3, CO_2) are shown in Appendix 7.3, p. 296. It is evident that for the 'inert' gases C (moles gas m^{-3} Pa^{-1}) does not exceed 10^{-2}, and that ΔH for the gas solution process is positive. For the 'active' gases C may exceed 10^{-2}, and ΔH is negative. The magnitude of the solubility of the inert gases and of the positive ΔH involved can be explained in terms of the energy required to produce a hole in the melt, of the size required to accommodate the gas molecule[33]. This quantity is given by

$$\Delta h \simeq 4\pi r_g^2 \mathbf{N}\sigma \tag{7.76}$$

where r_g is the radius of the gas molecule (if spherical, otherwise $4\pi r_g^2$ is its effective surface) and σ the surface tension of the salt (the macroscopic value assumed to apply also to the microscopic hole). This Δh should apply to the transfer of the gas molecule from the gas to the melt phase at constant concentration, the entropy effect of this step being negligible. For the dissolution process at unit pressure, the (ideal) gas is first compressed isothermally to its pressure at the concentration equalling that in the melt. The enthalpy effect for this step is zero, the entropy effect $\Delta s = \mathbf{R} \ln \mathbf{R}TC$. The value of gas solubility arising from these considerations is

$$C = (\mathbf{R}T)^{-1} \exp(-4\pi r_g^2 \sigma/\mathbf{k}T) \tag{7.77}$$

Fairly good fits for actual solubilities of inert gases were obtained from this simple derivation. The experimental $\Delta s = \Delta h/T$ (since for the solubility equilibrium $\Delta g = 0$) may be more negative than $\mathbf{R} \ln \mathbf{R}TC$ in those cases where rotational entropy is lost on dissolution at constant concentration.

a. Solutions of water and molten hydrated salts

The negative ΔH for the solution of 'active' gases in salt melts show that not only is the energy required for creating the hole in the melt recovered in the process, but that additional interaction energy is released. Most of the information is available (of the Appendix 7.3, p. 296) for water as the solute. The enthalpy change of solution has an absolute magnitude similar to the heat of vaporization of liquid water at the relevant temperatures. There is, however, a difference, and more energy is required to remove the water from the nitrate melts containing lithium cations, pointing to the considerable strength of binding of the water in these melts[34,35]. The increased energy of vaporization of water ($-\Delta H$ for dissolution) is even more evident when polyvalent cations are present, but here the data extend to higher water concentrations – comparable to those of the salt, and hence to lower temperatures.

At these higher water concentrations, say at ratios water-to-salt R_H from 1 to 10, the distinction between solutions of water in molten salts and concentrated solutions of salts in water becomes vague. In particular several systems were found to behave as molten hydrated salts, in which all the water is tightly bound to the cations. This occurs for cations of sufficiently high field strength, when R_H corresponds to the coordination number (primary hydration number, p. 215), and leads to large cations such as $Li(H_2O)^+_{3\text{ to }4}$ in $LiCl + H_2O$ or $LiNO_3 + H_2O$, $Na(H_2O)^+_n$ in molten $Na_2S_2O_3 \cdot 5H_2O$, $Mg(H_2O)^{2+}_6$ in molten $MgCl_2 \cdot 6H_2O$ or $Mg(NO_3)_2 \cdot 6H_2O$, $Ca(H_2O)^{2+}_4$ in molten $Ca(NO_3)_2 \cdot 4H_2O$, $Al(H_2O)^{3+}_{6\text{ to }9}$ in $AlCl_3 + H_2O$ systems, etc. In fact, low-melting salt hydrates provide excellent examples of molten salts with hydrated cations[36,37] since they may as a rule be considerably undercooled. The resulting hydrated cations have low field strengths, $z_+(r_+ + 2r_{H_2O})^{-1}$ (nm) of 2.7 and 3.0 for the sodium and lithium ion, 5.3 and 5.8 for the calcium and magnesium ion and 9.1 for the aluminium ion, compared with 5.9 for Cs^+, 7.5 for K^+, 16.7 for Li^+ and > 20 for the multivalent bare (unhydrated) cations. The low field is manifested by the formation of tetrahedral $NiCl_4^{2-}$ anions when low concentrations of nickel are introduced into hydrated chloride melts, in analogy with chloride melts of other low-field cations, and in the behaviour of these molten hydrated salts with a given anion, in common-anion mixtures with other cations, in respect of conductivity, viscosity, and nuclear magnetic resonance spectra[37].

There are, however, some indications that the anions do penetrate the hydration shell when $R_H < Z$, where Z is the coordination number. Contact of cation and anion has been demonstrated spectroscopically[38] in such cases. One approach to this problem is to compare the energies of interaction of the water and the anion with some solute cation. The systems studied in this way were dilute reciprocal systems where the solvent salt was a light-metal nitrate (alkali metal, alkaline earth, or mixtures) and the solute was an associating heavy-metal halide (silver, lead and cadmium chlorides and bromides). The specific Helmholz energy A_1 for the association of, say, $Cd^{2+} + Br^-$ or $Ag^+ + Cl^-$ (p. 279) depends on the composition of the solvent: type of cations, moles of cation per mole of nitrate (for mixtures of uni- and di-valent salts), and moles of water per mole of nitrate. In agreement with the behaviour of anhydrous melts, this parameter usually depends linearily on the cation composition and is independent of the temperature[35–36].

Table 7.3. Specific molar Helmholz energies for association in hydrous nitrate melts (Z = 5 is arbitrarily selected).

Salt	R_H	T (K)	$-A_1$ (kJ mol^{-1})	Ref.
Association $Ag^+ + Cl^-$				
NH_4NO_3	0	(extrap.)	21.0	a
$(0.81NH_4 + 0.19Na)NO_3$	0	423	21.3	a
$(0.89NH_4 + 0.11K)NO_3$	0	433	22.2	a
NH_4NO_3	0.4	383	19.9	b
	0.6	383	19.5	b
	1.0	383	19.0	b
	1.4	383	18.7	b
	2.0	358	17.7	c
	2.0	383	18.5	b
	2.0	328	16.6	c
Association $Cd^{2+} + Br^-$				
$(0.5Li + 0.5K)NO_3$	0	392	26.9	d
	0	441	27.1	d
	0.1	441	25.9	d
	0.26	392	24.5	d
	0.51	392	22.9	d
	1.0	392	21.4	d
$Ca(NO_3)_2$	0	(extrap.)	31.8	e
$(0.5Ca + 0.5K)NO_3$	1.33	323	18.1	f
$Ca(NO_3)_2$	1.5	323	20.0	e
$(0.67Ca + 0.33K)NO_3$	1.6	323	17.9	f
$Ca(NO_3)_2$	2.0	323	17.9	f,g
$(0.67Ca + 0.33K)NO_3$	2.4	323	15.2	f
$Ca(NO_3)_2$	3.0	323	15.4	e,f
	5.0	323	12.9	e
NH_4NO_3	1.5	343	13.7	h
	2.0	343	13.0	h
	3.0	343	12.4	h

[a] M. Peleg, *J. Phys. Chem.*, **75**, 3711 (1971).
[b] M. Peleg, *J. Phys. Chem.*, **75**, 2060 (1971).
[c] I. J. Gal, *Inorg. Chem.*, **7**, 1611 (1968); J. Braunstein and H. Braunstein, *Inorg. Chem.*, **8**, 1558 (1969).
[d] P. C. Lammers and J. Braunstein, *J. Phys. Chem.*, **71**, 2626 (1967).
[e] H. Braunstein, J. Braunstein and P. T. Hardesty, *J. Phys. Chem.*, **77**, 1907 (1973).
[f] J. Braunstein and H. Braunstein, *Inorg. Chem.*, **8**, 1528 (1969).
[g] J. Braunstein, A. Alvarez-Funes and H. Braunstein, *J. Phys. Chem.*, **70**, 2734 (1966).
[h] R. M. Nikolic and I. J. Gal, *J. Chem. Soc. Dalton Trans.*, **1972**, 162.

The dependence on the water content, however, is not simple (see Table 7.3). For a melt like lithium–potassium nitrate, where water activities are proportional to the mole ratio R_H $(= n_{H_2O}/n_{NO_3})$ up to $R_H \sim 1$, the quasi-lattice model[36] can be invoked. It is assumed that the solvent and solute cations mix randomly on the cation sites, but that the solvent and solute anions and the water molecules mix on the anion sites, subject to energy constrains. This model is thus restricted to cation hydration. The specific Helmholz energy $A_1 = -\mathbf{R}T \ln(1 + K_1/Z)$ obtained from the association constant K_1 is given according to this model as

$$A_1 = \mathbf{N}\epsilon_C + \mathbf{R}T \ln(1 + R_H \exp(-\epsilon_H/\mathrm{k}T)) \tag{7.78}$$

where ϵ_C and ϵ_H are the relative energies of a halide ion and a water molecule, respectively, compared to a nitrate ion, neighbouring a solute cation. The values $\mathbf{N}\epsilon_C = -18.4$ and $\mathbf{N}\epsilon_H = 0 \pm 2$ for $Ag^+ + Cl^-$ in NH_4NO_3 solvent, and $\mathbf{N}\epsilon_C = -27.1$ and $\mathbf{N}\epsilon_H = -5.4$ for $Cd^{2+} + Br^-$ in $(Li,K)NO_3$ solvent, all in kJ mol^{-1}, have been obtained. Since A_1 and ϵ_C are negative, and the second term on the right-hand side of (7.78) is positive, the role of water is seen to destabilize the solute association relative to the anhydrous melt. In other nitrate melts, however, Henry's law is not (even fortuitously) obeyed beyond, say, $R_H = 0.1$, and the dependence is no longer according to the simple quasi-lattice model, eq. (7.78). Rather, an equally simple mass-action expression is followed[35]

$$\ln K_1 = \ln K_1^0 - \ln(1 + K_H a_{H_2O}^n) \tag{7.79}$$

Equation (7.79) can be converted to the form of (7.78) by multiplying with $(-\mathbf{R}T)$ but it differs in its dependence in the last term on $a_{H_2O}^n$ (values of n from 0.8 to 2 have been found) instead of on R_H. The activity of water refers to pure water as the standard state, R_H to the anhydrous molten nitrate. The two models are, therefore, incompatible. Restriction to cation hydration is also debatable, but the results available at present are restricted to too few solvent salts (essentially only ammonium nitrate, lithium–potassium nitrate and systems involving calcium nitrate, see Table 7.3) to permit a critical comparison.

b. Mixtures of ionic and molecular salts

Whereas water cannot be properly called a salt, situations similar to those encountered in the previous section may occur also when ionic salts are dissolved in, or mixed with, certain salt-like compounds, such as in $LiF + BeF_2$, or $KBr + HgBr_2$ mixtures. In these cases, it is the anions which may be said to be solvated, to form BeF_4^{2-} from F^-, and $HgBr_3^-$ from Br^-. The cations Li^+ and K^+ in these examples are relatively free. There is however a marked difference in behaviour, according to whether the 'solvating' salt is monomeric, as is $HgBr_2$, or forms a polymeric network, as does BeF_2 (section 3C(c)). These salts and their mixtures will be discussed as typical for a large group of such mixtures. The scope of this book does not permit a discussion of the silica and oxide mixture analogues to the BeF_2 and fluoride mixtures, i.e. the large and important glass- and ceramic-forming oxide melts.

A monomeric 'molecular' salt, such as mercury(II) bromide (aluminium chloride, which is dimeric, still behaves analogously), will typically show very low electrical conductivity in the molten state between the melting point of 511 K and the boiling point of 592 K. It has a relatively large thermal expansibility, suggestive of coherence forces not originating from ion–ion coulombic interactions.

A potentiometric determination also shows very low concentrations of halide ions. The degree of dissociation in the melt has been estimated at $\alpha = 10^{-5}$ to 2×10^{-4}. A salt such as thallium (I) nitrate as a solute would dissociate in mercury bromide solvent, and increase the conductivity. The mercury bromide acts as a diluent but is capable of solvating the anions, i.e. the nitrate ions, and remains undissociated even at low concentrations in thallium nitrate (where the roles of solute and solvent have been exchanged). Conductometry, cryoscopy and Raman spectroscopy have given consistent information on these points[39].

A different case arises when an alkali metal bromide is dissolved in the mercury bromide. Again, there is a sharp increase in conductivity, but the evidence from cryoscopy and Raman spectroscopy is difficult to reconcile with the small degree of self dissociation of the mercury bromide solvent. The Raman evidence shows clearly the successive formation of $HgBr_3^-$ and $HgBr_4^{2-}$ when increasing amounts of solute are dissolved. Cryoscopy indicates two foreign ions added at high dilutions of the solute, but later evidence pointed to only one foreign ion added, when, say, KBr is dissolved in molten $HgBr_2$ (refs. 39, 40). This would be readily understandable if the solvent dissociated to an appreciable extent to yield bromide ions, but the evidence quoted above shows only very slight dissociation, and this most probably according to

$$2HgBr_2 \rightleftharpoons HgBr^+ + HgBr_3^- \tag{7.80}$$

i.e. the bromide is solvated by $HgBr_2$. The added bromide ions, even if solvated, should make themselves felt cryoscopically and several schemes have been suggested[40] in explanation of the contrary behaviour, but with little supporting evidence. For example

$$2K^+ + 2Br^- + HgBr_2 \rightleftharpoons K^+ + KHgBr_4^- \tag{7.81}$$

provides two moles of new ions per two moles of dissolved solute, as also does

$$2K^+ + 2Br^- + 2HgBr_2 \rightleftharpoons K^+ + K(HgBr_3)_2^- \tag{7.82}$$

The species $KHgBr_4^-$ as the major species formed in the relevant concentration range, is contrary to the Raman spectroscopic evidence, while $K(HgBr_3)_2^-$ is consistent with it, but both contain rather implausible attachments of potassium ions. An alternative explanation is in terms of considerable dissociation of the solvent according to (7.80), so that (solvated) bromide ions are no longer foreign ions. The low conductivity, in spite of an appreciable dissociation, is attributed to the low mobility of the ions, if each of the species in (7.80) is further solvated. This explanation, however, does not account for the relatively high mobility, and conductivity of other electrolytes, such as thallium (I) nitrate quoted above, in the mercury (II) bromide solvent. There remains, therefore, a real unresolved problem concerning these solutions.

Another salt which has very low conductivity in its molten state is beryllium fluoride. This is due not to an inherent appreciable covalency, as in mercury bromide, but rather, in spite of pronounced ionicity, to the cross-linked network character of the salt. In fact, beryllium fluoride behaves as a low-melting model for silica, and its mixtures with lithium fluoride as analogues to SiO_2 + MgO mixtures, with similar sized ions but a weaker field, owing to the lower charges. On the side of the mixtures rich in lithium fluoride, up to $x_{BeF_2} = 0.33$, corresponding to the composition Li_2BeF_4 (which is an almost congruently melting solid), the mixture behaves as an essentially ideal binary mixture with a common cation Li^+, and the anions F^- and BeF_4^{2-}. Raman spectroscopy concurs with e.m.f. and cryoscopic data that monomeric tetrahedral BeF_4^{2-} anions are the only beryllium-containing species[41]. The partial molar excess entropy of beryllium fluoride is positive, $\bar{s}^E_{BeF_2} \sim 14$ J K^{-1} mol^{-1} and changes little with composition, and is due to complete depolymerization of the network of beryllium fluoride to give the monomeric BeF_4^{2-} in the dilute solution[42].

The main interest in the system rests, however, in the composition range $0.22 < x_{BeF_2} < 1.00$, where polymeric species prevail (Fig. 7.6). In the range $0.33 < x_{BeF_2} < 0.50$, the only conducting ion (relative to fluoride) is lithium, and the activity coefficient of the component LiF becomes very small. The sole cation is, again, Li^+, and since presumably $a_{Li^+} \sim 1$, f_{F^-} becomes very small. There is no longer sufficient fluoride to remain free, or even as separate BeF_4^{2-} tetrahedra, but rather singly bonded —F (unshared corners of BeF_4 tetrahedra) and bridging —F— (shared corners). In this range dimeric $Be_2F_7^{3-}$ could be identified directly by Raman spectroscopy, and it yields, presumably, $Be_3F_{10}^{4-}$ by disproportionation and sharing a corner with BeF_4^{2-}, but edge-shared polymers, such as $Be_3F_8^{2-}$, cannot be ruled out completely. These latter polymers, however, do not play a major role in molten or glassy beryllium fluoride, and will therefore be disregarded. As the region rich in beryllium fluoride is attained, $x_{BeF_2} > 0.7$, polymerization increases rapidly, and with it the viscosity, but no abrupt change is observed up to the pure salt[43].

A polymerization model has been proposed which can rationalize detailed thermodynamic data[42,44] over wide composition and temperature ranges, and is consistent also with the spectroscopic and transport properties. The model assumes that the beryllium is always tetrahedrally coordinated, and that the BeF_4 tetrahedra may share zero to four corners with others. (Two tetrahedra that share the same two corners share an edge.) Three parameters describe the interactions: $G = \alpha^0 \mathbf{R}T + \alpha' \mathbf{R}$ and $H = \beta \mathbf{R}T$, where G is the molar Gibbs energy change associated with the reaction

$$\gt\!\!\text{Be—F—Be}\!\!\lt + F^- \rightleftharpoons 2[\gt\!\!\text{Be—F}]^{-1/2} \qquad (7.83)$$

and H is the enthalpy of a mole of 'contacts' of shared —F and free F^- anions relative to their contacts with shared —F—, i.e. bridging fluorides. (The small Li^+ cations and the beryllium ions are located in the interstices, and do not themselves

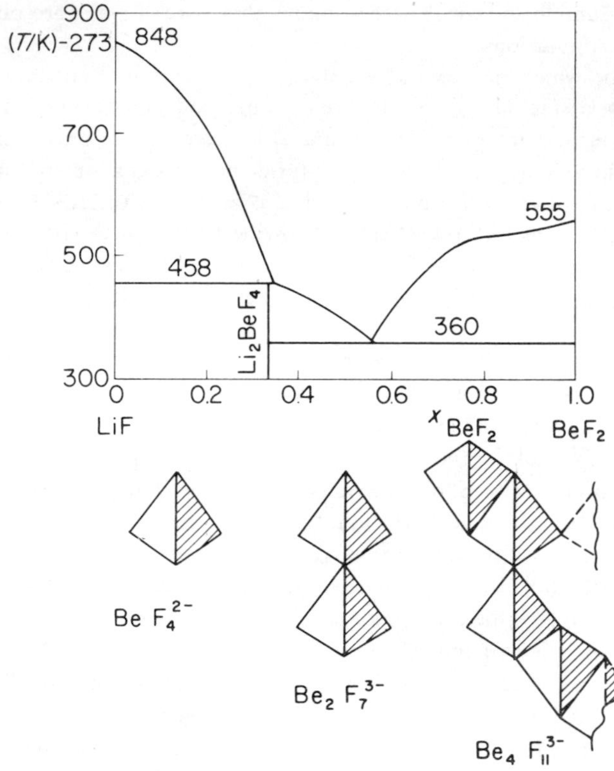

Figure 7.6. The phase diagram of the LiF + BeF_2 system. In the lower part of the Fig. are schematic representations of the monomeric tetrahedral BeF_4^{2-}, with four bound but unshared fluoride ions, the dimer $Be_2F_7^{3-}$ with one shared fluoride (corner sharing), and a fragment, $Be_4F_{11}^{3-}$, of a higher polymer with both corner and edge sharing.

contribute to the heat of mixing.) Because of the various sizes of the polymers, volume fraction statistics are employed. The model recognizes all the species Be_aF_b (omitting charges) with $2 \leqslant ba^{-1} \leqslant 4$, the number of shared corners being $s = 4a - b$ and of unshared corners $t = 2b - 4a$ in a given species. For the generalized polymerization reaction

$$a\mathrm{BeF}_4^{2-} \rightleftharpoons \mathrm{Be}_a\mathrm{F}_b + (4a - b)\mathrm{F}^- \tag{7.84}$$

$$\ln K_{ab} = -\Delta G_{ab}/\mathbf{R}T = (4a - b)G/\mathbf{R}T = (4a - b)(\alpha^0 + \alpha'/T) \tag{7.85}$$

as can be seen from counting the number of —F— bridges formed. The volume fraction sum $\Sigma\Sigma\phi_{ab} = 1 - \phi_{01}$ (the subscript 01 denotes F^-) since only the fluorides contribute to the volume, and $\Sigma\Sigma ab^{-1}\phi_{ab} = x_{BeF_2}/(1 + x_{BeF_2})$. The heat of mixing per mole of mixture is

$$h^M = H\phi_{01}(1 + x_{BeF_2})\Sigma\Sigma(2b - 4a)b^{-1}\phi_{ab} \tag{7.86}$$

where $(1 + x_{BeF_2})$ times the double sum is the volume fraction of the unshared fluorides. The molar entropy of mixing is

$$s^M = (1 + x_{BeF_2})\mathbf{R}[\phi_{01} \ln \phi_{01} + \Sigma\Sigma b^{-1} \phi_{ab} \ln \phi_{ab}] \qquad (7.87)$$

The activity of any species may be computed from (7.85), (7.86) and (7.87) by partial differentiation of $(h^M - Ts^M)$, and the activities of the components, $a_{LiF} = a_{F^-}$ (since $a_{Li^+} = 1$ has been assumed) and a_{BeF_2}, may be computed by an iterative prodedure, for given values of the parameters α^0, α' and β, by choosing trial values of ϕ_{01} in certain explicit equations involving those parameters, x_{BeF_2}, T and ϕ_{01} that were derived[44]. Limiting values of the activity coefficients are $\lim f_{LiF}(x_{BeF_2} \rightarrow 1) = e/2$ and $\lim f_{BeF_2}(x_{BeF_2} \rightarrow 0) = 4 \exp[2\alpha^0 + 2\alpha'/T + 4\beta - 3]$. The values of the parameters that fit e.m.f. and phase diagram activity data are $\alpha^0 = -1.71$, $\alpha' = -530$ K and $\beta = 0.44$. The average degree of polymerization $\langle b \rangle = \Sigma\Sigma\phi_{ab}/\Sigma\Sigma b^{-1}\phi_{ab}$ may also be calculated; this reaches two at approximately $x_{BeF_2} = 0.6$ and increases rapidly beyond that. By proper choice of the parameters, the model has been made consistent with all the data available.

It should be noted that this treatment of certain binary molten salt mixtures in terms of polymerized species is quite similar in concept to the treatment of self associating molecular mixtures, such as alcohols in inert solvents, dealt with in section 5B. The idea of volume fraction statistics is combined with a non-zero interaction between the species to give a more realistic consideration of polymerizing systems that each of the athermal and regular solution treatments alone. The present model for lithium–beryllium fluoride mixtures has its own restrictive assumptions, and requires several free parameters to fit the data (it has not been fitted to available[42] data of h^M at 1135 K, although it should fit). The main assumption of unvarying tetrahedral coordination of fluorides around the beryllium is reasonable; another, the independence of G, describing (7.83), from the nature (shared or not) of the other fluorides bound to the beryllium, can be modified[44] to take into account a reduced stability of strained $(-F-Be-)_n$ configurations relative to the hexagonal ring with $n = 6$. Again, disregard of contributions to the enthalpy of mixing from interactions of bridging –F– with bound –F and with free F^- may not be warranted, but their contribution will only add more adjustable parameters. It may be concluded that this model is useful for dealing with molten salt mixtures which involve polymerizing, non-ionizing components, but that further work is warranted.

E. Appendixes

(a) **Appendix 7.1.** Parameters describing the heat of mixing of common-ion binary systems (eq. 7.6, p. 258).

Common Anion Systems

A	B	a	b	c	Ref.
$X = F^-$					
Li	Na	−7.95	−0.25	0	a
	K	−15.73	−5.61	−3.64	a
	Rb	−17.15	−6.28	0	a
	Cs	−11.04	−7.45	−5.52	a
	Be	−8.4	−20.9	+43.9	b
	Ca	−5.27	+0.38		c
Na	K	−0.38			a
	Rb	+0.38			a
	Mg	−12.55			d
	Ca	−19.79	+0.38		c
	Sr	−1.67			d
	Ba	−1.05			d
	Cd	−9.00			d
	Pb	−7.32			d
K	Rb	+0.38			a
	Be	−54.4	−120.3	+153.8	b
	Ca	−34.65	+1.80		c
Be	Ca	−1.26			d
Mg	Ba	−4.18			d
$X = Cl^-$					
Li	Na	−4.69	0	0	e
	K	−17.57	+0.38	0	e
	Rb	−17.86	−4.81	−3.97	e
	Cs	−19.46	−7.44	−9.08	e
	Ag	+8.20	+1.26	−2.38	e,f
	Mg	−11.72	+9.20	0	g
	Ca	−1.01	−0.19	+0.60	h
	Mn	−4.44	+2.34	+1.30	i
	Fe	−4.33	+2.51	+3.81	i
	Co	−3.93	+2.47	+5.40	i
	Ni	−5.98	0		g
	Zn	+20.9	−27.2	−20.1	y
	Cd	−0.67	−4.06	−2.93	z
	Pb	−1.88	0		d
	La	−3.21	+1.40		j
	Ce	−3.07	+2.16		k
Na	K	−2.05	−0.27	0	e
	Rb	−3.22	−0.33	0	e
	Cs	−4.31	+0.42	0	e
	Ag	+3.85	+2.59	0	e,f
	Mg	−15.3	−8.7	+12.9	g
	Ca	−9.41	+3.35	0.0	d,h
	Sr	−3.13	+3.25	−1.66	d,h
	Ba	0.0			d
	Mn	−14.14	−8.28	+11.11	d,i
	Fe	−14.16	−11.17	+18.26	i
	Co	−16.51	−16.44	+29.00	i
	Ni	−12.13	−18.41	+32.63	q
$X = Cl^-$ (contd.)					
Na	Pb	−3.77			d
	La	−12.96	−2.73	+32.65*	j
	Ce	−11.37	−9.59	+41.0*	k
K	Rb	+0.08			e
	Cs	+0.79			e
	Ag	−7.99	−2.51	0	e,f
	Mg	−27.6	−32.2	+32.6	g
	Ca	−21.86	+6.82	0	h
	Sr	−11.96	+10.60	−5.23	h
	Ba	−4.18	0		d
	Mn	−28.68	−23.85	+27.09	i
	Fe	−28.30	−36.21	+43.68	i
	Co	−33.83	−39.69	+55.04	i
	Ni	−27.61	−60.25	+105.4	g
	Pb	−22.59			d
	La	−21.91	−25.90	+119.01*	j
	Ce	−23.93	−25.75	+124.70*	k
Rb	Cs	+0.33			e
	Ag	−12.93	−3.39	0	e,f
	Mg	−31.4	−41.1	+38.6	g
	Ca	−22.91	−7.36	+12.57	h
	Sr	−14.74	+11.05	−4.60	h
	Mn	−33.93	−38.05	+38.47	i
	Fe	−33.83	−49.98	+53.24	i
	Co	−40-90	−45.50	+60.44	i
	Ni	−33.47	−74.48	+133.05	g
	La	−29.61	−19.62	+120.89*	j
	Ce	−29.57	−38.12	+202.51*	k
Cs	Ag	−8.75	+3.10	−6.86	e,f
	Cu(I)	−31.06	−21.14	0	a'
	Mg	−35.56	−53.35	+47.07	g
	Ca	−21.30	−22.47	+41.90*	h
	Sr	−15.61	+4.06	+9.35*	h
	Mn	−39.98	−43.97	+44.46	i
	Fe	−40.10	−53.07	+59.91	i
	Co	−46.00	−60.25	+73.22	i
	Ni	−39.75	−84.73	+154.39	g
	Zn	−30.1	−58.2	−175.7*	y
	Cd	−85.9	+10.5	−192.5*	z
	La	−37.32	−30.42	+173.19*	j
	Ce	−36.48	−39.99	+249.10*	k
Cu(I)	Pb	−2.93			d
Ag	Tl	+10.75	−2.85	0	e
	Mg	+1.26			d,g
	Pb	0.00			d
	Zn	+17.6	−13.8	−13.6*	y
	Cd	−1.74	−1.26		l
Tl	Cd	−13.70	−3.12	−17.41	l
Mg	Ca	+0.13	+4.80	−2.72	m,n

Appendix 7.1. *(continued)*

Common Anion Systems

A	B	a	b	c	Ref.
X = Cl⁻ (contd.)					
Mg	Sr	−3.66	+9.33	−19.04	n
	Ba	−11.71	+12.15	−20.44	n
	Mn	+0.67			p
	Fe	+0.84			p
	Co	+0.84			p
	Sn	+0.42			d
	Pb	−0.84			d
Ca	Sr	−1.01	−0.01	0	n
	Ba	−4.35	+0.10	0	n
	Mn	+3.57	−3.18	−1.17	p
	Fe	+5.19	−4.33	−0.94	p
	Co	+9.82	−6.72	−3.77	p
	Zn	+0.84			d
	Sn	+1.67			d
	Pb	+0.84			d
Sr	Ba	−0.91	−0.01	0	n
	Zn	−2.93			d
	Pb	+1.26			d
Ba	Mn	−9.62			d
Mn	Fe	0.00			p
	Co	+0.71			p
	Zn	−1.26			d
	Cd	+0.17			p
	Sn	−1.26			d
	Pb	−3.14			d
Fe	Co	+0.33			p
	Cd	+0.17			p
	Sn	−0.84			d
	Pb	−2.51			d
Co	Sn	+0.84			d
	Pb	+0.84			d
Zn	Pb	−4.39			d
Cd	Sn	−1.67			d
	Pb	+1.26			d
X = Br⁻					
Li	Na	−2.97	−0.59	0	e
	K	−13.22	−1.76	−1.97	e
	Rb	−16.07	−4.02	−4.69	e
	Cs	−18.66	−6.49	−3.89	e
	Ag	+8.79			e
	Mg	−0.53			h
	Sr	+0.67			h
	Ba	−1.67			d
	Zn	+23.3	−29.3	−13.8	y
Na	K	−2.13	−0.25	0	e
	Rb	−3.45	−0.46	0	f
	Cs	−4.73	−0.21	0	f
	Ag	+1.26	+4.52		q
	Mg	−10.46			d,h
	Ca	−9.60	+3.47	0	h
	Sr	−3.35	+3.10	−1.66	h

A	B	a	b	c	Ref.
X = Br⁻ (contd.)					
Na	Ba	+0.48	−0.55	+0.55	h
	Pb	−8.37			d
K	Rb	0.00			e
	Cs	+0.38			e
	Ag	−7.20			e
	Mg	−24.9			h
	Ca	−19.38	+3.77		h
	Sr	−12.48	+10.72	−5.69	h
	Ba	−3.68	+1.90		h
	Pb	−25.94			d
Rb	Cs	+0.25			e
	Ag	−11.79			e
	Mg	−30.8			h
	Ca	−19.84	−22.81	+56.23*	h
	Sr	−16.15	+11.80	−5.31	h
	Ba	−6.38	+5.40	−2.06	h
Cs	Cu(I)	−26.14	−18.19	0	a′
	Mg	−36.8			h
	Ca	−19.13	−35.06	+60.84*	h
	Sr	−18.77	+0.24	+21.74*	h
	Ba	−0.13	+6.12	−1.68	h
	Zn	−20.1	−68.2	−125.5*	y
Ag	Tl	−10.33	−3.10	0	e
	Pb	0.00			d
X = I⁻					
Li	Na	−3.90	−0.54		r
	K	−11.39	−2.82		r
	Rb	−15.38	−4.35		r
	Cs	−17.36	−7.62		r
	Sr	+0.57			h
Na	K	−2.12	−0.17		r
	Rb	−3.57	−0.40		r
	Cs	−5.44	−0.56		r
	Sr	−3.04			h
K	Rb	−0.08			r
	Cs	−0.06			r
	Sr	−9.60			h
Rb	Cs	+0.09			r
	Sr	−13.4			h
Cs	Sr	−7.2			h
X = NO_3^-					
Li	Na	−1.87	−0.06		s
	K	−6.94	−0.55	−2.33	s
	Rb	−10.06	−0.85	−4.16	s
	Cs	−14.35	−1.00	−6.44	s
	Ag	+2.94	−0.39		s
	Ca	+0.84			d
	Cd	0.00			d
Na	K	−1.64	−0.28		s
	Rb	−2.96	−0.95	−0.99	s

Appendix 7.1. (*continued*)

Common Anion Systems

A	B	a	b	c	Ref.
$X = NO_3^-$ (contd.)					
Na	Cs	−4.36	−1.82	−0.39	s
	Ag	+2.18	+0.67		q,s
	Ca	−1.88			d
	Sr	−0.21			d
	Ba	+0.21			d
K	Rb	−0.25			s
	Ag	−2.33	+1.15		s
	Ca	−7.53			d
	Sr	−4.60			d
	Ba	−2.93			d
Rb	Ag	−5.09	+1.27	−1.10	s
	Ca	−11.30			d
	Sr	−7.32			d
	Ba	−5.02			d
Cs	Ag	−8.74	+3.10	−6.86	s
	Sr	−10.88			d
	Ba	−7.95			d
Ag	Tl	−5.31			e
	Ca	+1.26			d
	Cd	−0.63			l
Tl	Ca	−2.93			d
	Cd	−1.42	−7.72	−10.02	l
$X = ½SO_4^{2-}$					
Li	Na	−4.29	−1.11		t
	K	−10.71	−3.89		t
	Rb	−11.09	−7.51		t
	Cs	−9.75	−9.37		t
Na	K	−2.20			t
	Rb	−2.57	−0.79		t

A	B	a	b	Ref.
$X = ½SO_4^{2-}$ (contd.)				
Na	Cs	−2.15	−0.36	t
K	Rb	−0.02		t
	Cs	+0.46		t
Rb	Cs	+0.19		t
$A = PO_3^-$				
Li	Na	−3.84	+0.44	u
	K	−12.83	−1.42	u
	Rb	−17.26	−0.44	u
	Cs	−19.58	−2.74	u
Na	K	−2.53	−0.16	u
	Rb	−4.55	+0.21	u
	Cs	−5.92	+0.21	u

*Higher terms.

A	B	X	d	e	f
Na	Ce	Cl	−48.7	+20.4	
K	Ce	Cl	−156.3	+62.6	
Rb	Ce	Cl	−276.9	+119.4	
Cs	Ca	Cl	−11.0		
	Sr	Cl	−7.8		
	Zn	Cl	−385	+1607	−1950
	Cd	Cl	−661	+1130	−469
	Ce	Cl	−356.0	+161.1	
Ag	Zn	Cl	+34.3	−34.3	
Rb	Ca	Br	−35.14		
Cs	Ca	Br	−30.18		
	Sr	Br	−16.11		
	Zn	Br	−1046	+3607	−4096

Common Cation Systems

X	Y	a	b	Ref.
$A = Li^+$				
Cl	Br	+0.08	+0.12	v
	I	+1.39	+0.63	v
	NO_3	+0.88		w
Br	I	+0.81	+0.02	v
$A = Na^+$				
F	Cl	+1.05	+2.51	x
Cl	Br	+0.29	+0.10	v
	I	+1.62	+0.64	v
	NO_3	+1.67		q,w
Br	I	+0.63	+0.02	v
	NO_3	+1.51		q,w

X	Y	a	b	Ref.
$A = K^+$				
Cl	Br	+0.15	+0.11	v
	I	+1.24	+0.27	v
	NO_3	+0.88		v
Br	I	+0.45	−0.01	v
	NO_3	+0.50		w
$A = Rb^+$				
Cl	Br	+0.12	+0.05	v
	I	+1.14	−0.06	v
	NO_3	+0.50		
Br	I	+0.42	−0.06	v
	NO_3	+0.33		w

Appendix 7.1. *(continued)*

Common Cation Systems

X	Y	a	b	Ref.
$A = Cs^+$				
Cl	Br	+0.06	0	v
	I	+0.74	−0.17	v
	NO_3	+0.71		w
Br	I	−0.06	+0.01	v
	NO_3	+0.33		v
$A = Ag^+$				
Cl	NO_3	−1.92	+2.97	l
Br	NO_3	−9.08		q,w
$A = Tl^+$				
Cl	NO_3	+3.14	−0.46	l
Br	NO_3	+4.69		w
$A = \frac{1}{2}Mg^{2+}$				
Cl	Br	−0.33		m
$A = \frac{1}{2}Ca^{2+}$				
F	Cl	−1.26		d
	Br	+5.02		d
	I	+8.4		d
Cl	Br	−0.08		m
$A = \frac{1}{2}Pb^{2+}$				
Cl	Br	+4.18		d
	I	+7.95		d

[a]J. L. Holm and O. J. Kleppa, *J. Chem. Phys.*, **49**, 2425 (1968).
[b]J. L. Holm and O. J. Kleppa, *Inorg. Chem.*, **8**, 207 (1969).
[c]O. J. Kleppa and K. C. Hong, *J. Phys. Chem.*, **78**, 1478 (1974).
[d]J. Lumsden, *Thermodynamics of Molten Salt Mixtures*, Academic Press, London, 1966, pp. 183–255.
[e]L. S. Hersch and O. J. Kleppa, *J. Chem. Phys.*, **42**, 1309 (1965); S. V. Meschel, J. Toguri and O. J. Kleppa, *J. Chem. Phys.*, **45**, 3075 (1966).
[f]L. S. Hersch, A. Navrotsky and O. J. Kleppa, *J. Chem. Phys.*, **42**, 3752 (1965); P. Datzer and O. J. Kleppa, *J. Chim. Phys.*, **71**, 216 (1974).
[g]G. N. Papatheodorou and O. J. Kleppa, *J. Inorg. Nucl. Chem.*, **32**, 889 (1970); O. J. Kleppa and F. G. McCarty, *J. Phys. Chem.*, **76**, 1249 (1966).
[h]T. Østwold, *J. Phys. Chem.*, **76**, 1616 (1972).
[i]G. N. Papatheodorou and O. J. Kleppa, *J. Inorg. Nucl. Chem.*, **33**, 1249 (1971).
[j]G. N. Papatheodorou and T. Østwold, *J. Phys. Chem.*, **78**, 181 (1974).
[k]G. N. Papatheodorou and O. J. Kleppa, *J. Phys. Chem.*, **78**, 178 (1974).
[l]S. V. Meschel and O. J. Kleppa, *J. Chem. Phys.*, **46**, 1853 (1967).
[m]J. Toguri, H. Flood and T. Førland, *Acta Chem. Scand.*, **17**, 1502 (1963).
[n]G. N. Papatheodorou and O. J. Kleppa, *J. Chem. Phys.*, **47**, 2014 (1967).
[p]G. N. Papatheodorou and O. J. Kleppa, *J. Chem. Phys.*, **51**, 4624 (1969).
[q]S. V. Meschel and O. J. Kleppa, *J. Chem. Phys.*, **43**, 4160 (1965).
[r]N. E. Melnichak and O. J. Kleppa, *J. Chem. Phys.*, **52**, 1790 (1970).
[s]S. V. Meschel and O. J. Kleppa, *J. Chem. Phys.*, **48**, 5146 (1968).
[t]T. Østwold and O. J. Kleppa, *Acta Chem. Scand.*, **25**, 919 (1971).
[u]H. C. Ko and O. J. Kleppa, *Inorg. Chem.*, **10**, 771 (1971).
[v]N. E. Melnichak and O. J. Kleppa, *J. Chem. Phys.*, **57**, 5231 (1972).
[w]O. J. Kleppa and S. V. Meschel, *J. Phys. Chem.*, **67**, 668 (1963).
[x]K. Grjotheim, T. Halvorsen and J. L. Holm, *Acta Chem. Scand.*, **21**, 2300 (1967).
[y]G. N. Papatheodorou and O . J. Kleppa, *Z. Anorg. Allg. Chem.*, **401**, 132 (1973).
[z]G. N. Papatheodorou and O. J. Kleppa, *Inorg. Chem.*, **10**, 872 (1971).
[a']P. Datzer and O. J. Kleppa, *Inorg. Chem.*, **12**, 2699 (1973).

(b) **Appendix 7.2.** Specific Helmholz energies for the association of silver, cadmium and lead cations (A) with an anion X in B-nitrate melts, calculated from eq. (7.70) with Z = 5.

AX	B	T (K)	$-A_1$ (kJ mol^{-1})	10^3d$(-A_1)$/dT*	Ref.
AgCl	Li	713	22.8		a
	(0.43Li + 0.57K)	498	22.1		b
	(0.33Li + 0.67K)	513	22.1		c
	(0.30Li + 0.70K)	664	23.8		a
	Na	604–773	20.3	—	a
		643	20.2		d
		647	20.4		e
		658	20.2		e
	(0.83Na + 0.17K)	643	20.8		d
	(0.53Na + 0.47K)	643	22.2		d
		606–647	22.0	—	f
	(0.50Na + 0.50K)	506–752	22.5	—	g
	(0.23Na + 0.77K)	643	22.4		d
	K	623–709	24.6	—	a,h
		675	24.6		i
		658	24.6		j
		643	24.2		d
	(0.67K + 0.33Cs)	658	25.5		a
	Cs	713	27.4		a
	(0.80K + 0.20Ca)	593–623	22.9	—	k
	Ca	623	17		k
	(0.80K + 0.20Sr)	623	21.9		k
	Sr	623	12		k
	(0.89K + 0.11Ba)	623–663	22.6	—	l
AgBr	Na	675–773	27.6	—	m
	(0.53Na + 0.47K)	649–683	28.7	—	f
	K	675–773	29.9	—	i
AgI	Na	623	47		n
	(0.50Na + 0.50K)	623–648	42.0	—	n
	(0.20Na + 0.80K)	673	40.3		n
	K	623	40.2		n
		675	39.2		i
AgCN	(0.50Na + 0.50K)	519	46.4	79	o
$AgSO_4$	Na	595–684	0.9	—	p
	(0.50Na + 0.50K)	526–685	2.3	—	p
	K	622–715	6.6	—	p
		622	6.7	9	q
		636	6.4	15	r
$AgCrO_4$	Li	528–600	4.3	—	s
	(0.60Li + 0.40Na)	519–610	5.9	—	s
	(0.43Li + 0.57K)	429	6.9	3	s
	Na	584–673	9.8	—	s
	(0.50Na + 0.50K)	580–695	11.0	—	s
	K	615–675	12.5	—	s
$AgMoO_4$	Li	541–622	5.6	—	p
	(0.60Li + 0.40Na)	566–650	6.9	—	p
	(0.43Li + 0.57K)	549–653	7.7	—	p
	Na	584–690	10.6	—	p
	(0.50Na + 0.50K)	517–646	11.9	—	p
	K	619–714	14.3	—	p
CdCl	Li	523–563	22.0	—	t
	(0.75Li + 0.25K)	523–563	22.3	—	t
	(0.50Li + 0.50K)	523–563	22.5	—	t
		433–473	21.4	—	u
	(0.40Li + 0.60K)	453	21.5		u

Appendix 7.2. (*continued*)

AX	B	T (K)	$-A_1$ (kJ mol^{-1})	$10^3 d(-A_1)/dT$*	Ref.
CdCl	(0.39Li + 0.61K)	453	23.0		v
		453	19.6		w(v)
	Na	580	17.7		w(x)
	(0.50Na + 0.50K)	527	22.5		y
		523	17.1	50	w(z)
CdBr	Li		29.5		a′
	(0.50Li + 0.50Na)	513	26.9		a′
	(0.80Li + 0.20K)	513	26.2		u
	(0.50Li + 0.50K)	443–513	27.3	—	u
	(0.26Li + 0.74K)	513	26.2		u
	Na	604	24.2		b′
	(0.53Na + 0.47K)	523	16.5	48	z
		529–571	24.1	—	c′
	(0.50Na + 0.50K)	528	26.3		y
		513	24.5		u
		513	24.4	15	d′
		527	26.2		y
		531	25.2		b′
	(0.67Na + 0.33Ca)	533	27.4		e′
	K	631	25.6		b′
	(0.67K + 0.33Ca)	473–553	27.6	—	e′
	Ca	323	31.9		e′
CdI	(0.50Na + 0.50K)	513	29.8	10	d′
PbCl	(0.39Li + 0.61K)	453	17.2		v
		453	15.1		w(v)
	(0.50Li + 0.50K)	433–473	14.4	15	u
	Na	580	12.4		w(x)
	(0.50Na + 0.50K)	523	16.1	−70	w(z)
PbBr	(0.50Li + 0.50K)	433–473	18.9	12	u
	(0.75Na + 0.25K)	553–573	16.6	—	f′
PbBr	(0.53Na + 0.47K)	528–592	16.3	—	c′
		523–573	15.6	—	z
	(0.50Na + 0.50K)	513	16.8		u
		513–573	16.9	—	f′
	(0.25Na + 0.75K)	553–573	17.1	—	f′

*– indicates $10^3 d(A_1)/dT \simeq 0 \pm 1$, *i.e.* the data within the temperature range indicated show no temperature-dependence within their experimental precision.

[a]C. Thomas and J. Braunstein, *J. Phys. Chem.*, **68**, 957 (1964).
[b]G. W. Harrington and H. T. Tien, *Inorg. Chem.*, **3**, 215, 1333 (1964).
[c]J. Braunstein, H. Braunstein and A. S. Minano, *Inorg. Chem.*, **3**, 1334 (1964).
[d]H. M. Garfinkel and F. R. Duke, *J. Phys. Chem.*, **65**, 1629 (1961).
[e]D. G. Hill, J. Braunstein and M. Blander, *J. Phys. Chem.*, **64**, 1038 (1960).
[f]F. R. Duke and H. M. Garfinkel, *J. Phys. Chem.*, **65**, 461 (1961).
[g]D. G. Hill and M. Blander, *J. Phys. Chem.*, **65**, 1866 (1961).
[h]D. L. Manning, J. Braunstein and M. Blander, *J. Phys. Chem.*, **66**, 2069 (1962).
[i]A. Alvarez-Funes, J. Braunstein and M. Blander, *J. Am. Chem. Soc.*, **84**, 1538 (1962).
[j]J. Braunstein and M. Blander, *J. Phys. Chem.*, **64**, 10 (1960).
[k]J. Braunstein and J. D. Brill, *J. Phys. Chem.*, **70**, 1261 (1966).
[l]H. C. Gaur and R. S. Sethi, *Trans. Faraday Soc.*, **64**, 445 (1968).
[m]D. L. Manning, R. C. Bansal, J. Braunstein and M. Blander, *J. Am. Chem. Soc.*, **84**, 2028 (1962).
[n]J. Braunstein and R. E. Hagman, *J. Phys. Chem.*, **67**, 2881 (1962).
[o]D. L. Manning and M. Blander, *Inorg. Chem.*, **1**, 594 (1962).
[p]G. A. Sacchetto, C. Macca and G. G. Bombi, *J. Electroanal. Chem.*, **36**, 319 (1972).

Appendix 7.2. (*continued*)

[q]C. E. Vallet and J. Braunstein, *J. Phys. Chem*., **77**, 2672 (1973).
[r]W. J. Watt and M. Blander, *J. Phys. Chem.*, **64**, 729 (1960).
[s]G. A. Sacchetto, C. C. Bombi and C. Macca, *J. Electroanal. Chem.*, **36**, 47 (1972).
[t]T. P. Flaherty and J. Braunstein, *Inorg. Chim. Acta*, **1**, 335 (1967).
[u]J. Braunstein and A. S. Minano, *Inorg. Chem.*, **3**, 218 (1964).
[v]J. H. Cristie and R. A. Osteryoung, *J. Am. Chem. Soc.*, **82**, 1841 (1960).
[w]J. A. Braunstein, M. Blander and R. M. Lindgren, *J. Am. Chem. Soc.*, **84**, 1529 (1962).
[x]E. R. van Artsdalen, *J. Phys. Chem.*, **60**, 172 (1956).
[y]D. Inman, *Electrochim. Acta*, **10**, 11 (1965).
[z]F. R. Duke and M. L. Iverson, *J. Phys. Chem.*, **62**, 417 (1958).
[a′]J. Braunstein and A. S. Minano, *Inorg. Chem.*, **5**, 942 (1966).
[b′]H. Braunstein, J. Braunstein and D. Inman, *J. Phys. Chem.*, **70**, 2726 (1966).
[c′]F. R. Duke and H. M. Garfinkel, *J. Phys. Chem.*, **65**, 1627 (1961).
[d′]J. Braunstein and R. M. Lindgren, *J. Am. Chem. Soc.*, **84**, 1534 (1962).
[e′]H. Braunstein, J. Braunstein, A. S. Minano and R. E. Hagman, *Inorg. Chem.*, **12**, 1407 (1973).
[f′]D. L. Manning, M. Blander and J. Braunstein, *Inorg. Chem.*, **2**, 345 (1963).

(c) **Appendix 7.3.** Gas solubilities in molten salts, in concentration units, C (mole gas m^{-3} Pa^{-1}), and reciprocals of Henry's law constant, K (mole gas (mole salt)$^{-1}$ Pa^{-1}).

Salt	Gas	T (K)	10^2C	10^7K	H (kJ mol^{-1})	Ref.
LiF	HF	773 (extrap.)			−10	a
(0.66Li + 0.34Be)F	HF	773		0.033	−25.0	a
	DF	773		0.029	−26.9	a
	He	873	0.114			q
	Ne	873	0.046			q
	Ar	873	0.010			q
	Xe	873	0.0023			q
(0.46Li + 0.12Na + 0.42K)F	He	873	0.112	0.224	+33.5	b
	Ne	873	0.043	0.086	+37.2	b
	Ar	873	0.009	0.018	+51.9	b
(0.53Na + 0.47Zr)F	HF	873	12.1	39	−19.7	c
	He	873	0.213	0.69	+25.9	d
	Ne	873	0.112	0.36	+32.6	d
	Ar	873	0.050	0.162	+34.3	d
	Xe	873	0.019	0.062	+46.4	d
NaF	CO_2	1273	1.84			r
KF	CO_2	1173	2.15			r
LiCl	Cl_2	893		4.1		s
	Cl_2	1100	0.186			t
(0.5Li + 0.5K)Cl	HCl	763		4.06		u
	Cl_2	823		0.71		w
	H_2O	1013		0.296		v
	H_2O	663		2.25	−33	e
(0.6Li + 0.4K)Cl	H_2O	753		0.85	−46.0	e
	HCl	753		0.075		e
NaCl	CO_2	1100	0.51	1.93	+24.7	f
	Cl_2	1120	0.218	0.83	+48.5	g
	Cl_2	1113	0.228			t
	Cl_2	1093		1.30		s
	HCl	1203	0.99	3.75		x
KCl	CO_2	1100	0.75	3.7	+18.8	f
	Cl_2	1121	1.03	5.2	+91.6	g
	Cl_2	1096	1.31			t
	Cl_2	1093		5.2		s

Appendix 7.3. (*continued*)

Salt	Gas	T (K)	10^2C	10^7K	H (kJ mol^{-1})	Ref.
	HCl	1173		12.0		x
(0.5Na + 0.5K)Cl	Cl_2	1123	0.69		+67.4	g
	HCl	1023		8.9		x
RbCl	HCl	1103		23.2		x
CsCl	Cl_2	1195	4.7			t
$MgCl_2$	Cl_2	1106	0.67	1.94	+34.3	g
(0.5Na + 0.5Mg)Cl	Cl_2	1093	0.209		+37.7	g
(0.5K + 0.5Mg)Cl	Cl_2	1117	0.68		+57.7	g
$CaCl_2$	Cl_2	1073		4.1		s
AgCl	Cl_2	791		0.69		w
$PbCl_2$	Cl_2	786		0.29		w
(0.77Pb + 0.23K)Cl	Cl_2	908	1.39		+33.2	y
(0.30Pb + 0.70K)Cl	Cl_2	861	1.75		−15.4	y
LiBr	Br_2	890		6.02		w
NaBr	Br_2	1073		17.1		s
KBr	Br_2	1073		72.2		s
	CO_2	1100	0.90	5.2	+16.7	f
$CaBr_2$	Br_2	1073		22.7		s
AgBr	Br_2	1033		0.36		z
$PbBr_2$	Br_2	792		8.21		w
NaI	I_2	973		0.167		s
KI	I_2	973		0.406		s
	CO_2	973	1.88	0.135		f
$LiNO_3$	H_2O	530	1.34	5.2		h
	H_2O	541	4.5	17.4	−52.3	i
	H_2O	608	1.11	4.4	−39.1	j
	NH_3	533	2.42	9.4	−69.0	k
	He	543	0.151	0.59		l
	Ar	546	0.091	0.35	+14.0	l
	N_2	550	0.073	0.284		l
(0.5Li + 0.5Na)NO_3	H_2O	541		8.3		i
	H_2O	423		81.6	−43.5	n
(0.5Li + 0.5K)NO_3	H_2O	541		62		i
(0.75Li + 0.25K)NO_3	NH_3	515		3.2	−49.4	k
(0.43Li + 0.57K)NO_3	NH_3	523		0.79	−29.3	k
$NaNO_3$	H_2O	580	1.27	5.7		h
	H_2O	541 (extrap.)			−41.4	i
	H_2O	606	0.282	1.27	−34.1	j
	He	605	0.186	0.84	+13.4	l
	He	637	0.105	0.47	+13.5	m
	Ar	604	0.064	0.29	+15.8	l
	Ar	637	0.071	0.33	+14.7	m
	N_2	580	0.002	0.01		h
	N_2	604	0.050	0.23	+16.0	l
	N_2	637	0.22	1.01	+11.5	m
	CO_2	637	0.103	0.47	−11.2	m
	CO_2	595	1.24			r
KNO_3	H_2O	608	0.29	1.57		h
	H_2O	541 (extrap.)			−33.9	i
$RbNO_3$	Ar	604	0.13	0.78	+20.1	l
$CsNO_3$	H_2O	679	0.54	3.7		h
$AgNO_3$	Ar	507	0.019	0.082		l
(0.76Li + 0.24K)ClO_4	NH_3	524		0.63	−57.7	k
(0.5Li + 0.5Na)CO_3	O_2	1073	0.49			a′
(0.36$NaNO_2$ + 0.64KNO_3)	H_2O	423	8.45	39.3	−38.1	p
	H_2O	523	1.07	5.3	−35.1	p

Appendix 7.3. (*continued*)

[a]P. E. Field and J. S. Shaffer, *J. Phys. Chem.*, **71**, 3218 (1967).
[b]M. Blander, W. R. Grimes, N. V. Smith and G. M. Watson, *J. Phys. Chem.*, **63**, 1164 (1959).
[c]J. H. Shaffer, W. R. Grimes and G. M. Watson, *J. Phys. Chem.*, **63**, 1999 (1959).
[d]W. R. Grimes, N. V. Smith and G. M. Watson, *J. Phys. Chem.*, **62**, 862 (1958).
[e]W. J. Burkhard and J. D. Corbett, *J. Am. Chem. Soc.*, **79**, 6361 (1957).
[f]D. Bratland, K. Grjotheim, C. Krohn and K. Motzfeld, *Acta Chem. Scand.*, **20**, 1811 (1966).
[g]Yu. M. Ryabukhin, *Zh. Neorg. Khim.*, **7**, 1101 (1962).
[h]J. P. Frame, E. Rhodes and A. R. Ubbelohde, *Trans. Faraday Soc.*, **57**, 1075 (1961).
[i]G. Bertozzi, *Z. Naturforsch. A*, **22**, 1748 (1967).
[j]M. Peleg, *J. Phys. Chem.*, **71**, 4553 (1967).
[k]S. Alluli, *J. Phys. Chem.*, **73**, 1084 (1969).
[l]B. Cleaver and D. E. Mather, *Trans. Faraday Soc.*, **66**, 2469 (1970).
[m]P. E. Field and W. J. Green, *J. Phys. Chem.*, **75**, 821 (1971).
[n]T. B. Tripp and J. Braunstein, *J. Phys. Chem.*, **73**, 1984 (1969).
[p]H. S. Hull and A. G. Turnbull, *J. Phys. Chem.*, **74**, 1783 (1970).
[q]G. W. Warson, R. B. Evans, W. R. Grimes and N. V. Smith, *J. Chem. Eng. Data*, **7**, 285 (1962).
[r]D. Bratland and C. Krohn, *Acta Chem. Scand.*, **23**, 1839 (1969).
[s]H. v. Wartenberg, *Z. Elektrochem.*, **32**, 330 (1926).
[t]Yu. M. Ryabukhin and N. G. Bukun, *Z. Neorg. Khim.*, **13**, 597 (1968).
[u]J. D. Van Norman and R. J. Tivers, *J. Electrochem. Soc.*, **118**, 208 (1971).
[v]D. L. Maricle and D. N. Hume, *J. Electrochem. Soc.*, **107**, 354 (1960).
[w]J. D. Van Norman and R. J. Tivers, in *Molten Salts*, edited by G. Mamantov, Dekker, New York, 1969.
[x]V. N. Devyatkin and E. A. Ukshe, *Z. Prikl. Khim. (Leningrad)*, **38**, 1612 (1965).
[y]M. Kowalski and G. W. Harrington, *Inorg. Nucl. Chem. Lett.*, **3**, 121 (1967).
[z]A. Block-Bolten and S. N. Flengas, *Can. J. Chem.*, **49**, 2266 (1971).
[a']M. Schencke, G. H. J. Broers and J. A. A. Ketelaar, *J. Electrochem. Soc.*, **113**, 404 (1966).

Reviews on the solubilities of reactive gases in molten salts, by S. N. Flengas and A. Block-Bolten, and by P. Field appear in *Advances in Molten Salt Chemistry, Vol. 2* and *3*, edited by J. Braunstein, G. Mamantov and G. P. Smith, Plenum Press, New York. 1973 and 1975 respectively.

References

1. M. Temkin, *Acta Physicochim. USSR*, **20**, 411 (1945); R. Haase, *J. Phys. Chem.*, **73**, 1160, 4023 (1969); D. M. Moulton, *J. Phys. Chem.*, **73**, 4022 (1969).
2. (a) T. Østvold, *Acta Chem. Scand.*, **22**, 435 (1968); (b) T. Førland, *Norg. Tek. Vitenskapsakad*, **Ser. 2**, No. 4 (1957); (c) *Trans. Faraday Soc.*, **58**, 122 (1962); (d) *J. Phys. Chem.*, **59**, 152 (1955).
3. H. Reiss, J. L. Katz and O. J. Kleppa, *J. Chem. Phys.*, **36**, 144 (1962); M. Blander, *J. Chem. Phys.*, **37**, 172 (1962); H. T. Davis, *J. Chem. Phys.*, **41**, 2760 (1964); *J. Phys. Chem.*, **76**, 1629 (1972).
4. E. A. Guggenheim, *Proc. Roy. Soc., Ser. A*, **148**, 304 (1935); *Mixtures*, Clarendon Press, Oxford, 1952, pp. 38–52; T. Førland, in *Fused Salts*, edited by B. R. Sundheim, 1964, p. 86.
5. M. Gaune-Escard, J. C. Mathieu, P. Desré and Y. Doucet, *J. Chim. Phys.*, **1972**, 1390, 1397; **1973**, 1666.
6. H. T. Davis and S. A. Rice, *J. Chem. Phys.*, **41**, 14 (1964); K. D. Luks and H. T. Davis, *Ind. Eng. Chem., Fundam.*, **6**, 194 (1967); H. T. Davis, *J. Phys. Chem.*, **76**, 1639 (1972).
7. J. Lumsden, *Discuss. Faraday Soc.*, **32**, 138 (1961).
8. L. S. Hersh and O. J. Kleppa, *J. Chem. Phys.*, **42**, 1309 (1965).
9. G. N. Papatheodorou and O. J. Kleppa, *J. Chem. Phys.*, **51**, 4624 (1969).
10. O. J. Kleppa and L. S. Hersh, *Discuss. Faraday Soc.*, **32**, 99 (1961); *J. Chem. Phys.*, **35**, 175 (1961); **36**, 544 (1962).
11. F. H. Stillinger, *J. Chem. Phys.*, **35**, 1581 (1961).
12. Y. Marcus, *Israel J. Chem.*, **10**, 659 (1972).

13. H. Flood and S. Urnes, *Z. Elektrochem.*, **59**, 834 (1955); K. Grjotheim, *Kgl. Norske Videnskab. Selskabs Skrifter*, No. 5 (1956).
14. Y. Marcus, in *Advances in Molten Salt Chemistry, Vol. 1*, edited by J. Braunstein, G. Mamantov and G. P. Smith, Plenum Press, New York, 1971, p. 81; I. N. Belyaev, *Usp. Khim.*, **29**, 899 (1960); C. Sinistri, P. Franzosini and M. Rolla, *An Atlas of Miscibility Gaps in Molten Salt Systems*, University of Pavia, Italy, 1968; P. Franzosini, P. Ferloni and G. Spinola, *Molten Salts with Organic Anions*, University of Pavia, Italy, 1973.
15. J. Braunstein, *J. Chem. Phys.*, **49**, 3508 (1968).
16. A. D. Pelton and S. N. Flengas, *J. Electrochem. Soc.*, **117**, 1130 (1970).
17. H. Bloom, *Pure Appl. Chem.*, **7**, 389 (1963); K. Balasubrahmanyam and L. Nanis, *J. Chem. Phys.*, **40**, 2657 (1964); F. G. McCarty and O. J. Kleppa, *J. Phys. Chem.*, **68**, 3846 (1964); H. Bloom and J. W. Hastie, *J. Phys. Chem.*, **72**, 2361 (1964); G. J. Janz and Chr. G. M. Dijkhuis, *Nolten Salts, Vol. 2*, U.S. National Bureau of Standards, Washington, D.C., NSRDS-NBS 28, 1969.
18. G. Perkins, R. B. Escue, J. F. Lamb and T. A. Tidwell, *J. Phys. Chem.*, **64**, 495 (1960); M. A. Bredig, in *Molten Salts*, edited by G. Mamantov, Dekker, New York, 1968, p. 55; B. Vos, Ph.D. Thesis, University of Amsterdam, 1970.
19. S. V. Meschel and O. J. Kleppa, *J. Chem. Phys.*, **48**, 5146 (1968).
20. J. Guion, *J. Chim. Phys.*, **64**, 1635 (1967); J. Guion, M. Blander, D. Hengstenberg and K. Hagemark, *J. Phys. Chem.*, **72**, 2086 (1968).
21. K. Hagemark, *J. Phys. Chem.*, **72**, 2316 (1968).
22. H. Flood, T. Førland and K. Grjotheim, *Z. Anorg. Allg. Chem.*, **276**, 789 (1954).
23. O. J. Kleppa and J. M. Toguri, in *Selected Topics in High Temperature Chemistry*, Universitetsforlaget, Oslo, 1966, p. 15; S. V. Meschel and O. J. Kleppa, *J. Chem. Phys.*, **46**, 1853 (1967).
24. (a) M. Blander and L. E. Topol, *Inorg. Chem.*, **5**, 1641 (1966); (b) M. L. Saboungi, *J. Phys. Chem.*, **77**, 1699 (1973); **78**, 1091 (1974).
25. M. Blander and S. J. Yosim, *J. Chem. Phys.*, **39**, 2610 (1963).
26. M. Blander and J. Braunstein, *Ann. N.Y. Acad. Sci.*, **79**, 838 (1960); cf. also T. Førland, in *Fused Salts*, edited by B. R. Sundheim, McGraw-Hill, New York, 1964, p. 89.
27. M. Blander, *J. Phys. Chem.*, **63**, 1262 (1959).
28. Y. T. Hsu, R. B. Escue and T. H. Tidwell, Jr., *J. Electroanal. Chem.*, **15**, 245 (1967).
29. D. G. Hill, J. Braunstein and M. Blander, *J. Phys. Chem.*, **64**, 1038 (1960); M. Blander, *J. Chem. Phys.*, **34**, 432 (1961).
30. E. Q. Adams, *J. Am. Chem. Soc.*, **38**, 1503 (1916); N. Bjerrum, *Z. Phys. Chem. Stoechiom. Verwandschaftslehre*, **106**, 219 (1923).
31. M. Blander, in *Molten Salts*, edited by M. Blander, Wiley, New York, 1964, p. 228.
32. D. L. Manning, R. C. Bansal, J. Braunstein and M. Blander, *J. Am. Chem. Soc.*, **84**, 2028 (1962).
33. M. Blander, W. R. Grimes, N. V. Smith and G. M. Watson, *J. Phys. Chem.*, **63**, 1164 (1959).
34. J. Braunstein, L. Orr, and W. Macdonald, *J. Chem. Eng. Data*, **12**, 415 (1967); T. B. Tripp and J. Braunstein, *J. Phys. Chem.*, **73**, 1984 (1969).
35. H. Braunstein, J. Braunstein and P. Hardesty, *J. Phys. Chem.*, **77**, 1907 (1973).
36. J. Braunstein, *J. Phys. Chem.*, **71**, 3402 (1967); in *Ionic Interactions, Vol. 1*, edited by S. Petrucci, Academic Press, New York, 1971, pp. 240–254.
37. C. A. Angell, *J. Electrochem. Soc.*, **112**, 1224 (1965); *J. Phys. Chem.*, **69**, 2137 (1965); C. A. Angell and D. M. Gruen, *J. Am. Chem. Soc.*, **88**, 5192 (1966); C. A. Angell and E. J. Sare, *J. Chem. Phys.*, **49**, 4713 (1968); C. T. Moynihan and A. Fratiello, *J. Am. Chem. Soc.*, **89**, 5546 (1967); D. E. Irish, D. L. Nelson and M. H. Brooker, *J. Chem. Phys.* **54**, 654 (1971).
38. R. E. Hester and R. A. Plane, *J. Chem. Phys.*, **40**, 411 (1964); D. E. Irish, A. R. Davis and R. A. Plane, *J. Chem. Phys.*, **50**, 2262 (1969); V. S. Ellis and R. E. Hester, *J. Chem. Soc. A*, **1969**, 607; M. Peleg, *J. Phys. Chem.*, **76**, 1019 (1972).
39. G. Jander and K. Brodersen, *Z. Anorg. Allg. Chem.*, **264**, 57, 92 (1951); G. J. Janz and J. D. E. McIntyre, *Ann. N.Y. Acad. Sci.*, **79**, 790 (1960); *J. Electrochem. Soc.*, **109**, 842 (1962); M. Rolla, P. Franzosini and R. Riccardi, *Discuss. Faraday Soc.*, **32**, 84 (1961); G. J. Janz and T. R. Kozlowski, *J. Chem. Phys.*, **39**, 843 (1963); G. J. Janz A. Timedei and F. W. Damper, *Electrochim. Acta*, **15**, 609 (1970).

40. G. J. Janz and J. Goodkin, *J. Phys. Chem.*, **64**, 308 (1960); G. J. Janz and D. W. James, *J. Chem. Phys.*, **38**, 902, 905 (1963); J. Lumsden, *Thermodynamics of Molten Salt Mixtures*, Academic Press, London, 1966, pp. 253, 254.
41. J. Braunstein, K. A. Romberger and R. Ezell, *J. Phys. Chem.*, **74**, 4383 (1970); A. S. Quist, J. B. Bates and G. E. Boyd, *J. Phys. Chem.*, **76**, 78 (1972); R. A. Romberger, J. Braunstein and R. E. Thomas, *J. Phys. Chem.*, **76**, 1154 (1972).
42. J. L. Holm and J. O. Kleppa, *Inorg. Chem.*, **8**, 207 (1969).
43. J. D. Mackenzie, *J. Chem. Phys.*, **32**, 1150 (1960); S. Cantor, W. T. Ward and C. T. Moynihan, *J. Chem. Phys.*, **50**, 2874 (1969); A. H. Narten, *J. Chem. Phys.*, **56** 1905 (1972); J. B. Bates, *J. Chem. Phys.*, **56**, 1910 (1972); A. S. Quist, J. B. Bates and G. E. Boyd, *Spectrochim. Acta, Part A*, **28**, 1103 (1972).
44. C. F. Baes, Jr., *J. Solid State Chem.*, **1**, 159 (1970); B. F. Hitch and C. F. Baes, Jr., *Inorg. Chem.*, **8**, 201 (1969); R. E. Thomas, H. Insley, H. A. Friedman and G. M. Hebert, *J. Nucl. Mater.*, **27**, 166 (1968).

Chapter 8

Mixtures of Liquid Metals

A. Liquid Metal Alloys

Although solid solutions of metals, i.e. alloys, have been in use for millenia, and have been well characterized long ago, interest in liquid alloys has awakened mainly in the last three decades. Much structural and thermodynamic information is now available on liquid alloys, but their interpretation is still hampered by the lack of suitable theories. Liquid alloys differ from the liquid mixtures dealt with in the previous chapters mainly in two respects. The first is the fluctuating potential field produced by the free electrons, which provides a continuous background to the potentials arising from individual atoms or ions. The second is the strong tendency to form compounds, with strong directed bonding and (negative) energies many times the thermal energy. Because of the high melting points of metals generally, most of the information on liquid alloys[1,2] deals with a restricted group of metals, those having melting points below 1000 K, copper, silver and gold being notable exceptions. The low melting metals are mainly the alkali metals and post-transition metals (B subgroup elements of the Periodic Table). Information on mixtures of the refractory metals (mainly the transition metals) is limited to their phase diagrams, to a large extent.

Phase diagrams are a much more important source of information on liquid alloys than on mixtures of other substances. The generalizations that have been derived for liquid alloys, on the other hand, serve to predict liquidus temperatures, or to correct and supplement imperfectly established liquidus curves.

a. Structural information

Much of the effort made on liquid alloys has been directed at establishing any residual order that may have remained in the melt. Most of the information has been obtained at temperatures not too far above the liquidus temperatures. It is reasonable therefore to apply quasi-lattice models to the liquid alloys, and to look for remnants of the superlattice (p.158), or of the structure of congruently melting compounds, and even of peritectics, in the melt. The atomic nature of the metals (excluding metalloids, such as silicon, arsenic or selenium) facilitates the use of diffraction methods, both X-ray and neutron, for elucidating the structure. As for pure liquid metals (Chapter 3) the one-dimensional radial correlation function $g(r)$ is but a poor representation of three-dimensional structure, even at short-range. Usually it does give information, however, on the distances of nearest- and next-nearest-neighbours and on the corresponding coordination numbers. From one kind of measurement, however, it is impossible to identify the atoms which correspond to these distances. Three coordinated measurements from both X-ray and neutron diffraction, utilizing different isotopes of the metals, are required for this purpose in liquid binary alloys. The coherent scattering intensity is given by

$$I = N(x_1 f_1^2 + x_2 f_2^2 + x_1^2 f_1^2 S_{11} + x_2^2 f_2^2 S_{22} + 2x_1 x_2 f_1 f_2 S_{12}) \tag{8.1}$$

where N is the total number of atoms in the scatterer of volume V, f_i is the atomic coherent scattering amplitude of component i, and S_{ij} is the structure factor (cf. eqs. (2.23) and (2.18)). This situation is analogous to that in ionic liquids, where the two kinds of ions also require three independent measurements in order to obtain structural information (p. 120). From $S_{ij}(k)$, $g_{ij}(r)$, the pair correlation function is obtained: the three measurements with differing f_i and f_j thus yield $S(k, x)$ values, from which $g(r, x)$ values may be calculated. An example of structure factors $S(k)$ obtained[2] for a liquid copper+tin alloy ($x_{Cu} = 0.54$) is shown in Fig. 8.1. The first Cu–Sn peak is seen to nearly coincide in position with the Cu–Cu peak, and in genral S_{Cu-Sn} is not midway between S_{Cu-Cu} and S_{Sn-Sn}, as expected for random distribution (at $x_{Cu} = 0.5$), so that some ordering is indicated. The three-dimensional arrangements cannot, however, be derived from these results.

About a dozen binary systems have by now been examined, although only very few in a manner that permits the separation of the three $S_{ij}(k, x)$ structure factors, or the $g_{ij}(r, x)$ pair correlation functions[3]. In all, the results available from these studies have not been as informative as expected. In most systems random distribution is observed, but there are notable exceptions, as for the copper–tin system referred to above.

Perhaps the most interesting conclusion from the diffraction studies applies to the average coordination numbers in the liquids. Whereas the interatomic distances

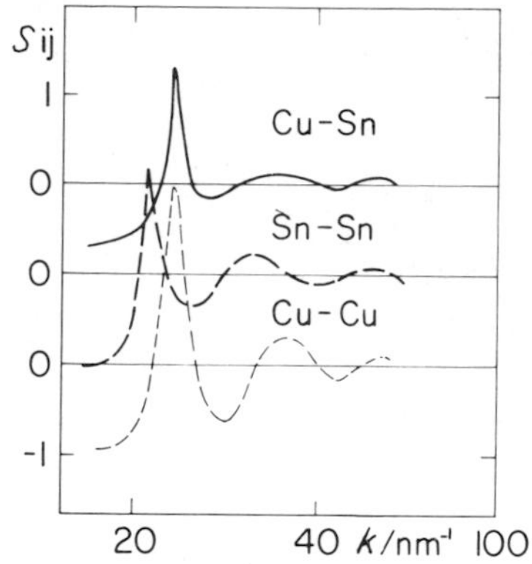

Figure 8.1. Structural factors $S_{ij}(k)$ for a liquid copper–tin alloy ($x_{Cu} = 0.54$) obtained from X-ray and neutron diffraction experiments[2]. The ordinates of the three curves are displaced by one unit from each other.

are more or less additive in terms of metallic radii, and therefore not sensitive indicators of structural peculiarities, the coordination numbers Z vary in characteristic manners. For instance, in the bismuth–lead system Z remains near the value for liquid lead, 11.9 ± 0.2, up to a substitution of $x_{Bi} = 0.60$, and only above this does it fall gradually to the value in liquid bismuth, 7.7 ± 0.1. Similarly $Z = 11.5 \pm 0.1$ for $0 < x_{Hg} < 0.60$ in the mercury+thallium system, and only at higher mercury concentrations does it reach the value for liquid mercury, 10.0. In other systems, e.g. silver+tin, Z varies nearly linearly between the values for the pure liquid metals[4]. The non-equality of the average coordination numbers of the liquid components themselves, and the non-linearity of $Z(x)$, discredit attempts to apply quasi-lattice methods to liquid alloys, a point often disregarded.

b. Solid-solution- and eutectic-forming liquid alloys

Liquid alloys may be classified in several ways; because of the importance of the phase diagrams to the elucidation of their properties, referred to above, it is convenient to use these diagrams as a basis for their classification[3]. On the one hand there are systems with very strong mutual interactions of the components, which form congruently melting compounds (Fig. 8.2(d)), or at least well characterized stoichiometric peritectics, on freezing. On the other hand, there are systems where the compounds form solid solutions on freezing (Fig. 8.2(a)), with a relatively large mutual solubility, or where eutectic points are found on the liquidus (Fig. 8.2(b)), anywhere from nearly stoichiometric compositions to nearly the pure components, including systems which show liquid–liquid immiscibility (Fig. 8.2(c)).

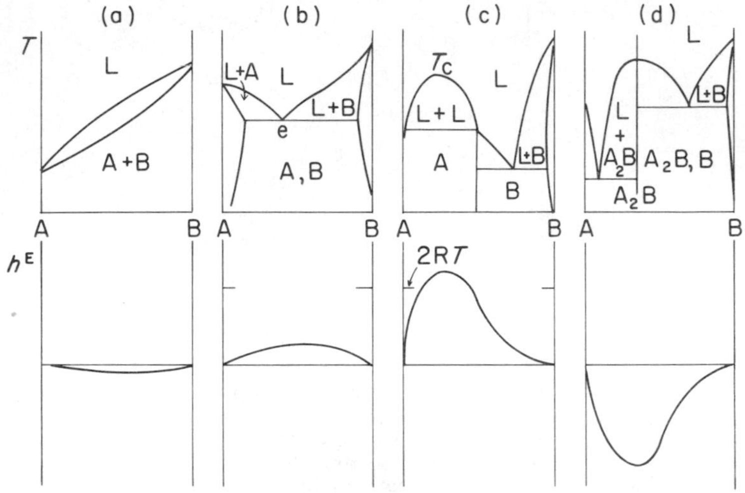

Figure 8.2. Schematic diagrams representing several types of liquid alloys. The upper part represents phase diagrams, and the lower part the heat of mixing curves $h^E(x)$. (a) Solid-solution system (with complete miscibility in the solid); (b) eutectic system with one liquidus inflected (some mutual solubility in the solid); (c) miscibility-gap system in the A-rich region (note the high positive h^E); (d) compound (A_2B)-forming system (note the large negative h^E).

In most metallic systems, the interaction energies of tens of kilojoules per mole between like or between unlike kinds of atoms are very large, compared to the modest energies of molecular systems. However, solid-solution- and eutectic-forming alloys have comparable self and mutual interaction energies, while compound-forming systems have considerably larger (more negative) mutual than self interaction energies. The thermodynamic properties of liquid alloys of the former group of systems will now be discussed, while the latter group is discussed in section 8A(b).

The difference in atomic volumes of the two metals forming the liquid alloy is an important factor in describing their properties. The factor $\sigma = 2 \mid v_1^* - v_2^* \mid / (v_1^* + v_2^*)$ can be used, and usually the *size factor* $\rho = 2 \mid r_1 - r_2 \mid / (r_1 + r_2)$ is employed. A further notable influence on the properties of the alloys arises from the *electronegativity factor,* η (kJ mol^{-1}) $= (\chi_1 - \chi_2)^2$, where χ_i(kJ mol^{-1})$^{1/2}$ is the electronegativity (the quantity χ_i is $96.487^{1/2} \simeq 10$ times the electronegativity on the (electron-volt)$^{1/2}$ scale, i.e. on the ordinary Pauling scale).

The classification of the liquid alloys dealt with here according to the phase diagrams correlates[3] with that according to the size factor ρ and the electronegativity factor η. Systems with solid solutions, or eutectics with no more than one liquidus curve having an inflection point, have η (kJ mol^{-1}) values of 1 to 4, and ρ values of 0.06 for solid-solution systems, 0.10 for eutectics with no inflection, and 0.17 for eutectics with one inflection. These are mean values, but the deviations of individual systems are considerable. On the other hand, eutectic systems with two inflected liquidus curves have η (kJ mol^{-1}) of 9 and $\rho \sim 0.20$, and systems where the eutectic point nearly coincides with the melting point of the lower melting component have η (kJ mol^{-1}) of 12 and $\rho \sim 0.30$, again on average. Systems showing miscibility gaps in the liquid state tend to have, on average, even larger size and electronegativity factors.

The thermodynamic excess functions follow very roughly the same classification (Fig. 8.2(a) to (c)): from negative or small positive values of g^E and h^E for solid-solution systems to large positive values for doubly inflected eutectic systems, or those with miscibility gaps. Deviations from regular-solution behaviour (section 5A(c)) are also noted. In fact, the partial molar excess Gibbs energies are better expressed by $a + b(1 - x_i)^2$ than by the regular solution expression. These quantities are obtained with good precision from e.m.f. data, from cells where the opposing electrodes are the pure (liquid) metal and its liquid alloy, and the electrolyte a molten salt of this metal, in which it is practically insoluble. Another source for precise values are vapour pressure measurements, carried out with suitable Knudsen cells and if necessary a mass-spectrometric analysis of the vapours. Less precise values are obtained from the liquidus curves of phase diagrams and from liquid–liquid distribution measurements. The partial molar excess entropies are obtained from the temperature coefficients of the e.m.f. data, or from combined μ_i^E and h_i^E data, the latter obtained calorimetrically.

The partial excess functions are quite sensitive to structural effects. Various correlations of the excess functions with structural peculiarities have been noted, e.g. h^E correlates with the size factor σ. It is difficult however to relate these quantities uniquely to the properties of the pure liquid metals.

The difficulties in the interpretation of the thermodynamic functions of mixing of liquid alloys arise from their reluctance to conform to simple models that have been found successful for other systems, e.g. the regular-solution model or the quasi-chemical, quasi-lattice model. The non-constancy of Z has already been commented on above, the non-vanishing of v^E and of s^E is demonstrated in Table 8.1. Many more systems where only some of the data ($\hat{v}^E$ or $\hat{s}^E$) are known[3], are not shown in Table 8.1, but also show finite values of these quantities. Thus, the premises for the use of most of the simpler models do not exist.

On the other hand, it is true that the absolute values of $\hat{v}^E$ or $\hat{s}^E$ are not large. The relative values of $\hat{v}^E/\hat{v}_m$ are of the order of 1–2%. Since for the heat capacities of these systems the Neumann–Kopp law, $c_P^E = c_{Pm} - (x_1 c_{P1} + x_2 c_{P2}) \simeq 0$, is usually obeyed[3], excess entropies are also quite small. Furthermore, the maximal values of h^E are obtained at $0.4 < x_B < 0.6$, so that the near-symmetric curves required for regular behaviour result, and $\hat{h}^E$ is a fair representation of h^E_{max}.

There is a good correlation between the signs of v^E and s^E. A contribution to the excess entropy arising from the excess volume, $s^E_{vol} = \alpha\kappa^{-1} v^E$, where α is the isobaric thermal expansivity and κ the isothermal compressibility, may account for *ca.* 70 % of the total $\hat{s}^E$ in many systems[5]. This can be put on a different basis, by considering the lattice-perturbation energy caused by mixing metals with differing atomic volumes

$$Ts^E_{vol} = H^E_{vol} = 0.4 x_1 x_2 \mid v_1^* - v_2^* \mid [x_1 T_{m1} \alpha_1 \kappa_1^{-1} + x_2 T_{m2} \alpha_2 \kappa_2^{-1}] \tag{8.2}$$

Table 8.1. Excess volumes, entropies and heats of mixing of some equimolar binary liquid alloys, forming solid solutions or eutectics.

A	B	$\hat{v}^E$ (cm³ mol⁻¹)	$\hat{s}^E$ (J K⁻¹ mol⁻¹)	$\hat{h}^E$ (kJ mol⁻¹)	T (K)*	Ref.
Ag	Au	<0.1	−1.42	−4.60	1350	a
	Bi	>0	+1.88	+1.88	1000	a
	Sn	−0.67	<0	+0.25	1248	b
	Pb	>0	+2.72	+4.23	1000	a, v
Al	Mg	−0.36	−2.1	−3.8	1000	a, q
	Zn	~0	−3.2	−4.42	1050	a*, i
Au	Cu	−0.05	~0	−4.60	1300	a, w
Bi	Cd	+0.32	+1.59	+0.92	773	c, j, k, w
	Pb	+0.06	+0.21	+1.09	700	a*, d
	Sn	+0.21	−0.3	+0.08	608	a, k, w
Cd	Hg	−0.39	−1.09	−2.72	600	a*, d
	In	+0.11	+0.59	+1.46	723	c*, l
	Pb	+0.12	+0.92	+2.85	773	c*, k, w
	Sb	+0.49	+1.9	−2.13	773	a, m
	Sn	+0.24	+1.17	+1.88	773	c*, a, e, w
	Tl	+0.16	+0.96	+2.30	673	c*, n
	Zn	+0.06	~0	+2.13	673	d*, n, w
Cu	Ni	~0	<0	+2.1		f
	Sb	−0.58	>0	−5.77	1215	a
	Sn	−0.92	<0	−5.0	1423	f
	Zn	−0.32	−1.76	−7.07	1200	f, p
Hg	In	−0.28	−0.46	−2.26	433	d, r
	Sn	−0.18	−0.2	+0.9	423	d, r
	Zn	−0.30	+0.4	−0.96	608	d*, a
In	Zn	+0.07	+1.26	+3.26	700	d*, k, s
Mg	Pb	−1.4	−3.14	−9.6	823	a
	Zn	−0.80	+2.51	−5.9	923	a, t
Pb	Sn	~0	+0.59	+1.38	1050	e*, x
	Zn	+0.15	+1.1	+5.69	926	g, l
Sb	Zn	+0.06	+1.05	−3.56	823	a, d
Sn	Zn	+0.18	+2.34	+3.43	700	d*, h*, e, u, w

*For systems marked with an asterisk, T refers to $\hat{s}^E$ and $\hat{h}^E$, but not to $\hat{v}^E$, which was determined at another temperature.

a J. R. Wilson, *Met. Rev.*, **10**, 381 (1965); *cf.* also O. Kubachewski and J. A. Catterall, *Thermochemical Data of Alloys*, Pergamon Press, London, 1956; R. Hultgren R. L. Orr, P. D. Anderson and K. K. Kelley, *Selected Values of Thermodynamic Properties of Metals and Alloys*, Wiley, New York, 1963.

b F. E. Wittig and E. Gehring, *Z. Naturforsch.*, **18**, 351 (1963).

c O. J. Kleppa, *J. Phys. Chem.*, **64**, 1542 (1960).

d O. J. Kleppa, M. Kaplan and C. E. Thalmayer, *J. Phys. Chem.*, **65**, 843 (1961); O. J. Kleppa and M. Kaplan, *J. Phys. Chem.*, **61**, 1120 (1957).

e A. F. Crawley, *Met. Trans.*, **3**, 971 (1972).

f M. G. Benz and J. F. Elliott, *Trans. Met. Soc. AIME*, **230**, 706 (1964).

g D. D. Todd and W. A. Oats, *Trans. Met. Soc. AIME*, **230**, 244 (1964).

h E. Ubelacker and L. D. Lucas, *Compt. Rend.*, **254**, 1622 (1962).

i G. J. Lutz and A. F. Voigt, *J. Phys. Chem.*, **67**, 2795 (1963).

j E. Wachtel and S. Nazareth, *Z. Metallk.*, **56**, 20 (1965).

k J. Terpilowski and E. Przezdziecka, *Arch. Hutn.*, **3**, 315 (1957).

l D. D. Todd, W. A. Oats and E. O. Hall, *J. Inst. Met.*, **93**, 302 (1964).

m M. F. Lantratov, *Zhur. Prikl. Khim (Leningrad)*, **34**, 130 (1961).

n F. E. Wittig, E. Muller and W. Schilling, *Z. Elektrochem.*, **62**, 529 (1958).

where T_m is the melting point, and only data for the pure components are used (the numerical factor 0.4 is empirical)[5].

Other contributions to the excess entropies arise from changes of the coordination number which are non-linear with composition, and to changes in the vibration frequencies and the electronic heat capacities. Some discussion was devoted to a possible correlation of s^E with the valency difference between the two alloying metals. The correlation found for zinc alloys (where often $s^E > 0$) is not encountered for mercury alloys, liquid amalgams ($s^E < 0$ is the rule) or with lead alloys. Positive partial entropies would result in alloys rich in a metal which has more tendency to covalent bonding, hence to clustering, than the second metal (cf. p. 186).

If, as a rough approximation, the finite values of v^E and s^E are disregarded, the regular-solution model may be applied to liquid alloys of the types discussed here. This approach is useful at least for the purpose of predicting the occurrence of miscibility gaps and of determining the course of the liquidus curve in phase diagrams. In fact, phase diagrams are not very sensitive to small errors in the excess entropy, and the assumption of the ideal entropy of mixing, which may be based on volume fractions (p. 142) in the case where the atomic volumes of the components are very disparate, does not lead to serious error. On the other hand, this insensitivity does not permit the deduction of excess entropies from phase diagrams with reasonable accuracy.

Whether the mixture obeys the regular solution expression $\ln f_2 = (b/\mathrm{R}T)(1 - x_2)^2$, or not, it is possible to describe the activity coefficient of a solute at trace concentrations as $\ln f_2^\infty = \ln f_2^{\infty 0} + \epsilon_{22} x_2$, where ϵ_{22} is a self interaction parameter. This may alternatively be defined as $\epsilon_{22} = \lim(x_2 \to 0)\mathrm{d}(\ln f_2)/\mathrm{d}x_2$, and is seen to equal $-b/\mathrm{R}T$, if the regular-solution expression applies. Values of this self interaction parameter for several dilute binary liquid alloys are shown in Table 8.2. Both positive and negative values are found, whereas positive b values are expected for the systems obeying the regular-solution law over the whole composition range. Furthermore, if the roles of solute and solvent are interchanged, different values of ϵ_{22} and ϵ_{11} (which relates to trace solute 1 in solvent 2) are as a rule found[6], contrary to the expectation for a single b value to apply for regular solutions (e.g. the Cd–Sb system).

---

[p] D. B. Downie, *Acta Met.*, **12**, 875 (1964).
[q] V. N. Eremenko and G. M. Lukaschenko *Ukr. Khim. Zh.*, **28**, 462 (1962).
[r] F. E. Wittig and P. Scheidt, *Naturwissenschaften*, **47**, 250 (1960).
[s] W. J. Swirbely and S. M. Read, *J. Phys. Chem.*, **66**, 658 (1962).
[t] P. Chiotti and E. R. Stevens, *Trans. Met. Soc. AIME*, **233**, 198 (1965).
[u] E. Scheil and F. Wolf, *Z. Metallk.*, **50**, 229 (1959); E. Scheil and E. D. Muller, *Z. Metallk.*, **53**, 389 (1962).
[v] A. T. Aldred and J. N. Pratt, *Trans. Faraday Soc.*, **57**, 611 (1961).
[w] O. J. Kleppa, *J. Am. Chem. Soc.*, **73**, 385, 885 (1951).
[x] S. K. Das and A. Ghosh, *Met. Trans.*, **3**, 803 (1972).

Table 8.2. Self interaction parameters ϵ_{22} for dilute binary liquid alloys[a].

Solvent	Solute	T (K)	ϵ_{22}	Solvent	Solute	T (K)	ϵ_{22}
Ag	Pb	1273	−0.4	Cu	Zn	877	+1.4
Au	Bi	973	+0.1	Pb	Ag	1273	−0.9
	Pb	873	+5.		Au	873	+3.2
	Tl	973	+0.1		Bi	773	+2.6
Bi	Au	973	+2.1		Cd	773	−2.6
	Cd	773	−1.2		Sn	773	−1.2
	Pb	773	+1.4		Zn	880	−5.3
	Sn	603	−0.1	Sb	Cd	773	+1.5
	Zn	873	−3.3	Sn	Cd	773	−1.1
Cd	Bi	773	−6.5		Pb	773	+0.7
	Hg	600	+3.4		Tl	625	−3.2
	Pb	773	−4.6		Zn	957	−0.6
	Sb	773	−6.5	Tl	Au	973	+2.6
	Sn	773	−1.5		Sn	625	−1.7
	Zn	955	−1.8	Zn	Cd	955	−3.3

[a]J. M. Dealy and R. D. Pehlke, *Trans. Met. Soc. AIME*, **227**, 88 (1963).

For regular solutions, obeying the law $g^E \simeq h^E \simeq bx_1x_2$, or $RT \ln f_1 = bx_2^2$ (pp. 188) in the region where the solution is in equilibrium with pure solid component 1, the liquidus temperature T is related to the mole fraction x_2 approximately by

$$T = [h_1^F - T_{m1}\Delta c_{p1} + bx_2^2]/[-R\ln(1-x_2) + (h_1^F/T_{m1}) + \Delta c_{p1}] \qquad (8.3)$$

where the parameters for fusion refer to pure component 1, and Δc_{p1} is the difference between the heat capacities of pure liquid and solid component 1 at the melting point. This latter quantity is often neglected in comparison with h^F/T_m. The properties of the pure components (eq. (8.3) applies, of course, *mutatis mutandis* to the composition region where the liquid is in equilibrium with pure component 2), and the single parameter b therefore suffice for fixing the liquidus curve.

The parameter b is also sufficient, provided that the regular solution model applies, for determining the critical consolute temperature for eventual liquid immiscibility. Equations (4.69) and (4.90) can be used, recalling (p. 148) that for regular solutions $b = \alpha$ and $\beta = 0$, so that $T_c = b/2R$. This condition may be also rephrased to state that phase separation will occur for b values at least so large that $b/2R$ is well above the liquidus temperature of the mixture at $x = 0.5$ (since composition-symmetric behaviour is assumed).

The parameter b may be estimated by applying the solubility parameter concept. It has become clear, however, that the simple expression along the lines used for non-polar molecular liquids (pp. 167 to 172) is not applicable, and that the size effect and the electronegativity effect are two major factors that should be included in the considerations. The former is taken care of to some extent by using volume fractions rather than mole fractions, applying (5.32) instead of (5.27), i.e. letting $\ln f_i$ be proportional to $(1 - \phi_i)^2$, instead of to $(1 - x_i)^2$. The solubility parameter criterion for immiscibility then becomes

$$T_c = b/2R = (v_1^* + v_2^*)(\delta_1 - \delta_2)^2/4R \tag{8.4}$$

which should be larger than the liquidus temperature at $\varphi = 0.5$, rather than at $x = 0.5$.

However, the Berthelot geometric mean rule (5.29) is not a good approximation in the heteroatomic liquid alloys, since the modes of bonding in the liquid-metal components may differ considerably. Cognizance of this is taken by including the electronegativity effect[7]. It is not the quantity $\frac{1}{2}(v_1^* + v_2^*)(\delta_1 - \delta_2)^2$ that should be compared with $2RT_c$, as in (8.4), but the criterion for immiscibility is

$$[\tfrac{1}{2}(v_1^* + v_2^*)(\delta_1 - \delta_2)^2 - 2RT] > n\eta \tag{8.5}$$

where n is the number of metallic bonds that can be formed between the two metals. According to the normal valency rules for metals, the maximal value of n is 6 (p. 95). Therefore, if at a given temperature T the left-hand side of (8.5) is larger than 6η, immiscibility should occur. Conversely, if 2η is larger than the left-hand side of (8.5), so that only two bonds are necessary, the liquid metals are completely miscible at the given temperature[8]. For $2 < n < 6$, the size factor becomes decisive, and immiscibility was found for systems with $0.12 < \rho < 0.60$.

A list of solubility parameters and of electronegativities is given in Table 8.3. It should be noted that the solubility parameters refer to the melting points of the metals T_m. Whereas for molecular liquids $\delta(T)$ over their liquid range differs little from δ(298 K), the standard value, there occurs a considerable decrease of the solubility parameters of the metals as the temperature increases to above the melting point. For the liquid metal

$$u^V(T) = h^S(298) + (3R/2)(T\,(K) - 298) - \bar{c}_P(T\,(K) - 298) - \Delta h^F(T_m) \tag{8.6a}$$

$$v(T) = v(298)(1 + \alpha^s(T_m\,(K) - 298)) + v^F + v^l(T_m)\alpha^l(T - T_m) \tag{8.6b}$$

where h^S is the heat of sublimation of the solid and $\bar{c}_P$ is the mean molar heat capacity over the temperature range. Here from the approximate relationship

$$\delta(T) = \delta(298)[0.97 - (T\,(K) - 298)(3.10^{-6}\,v(298)\,(cm^3) + R/h^S)] \tag{8.7}$$

is derived, which holds for metals, in view of their average values of heat capacity, heat of fusion Δh^F, volume change on fusion Δv^F and coefficients of thermal expansion. With the mean values of h^S and $v(298)$ of the metals, (8.7) expresses a -1 % change in δ per 100 K increase in temperature.

The description of the systems with the single parameter b, which depends on v_i^*, δ_i and χ_i of the pure metals, and with their volume fractions, in terms of the regular-solution model, is only a rough approximation. It can, of course, say nothing about s^E, which it assumes to equal zero, or about v^E either. Attempts have been made to use other models, employing perturbation methods (section 4B(b)), such as the average potential model (section 4B(c)). In one attempt[9], additivity of metallic radii in the mixture is assumed, i.e. $\sigma = 0$

Table 8.3. Solubility parameters of liquid metals at the melting point and their electronegativities[a].

Metal	T_m (K)	δ ($J^{1/2}$ $cm^{-3/2}$)	χ ($kJ^{1/2}$)	Metal	T_m (K)	δ ($J^{1/2}$ $cm^{-3/2}$)	χ ($kJ^{1/2}$)
Li	454	105.9	10	Mn	1517	173.8	15
Na	371	64.4	9	Re	3453	229[b]	19
K	336	42.3	8	Fe	1809	208.3	18
Rb	312	37.3	8	Co	1768	221.2	18
Cs	302	32.0	7	Ni	1725	229.4	18
Be	1556	236.0	15	Ru	2700	232[b]	22
Mg	923	91.1	12	Rh	2239	221[b]	22
Ca	1123	76.5	10	Pd	1823	180.7	22
Sr	1043	65.2	10	Os	3300	228[b]	22
Ba	983	71.4	9	Ir	2727	231[b]	22
Al	932	159.5	15	Pt	2043	227.3	22
Sc	1812	129.1	13	Cu	1357	196.5	19
Y	1803	133.6	12	Ag	1234	145.3	20
La	1193	109[b]	11	Au	1336	173.9	24
Ce	1077	118[b]	11	Zn	693	107.7	16
Ti	1940	156[b]	15	Cd	594	83.8	17
Zr	2125	152[b]	14	Hg	234	62.1	19
Hf	2500	180[b]	13	Ga	303	153.3	16
Th	1968	137[b]	13	In	429	121.7	17
V	2190	202[b]	16	Tl	577	96.7	19[c]
Nb	2740	206[b]	16	Ge	1210	159.3	18
Ta	3269	213[b]	15	Sn	505	130.4	18
Cr	2176	197.3	15[c]	Pb	601	97.0	18
Mo	2890	205[b]	18	Sb	903	111.3	21[c]
W	3650	223[b]	16[c]	Bi	545	95.4	19
U	1405	199.0	17	Te	723	81.4	21
Pu	913	147.2	13				

[1] The solubility parameters were calculated from h^S and v^l data reported by Wilson[3]. Electronegativities are those reported by L. Pauling, *Nature of the Chemical Bond*, Cornell University Press, Ithaca, N.Y., 3rd edn., 1960.

[b] Calculated from $\delta(298)$, taken from J. H. Hildebrand and R. L. Scott, *The Solubility of Nonelectrolytes,* Reinhold, New York, 3rd edn., 1950, p. 323, by means of eq. (8.6).

[c] From M. Haissinsky, *J. Phys. Radium,* **7**, 12 (1946). The following unusual valency states give[5] the most consistent results: Cr(II), W(IV), Tl(II) and Sb(V).

(eq. 4.106), and second-order and higher-order terms in the perturbation parameters (i.e. δ^2, θ^2, $\rho\delta$, $\rho\theta$, $\delta\theta$, but not ρ^2) are neglected. The results are not sensitive to ρ, the size difference, and give approximately the right magnitude and shape of the $h^E(x)$ curve, but often a wrong sign of the s^E function. More sophisticated treatments are therefore necessary in order to understand the source of the excess entropies in these systems.

c. Compund-forming liquid alloys

Many intermetallic compounds are well characterized with respect to stoichiometry, structure and bonding in the solid state. Many other intermetallics are better described as solid phases than as compounds, since they may not have constant stoichiometries. In any case, the particular compounds and phases formed are uniquely affected by the requirements of the crystalline lattice in which the alloy is solidified. This particular structure represents a local minimum of Gibbs energy in the temperature–composition space, dictated mainly by the relative atomic sizes of the metals and the orientation of their valences, i.e. bonding requirements.

In the liquid state, these structural constraints are absent. Compound formation does persist into the molten alloy, if the bonding between unlike atoms is so much stronger than that between like atoms that it can overcome the disorder caused by thermal agitation and produce short-range order. The systems belonging to this category have electronegativity factors $\eta > 16$, on average, and usually but not always, exhibit a maximum in the liquidus curve (Fig. 8.2(d)). These have been classified[3] as electron-compounds when their solid compositions are determined primarily by the Hume-Rothery rules, or as liquidus maximum systems which are non-electron-compounds. These latter have an average $\eta \sim 30$, while $\eta \sim 45$ characterizes electron-compounds, which may also show a liquidus maximum.

The Hume-Rothery rules specify certain ratios (3/2, 7/4, 21/13, etc.) between the number of electrons and the number of atoms in alloys forming stable solid phases[10]. The number of electrons is given by the valency, which is taken as the group number in the Periodic Table, except for metals of group VIII, which are assigned zero valency. The phase $Cu_{31}Sn_8$ thus corresponds to an electron-to-atom ratio of $(31 \times 1 + 8 \times 4)/(31 + 8) = 21/13$. Arguments have, however, been presented (p. 95) by Pauling according to which the metals of groups VII and VIII have a valency of 6, and the B-group metals in the Periodic Table should have valencies equal to 6.5 – (group number), e.g. 5.5 for copper, 3.5 for indium and 1.5 for bismuth[10]. Corresponding electron-to-atom ratios, differing from those of Hume-Rothery, are obtained on this basis, e.g. 127/26 for Cu_5Zn_8, Cu_9Ga_4 or $Cu_{31}Sn_8$. Since these stoichiometries are forced on the alloys by the crystal structures, they have no meaning for the liquid mixtures. The stoichiometries in the liquid become discernible if the composition-dependencies of excess thermodynamic functions are examined, together with other peculiarities, e.g. of pair correlation curves (section 8A(a)) or transport properties, especially conductivities.

Ideal associated mixtures (section 5B(a)) are very uncommon, and although the concept has been applied to alloys (e.g. the Bi + Cd, Hg + Tl and Pb + Tl systems[11]), a closer examination showed serious deviations. Regular associated solutions, as discussed below, seem also to be rare, and most systems show finite excess entropies, as for the mercury + thallium or gold + tin systems[11]. The latter shows maxima and minima in the $s^E(x)$ curve, the most pronounced minimum at $x_{Au} = 0.75$, corresponding to Au_3Sn which is unknown as a solid complex, but persists in the liquid to high temperatures. Another minimum, corresponding to a mixture of the compounds AuSn and $AuSn_2$, known in the solid phase, disappears at higher temperatures.

The excess properties at equimolar compositions of some binary compound-forming alloys are given in Table 8.4. The compounds listed refer to the liquidus maximum (the molten silver compounds show no such maximum, and also have low η values). There is a rough negative correlation between the electronegativity factor η and the heat of mixing $\hat{h}^E$, and practically all systems with $\eta > 10$ also have $\hat{h}^E$

Table 8.4. Excess volumes, entropies and heats of mixing of some equimolar compound-forming binary alloys (unmarked data are from ref. 3).

A	B	Compounds	η (kJ mol^{-1})	$\hat{v}^E$ (%)	$\hat{s}^E$ (J K^{-1} mol^{-1})	$\hat{h}^E$ (kJ mol^{-1})	Ref.
Ag	Al		25		+2.8	−4	a
	Cd		9		−2		
	In		9			−3.3	b
	Mg	AgMg, Ag_3Mg_2	64			−12.1	
Al	Cu	AlCu	16	−7		−18.4	
Au	Cd	AuCd	49		+4	−18.8	
	Ga	$AuGa_2$	64			−13.2	c
	In	$AuIn_2$	49			−17	
	Sn	AuSn, $AuSn_2$	36		+2.6	−10.8	
Bi	In	BiIn	4		+0.4	−1.9	d
	Mg	$BiMg_2$, Bi_2Mg_3	49	+2	+0.8	−22.2	
	Tl	Bi_5Tl_3	0		+0.8	−4.6	
Cd	Cu	Cd_3Cu_2	4		−3.3	−3.3	e
	Na	Cd_2Na	64		+2.9	+1.9	
Cu	Mg	Cu_2Mg	49		−2.9	−10.0	
Hg	K	Hg_2K	121	−25	−16.3	−22.2	f
	Na	Hg_2Na	100	−20	−22.2	−28.0	f
	Tl	Hg_5Tl	0		−0.5	−1.0	g
K	Pb		100		−9.3	−20.3	h
	Tl	KTl	121		+9	−11.1	f
Mg	Pb	Mg_2Pb	36	−6.5	−3.2	−9.6	
	Sn	Mg_2Sn	36	−4	−3.4	−14.4	i
Na	Pb	Na_5Pb_2	81		−7.7	−17.6	f
	Sn	NaSn	81		−11.3	−22.2	f
	Tl	NaTl	100		−3.4	−11.3	f
Pb	Tl	Pb_3Tl_5	1		−0.9	−1.7	d

[a] F. E. Wittig and J. F. Elliot, *J. Electrochem. Soc.*, **107**, 628 (1960).
[b] C. B. Alcock, R. Sridhar and R. C. Svedberg, *Acta Met.*, **17**, 839 (1969).
[c] B. Predel and D. W. Stein, *Acta Met.*, **20**, 515 (1972).
[d] J. Terpilowski, *Arch. Hutn.*, **3**, 226 (1958).
[e] A. V. Nikolskaya, P. P. Otopkov and Ya. I. Gerasimov, *Zh. Fiz. Khim.*, **31**, 1007 (1957).
[f] F. E. Wittig and T. Klemstuber, *Proceedings of XVIII International Congress of IUPAC*, 1961.
[g] F. E. Wittig and P. Scheidt, *Naturwissenschaften*, **47**, 250 (1960).
[h] M. F. Lantratov, *Zh. Fiz. Khim.*, **34**, 782 (1960).
[i] A. Steiner, E. Miller and K. L. Komarek, *Trans. Met. Soc. AIME*, **230**, 1361 (1964).

more negative than −10 kJ mol^{-1}. The curves $h^E(x)$ are often very asymmetric, and the maximal value of $-h^E$ often coincides with the compound composition, rather than with equimolarity, characteristic for eutectic or solid-solution systems. Excess volume data for these systems are scarce, but large negative v^E values have been observed for them (especially for many systems involving silicon, not included in Table 8.4) compared with small positive values ⩽ 2 %, characteristic for eutectic and solid-solution systems. The excess entropies do not follow systematic trends, but, as already mentioned, inflected curves $s^E(x)$, with both negative and positive portions, are common.

Highly asymmetrical, and in extreme cases inflected $h^E(x)$ curves are also observed in systems involving zinc, cadmium and metals of groups IVB and VB. The partial heats of mixing of the 'solvent', when zinc or cadmium are present at a large excess, may be positive, in spite of strong compound formation and generally negative heats of mixing. Similar effects occur with group IVB and group VB solvents and copper or silver solutes, e.g. in the silver–tin system. This effect has been explained by strong self-bonding of the solvent, although the type of bonding is different in the two groups of systems. In group IVB and VB solvents, the breaking of homopolar bonds between the solvent metal atoms by the solute is involved, while for zinc and cadmium solvents an increase in the Fermi level of these metals is produced by the solute. Phenomenologically, the excess functions follow closely the behaviour of self associated molecular liquids mixing with inert liquids (cf. sections 5B(d) and 5B(e), Fig. 5.12).

Compound formation may change the nature of the bonding from metallic to ionic. In systems such as silver + tellurium or bismuth + magnesium, there is little doubt of a more or less complete electron transfer, to form anions (telluride or bismuthide) and cations (silver or magnesium). This process involves, however, large negative enthalpy changes. On the other hand, these ionic compounds may not be

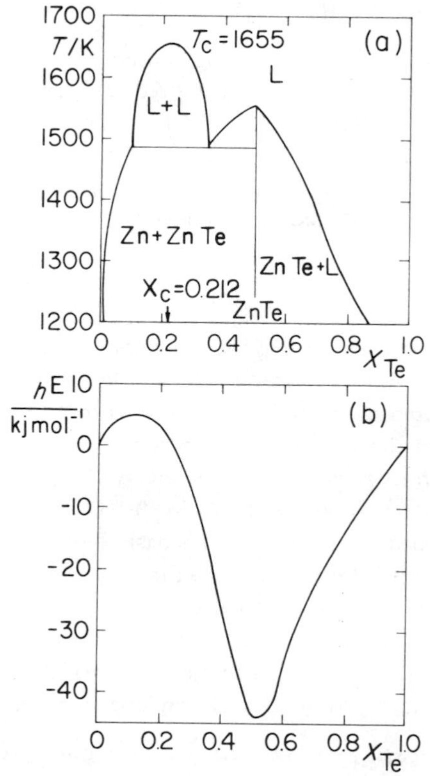

Figure 8.3. The zinc–tellurium system: upper part, phase diagram[1 2]; lower part, heat of mixing curve (calculated from the data in reference 11).

compatible with the metallic bonding that prevails at compositions near the pure metals, so that positive enthalpies of mixing result in these composition regions.

The tellurium + zinc system is also typical for such cases, and has been studied[12] (Fig. 8.3) with reference to the concept of associated regular solutions (section 5B(c)). The compound formed is TeZn, and it is assumed to interact in a regular manner with zinc, in the region $x_{Te} < 0.5$, and with tellurium in the region $x_{Te} > 0.5$. Two independent interaction parameters $\alpha_{Zn} = 71.1$ kJ mol^{-1} and $\alpha_{Te} = 3.8$ kJ mol^{-1} are said to describe these interactions, and the fraction associated at the equimolar composition, the parameter $\hat{\zeta} = 0.94$, is directly related to the equilibrium constant (eq. 5.97). The excess enthalpy is given by

$$\begin{aligned} h^E &= \alpha_{Zn}(x_{Te} - \zeta)\zeta + \zeta(1+\zeta)^{-1}\Delta h^0_{TeZn} \qquad \text{for} \quad x_{Te} < 0.5 \\ &= \alpha_{Te}(1 - x_{Te} - \zeta)\zeta + \zeta(1+\zeta)^{-1}\Delta h^0_{TeZn} \qquad \text{for} \quad x_{Te} > 0.5 \end{aligned} \tag{8.8}$$

where ζ is the fraction associated, and Δh^0_{TeZn} is obtained from thermochemical data. The liquidus temperature T is related to the parameters α and ζ by

$$T = [h^F + \alpha(x - 0.5)^2]/[(h^F/T_m) + 0.5\,\mathbf{R}\ln x(1-x)^{-1} - \mathbf{R}(1 - \hat{\zeta})] \tag{8.9}$$

where h^F and T_m refer to the crystallizing component, and α and x to the solute, Phase separation in the zinc-rich region is predicted, since the critical consolute temperature $T_c = \alpha_{Zn}\sqrt{3}/9\mathbf{R}\hat{\zeta}(2 - \hat{\zeta}) = 1655$ K is above the liquidus at the critical composition $x_{cTe} = 0.212$. Because of the low α_{Te}, the corresponding critical consolute temperature for the tellurium-rich region is much below the liquidus, and no liquid immiscibility is predicted there, nor found experimentally.

d. Ternary liquid alloys

Not much information concerning the behaviour of ternary or multicomponent liquid alloys is available. Darken[13] has shown that properties of binary liquid alloys may be established from data on dilute solutions of a third component in them, used as a probe. Thus the excess Gibbs energy of a B+C alloy may be obtained from e.m.f. measurements using A as an electrode, a molten salt of A as an electrolyte, and a dilute solution of A in B + C as the other electrode. The expressions that have been developed on this basis[14] are shown in section 7A(d), in the discussion of ternary additive molten salts. The composition variables n_A, n_B and n_c are replaced by $n = n_A + n_B + n_C$, $x = n_B/(n_B + n_C)$ and $y = (n_B + n_C)/n$, so that the molar excess Gibbs energy is a function of the two composition variables x and y only, and the condition $y \to 1$ is applied. The deviation of the excess chemical potential of the probe metal from linear dependence is given by (7.43)

$$\Delta\mu^E_A = \lim_{y\to 1} \mu^E_A - x\mu^E_{A(B)} - (1-x)\mu^E_{A(C)} = -g^E_{BC} + x(1-x)f(x, y \to 1) \tag{8.10}$$

where the leading term on the right-hand side is the negative of the excess Gibbs energy of the binary system B—C. This would mean that (8.10) predicts the same deviation $\Delta\mu^E_A$ for all probe metals A, if it were not for the correction term

$f(x,y \to 1)$ which depends specifically on the interaction of A with B and with C. For example[14], in dilute solutions of cadmium in the binary Bi + Pb at $x = 0.5$, $x(1-x)f = -0.13$ kJ mol^{-1} while for aluminium as the probe the value is -0.34 kJ mol^{-1}. Compared with $-\hat{g}^E_{BiPb} = 1.26$ kJ mol^{-1}, these values of f are not large, but certainly not negligible, and equating $\Delta\mu^E_A$ to $-g^E_{BC}$ is only an approximation.

The function $f(x,y \to 1) = g(x,y \to 1) + g^E_{BC}/x(1-x)$ has been derived according to the quasi-chemical model (section 4B(d)), where only nearest-neighbour interactions are considered. It is a complicated function of the three $\Delta\epsilon_{ij} = \epsilon_{ii} + \epsilon_{jj} - 2\epsilon_{ij}$ (where i and j are A, B or C) and the temperature, in addition to the composition variables x and y. Since its effect is most noticeable at compositions near $x = 0.5$

$$f(x = 0.5, y \to 1) = 2ZRT[\ln(1 + \exp(\Delta\epsilon_{AB} - \Delta\epsilon_{AC})/2kT) + \ln(1 + \exp(\Delta\epsilon_{AC} - \Delta\epsilon_{AB})/2kT) - 2\ln 2] \tag{8.11}$$

which may be written as (7.44), except that for the liquid alloys it is the nearest-neighbour coordination number, $Z \sim 10$, rather than the next-nearest-neighbour coordination number, Z' (also ~ 10), which applies for the additive ternary molten salts. In any case, f is a negative quantity, and counteracts a large negative g^E_{BC}.

Only a few systems have been treated according to this manner, e.g. the Cd + Bi + Pb and Al + Bi + Pb systems, where Cd and Al are the probe metal A in the above treatment, and the Cd + Sn + Pb and Cd + Sn + Bi systems. Agreement with the model was found, so that the ternary systems can be described in terms of the properties of the binary ones.

A different formulation[15] for the function $f(x,y \to 1)$ is obtained formally if a two-parameter Redlich–Kister equation (section 4A(d), eq. (4.65)) is applied to the binary alloys. It is found that f contains only the second-order terms, b_2, of the three binary alloys

$$f(x,y \to 1) = x^2(1-x)^{-1}b_{2\,AB} - (1-x)^2x^{-1}b_{2\,AC} - (1-2x)b_{2\,BC} \tag{8.12}$$

This expression is completely empirical, but seems to express the data for ternary combinations from among Bi, In, Sn and Zn[15], from data on the binary systems.

Dilute ternary systems are also often described in terms of the parameters $\epsilon_{AA} = \lim(x_A, x_B, \ldots \to 0)\ d(\ln f_A)/dx_A$ for self interaction (cf. p. 307) and Table 8.1) and $\epsilon_{AB} = \lim(x_A, x_B, \ldots \to 0)\ d(\ln f_A)/dx_B$ for the cross interaction. For the latter $\epsilon_{AB} = \epsilon_{BA}$ applies of course. It can be shown that $\epsilon_{AB} = \ln f_{A(B)} - \ln f_{A(S)} - \ln f_{B(S)}$, where the first term refers to dilute solutions of the solutes A and B in the common solvent S. This case of dilute solutions of solutes A,B, . . . in a solvent S can also be described in terms of the quasi-chemical model, to the second-order terms, as[16]

$$g^E/RT = (Z/2)\Sigma_i x_i \ln(1 + \lambda_{iS}) - (Z/2)\Sigma_i x_i^2 \lambda_{iS} - (Z/2)\sum_{i \neq j} x_i x_j(\lambda_{iS} + \lambda_{jS} - \lambda_{ij}) \tag{8.13}$$

where $\lambda_{ij} = \exp(2\epsilon_{ij}/kT) - 1$ and similarly for λ_{iS} and λ_{jS}, and where the indices i,j refer to the solutes A,B, ... Double differentiation of (8.13) with respect to x_i at $x_{j \neq i} = 0$ and $x_i \to 0$ (i.e. for dilute binary solutions) shows that $\lambda_{iS} = -\epsilon_{ii}/Z$, so that the binary terms of (8.13) can be obtained from tabulated ϵ_{ii} data (e.g. Table 8.2). The cross-term, $\lambda_{ij} = Z^{-1}(2\epsilon_{ij} - \epsilon_{ii} - \epsilon_{jj})$, however, must be determined, from the effect of one solute on the activity coefficient of the other.

Another aspect of ternary systems is the distribution of one metal between two other, partly immiscible, liquid metals. Since the activities of the distributing metal are equal in the two phases, the distribution ratio $D_A = x'_A/x''_A$, where A distributes between the (partly) immiscible liquid metals B and C, forming the phases $'$ and $''$, equals the inverse ratio of the activity coefficients f''_A/f'_A. Hence, if the one is known, the other can be calculated, as can the standard Gibbs energy of transfer and also the enthalpy and entropy of transfer from the temperature-dependencies. The mutual miscibility of B and C makes the distribution method less useful for obtaining information on the binary systems (say A + C from known A + B), but corrections can be applied. The application of Raoult's law to the immiscible metals in each phase[17] is erroneous (pp. 150, 192), but eq. (8.10) may be applied, with the approximation that $f(x,y \to 1) = 0$, which should hold for immiscible metals since x is then very far from 0.5 in each phase. The expression that is derived from (8.10)

$$\mathbf{R}T \ln D_A = (x'' - x')[(\mu^E_{A(B)} - \mu^E_{A(C)}) + (1 - x'' - x')b_{BC}] \tag{8.14}$$

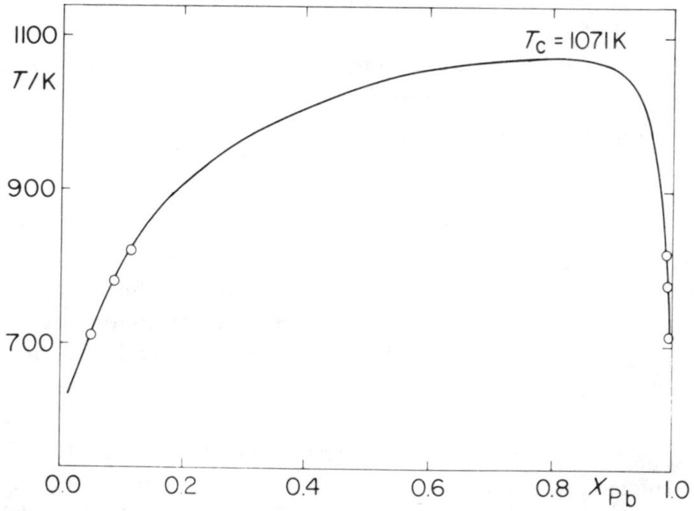

Figure 8.4. The experimentally established coexistence curve for the immiscible pair lead+zinc. The critical consolute temperature is 1071 K, and is increased by the addition of silver. The distribution of trace silver quantities between the phases at equilibrium at 711, 781 and 821 K (the points indicated), D_{Ag}, is 55.7, 27.5 and 20.1 in favour of the zinc–rich phase[18].

relates the distribution ratio to the properties of the pure binaries A + B, A + C and B + C. For the latter, b_{BC} is to be taken as potentially both temperature- and composition-dependent, while $\mu^E_{A(B)}$ and $\mu^E_{A(C)}$ apply to the limit of low concentrations of A and should be composition-independent. The more immiscible B and C are, the more does $(x'' - x')$ tend to unity and $(1 - x'' - x')$ to zero, while as the critical consolute temperature T_c for B and C is approached, the smaller does $(x'' - x')$ become, and the more does D_A approach unity[17].

The lead+zinc system has been studied by several workers with respect to the distribution of other metals[17], and in particular of silver[18] between the two immiscible phases. The function $b_{Pb,Zn}(x,T)$ and the coexistence curve $T(x',x'')$ (Fig. 8.4) are known with good precision, as well as the activity coefficients of silver in each of the other liquid metals[19], so that (8.14) can be tested. Any deviation found may be ascribed to the neglect of the function f, (8.10). Silver shows positive deviations from ideality in lead, and negative ones in zinc. The term in square brackets in (8.14) is found to contribute *ca.* 89 % of $(RT \ln D_A)/(x'' - x')$, the term in $b_{Pb,Zn}$ alone only *ca.* 7%, so that a sizeable f term should be important.

e. Electron localization in liquid alloys

An interesting feature of liquid metals and alloys, which makes them different from most other liquid systems, is the relative freedom which some of the electrons (of the order of one per atom) have. Some of the valence electrons are localized and participate in the bonding, while others are free to occupy the whole volume of the metallic system, on a time-average basis. The electrons may be considered to move among potential-wells which may trap them, if they are deep enough. The trapping is more or less permanent if the potential-well arises from a covalent, directional homopolar bond. Mixing of two metals provides then new types of electron traps, produced by electron transfer from one metal to the other which provides for ionic bonding, and even if this is absent, the state of delocalization in the mixture need not be linear with the composition.

Certain optical, magnetic and electrical properties of the liquid alloys shed light on the problem of the localization of the electrons in them, but here only the results from electric conductance will be discussed. These have been studied most systematically, and added to the structural and thermodynamic properties discussed above lead to a detailed picture of liquid metal alloys.

The resistivity of a liquid metal may be expressed as $\rho = C \langle S(K)U^2(K) \rangle$, where C is a constant that depends directly on the number density of the metal atoms and inversely on the Fermi energy of the electrons, $S(K)$ the structure factor (cf. p. 50), $U(K)$ the atomic scattering potential (also a function of the Fermi energy), and $K = 2k_F \sin(\theta/2)$ depends on the Fermi wave number k_F of the electrons. The averaging is equivalent to integrating over the angle θ

$$\langle S(K)U^2(K) \rangle = \int_0^{\pi} S(K)U^2(K)(1 - \cos\theta) \sin\theta \, d\theta \tag{8.15}$$

For a liquid alloy the corresponding expression for the resistivity is[20]

$$\rho(x) = C\{\langle xSU_A^2 + (1-x)SU_B^2\rangle + x(1-x)(1-S)(U_A - U_B)^2\} \qquad (8.16)$$

provided that the two metals have the same valency and nearly the same atomic volume, so that $S_{AA} \simeq S_{AB} \simeq S_{BB} = S$. The term in $x(1-x)$ corresponds to an empirical rule (Nordheim's) that is found to hold for solid alloys, and expresses the composition-dependence of their resistivities. If the two metals differ in valency, Nordheim's rule still holds for solid alloys, but generally not for liquid ones. In the latter case, for dilute solid alloys, another empirical rule, Linde's, holds, namely that $(d\rho/dx) = const.\ (n_A - n_B)^2$ where n is the valency. This expression is found to be obeyed for some dilute liquid alloy systems (e.g. with copper or aluminium as the solvent) but not in others (e.g. with zinc as the solvent)[3].

Very detailed composition-dependencies $\rho(x)$ have been obtained for a number of solid-solution- or eutectic-forming systems[21]. These curves show anomalies (local maxima and minima) near the compositions 1:2 and 2:1 (e.g. for Ag +Au, Ag + Cu, Au + Cu, Bi + Sb, Cd + Hg, Cd + Zn, Ga + In and Pb + Sn) in systems of equal valency. Some correlation with similar anomalies in the structure factors $S(K,x)$ is expected. For systems of different valencies, where the electron-to-atom ratio depends on the composition, the anomaly occurs nearer the mole fraction 0.4 of the higher-valent metal (e.g. for Ag (or Au) + Pb (or Sn)). These anomalies have been interpreted in terms of partial ordering of the Fermi (free) electron-gas in the disordered atomic liquid, i.e. in terms of partial binding of the electrons[3,22].

More definite information on electron binding is obtained from resistivity data in compound-forming systems. The copper + tin system, for instance, shows maximal resistivity, and an uncommon negative temperature coefficient of the resistivity, at the same composition, where there occurs a maximum in the (absolute) value of the heat of mixing, namely at $x_{Cu} \simeq 0.75$, the composition corresponding to Cu_3Sn. A solid phase at this composition corresponds to an electron-to-atom ratio of 7:4 according to the Hume-Rothery rules (cf. p. 311), which is dictated by the lattice requirements. That this composition has a special significance also in the liquid state is coincidental, but electron localization in bonds between copper and tin is indicated, and substantiated by the minimum in the (electronic) magnetic susceptibility, which also occurs at this composition. The similar behaviour of the silver + tin system in respect of the maximum in resistivity and vanishing temperature coefficient at compositions near Ag_3Sn in the liquid alloys points to similar bonding, although the negative enthalpy of mixing is smaller (h^E -3.0 kJ mol^{-1} at $x_{Ag} = 0.75$, compared with $+0.25$ kJ mol^{-1} at $x_{Ag} = 0.50$, and with -5.0 kJ mol^{-1} at $x_{Cu} = 0.75$), and in spite of the systems being classified according to the phase diagrams as eutectics rather than compound-forming systems.

With more electropositive metals, such as magnesium, the electron localization on the more electronegative metallic partner is clearer, and more or less ionic liquid structures are obtained. In Mg_2Sn, or Mg_3Bi_2, the alloys have reached full ionicity. Addition of magnesium to the stoichiometric alloy provides additional magnesium ions, which mix almost ideally with those already present (hence the low partial enthalpy of mixing of magnesium h^E_{Mg})[23], and free electrons, which contribute strongly to the conduction beyond the sharp minimum[23]. The addition of bismuth, on the other hand, to the stoichiometric liquid alloy provides positive hole conduction. The mole fraction of these holes at $x_{Mg} = 0.6$ (corresponding to Mg_3Bi_2) has been estimated at 0.046 and the partial excess chemical potential of magnesium relative to this composition has been given[23] as $\mu^E_{Mg} - \mu^E_{Mg}\ (x_{Mg} = 0.6) = 2RT \sinh^{-1}\ [(x_{Mg} - 0.6)/0.046]$ with good agreement with experiment.

B. Mixtures of Metals and Molten Salts

Many important metallurgical processes involve metals and molten salts (slags). The equilibria between a pair of metals A and B, and the corresponding salts AX_{z_A} and BX_{z_B}, where X represents one equivalent of the anion, e.g. oxide, silicate, halide, etc., have been discussed in detail many years ago. At that time there was little systematic information, not to mention prediction ability, on liquid alloys and on molten salts. Today, knowledge of these fields is much wider and deeper, so that if the standard thermodynamic functions for the equilibrium

$$AX_{z_A} + (z_A/z_B)B \rightleftharpoons (z_A/z_B)BX_{z_B} + A \qquad (8.17)$$

as well as the behaviour of the binary mixtures, the liquid alloy A + B and the molten salts $AX_{z_A} + BX_{z_B}$, are known, one requires only the properties of the binary mixtures $A + AX_{z_A}$ and $B + BX_{z_B}$ to obtain a good approximation to the behaviour of the system. A correction for the simultaneous presence of all the four constituents (virtual components) may modify this approximate description, but the main features will be given if knowledge of the properties of mixtures of liquid metals in their own salts is added to that of the liquid alloy and salt binaries.

Some metals show considerable solubility, even complete miscibility, in some of their salts; in other cases, the solubility is very limited, Fig. 8.5. Metallic properties may be conferred to a varying extent on the solutions. It is possible to classify these solutions into a group where the metallic electrons are delocalized to a large extent, or into another where they are localized, so that the metal ions of the salt are reduced to a lower valency. The boundary between these groups is certainly not sharp, but mixtures of alkali metals with their halides are a typical example of the first group, and bismuth and its halides of the second.

Detailed information on many metal–metal halide systems has by now been amassed. This is not the case for other salts: information on oxides, sulphides, silicates, etc. is not as systematic. The present discussion will therefore emphasize the halide systems.

a. Metal-salt systems with delocalized electrons

Alkali metals, alkaline-earth metals, and some of the rare-earth metals form with their halides liquid mixtures with more or less delocalized electrons. These consist

Table 8.5. Mutual liquid solubilities, metal in salt ($x_{M(S)}$) and salt in metal ($x_{S(M)}$) of group IA to IIIA metals and their halides.

M	X	$x_{M(S)}$	$x_{S(M)}$	x_{cM}	T_c (K)	T (K)[a]	T_{mM} (K)	T_{mMX_n} (K)	Ref.
Li	H	0.0045	0.224	0.394	1305	958M	452	961	b
	F	0.01	0.03	0.40	1603	1120M	452	1121	c
	Cl	0.005				882M	452	883	c
	I	0.01				741M	452	742	c
Na	F	0.03	0.17	0.28	1453	1263M	370	1268	c
	Cl	0.021	0.023	0.50	1353	1068M	370	1073	c
	Br	0.029	0.034	0.52	1299	1013M	370	1020	c
	I	0.016	0.014	0.59	1306	929M	370	933	c
K	F	0.049	0.483	0.20	1177	1122M	337	1131	c
	Cl	0.105	0.250	0.39	1063	1024M	337	1043	c
	Br	0.190	0.308	0.44	1001	981M	337	1007	c
	I	0.135	0.175	0.50	990	931M	337	954	c
Rb	F	0.09	0.60	0.21	1063	1046M	312	1068	c
	Cl	0.18	0.43	0.37	979	969M	312	995	c
	Br			{miscible}			312	965	c
	I	0.22	0.27	0.51	907	888M	312	920	c
Cs	F			{miscible}			302	976	c
	Cl			{miscible}			302	918	c
	Br			{miscible}			302	909	c
	I			{miscible}			302	899	c
Mg	F	0.004					923	1529	d
	Cl	0.002				~987M	923	987	c
Ca	F			{miscible}			1110	1689	d,e
	Cl	0.027	0.005	0.62	1610	1093M	1110	1045	c
	Br	0.023	0.004	0.64	1610	1100M	1110	1015	c
	I	0.038	0.003	0.74	1650	1104M	1110	1053	c
Sr	Cl	0.055	0.01	0.52	1379	1112M	1042	1145	c,f
	Br	0.02	0.015	0.62	1337	1018M	1042	931	f
	I	0.02	0.01	0.66	1329	1024M	1042	810	f
Ba	H			{miscible}			998	1473	g
	F			{miscible}			998	1627	d
	Cl	0.150	0.10	0.55	1293	1163M	998	1235	c,f
	Br	0.16	0.15	0.65	1233	1073M	998	1128	f
	I	0.17	0.20	0.75	1187	957M	998	985	f
Sm	F	0.32	0.015			1487E	1359	1698	d
Yb	F			{miscible}			1116	1680	d
Sc	Cl	0.185				1076E	1812	1240	h
Y	Cl	0.023				988E	1803	994	h
	I	0.118				1221E	1803	1270	h
La	Cl	0.09				1099E?	1193	1135	h
	Br	0.14				1001E	1193	1059	i
Ce	Cl	0.093				1050E?	1077		h
	Br	0.12				960E	1077	1005	i
Pr	Cl	0.188				1043M	1200	1058	h
	Br	0.17				874P	1200	966	i
Gd	Cl	0.02				905P	1623	878	h
Er	Cl	0.048				1019E	1798	1050	h
	I	0.110				1206E	1798	1287	h

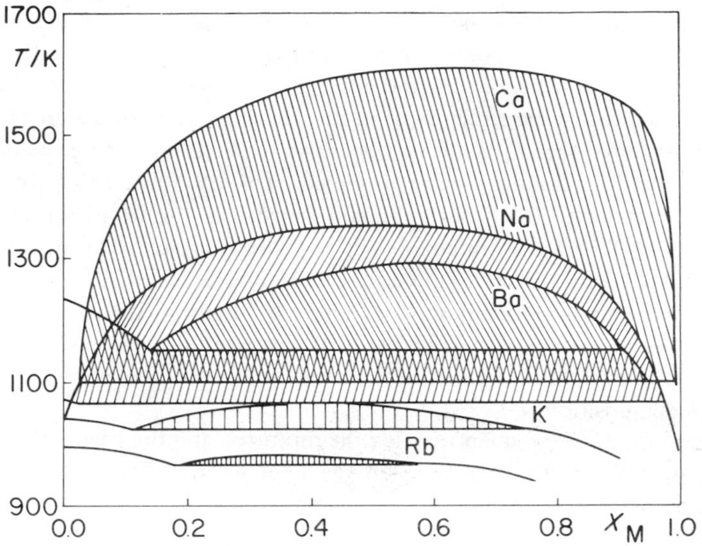

Figure 8.5. Miscibility gaps in some metal+metal chloride systems. For details of temperatures, compositions and references see Table 8.4. Caesium is miscible with its chloride at all temperatures above the liquidus.

usually of two phases, one rich in metal and the other rich in salt. The metals can be considered to dissociate in these mixtures to the metal cations of normal valency, and 'electronide' anions. The metal cations are indistinguishable, whether originating from the metal or the salt, and produce as in ordinary molten salts the cationic quasi-sublattice. The anions mix with the electrons on the anionic quasi-sublattice, with more or less severe deviations from ideality. Partial association to configurations involving one or several of the cations, anions and electrons with life-times long compared with thermal vibrations also contribute to the thermodynamic, transport and other properties. The situation as it was known a decade ago has been reviewed in great detail by Bredig[25] and by Corbett[26].

A summary of the phase relationships (compositions and temperatures) of these systems is shown in Table 8.5. Complete miscibility is seen to be the exception, limited solubility the rule, with higher salt solubility in molten univalent metals, and lower salt solubility in the more tightly bonded molten multivalent metals,

---

[a] M = monotectic; E = eutectic, P = peritectic: temperature at which compositions $x_{M(S)}$ and $x_{S(M)}$ apply.
[b] E. Veleckis, E. H. van Deventer and M. Blander, *J. Phys. Chem.*, **78**, 1933 (1974).
[c] M. Bredig, in *Molten Salts*, edited by M. Blander, McGraw-Hill, New York, 1964, pp. 367.
[d] A. S. Dworkin and M. A. Bredig, *J. Phys. Chem.*, **75**, 2340 (1971).
[e] B. D. Lichter and M. A. Bredig, *J. Electrochem. Soc.*, **112**, 506 (1965).
[f] A. S. Dworkin, H. R. Bronstein and M. A. Bredig, *J. Phys. Chem.*, **72**, 1892 (1968).
[g] D. T. Peterson and M. Indig, *J. Am. Chem. Soc.*, **82**, 5645 (1960).
[h] J. D. Corbett, D. L. Pollard and J. E. Mee, *Inorg. Chem.*, **5**, 761 (1966).
[i] R. A. Sallach and J. D. Corbett, *Inorg. Chem.*, **2**, 457 (1963).

then *vice versa*, metal in the molten salt. For many systems, however, an upper critical consolute temperature T_c can be determined, and the extent of the miscibility gap is given mainly by the difference between T_c and the monotectic temperature, which, in turn, depends on the melting points of the metal and the salt. It may be noted that T_c is not very sensitive to the anion for a given metal: the average for the four halides within a spread of $< \pm 8\,\%$ is 1353 K for sodium, 1058 K for potassium, 983 K for rubidium (excluding the bromide), 1623 K for calcium and 1348 K for strontium (excluding the fluoride). The miscibility gap may, therefore, vanish if the monotectic temperature is raised above T_c, when one of the salts has a high melting point (calcium or barium fluoride). The critical mole fraction of the metal x_{cM}, at T_c, is seen to increase with increasing size of the anion. The corresponding volume fractions are more nearly constant, at nearly the equivolume composition.

Excess thermodynamic quantities for the mixtures, and the partial functions for the metal and for the salt, have been obtained for several of these systems. Excess volumes, which are usually quite small in molten salts and in liquid alloys, are very difficult to obtain in these high-temperature two-phase systems and have not been reported. Heats of mixing have also been deduced mainly from the phase diagrams (cf. discussion on p. 308), with low precision. Most of the following therefore applies to the excess Gibbs energies and chemical potentials (activity coefficients). Regular-solution-type behaviour has been noted in several systems, both dilute and up to saturation[27]. Cryoscopically, z_A foreign particles are expected per atom of metal A dissolving in the salt AX_{z_A}. A deviation from this number, as well as the interaction parameter b obtained from the activity coefficients, should be explained by the models and the theories that have been proposed for these systems.

The regular-solution theory provides for a composition-independent parameter (eq. 5.33) $b_i/v_i^* = \mu_i^E(1-\varphi_i)^{-2}\,v_i^{*\,-1}$, which depends only on the properties of the two components. The values of this parameter found[28] for potassium + potassium halide mixtures are 0.46, 0.38, 0.34 and 0.29 kJ cm^{-3}, respectively, for the fluoride, chloride, bromide and iodide. The corresponding values for sodium, rubidium and caesium are, respectively, uniformly 2.21, 0.77, and 0.61 times as large as those for potassium. It is futile, however, to go on and try to interpret these data in terms of the solubility parameters (eq. 5.33) since no independent estimates for the molten salts are available; The following additional reasons also make such an interpretation pointless. For the systems involving sodium, for which the solubility gap extends over a sufficiently large temperature range to enable the extraction of reliable $b_i(T)$ functions from the phase diagram, the actual dependence found is larger than expected from the temperature-variation of v_i^* and of the solubility parameters (eq. 8.7). This, and the sign of the dependence, points to a sizable positive excess entropy, contrary to the requirements of the theory. Furthermore, b_i/v_i^* is not the same for i being the salt or the metal, again contrary to theory, unless the temperature is much in excess of T_c, and it is doubtful whether the data are valid to these higher temperatures. On the contrary, deviations from regular solution-solubility parameter theory and the finding of a positive excess entropy are expected[28]. In the salt-rich phase, this would arise from the uncoupling of spins of the metallic electrons, and from the smaller rigidity of the F-centre-like cavities compared with the halide anions. In the metal-rich phase, it would arise from a decreased restraint on the vibration of the metal atoms by halide anions compared with metallic bonding. Thus, instead of relating the interaction parameters to the densities of cohesive energies of the components, the model for these systems considers the conversion energy of the metal to an ionic liquid, the 'metal electronide', with electrons localized but not in metallic-type orbitals as the major contribution. The delocalization of the electrons, i.e. their mixing with the anions, is only a minor contribution to the interaction parameter. This explains the much larger variation, noted above, between the several alkali metals then between the halide salts. The ratio of conversion energy to molar volume (at T_m) of the alkali metals is: Na 2.28, K 1.10, Rb 0.84 and Cs 0.68, all in kJ cm^{-3}. These are almost in the same ratios to one another as the interaction energies per unit volume, b_i/v_i^* for the metal + salt systems (irrespective of the anion) listed above. These conversion energy densities are, in absolute terms, considerably higher (two to four times) than the interaction energy densities, so that the mixing energy must be not only negative but very appreciable, rather than a minor factor. Lower conversion energies, to a hypothetical state of the liquid metal nearer its actual state than the imagined 'metal electronide', would be more compatible with the experimental results.

Cryoscopic measurements on dilute solutions of the metals in the salts indicate m foreign particles for an m-valent metal, but deviations therefrom have been noted. The dissolution of the metal can be written as

$$M \rightarrow m(x-y)M^{m-1} + yM_2^{2m-2} + (1-mx)M^{m+} + m(1-x)e^- \qquad (8.18)$$

with $0 \leqslant y \leqslant x \leqslant 1$, and where y expresses the deviation from the expected m foreign particles per metal atom dissolved, and x the trapping of electrons and their localization on monomeric or dimeric ions[3]. For a monovalent ion like potassium, $x \sim y \sim 0$ and the reaction is $K \rightarrow K^+ + e^-$, and a large increase in conductivity is observed when potassium metal is dissolved in its halides. This is due to increasing overlap of the orbitals of the free electrons as the metal concentration is increased, and a more metal-like mechanism of conduction. For sodium, $0 < x \sim y \leqslant 1$ indicates that as the metal concentration is increased, some dimerization to Na_2 occurs, where the electrons are trapped in pairs on the metal dimer, which is known from the gas phase. For divalent metals[29], such as calcium, $0 < y < x \leqslant 1$, which corresponds to the observed deviation from the cryoscopically expected 2 foreign particles per calcium atom dissolved, as well as a not very steep increase in conductivity on addition of metal, owing to electron trapping. Thus both Ca^+ (perhaps 'solvated' by calcium Ca^{2+} ions to Ca_2^{3+}) and Ca_2^{2+} are produced, in addition to the free electrons e^-. For praseodymium and neodymium dissolved in their chlorides, the increase in conductivity is so small that x is no longer small compared with $m^{-1} = 1/3$, and the strong localization of the electrons leads to their classification in the second group of systems discussed below.

b. Metal-salt systems with reduced metal ions

Little is known about the mutual solubility of other salts and metals belonging to A-groups of the Periodic Table. Of the B-group metals, however, the IB metals do not have reduced metal species. The solubility of liquid bivalent and trivalent B-group metals in their molten salts, however has been better documented, Table 8.6, and very detailed studies of the cadmium and the bismuth systems have been made. The temperature selected for presenting the data, 773 K (500 °C) is usually within the liquid range of the salt, otherwise a short extrapolation of the data has been made. Plots of log x_M vs $(1000/T)$ are linear, and $-\mathbf{R}\ \ln 10$ times their slope has been given as h_{soln}. These enthalpies of solution are all positive in the systems studied, and the mutual solubility (at least that of metal in salt) increases in the range examined with increasing temperatures.

Notable exceptions are the bismuth–bismuth chloride and bromide systems, where retrograde solubility is noted at intermediate temperatures, before the solubility gap is closed at higher ones. The critical consolute temperatures T_c are 1053 K, 811 K and 731 K, respectively, for the $Bi + BiCl_3$, $Bi + BiBr_3$, and $Bi + BiI_3$ systems, while the monotectic (or syntectic) temperatures T_M are 595 K, 567 K and 609 K, respectively. The height of the miscibility gap in the bismuth systems is thus seen to diminish with increasing size of the anion. (The data in the chloride system are not isobaric, since a high pressure of the salt vapours, 8×10^6 Pa, is required for attaining liquid–liquid equilibria near the consolute temperature.)

Many experimental methods were applied to the bismuth–bismuth halide systems in order to elucidate the species formed and the mechanism for their formation. These include cryoscopy, density, viscosity, spectrophotometry, e.m.f. vapour pressures, magnetic susceptibility, electrical conductivity measurements, etc.[30]. The accummulated evidence points to dissolution of bismuth to very dilute solutions in BiX_3 as Bi^+ which polymerizes to species Bi_m^{n+}, solvated by halide, with increasing tendency to do so from iodide to bromide to chloride. The data are more or less consistent with values $n = m = 2$, 3 or 4, but in similar solutions in a more acidic molten salt solvent, sodium tetrahaloaluminate, species with $m = 5$ and $n = 3$, and $m = 8$ and $n = 2$, have been found. Polymerization is evidently very slight in the iodide system, and Bi^+ ions are the major species. In the metal-rich composition region, metallic conductivity prevails, and BiX_3 contributes three foreign particles and has a much decreased partial molar volume. This points to the bismuth behaving as the metal, apparently present as Bi^{3+} with delocalized electrons, and the halide ions finding interstitial holes to accommodate them. In the salt-rich region, the polymeric Bi_m^{n+} ions conduct much less than Bi^+, leading to decreasing conductivities with increasing x_{Bi} in the chloride and bromide systems, but not in the iodide one. This effect diminishes with increasing temperatures, where the polymeric ions dissociate, more so for the bromide than for the chloride systems. No *solid* stoichiometric compounds BiX have been found. The 'subchloride' and 'subbromide' contain polymeric cations and a stoichiometry deviating from 1:1, while the iodide is not stable when heated to the liquid. Therefore, the systems are

Table 8.6. The solubilities of transition- and B-group metals in their halides at 773 K (unless otherwise stated) and the corresponding molar enthalpies of solution h_{soln}.

Metal	Halide	$10^3\ x_M$	h_{soln} (kJ mol^{-1})	Ref.
Al	$AlCl_3$	10^{-6} (503 K)		a
	AlI_3	3 (696 K)		b
Cr	$CrCl_2$	35 (1191 K)		c
U	UCl_3	19	16.1	c
	UBr_3	17	16.5	c,d
	UI_3	23	20.7	c
Mn	$MnCl_2$	⩽10(925 K)		c
Co	CoI_2	⩽20		d
Ni	$NiCl_2$	91 (1251 K)		e
Ag	AgCl	0.30	17.2	f,b
	AgBr	0.12		f
Zn	$ZnCl_2$	1.7	69.6	a,g
	ZnI_2	3.0	60.1	a
Cd	$CdCl_2$	120	8.4	h,i
	$CdBr_2$	120	10.3	h,i
	CdI_2	68	21.7	a,c
Ga	$GaCl_2$	19 (453 K)		b
	$GaBr_2$	93 (443 K)		a
	GaI_2	400 (503 K)		j
Tl	TlCl	0.09	~0	b
Sn	$SnCl_2$	0.032		b
	$SnBr_2$	0.09		a
Pb	$PbCl_2$	0.040	78.5	b,g,k
	PbI_2	0.68	70.8	a,k
Sb	$SbCl_3$	0.18 (546 K)		a
	SbI_3	86	16.8	a
Bi	$BiCl_3$	370	8.0	l,m
	$BiBr_3$	488		l,n
	BiI_3	miscible		n

[a] J. D. Corbett, S. von Winbush and F. C. Albers, *J. Am. Chem. Soc.*, **79**, 3020 (1957).
[b] J. D. Corbett and S. von Winbush, *J. Am. Chem. Soc.*, **77**, 3964 (1955).
[c] J. D. Corbett, in *Fused Salts*, edited by B. R. Sundheim, McGraw-Hill, New York, 1964, p. 359.
[d] J. D. Corbett, R. J. Clark and R. F. Munday, *J. Inorg. Nucl. Chem.*, **25**, 1286 (1963); *cf.* also ref. c.
[e] J. W. Johson, D. Cubicciotti and C. M. Kelley, *J. Phys. Chem.*, **62**, 1107 (1958).
[f] J. D. Van Norman, *J. Electrochem. Soc.*, **112**, 1126 (1965).
[g] J. D. Van Norman, J. S. Bookless and J. J. Egan, *J. Phys. Chem.*, **70**, 1276 (1966).
[h] G. von Hevesy and E. Lowenstein, *Z. Anorg. Allg. Chem.*, **187**, 266 (1930).
[i] L. E. Topol and A. L. Landis, *J. Am. Chem. Soc.*, **82**, 6291 (1960).
[j] J. D. Corbett and R. K. McMullan, *J. Am. Chem. Soc.*, **77**, 4217 (1955).
[k] L. E. Topol, *J. Phys. Chem.*, **67**, 2222 (1963).
[l] B. Eggink, *Z. Phys. Chem.*, **64**, 449 (1908).
[m] S. J. Yosim, A. J. Darnell, W. Gehman and S. W. Mayer, *J. Phys. Chem.*, **63**, 230 (1959).
[n] S. J. Yosim, L. D. Ransom, R. A. Sallach and L. E. Topol, *J. Phys. Chem.*, **66**, 28 (1962).

properly classified as liquid metal-salt solutions with reduced metal species, but not as liquid subhalide systems.

It is interesting to note that in the analogous bismuth–bismuth sulphide system[31], the sulphur behaves in one way like a halide ion larger than iodide, but on the other hand also as if the sulphide is oxidized, in addition to the bismuth being reduced. There is complete miscibility of liquid $Bi + Bi_2S_3$ mixtures, and bismuth behaves ideally in the metal-rich region: $\mu^E_{Bi} = s^E_{Bi} = h^E_{Bi} = v^E_{Bi} \sim 0$ for $0.5 < x_{Bi} < 1.0$. The sulphide ions may provide their electrons to the pooled free-electron-gas of the metal, and exist as atoms, as the behaviour of its excess functions suggests. The species existing in this system thus seem to be Bi^{3+}, S^0 and e^- at all compositions.

For the cadmium–cadmium halide systems, again, appreciable solubilities have been noted[32], but complete miscibility has not been attained, contrary to the bismuth systems discussed above, the critical temperatures being apparently too high ($\sim$ 1800 K). There is much accumulated evidence in favour of dissolution of cadmium to form $Cd(Cd^{2+})_n$ species, with n values of 1 to 3, but there is no conclusive evidence for the state of the reduced cadmium. The cryoscopic, magnetic susceptibility and density data cannot determine whether the reduced cadmium in, say, Cd_2^{2+} ($n = 1$), is dimeric $(Cd^+)_2$ or atomic Cd^0 'solvated' by a Cd^{2+} ion. The difference lies in the presence or absence of charge-symmetry, but the localization of the electrons somewhere in this species, and the absence of free electrons of the metallic type, are accepted. Thus the addition of cadmium to its chloride decreases the conductivity, as with bismuth and contrary to the alkali metals. The excess volume is positive for the chloride, $v^E = 15.3x_{Cd}(1 - x_{Cd})$ cm^3 mol^{-1}, nearly zero for the bromide and negative for the iodide, $v^E = -4.6x_{Cd}(1 - x_{Cd})$ cm^3 mol^{-1}, at temperatures 20% above the melting points of the salts.

As has been noted above, the values of $v^E/x_{Bi}(1 - x_{Bi})$ for the bismuth systems are all negative, becoming more so from chloride to iodide to sulphide. The excess heats, on the contrary, are all positive; as seen, the systems all have miscibility gaps. The opposing signs of v^E and h^E (except for $Cd + CdCl_2$) require the system to be described by a model more complicated than first-order conformal solutions, perhaps by the single-liquid second-order model (section 4B(b)) or better the two-liquid approximation (section 4B(c)), since g^E is quite asymmetric with respect to composition.

For cadmium–cadmium chloride solutions all the excess functions are positive[32]. The data may be interpreted in terms of the reaction $Cd(l) + CdCl_2(l) \rightleftharpoons Cd_2Cl_2(l)$, and subsequent ideal mixing of the salts $CdCl_2$ and Cd_2Cl_2. The standard functions at 873 K for this reaction are $\Delta v^0 = 16.1$ cm^3 mol^{-1}, $\Delta h^0 = 31$ kJ mol^{-1} $\Delta s^0 = 18$ J K mol^{-1}, leading to $\Delta g^0 = 15$ kJ mol^{-1}, compared with a value near zero at this temperature for the corresponding reaction of mercury.

Metals such as mercury form stable solid halides of lower valency, which may remain stable also at elevated temperatures. In the systems $Hg + HgX_2$, $Ga + Ga[GaX_4]$, $In + In[InX_4]$, $Tl + TlX_3$, $Cu + CuX_2$, etc., the species of intermediate valency are formed. Not very much is known about the stability of these species in the liquid state. Mercury (I) chloride, Hg_2Cl_2, is known to dissociate on melting at the syntectic temperature of 798 K to mercury metal and a salt-rich phase of nearly, but not quite, the composition Hg_2Cl_2. On the other hand, copper (I) chloride is stable to disproportionation in the molten state.

C. Liquid Mixtures of Metals and Molecular Compounds

Several categories of liquid mixtures that involve metals and molecular compounds may be distinguished. One includes solutions of gases in liquid metals, where they occupy interstitial positions. Another includes solutions of metals in non-polar liquids, illustrated by mercury in hydrocarbons. The third, and of major importance, includes the solution of reactive metals in polar liquids, illustrated by alkali metals in ammonia. These latter solutions have been investigated from many aspects (spectroscopic and transport properties, for example), which are outside the scope of this book. Here only the problem of the solubility, or the mutual miscibility of the two liquid phases, will be discussed.

a. Solutions of liquid metals and non-polar substances

Liquid metals are very poorly miscible with non-polar substances, contrary to the appreciable solubility in molten salts noted above, and in certain polar liquids discussed further below. The ideal entropy of mixing cannot provide, at reasonable temperatures, the energetic equivalent of the strong metallic bonds that have to be broken, uncompensated by other strong interactions. Therefore, only very dilute solutions of non-polar liquids or gases in metals, or metals in liquids, are obtained.

The solubility of inert gases in liquid metals and liquid alloys should provide a relatively straightforward testing of theories of the structure and interactions in these metallic liquids. It is therefore unexpected that relatively little work has been done in this direction. The solubility in liquid alloys A + B of a gas C should be given to a good approximation by Wagner's relationship[24]

$$\ln x_C = z \ln[x_A x_{C(A)}^{1/z} + x_B x_{C(B)}^{1/z}] \simeq x_A \ln x_{C(A)} + x_B \ln x_{C(B)} \tag{8.19}$$

provided z is fairly large and constant and the solubilities of C in liquid A and in liquid B do not differ greatly. The main interest, to date, is in the solubility of the gas C in a given liquid metal A.

The solubility of many kinds of interstitial atoms, such as carbon, nitrogen, oxygen or hydrogen in solid metals is known, but in liquid metals, data are available mainly concerning hydrogen and the inert gases. Molecular hydrogen is thought to dissolve by dissociating to atoms. These dissociate further to protons and electrons, which join the free electron gas in the metal, or else pick up free electrons to form hydride anions. The data can be expressed as Ostwald coefficients, i.e. as the ratio of the volume concentration in the metal and in the gas, or as the volume concentration at unit pressure C (moles gas m^{-3} Pa^{-1}) (cf. p. 282), as given in Table 8.7. The heats of solution are $-\mathbf{R}$ times the factor of T^{-1} in the expressions given for $\ln C$ (with the addition of $\mathbf{R}T$ for those that include the term $\ln T$), and

Table 8.7. Solubility of gases in liquid metals in volume concentrations.

Gas	Metal	$\ln[C$ (moles gas m^{-3} Pa^{-1})]	Ref.
H_2	Fe	$-5.177-4390T^{-1}$	a
	Co	$-4.959-4940T^{-1}$	a
	Ni	$-5.649-2420T^{-1}$	a
	Cu	$-4.982-5460T^{-1}$	a
	Na	$+12.705-12210T^{-1}$	b
He	Na	$-0.930-7090T^{-1} - \ln T$	c
Ar	Na	$+0.369-10270T^{-1} - \ln T$	c
Kr	Cd	$-1.712-25330T^{-1} - \ln T$	d
	In	$-1.712-23310T^{-1} - \ln T$	d
	Sn	$-2.433-21990T^{-1} - \ln T$	d
	Pb	$-4.158-16370T^{-1} - \ln T$	d
Xe	Bi	-24.7 (at 773 K)	e

[a]T. Emi and R. D. Pehlke, *Met. Trans.*, **1**, 2733 (1970).
[b]R. J. Newcombe and J. Thompson, *J. Polarogr. Soc.*, **14**, 104 (1968).
[c]G. W. Johnson, *Phil. Mag.*, **6**, 943 (1961).
[d]E. Veleckis, S. K. Dhar, F. A. Cafasso and H. M. Feder, *J. Phys. Chem.*, **75**, 2832 (1971).
[e]G. F. Hewitt, J. A. Lacey and E. Lyall, *J. Nucl. Energy*, **B1**, 167 (1960).

are all positive, ranging to above 210 kJ mol^{-1} for the dissolution of krypton in liquid cadmium. The solubilities of the inert gases are all very small and at a temperature 10 % above T_m at least seven orders-of-magnitude lower than that of hydrogen.

The solubility of non-reacting gases in liquid metals has been rationalized in a manner similar to that for the corresponding molten-salt systems (p. 282). Most of the heat of solution has been ascribed to the hole-creation energy, but $\Delta h_{soln}/\sigma$, which should be constant for a given dissolving gas according to (7.76), is found to vary with the liquid metal solvent. Instead of using the macroscopic surface tension, the gas solubility may be computed also from the perturbation of the packing of the solvent molecules, using a hard-sphere model[33]. The radius of the cavity formed, i.e. half the sum of the diameters of the gas and metal atoms, considered as hard spheres, can be defined in a way that takes into account also a polarization interaction energy. Whereas the metal diameters vary only from 0.123 to 0.224 nm for the metallic solvents of hydrogen, the diameter of the (screened) hydrogen ions is made to vary from 0.224 to 0.322 nm, to give agreement with the measured heats of solution. For the solution of helium and argon in sodium, however, this approach yielded solubilities too high by two to four orders-of-magnitude. The electronic work involved in cavity formation must, apparently, also be taken into account, but no satisfactory general model for the solubility of gases in liquid metals exists as yet.

Whereas only a few metallic bonds are broken in order to provide for the cavity, and the metallic nature of the liquid is not altered when a gas dissolves in a liquid metal, the situation is quite different when a metal dissolves in a non-polar liquid. Here the metal is atomized, and only weak dispersion forces compensate for the broken bonds. Most metals either react chemically with non-polar liquids, or do not dissolve at all (consider, for example, the use of sodium metal to dry organic solvents). Solubilities, indeed, have been measured only for mercury, the metal liquid at room temperature.

However, it is not the low melting point of mercury which is directly responsible for this situation. If the solubility parameters of the metals (Table 8.3) are examined, it is seen that apart from the reactive alkali metals, mercury has indeed the lowest solubility parameter $\delta = 62.1\ J^{1/2}\ cm^{3/2}$. The highest solubility parameter of non-polar liquids (Table 5.2) is *ca.* 20. Since it is the square of the difference of solubility parameters that affects the heats of mixing or excess Gibbs energy, the metal with the next lowest solubility parameter, cadmium with $\delta = 83.8\ J^{1/2}\ cm^{-3/2}$, is at a disadvantage of a factor of 2.3 in this respect (the molar volumes are nearly equal). The melting point of cadmium, however, is too high for most non-polar liquids, while lower melting metals (gallium, indium, tin, bismuth) have all considerably higher solubility parameters. Since solubility parameter theory has been applied quite successfully to solutions of mercury, and since solubilities should increase at increasing temperatures, it would be of interest to test the theory with another metal, the best candidate being, seemingly, bismuth.

The solubility of mercury has been studied most recently by Voigt and co-workers[34], who considered also the few earlier determinations. The solubility c ranges from 4.2 to 6.7 μmol l^{-1} for non-cyclic paraffin solvents, from 8.2 to 12.1 μmol l^{-1} for cyclic paraffins, from 9.6 to 12.0 μmol l^{-1} for aromatic hydrocarbons, and has considerably lower values for polar solvents: ethers (4.8 μmol l^{-1} for isopropyl ether) and alcohols (2.7 μmol l^{-1} for isopropanol). The lowest solubilities have been noted for a perfluorinated hydrocarbon (0.38 μmol l^{-1} for perfluorodimethylcyclobutane) and for water (0.28 μmol l^{-1}, see below). For the paraffins and ethers, very good agreement is obtained between the observed solubilities and those calculated from eqs. (5.12) and (5.50), the first correcting for the discrepancy between molar volumes of solute (mercury) and solvent, and χ in the second being given the value $v^*_{Hg}(\delta^2_{Hg} + \delta^2_{solvent} - 2\lambda_{Hg}\delta_{solvent})\ (RT)^{-1}$ with $\lambda_{Hg} = 55.2\ J^{1/2}\ cm^{-3/2}$ defined in (5.54). The predicted solubilities in the aromatic solvents, however, are more than twice as high as the observed ones. The partial entropies of solution for both classes of solvents are nearly ideal, and conform closely to eq. (5.10), which takes into account the size discrepancy. Thus the premises for the application of solubility parameter theory are fulfilled.

b. Solutions of metals in polar solvents

As noted above, the solubility of mercury in alcohols and in water is low, much lower than predicted using the solubility parameters $\delta = (h^V - RT)^{1/2} v^{-1/2}$. This is obviously due to the breaking of hydrogen bonds in the solvent, and, in the case of aqueous solutions, also to the negative entropy effect of building up the water structure around the dissolved atoms of mercury (p. 199). The mole-fraction solubility is[35] $\log x_{Hg(H_2O)} = -122.81 + 4475T^{-1} + 40.22 \log T$, and for the process $Hg(l) \rightarrow Hg(in\ H_2O,\ x_{Hg} = 1)$ $\Delta h = 13.93$ kJ mol^{-1} and $\Delta s = -111.7$ J K^{-1} mol^{-1} at 298 K. Mercury is also much more volatile from its aqueous solutions than from those in hydrocarbon solvents, and more volatile from these latter, in turn, than from pure liquid mercury where metallic bonds have to be broken.

Of far greater interest than the above, however, are the well known solutions of the alkali metals and the alkaline-earth metals in liquid ammonia and amines or ethers. The solutions are metastable in the sense that on prolonged standing or heating an irreversible chemical reaction takes place, the evolution of hydrogen and the formation of amide anions. The solutions are sufficiently stable, however, to permit accurate determination of their properties, and for the recovery of the unchanged metal on evaporation of the ammonia. The subject has been reviewed in many recent publications[36], and here only certain aspects of the solutions of alkali metals and alkaline-earth metals in liquid ammonia will be discussed.

Dilute solutions of metals in ammonia are inky-blue, and concentrated ones bronze-red with a metallic lustre, the transition region from 'dilute' to 'concentrated' occuring at 2–5 mole-percent metal. Electric conduction is ionic in nature in dilute solutions, becoming metallic in the concentrated ones, except for the anomalous temperature coefficient of resistivity, $d\rho/dT < 0$. The transition from non-metallic to metallic character has been the subject of much research and speculation. This transition is also connected with a liquid–liquid phase separation, which occurs in this region of 2–5 mole-percent metal for several of the metals dissolving in liquid ammonia, Fig. 8.6. Some other metals, however, notably caesium and possibly also rubidium, do not show a miscibility gap[37], while for some that do dissolve, such as strontium, barium and ytterbium, this has not yet been ascertained. The situation is analogous in some respects to the mutual solubilities of metals in their molten salts, see Fig. 8.5 and Table 8.4. The notable difference, apart from that in temperature, is the near-symmetry of the gaps in the metal-salt systems (becoming more so when plotted against φ_M rather than x_M) compared with the very one-sided bias in the metal–ammonia systems.

The sodium–ammonia system has been examined in more detail than the others. From the densities, the apparent molar volume of the metal in dilute solutions is 65 cm^3, that of the solvent, ammonia, being 25 cm^3 at 238 K. Volume fractions may be calculated according to a solvation model[38], where the components are A = ammonia and M = metal, but the species are A = ammonia and B = MA_n. In the solvated solute only the stoichiometry n is specified, not the nature of 'M', which may be an 'ion-pair', M^+e^-. Then $v_B = 65 + 25n$ cm^3 and $\varphi_A = (1 - (n + 1) x_M)v_A/[(1 - (n + 1)x_M)v_A + x_M v_B]$. The vapour pressure p_A of ammonia above the solutions has been measured, and leads to $\mu_A^E = RT \ln(p_A/(1 - x_M)p_A^*) = b(1 - \varphi_A)^2$ (cf. eq. (5.33)) with a constant value of b (independent of x_M), provided the correct value of n is selected, which occurs at $6.5 < n < 7.0$. The vapour pressures have been obtained at 238 K, which is above the consolute temperature, and for $0.01 < x_M < 0.09$. The condition for phase separation

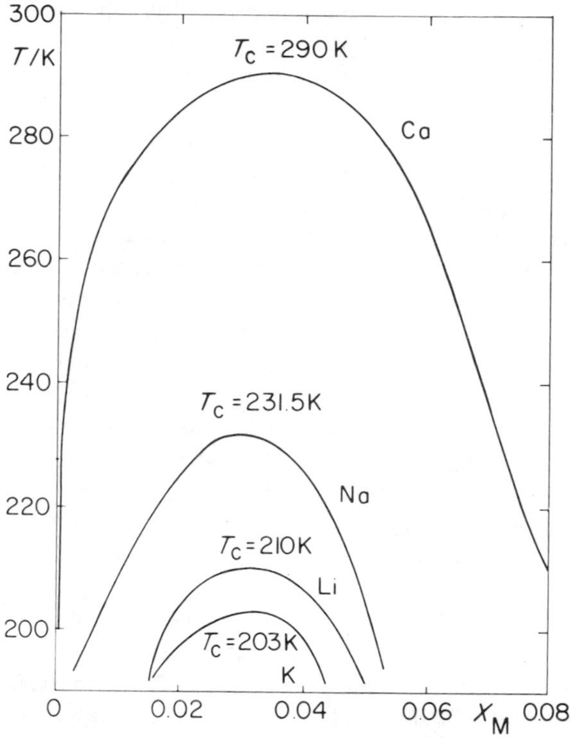

Figure 8.6. Miscibility gaps in metal+ammonia systems. Rubidium may show a very small gap, while caesium is known to be miscible at all temperatures, above the liquidus[37]. The abscissa for the calcium curve has been doubled so that 'equivalent fraction' is represented by x_M rather than mole fraction, the true maximum occurring at a mole fraction of calcium = 0.0168.

$(\partial\mu_A^E/\partial(1-x_M))_T < 0$ can now be applied, and the upper critical consolute temperature calculated from eq. (4.87) or equivalently from $(\partial\mu_A/\partial x_M)_{T_c} = 0$, where $\mu_A = \mu_A^*(P,T) + \mathbf{R}T\ \ln(1-(n+1)x_M)(1-nx_M)^{-1} + \mu_A^E$ to give $T_c = 2\mathbf{R}^{-1} b_A(1-\varphi_A)^2 x_M^{-1} = 232$ K, in good agreement with the experimental value, provided, however, that the solvation number is $6.5 < n < 7.0$. The precision of available vapour pressure data for solutions of lithium and potassium is sufficient only for establishing the wide limit $5 < n < 9$, and the correct order $T_c(\mathrm{K}) < T_c(\mathrm{Li}) < T_c(\mathrm{Na})$.

There are some puzzling features about the coexistence curve, especially the almost square dependence[37] of $(T_c - T)$ on $(x_M' - x_M'')$, where $'$ and $''$ denote the dilute and concentrated phases, compared with the nearly cube-dependence observed ordinarily for miscibility gaps. The correspondence of the gap with the steep increase in conductivity (for homogeneous solutions at $T > T_c$) implies that the same forces responsible for the non-ideality (the size of the parameter b) are responsible for the appearance of metallic properties. This has been explained by the concept of clustering.

In very dilute solutions of metals in liquid ammonia, say at $x_M < 10^{-4}$, the solute dissociates to metal cations M^+ (divalent metals presumably to M^{2+}) and electrons, which become solvated. The solution behaves as a typical dilute solution of a strong electrolyte. The electron finds itself trapped in a cavity in the solvent, the surrounding ammonia molecules being strongly polarized, and the solvated electron being characterized by its absorption spectrum and large apparent molar volume. As the concentration increases, and in view of the relatively low dielectric constant of liquid ammonia, $\epsilon = 22$ at 239 K, 'ion pairing' between solvated cations M^+ and electrons occurs, to give a solvated 'M', the species called MA_n above. The association constants have been evaluated from conductivity and magnetic susceptibility measurements and $\log K_1 = \log [M]/[M^+]^2 y_{\pm}^2$ has been found between 2.2 and 2.5 for sodium, potassium and caesium near 240 K (above T_c) and near 2.8 for lithium at 202 K (below T_c)[38]. In all cases, association to the ion-pair, the so-called M-monomer, is complete by $x_M \sim 0.01$, so that the vapour pressure data discussed above indeed apply for these species.

As the metal concentration increases, further association occurs, which leads to dimer formation, 'M_2', and clustering. The species 'M_2', however, is not the simple metal dimer known from the gas phase, and presumably occurring also in the metal solutions in molten salts (cf. p. 325). The exact nature of this species is not known, the following interpretations having been offered to explain certain of the solution properties: $M_2^+e^-$ pairs, M^+M^- pairs, $2M^+ \cdot 2e^-$ quadrupoles, among others. Formal association constants $\log K_2 = \log [M_2]^{1/2}/[M]$, varying from 0.6 for lithium to 1.6 for potassium with that for sodium in between, have been calculated. At concentrations above $x_M = 0.01$, because of the large solvation number n, the array of charges in these monomers and dimers approaches that in a molten salt. As the electron density becomes about one tenth that of the liquid metal, the solution starts to show metallic conduction because of the appreciable probability of orbital overlap of the electrons, and phase instability sets in. At still higher concentrations, $x_M > 0.08$, the overlap is so extensive, that the electrons are no longer trapped in their cavities but may be considered free. The metallic solution ensuing then consists of solvated M^+ and of free electrons.

References

1. *The Properties of Liquid Metals*, edited by S. Takeuchi, Taylor and Francis, London, 1973.
2. J. E. Enderby, *Adv. Struct. Res. by Diffraction Methods,* **5**, 65 (1971); J. E. Enderby, D. M. North and P. A. Egelstaff, *Phil. Mag.*, **14**, 961 (1966).
3. J. R. Wilson, *Met. Rev.*, **10**, 381 (1965).
4. P. C. Sharrah, J. I. Petz and R. F. Kruh, *J. Chem. Phys.*, **32**, 241 (1960); N. C. Halder, R. J. Metzger and C. N. J. Wagner, *J. Chem. Phys.*, **45**, 1259 (1966); N. C. Halder and C. N. J. Wagner, *J. Chem. Phys.*, **47**, 4385 (1967).
5. O. J. Kleppa, *J. Phys. Chem.*, **64**, 1542 (1960); T. Heumann, *Z. Elektrochem.*, **57**, 724 (1953); B. Predel and D. W. Stein, *Acta Met.*, **20**, 515 (1972).
6. L. S. Darken, *Trans. Met. Soc. AIME*, **239**, 80 (1967).
7. B. W. Mott, *Phil. Mag.*, **2**, 259 (1957).
8. R. Kumar, in ref. 1. p. 467.

9. R. A. Oriani, *Advances in Chemical Physics,* Vol. 2, edited by I. Prigogine, Wiley, New York, 1960, p. 119.
10. W. Hume-Rothery, *J. Inst. Metals,* **35**, 295 (1926); W. Hume-Rothery and G. V. Raynor, *Structure of Metals and Alloys,* Institute of Metals, London, 4th edn., 1962; L. Pauling, *Nature of the Chemical Bond,* Cornell University, Press, Ithaca, N.Y., 3rd edn., 1960.
11. E. Högfeldt, *Rec. Trav. Chim. Pays-Bas,* **75**, 790 (1956); R. G. Ward and J. R. Wilson, *Nature (London),* **182**, 334 (1958); A. K. Jena and T. R. Ramachandran, *Met. Trans.,* **2**, 2958 (1971).
12. A. S. Jordan, *Met. Trans.,* **1**, 239 (1970).
13. L. S. Darken, *J. Am. Chem. Soc.,* **72**, 2909 (1950).
14. J. Guion, M. Blander, D. Hengstenberg and K. Hagemark, *J. Phys. Chem.,* **72**, 2086 (1968); K. Hagemark, *J. Phys. Chem.,* **72**, 2316 (1968).
15. T. Yokokawa, A. Doi and K. Niwa, *J. Phys. Chem.,* **65**, 202 (1961).
16. C. H. P. Lupis and J. F. Elliott, *Acta Met.,* **14**, 1019 (1966).
17. F. A. Cafasso, H. M. Feder and I. Johnson, *J. Phys. Chem.,* **66**, 1028 (1962); **68**, 1944 (1964).
18. W. Seith and H. Johnen, *Z. Elektrochem.,* **56**, 140 (1952); D. T. Peterson and R. Kontrimas, *J. Phys. Chem.,* **64**, 362 (1960).
19. J. Lumsden, *Discuss. Faraday Soc.,* **4**, 60 (1948); O. J. Kleppa, *J. Phys. Chem.,* **60**, 446 (1956); O. J. Kleppa, *J. Phys. Chem.,* **63**, 1953 (1959); R. D. Pehlke and K. Okajima, *Trans. Met. Soc. AIME,* **239**, 1351 (1967).
20. T. E. Faber and J. M. Ziman, *Phil. Mag.,* **11**, 153 (1965); N. C. Halder and C. M. J. Wagner, *Phys. Lett. A,* **24**, 345 (1967).
21. A. Roll and H. Motz, *Z. Metallk.,* **48**, 435, 495 (1957); A. Roll and N. K. A. Swamy, *Z. Metallk.,* **52**, 111 (1961); A. Roll and G. Fees, *Z. Metallk.,* **51**, 540 (1960); A. Roll and E. Uhl, *Z. Metallk.,* **50**, 159 (1959); A. Roll and P. Basu, *Z. Metallk.,* **54**, 511 (1963).
22. S. F. Edwards, *Proc. Roy. Soc., Ser. A,* **267**, 518 (1961); M. Watabe and M. Tanaka, *Progr. Theor. Phys.,* **31**, 525 (1964); L. E. Ballentine, *Can. J. Phys.,* **44**, 2533 (1966).
23. J. B. Wagner and C. Wagner, *J. Chem. Phys.,* **26**, 1602 (1957); B. R. Ilscher and C. Wagner, *Acta Met.,* **6**, 712 (1958); J. J. Egan, *Acta Met.,* **7**, 560 (1959); R. J. Heus and J. J. Egan, *Z. Phys. Chem. (Frankfurt am Main),* **74**, 108 (1971).
24. C. Wagner, *Thermodynamics of Alloys,* Addison-Wesley, Reading, Mass., 1952; this refers to early work of R. Lorentz et al., K. Jellinek et al., and many others.
25. M. A. Bredig, in *Molten Salt Chemistry,* edited by M. Blander, Interscience, New York, 1964, pp. 367–425.
26. J. D. Corbett, in *Fused Salts,* edited by B. R. Sundheim, McGraw-Hill, New York, 1964, pp. 341–408.
27. J. Lumsden, *Thermodynamics of Molten Salt Mixtures,* Academic Press, London, 1966, pp. 293–317.
28. K. S. Pitzer, *J. Am. Chem. Soc.,* **84**, 2025 (1962).
29. J. D. Van Norman and J. J. Egan, *J. Phys. Chem.,* **67**, 2460 (1963); A. S. Dworkin, H. R. Bronstein and M. A. Bredig, *J. Phys. Chem.,* **70**, 2384 (1966); M. Krumplet, J. Fischer and I. Johnson, *J. Phys. Chem.,* **72**, 506 (1968); R. A. Sharma, *J. Phys. Chem.,* **74**, 3896 (1970).
30. Ref. 25, pp. 418–420, and ref. 26, pp. 382–385, and refs. therein; L. E. Topol and F. Y. Lieu, *J. Phys. Chem.,* **68**, 851 (1964); L. F. Grantham, *J. Chem. Phys.,* **43**, 1415 (1965); K. Ichikawa and M. Shimoji, *Trans. Faraday Soc.,* **62**, 3543 (1966); J. D. Kellner, *J. Phys. Chem.,* **71**, 3254 (1967); **72**, 1737 (1968); C. R. Boston, *Inorg. Chem.,* **9**, 389 (1970).
31. D. Cubicciotti, *J. Phys. Chem.,* **66**, 1205 (1962); **67**, 118 (1963); **68**, 537 (1964).
32. L. E. Topol and A. L. Landis, *J. Am. Chem. Soc.,* **82**, 6291 (1960); N. H. Nachtrieb, *J. Phys. Chem.,* **66**, 1163 (1962); G. A. Crawford and J. W. Tomlinson, *Trans. Faraday Soc.,* **62**, 3046 (1966); L. Suski and J. Mascinski, *J. Phys. Chem.,* **75**, 3620 (1971).
33. R. A. Pierotti, *J. Phys. Chem.,* **67**, 1840 (1963); T. Emi and R. D. Pehlke, *Met. Trans.,* **1**, 2733 (1970); E. Veleckis, S. K. Dhar, F. A. Cafasso and H. M. Feder, *J. Phys. Chem.,* **75**, 2832 (1971).
34. J. N. Spencer and A. F. Voigt, *J. Phys. Chem.,* **72**, 464, 471, 1913 (1968).
35. D. N. Glew and D. A. Hames, *Can. J. Chem.,* **49**, 3114 (1971).
36. G. Lepoutre and M. J. Sienko (Eds.), *Metal Ammonia Solutions, Colloq. Weyl I, Lille, 1963,* Benjamin, New York, 1964; J. C. Thompson, *Rev. Mod. Phys.,* **40**, 704 (1968); U.

Schindewolf, *Angew. Chem., Int. Ed. Eng.*, **7**, 190 (1968); *Metal Ammonia Solutions, Colloq. Weyl II, Ithaca, 1969,* edited by J. J. Lagowski, Butterworth, London, 1971.

37. P. D. Schettler, Jr. and A. Patterson, Jr., *J. Phys. Chem.*, **68**, 2865 (1964); H. Teoh, P. R. Antoniewicz and J. C. Thompson, *J. Phys. Chem.*, **75**, 399 (1971); R. L. Schroeder, J. C. Thompson and P. L. Oertel, *Phys. Rev.*, **178**, 289 (1969).
38. P. Damay and G. Lepoutre, *J. Chim. Phys.*, **69**, 1276 (1972); A. Demortier, M. DeBacker and G. Lepoutre, *J. Chim. Phys.*, **69**, 380 (1972).

Author Index

(The numbers in parentheses denote the reference number or letter in *italics*, and the page on which the complete reference is found).

Subject Index

(Substances are generally in their liquid state)